Physical Geology of High-Level Magmatic Systems

Geological Society Special Publications
Society Book Editors
R. J. PANKHURST (CHIEF EDITOR)
P. DOYLE
F. J. GREGORY
J. S. GRIFFITHS
A. J. HARTLEY
R. E. HOLDSWORTH
J. A. HOWE
P. T. LEAT
A. C. MORTON
N. S. ROBINS
J. P. TURNER

Special Publication reviewing procedures

The Society makes every effort to ensure that the scientific and production quality of its books matches that of its journals. Since 1997, all book proposals have been refereed by specialist reviewers as well as by the Society's Books Editorial Committee. If the referees identify weaknesses in the proposal, these must be addressed before the proposal is accepted.

Once the book is accepted, the Society has a team of Book Editors (listed above) who ensure that the volume editors follow strict guidelines on refereeing and quality control. We insist that individual papers can only be accepted after satisfactory review by two independent referees. The questions on the review forms are similar to those for *Journal of the Geological Society*. The referees' forms and comments must be available to the Society's Book Editors on request.

Although many of the books result from meetings, the editors are expected to commission papers that were not presented at the meeting to ensure that the book provides a balanced coverage of the subject. Being accepted for presentation at the meeting does not guarantee inclusion in the book.

Geological Society Special Publications are included in the ISI Index of Scientific Book Contents, but they do not have an impact factor, the latter being applicable only to journals.

More information about submitting a proposal and producing a Special Publication can be found on the Society's web site: www.geolsoc.org.uk.

It is recommended that reference to all or part of this book should be made in one of the following ways:

BREITKREUZ, C. & PETFORD, N. (eds) 2004. *Physical Geology of High-Level Magmatic Systems.* Geological Society, London, Special Publications, **234**.

MACHOWIAK, K., MUSZYNSKI, K. & ARMSTRONG, R. 2004. High-level volcano-granodioritic intrusions from Zelezniak Hill (Kaczawa Mountains, Sudetes, SW Poland). *In*: BREITKREUZ, C. & PETFORD, N. (eds) 2004. *Physical Geology of High-Level Magmatic Systems.* Geological Society, London, Special Publications, **234**, 67–74.

GEOLOGICAL SOCIETY SPECIAL PUBLICATION NO. 234

Physical Geology of High-Level Magmatic Systems

EDITED BY

C. BREITKREUZ
Institut für Allgemeine Geologie, Germany

and

N. PETFORD
Kingston University, UK

2004
Published by
The Geological Society
London

THE GEOLOGICAL SOCIETY

The Geological Society of London (GSL) was founded in 1807. It is the oldest national geological society in the world and the largest in Europe. It was incorporated under Royal Charter in 1825 and is Registered Charity 210161.

The Society is the UK national learned and professional society for geology with a worldwide Fellowship (FGS) of 9000. The Society has the power to confer Chartered status on suitably qualified Fellows, and about 2000 of the Fellowship carry the title (CGeol). Chartered Geologists may also obtain the equivalent European title, European Geologist (EurGeol). One fifth of the Society's fellowship resides outside the UK. To find out more about the Society, log on to www.geolsoc.org.uk.

The Geological Society Publishing House (Bath, UK) produces the Society's international journals and books, and acts as European distributor for selected publications of the American Association of Petroleum Geologists (AAPG), the American Geological Institute (AGI), the Indonesian Petroleum Association (IPA), the Geological Society of America (GSA), the Society for Sedimentary Geology (SEPM) and the Geologists' Association (GA). Joint marketing agreements ensure that GSL Fellows may purchase these societies' publications at a discount. The Society's online bookshop (accessible from www.geolsoc.org.uk) offers secure book purchasing with your credit or debit card.

To find out about joining the Society and benefiting from substantial discounts on publications of GSL and other societies worldwide, consult www.geolsoc.org.uk, or contact the Fellowship Department at: The Geological Society, Burlington House, Piccadilly, London W1J 0BG: Tel. +44 (0)20 7434 9944; Fax +44 (0)20 7439 8975; E-mail: enquiries@geolsoc.org.uk.

For information about the Society's meetings, consult *Events* on www.geolsoc.org.uk. To find out more about the Society's Corporate Affiliates Scheme, write to enquiries@geolsoc.org.uk.

Published by The Geological Society from:
The Geological Society Publishing House
Unit 7, Brassmill Enterprise Centre
Brassmill Lane
Bath BA1 3JN,
UK
(*Orders:* Tel. +44 (0)1225 445046
        Fax +44 (0)1225 442836)
Online bookshop: http://bookshop.geolsoc.org.uk

The publishers make no representation, express or implied, with regard to the accuracy of the information contained in this book and cannot accept any legal responsibility for any errors or omissions that may be made.

© The Geological Society of London 2004. All rights reserved. No reproduction, copy or transmission of this publication may be made without written permission. No paragraph of this publication may be reproduced, copied or transmitted save with the provisions of the Copyright Licensing Agency, 90 Tottenham Court Road, London W1P 9HE. Users registered with the Copyright Clearance Center, 27 Congress Street, Salem, MA 01970, USA: the item-fee code for this publication is 0305-8719/04/$15.00.

British Library Cataloguing in Publication Data
A catalogue record for this book is available from the British Library.

ISBN 1-86239-169-6

Typeset by Type Study, Scarborough, UK
Printed by MPG Books Ltd, Bodmin, UK

Distributors

USA
AAPG Bookstore
PO Box 979
Tulsa
OK 74101-0979
USA
Orders: Tel. +1 918 584-2555
        Fax +1 918 560-2652
        E-mail bookstore@aapg.org

India
Affiliated East–West Press PVT Ltd
G-1/16 Ansari Road, Daryaganj,
New Delhi 110 002
India
Orders: Tel. +91 11 2327-9113
        Fax +91 11 2326-0538
        E-mail affiliat@nda.vsnl.net.in

Japan
Kanda Book Trading Company
Cityhouse Tama 204
Tsurumaki 1-3-10
Tama-shi
Tokyo 206-0034
Japan
Orders: Tel. +81 (0)423 57-7650
        Fax +81 (0)423 57-7651
        E-mail geokanda@ma.kcom.ne.jp

Contents

Preface	VII
BREITKREUZ, C. & PETFORD, N. Introduction	1
AWDANKIEWICZ, M. Sedimentation, volcanism and subvolcanic intrusions in a late Palaeozoic intramontane trough (the Intra-Sudetic Basin, SW Poland)	5
BREITKREUZ, C. & MOCK, A. Are laccolith complexes characteristic of transtensional basin systems? Examples from Permo-Carboniferous Central Europe	13
MARTIN, U. & NÉMETH, K. Peperitic lava lake-fed sills at Ság-hegy, western Hungary: a complex interaction of a wet tephra ring and lava	33
AWDANKIEWICZ, M., BREITKREUZ, C. & EHLING, B.-C. Emplacement textures in Late Palaeozoic andesite sills of the Flechtingen–Roßlau Block, north of Magdeburg (Germany)	51
MACHOWIAK, K., MUSZYŃSKI, A. & ARMSTRONG, R. High-level volcanic–granodioritic intrusions from Zelezniak Hill (Kaczawa Mountains, Sudetes, SW Poland)	67
LORENZ, V. & HANEKE, J. Relationship between diatremes, dykes, sills, laccoliths, intrusive–extrusive domes, lava flows, and tephra deposits with unconsolidated water-saturated sediments in the late Variscan intermontane Saar–Nahe Basin, SW Germany	75
BONIN, B., ETHIEN, R., GERBE, M. C., COTTIN, J. Y., FÉRAUD, G., GAGNEVIN, D., GIRET, A., MICHON, G. & MOINE, B. The Neogene to Recent Railler-du-Baty nested ring complex, Kerguelen Archipelago (TAAF, Indian Ocean): stratigraphy revisited, implications for cauldron subsidence mechanisms	125
MAZZARINI, F., CORTI, G., MUSUMECI, G. & INNOCENTI, F. Tectonic control on laccolith emplacement in the northern Apennines fold–thrust belt: the Gavorrano intrusion (southern Tuscany, Italy)	151
HABERT, G. & DE SAINT-BLANQUAT, M. Rate of construction of the Black Mesa bysmalith, Henry Mountains, Utah	163
CORAZZATO, C. & GROPPELLI, G. Depth, geometry and emplacement of sills to laccoliths and their host-rock relationships: Montecampione group, Southern Alps, Italy	175
WESTERMAN, D. S., DINI, A., INNOCENTI, F. & ROCCHI, S. Rise and fall of a nested Christmas-tree laccolith complex, Elba Island, Italy	195
MALTHE-SØRENSSEN, A., PLANKE, S., SVENSEN, H. & JAMTVEIT, B. Formation of saucer-shaped sills	215
THOMSON, K. Sill complex geometry and internal architecture: a 3D seismic perspective	229
JAMTVEIT, B., SVENSEN, H., PODLADCHIKOV, Y. Y. & PLANKE, S. Hydrothermal vent complexes associated with sill intrusions in sedimentary basins	233
VINCIGUERRA, S., XIAO, X. & EVANS, B. Experimental constraints on the mechanics of dyke emplacement in partially molten olivines	243
Index	251

Preface

This book is the outcome of a two-day international workshop on the physical geology of subvolcanic systems, held at TU Bergakademie Freiberg in Germany between 12 and 14 October 2002. Christened LASI by the conference organizers and participants (Laccoliths and Sills), the workshop was supplemented by a one-day field trip to visit quarries that expose Late Palaeozoic subvolcanic systems. In all, the meeting attracted 40 participants from 10 countries, who presented papers covering a wide range of topics relevant to the geology and emplacement of high-level intrusions, 14 of which are included in this volume. We make no apologies for the strong European bias, and we are especially pleased that a number of contributors are from the former Soviet bloc countries.

To our knowledge, nothing similar or as significant in its breadth has been published specifically on high-level intrusions since the now-classic 1970 volume *Mechanism of Igneous Intrusion* (Geological Journal Special Issue **2**, edited by G. Newall & N. Rast), and we hope that this volume fills a much-needed gap in the market. As well as appealing to igneous petrologists, volcanologists and structural geologists, we hope that the book will provide an important source of reference for petroleum geologists and engineers working in sedimentary basins where minor intrusions contribute to basin architecture, or act as hydrocarbon reservoirs or seals.

Funding for the workshop came from the Saxonian Ministry of Science and Art, Land Sachsen, Deutsche Forschungsgemeinschaft and The Volcanic and Magmatic Studies Group (Geological Society, London, and Mineralogical Society of Great Britain and Ireland). We are grateful to all, and would also like to thank the authors and reviewers for their patience, A. Mock for much hard work and effort in the organization of the meeting, and C. Iverson for help with artwork. NP would like to thank The Shed.

Christoph Breitkreuz
Nick Petford

Physical geology of high-level magmatic systems: introduction

CHRISTOPH BREITKREUZ[1] & NICK PETFORD[2]

[1] *Institut für Allgemeine Geologie, Bernhard-von-Cotta-Str. 209599 Freiberg, Germany*
(e-mail: cbreit@geo.tu-freiberg.de)
[2] *Geodynamics and Crustal Processes Group, Kingston University, Surrey KT1 2EE, UK*

Despite their wide occurrence and structural importance for the development of the upper continental crust, the physical geology of high-level dykes, sills and laccoliths (so-called minor intrusions) has not received the level of detailed attention that it deserves. Factors determining the final emplacement level of subvolcanic intrusions are complex, and depend upon a range of physical parameters, including magma driving pressure, the local (and regional) stress field, and the physical properties (viscosity and density) of the intruding material (Breitkreuz *et al.* 2002). SiO$_2$-poor magmas rise through tabloid or ring-shaped dykes, acting as feeder systems for Hawaiian to strombolian eruptions or for their phreatomagmatic to subaquatic equivalents. The ascent of silica-rich magmas leads to explosive eruptions, extrusion of lava or emplacement of subvolcanic stocks and laccoliths. The main reason for this variation in emplacement style appears to be the initial volatile content of the rising magma (e.g. Eichelberger *et al.* 1986). Despite this, and as shown in this volume, the resulting emplacement geometries are surprisingly limited in range, suggesting that interactions between magma pressures and local (and regional) stress fields act to minimize the degree of freedom available for space creation, irrespective of initial composition.

Interaction between magmas and sediments is an important process in high-level intrusive complexes, and a number of papers address this topic. In the field, the distinction between subvolcanic intrusions and lavas, and even some high-grade rheomorphic ignimbrites, is not always clear cut, especially in the case of ancient units exposed in limited outcrop or in drill cores. In particular, the distinction between very shallow-level intrusions and subaerial or subaquatic lavas can be made very difficult, due to their textural similarity (Orth & McPhie 2003), and careful analysis and modelling of rock textures remains an important task.

One high-level intrusion type in particular – laccoliths – serves as an important link between lava complexes and plutons (Fig. 1), and several papers in this volume deal with this relationship

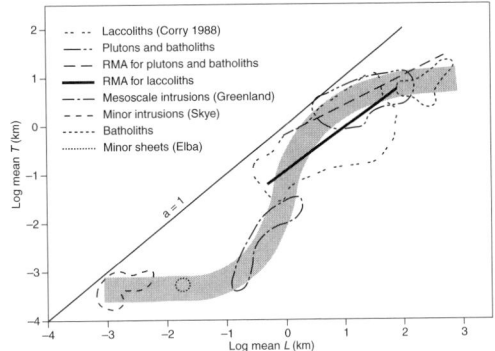

Fig. 1. Summary diagram showing the scaling relationships between minor intrusions, sills, laccoliths and plutons (from McCaffrey & Cruden 2002). The data suggest a genetic growth law linking each individual geometry over a length scale of several orders of magnitude. However, the one-size-fits-all power-law relationship originally used to explain these data (McCaffrey & Petford 1997) appears not to hold over the entire range of natural length-scales presented by igneous intrusions of roughly six orders of magnitude. Instead, the available data suggest an open S-curve with the power-law slopes tangential to the overall shape.

in some detail. Field descriptions of laccoliths and models for their emplacement reach back to the classic work of Gilbert (1877). Although analogue and numerical modelling of laccolith emplacement and host-rock rheology is quite advanced (e.g. Corry 1988; Jackson & Pollard 1988; Roman-Berdiel *et al.* 1995), internal processes and their controlling parameters still require further investigation. For example, many large laccoliths and sills cause remarkably little thermal overprint on the host rock. Emplacement and cooling textures that develop in the magmatic body and at its margins, such as flow foliation, vesiculation, brecciation, crystallization and jointing, are still poorly understood processes in this context. Given that high-level intrusions can also act as reservoirs for hydrocarbons, a better understanding of factors that

From: BREITKREUZ, C. & PETFORD, N. (eds) 2004. *Physical Geology of High-Level Magmatic Systems.*
Geological Society, London, Special Publications, **234**, 1–4. 0305-8719/04/$15.00
© The Geological Society of London 2004.

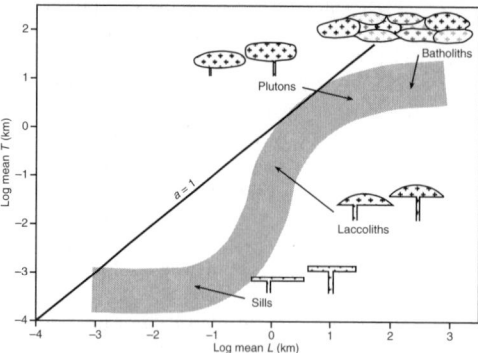

Fig. 2. Synthesis of Figure 1, indicating the possible relationship between sills, laccoliths, plutons and batholiths for typical intrusion geometries at different scales (after McCaffrey & Cruden 2002). In this model, plutons (and, by implication, pulsed batholiths), grow mostly via vertical inflation, but the final length/thickness ratio is arrived at via a more complex (S-type) path (Cruden & McCaffrey 2001).

contribute to the permeability and transport properties of sills and laccoliths may be crucial for future exploration and production from non-traditional oil and gas reservoirs (Petford & McCaffrey 2003). Indeed, several of the papers in this book deal with aspects of sill geometry and subsurface visualization from an industry perspective.

Recent studies of laccolith dimensions from the Henry Mountains Group, Utah, and other regions, show a power-law distribution of the form $N(\geq T) = kT^D$, where N = number of intrusions equal to or greater than a thickness T, and D is the power-law exponent. Significantly, similar power-law relationships have also been found in the measured dimensions of granitic plutons, suggesting a genetic link between sills, laccoliths and plutons. Despite uncertainty in the exact values of the exponent D (Cruden & McCaffrey 2001), an increasing number of studies appear to support the power-law growth model, over at least part of the emplacement history corresponding to a finite length-scale (Fig. 1). It will be interesting to see how this model develops as more data on intrusion dimensions are gathered (Fig. 2).

Compilations like the one of Corry (1988) on the Tertiary laccolith complexes of Utah reveal the importance of subvolcanic bodies in the magmatic continental systems throughout Earth's history. They also hint that some geotectonic settings such as intra-continental transtension in Late Palaeozoic Europe appear to favour the formation of sill and laccolith complexes. Could there be a preferred tectonic setting for laccolith emplacement? More provocatively, are laccoliths and high-level sills forming today, and what is the nature of the relationship (if any) between high-level intrusions and volcanic activity? Presumably, laccolith textures represent a frozen-in early stage of cooling plutons. To this end, Henry et al. (1997) have coined the term 'lacco-caldera', and suggested that many caldera-feeding, supracrustal magma chambers might have a laccolith geometry. Geophysical and in particular geodetic (interferometric) surveys in magmatically active zones like the Andes may help to provide new insight into these and related problems. It is hoped that with the publication of this volume, a consensus will emerge that will help to advance our understanding of the governing factors controlling the emplacement of high-level intrusions in the continental crust, and their geotectonic implications.

This volume contains 16 papers that cover a wide spectrum of topics relating to the physical geology of high-level magmatic systems. The structure of the book has been grouped broadly into three general themes: regional studies and magma sediment interaction, field constraints on the emplacement of laccoliths, and sills in sedimentary basins.

The geological complexity found in depositional, subvolume environments is documented by **Awdankiewicz** in a study of the Late Palaeozoic Intra-Sudetic Basin, southern Poland.

Breitkreuz & Mock describe multi-vent laccolith systems from the Permo-Carboniferous of central Europe, and suggest that they formed during transtensional basin development related to dextral strike-slip tectonics. It appears that the laccolith systems grow from multi-feeder systems that require the existence of a large lower- to mid-crustal magma chamber.

Martin & Nemeth address the topic of magma–wet sediment interaction in their study of the small Neogene Ság-hegy volcanic complex, western Hungary. Here, intense interaction and mixing of lava with the host tephras led to peperite formation along the outer rim of a lava lake. Fluidization of wet tephra allowed basaltic magma to invade and mix with phreatomagmatic tephra.

Awdankiewicz et al. discuss the origin of emplacement textures in Late Palaeozoic andesite sills of the Flechtingen–Roßlau Block, Germany, which intruded into a 100-m thick sequence of lacustrine to alluvial sediments. The resulting intrusive complex, considered as a lava by previous authors, comprises a range of structures, including domes, sills, dykes and failed sills of varying lateral thickness. Magma–sediment interactions include quench-clastic brecciation and post-emplacement hydrothermal breccias.

Machowiak et al. present new petrological and geochemical data on high-level silicic lava domes and laccoliths from the Kaczawa Mountains, SW Poland. The system comprises a carapace facies of exposed ignimbrites and spherulitic rhyolites. Recovered core (drilled to 55 m) includes volcanic rocks ranging in composition from andesite to rhyodacite, and a plutonic facies of microgranite and granodiorite. The ^{206}Pb–^{238}U zircon mineral ages from the volcanic and granitic rocks yield ages of 315 to 316 Ma, making the intrusions from Zelezniak Hill the oldest volcanic rocks dated so far in Armorica.

In a detailed review of the late-Variscan intermontane Saar–Nahe Basin, Germany, **Lorenz & Haneke** show that during intensive volcanism, basic to acidic maar–diatremes formed on hydraulically active faults or fault intersections with basic to intermediate sills emplaced at depths ranging from 2500 m to the palaeosurface. Ongoing inflation of some laccoliths led to large intrusive–extrusive domes, block-and-ash flows and lava extrusion. Some domes show evidence for magma mingling. Formation of maar–diatremes, sills, laccoliths, and most tephra deposits is related to magma intrusion into weak, water-saturated sediments. However, magmas of basic to intermediate composition in some areas reached the surface without being hampered by unconsolidated water-saturated sediments.

Bonin et al. revisit the stratigraphy of nested ring complexes in the Kerguelen Archipelago. The growth mechanism of the caldera-related ring structures are explained as the result of episodes of hydrofracturing followed by cauldron subsidence into the degassed magma chamber, resulting in the intrusion of discrete sheets with average volumes of $c.$ 200 km^3. They estimate a net crustal growth of $c.$ 10^5 m^3 yr^{-1}.

Mazzarini et al. present a compilation of historical mining exploration data, together with more modern structural and metamorphic work on the Gavorrano pluton, Tuscany, Italy. Using these data, they make a case that the pluton is laccolithic and that it grew by roof uplift after initial emplacement was localized at the subhorizontal interface between basement and autochthonous cover rocks. Some of the contact relationships of the pluton were modified by later post-emplacement faults.

Habert & Saint-Blanquat have re-investigated the emplacement of the Black Mesa bysmalith (Colorado Plateau, Utah), first studied by Pollard & Johnson (1973), and show how the emplacement rate can be constrained by a combination of textural studies of the pluton interior and numerical simulation of its thermal evolution. They propose that emplacement was a geologically very short event, with a maximum duration in the order of 60 years, implying minimum vertical displacement rates of the wallrocks above the pluton of around 4 metres per year. Their model suggests a characteristic thickness of 30 metres for magma pulses injected every six months, which is compatible with field constraints including lack of solid-state deformation around internal contacts and no significant recrystallization of the wall-rock at the contact.

Corazzato & Groppelli discuss the relationship between subvolcanic intrusions and their Upper Permian–Triassic sedimentary host rock from the southern Italian Alps. The magmatic rocks show large variations in thickness that can be related to mode of emplacement, which was close to 1 km from the palaeosurface. New lithostratigraphic units are defined for the area, and a combined geological map and GIS analysis is used to estimate the volume of intruded material at $c.$ 1 km^3.

The geometry of laccolith complexes is also investigated by **Westerman et al.** in a detailed field study on Elba Island, Italy. Here, Late Miocene granite porphyries are shown to be a layered series of intrusions that together comprise a nested Christmas-tree laccolith. Structural data suggest that the layers were originally part of a single sequence that was split by deformation. Magma traps controlled emplacement, with neutral buoyancy playing a negligible role.

The more applied aspects of intrusion geometry are taken up in complementary papers that address both the formation of sills from a mechanical perspective, and in 3D seismic imaging of subsurface geometry. **Malthe-Sørenssen et al.** present a model for sill emplacement in sedimentary basins, where the intruding magma is approximated as a non-viscous fluid. Their numerical model, using the discrete element method, shows that saucer-shaped sills occur in the simple setting of homogeneous basin fill and initial isotropic stress conditions. Anisotropic stresses lead to the formation of transgressive sill segments. The paper is illustrated with field examples from the Karoo Basin, South Africa, and seismic images from offshore Norway.

The seismic imaging theme is taken up by **Thompson**, who shows how seismic data can be used to good effect to image the 3D geometry and internal architecture of a sill complex from the North Rockall Trough. In particular, advances in seismic volume techniques allow features such as magma flow patterns to be visualized at a spatial resolution of c. 25 metres. Such models are important in hydrocarbon exploration, where igneous rocks contribute to overall basin architecture.

Jamtveit et al. address the role of hydrothermal vent complexes associated with deep-seated sill intrusions in sedimentary basins. Hydrothermal vents issuing from the tips of transgressive sills are observed in seismic profiles and, depending upon the extent of fluid pressure build-up, can result in explosive release of fluids. Associated high fluid-fluxes can help to alter the permeability structure of sedimentary basins, and may have long-term effects on their hydrogeological evolution.

Finally, **Vinciguerra et al.** have investigated experimentally the mechanics of basalt dyke emplacement during deformation at confining pressures of 300 MPa and temperatures of 1200 °C. Significant diffusion of basalt melt into the matrix was found when a deviatoric stress was applied. Creep and strain rate experiments induced melt propagation at up to 50% of the initial 'dyke' length/width ratio. The kinematics of deformation is essentially plastic and shows a strong dependence on the load applied and on the dyke geometry. Local pressure-drops due to dilatancy may have enhanced melt migration.

References

BREITKREUZ, C., MOCK, A. & PETFORD, N. (eds) 2002. *First International Workshop: Physical Geology of Subvolcanic Systems – Laccoliths, Sills and Dykes (LASI)*. Wissenschaftl. Mitt. Inst. Geol., TU Freiberg, 20/2002, Freiberg, 75 pp.

CORRY, C.E., 1988. Laccoliths: mechanics of emplacement and growth. *Geological Society of America Special Paper*, **220**, 1–110.

CRUDEN, A.R. & MCCAFFREY, K.J.W. 2001. Growth of plutons by floor subsidence: implications for rates of emplacement, intrusion spacing and melt extraction mechanisms. *Physics and Chemistry of the Earth*, **26(4–5)**, 303–315.

EICHELBERGER, J.C., CARRIGAN, C.R., WESTRICH, H.R. & PRICE, R.H. 1986. Non-explosive silicic volcanism. *Nature*, **323**, 598–602.

GILBERT, G.K. 1877. *Geology of the Henry Mountains, Utah*. US Geographical and Geological Survey of the Rocky Mountain Region, Washington, D.C., 1–170.

HENRY, C.D., KUNK, M.J., MUEHLBERGER, W.R. & MCINTOSH, W.C. 1997. Igneous evolution of a complex laccolith–caldera, the Solitario, Transpecos Texas. Implications for calderas and subjacent plutons. *Geological Society of America Bulletin*, **109**, 1036–1054.

JACKSON, M.D. & POLLARD, D.D. 1988. The laccolith–stock controversy. New results from the southern Henry Mountains, Utah. *Geological Society of America Bulletin*, **100**, 117–139.

MCCAFFREY, K.J.W. & CRUDEN, A.R. 2002. Dimensional data and growth models for intrusions. In: BREITKREUZ, C., MOCK, A. & PETFORD, N. (eds), *First International Workshop: Physical Geology of Subvolcanic systems – Laccoliths, Sills, and Dykes (LASI)*. Wissenschaftl. Mitt. Inst. Geol., TU Freiberg, 20/2002, Freiberg, 37–39.

MCCAFFREY, K.J.W. & PETFORD, N. 1997. Are granitic plutons scale invariant? *Journal of the Geological Society of London*, **154**, 1–4.

ORTH, K. & MCPHIE, J. 2003. Textures formed during emplacement and cooling of a Palaeoproterozoic, small-volume rhyolitic sill. *Journal of Volcanology and Geothermal Research*, **128**, 341–362.

PETFORD, N. & MCCAFFREY, K.J.W. 2003. *Hydrocarbons in Crystalline Rocks*. Geological Society, London, Special Publications, **214**.

POLLARD, D.D. & JOHNSON, A.M. 1973. Mechanics of growth of some laccolithic intrusions in the Henry Mountains, Utah II, Bending and failure of overburden layers and sill formation. *Tectonophysics*, **18**, 311–354.

ROMAN-BERDIEL, T., GAPAIS, D. & BRUN, J.P. 1995. Analogue models of laccolith formation. *Journal of Structural Geology*, **17**, 1337–1346.

Sedimentation, volcanism and subvolcanic intrusions in a late Palaeozoic intramontane trough (the Intra-Sudetic Basin, SW Poland)

MAREK AWDANKIEWICZ

University of Wrocław, Institute of Geological Sciences, Department of Mineralogy and Petrology, ul. Cybulskiego 30, 50-205 Wrocław, Poland (e-mail: mawdan@ing.uni.wroc.pl)

Abstract: The Intra-Sudetic Basin is a Late Palaeozoic intramontane trough, situated in the eastern part of the European Permo-Carboniferous Basin and Range Province. Within the basin, tectonics, sedimentation and volcanic/subvolcanic activity were intimately related. Tectonics controlled the location of the depositional and volcanic centres. Many volcanic centres with subvolcanic intrusions of rhyodacitic, rhyolitic and trachyandesitic composition were located close to the intra-basinal depositional troughs, where thick accumulations of sedimentary rocks partly obstructed the movement of magma to the surface. Differences in the structure and geometry of intrusions at separate subvolcanic complexes reflect the influence of different discontinuities, faults, margins of collapse structures, boundaries of contrasting lithologies in the country rocks and the volcanic structures.

The Intra-Sudetic Basin (Fig. 1) is a Late Palaeozoic intramontane trough, situated at the NE margin of the Bohemian Massif, in the eastern part of the Variscan belt of Europe (e.g. Wojewoda & Mastalerz 1989; Dziedzic & Teisseyre 1990 and references therein). The Carboniferous–Permian, volcanic–sedimentary infill of the basin records interrelated tectonic, sedimentary and volcanic processes in a late- to post-orogenic, extensional intra-continental setting, possibly similar to the Tertiary–Recent Basin and Range Province of the SW USA (Lorenz & Nicholls 1976, 1984; Menard & Molnar 1988). During the Carboniferous and Permian, substantial volumes of magmas were both erupted within the Intra-Sudetic Basin, and also emplaced as subvolcanic intrusions in the sedimentary sequence (e.g. Bubnoff 1924; Dathe & Berg 1926; Petrascheck 1938; Hoehne 1961; Grocholski 1965; Nemec 1979; Dziedzic 1980; Awdankiewicz 1999a, b). This paper considers the subvolcanic complexes and particularly their structure and distribution relative to the depositional and volcanic centres. These data illuminate some of the processes that control emplacement of subvolcanic intrusions within intra-continental, sediment-filled basins.

Outline geology and volcanic evolution of the Intra-Sudetic Basin

The Intra-Sudetic Basin, *c.* 60 km long and 30 km wide, is a NW–SE-aligned, fault-bounded, complex syncline. Its basement is mainly composed of lithologically variable, usually strongly deformed and metamorphosed, Upper Precambrian–Palaeozoic rocks of the West Sudetes (e.g. Franke & Żelaźniewicz 2000). Within the basin, the Carboniferous–Permian succession is up to 11 km thick and consists mostly of siliciclastic alluvial and lacustrine deposits, but some deltaic to marine deposits (of Late Viséan age), and volcanic and volcaniclastic rocks also occur. This sequence is overlain by Triassic and Upper Cretaceous deposits. The Triassic deposits are sandstones of alluvial origin, and the Upper Cretaceous deposits are shallow-marine sandstones and marls.

The Carboniferous–Permian succession includes *c.* 7 km of Lower Carboniferous (starting with the uppermost Tournaisian or lowermost Viséan), some 2.5 km of Upper Carboniferous and *c.* 1.5 km of Permian. However, the distribution of the deposits is highly asymmetrical, with thicker accumulations of older deposits in the NW and thinner, younger deposits in the SE, reflecting a progressive southeastward migration of the main depositional centres and decreasing subsidence rates. Possibly, the evolution of the basin was controlled by the main Intra-Sudetic Fault (e.g. Mastalerz 1996), a major NW-trending strike-slip dislocation zone.

The Early Carboniferous depositional centre, a west–east aligned graben, was located in the northern part of the present basin (Dziedzic & Teisseyre 1990). In Late Carboniferous times

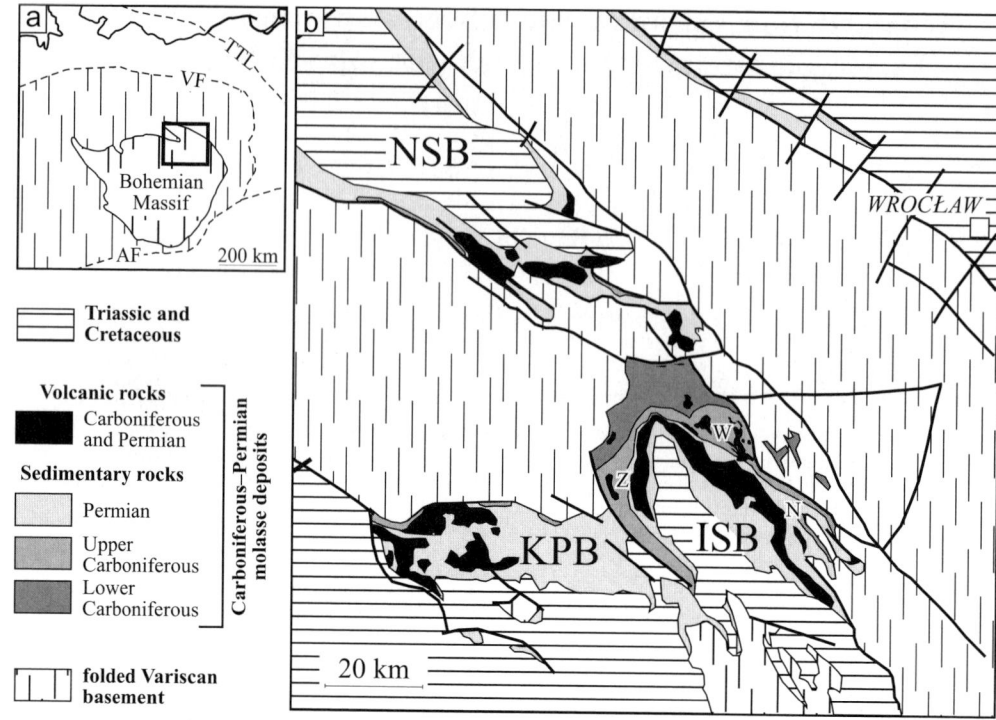

Fig. 1. (a) Location of the study area (frame) within the eastern part of the Variscan belt of Europe. TTL, Teisseyre–Tornquist line; VF, Variscan Front; AF, Alpine Front. (b) Carboniferous–Permian intramontane troughs at the NE margin of the Bohemian Massif. NSB, North Sudetic Basin; KPB, Krkonoše Piedmont Basin; ISB, Intra-Sudetic Basin (intra-basinal depositional troughs: Z, Žacleř Trough; W, Wałbrzych Trough; N, Nowa Ruda Trough).

the basin expanded to the south and became more complex, with slower subsidence close to the axis, surrounded by smaller depositional centres (Mastalerz 1996), including the Żacler Basin to the west, the Wałbrzych Basin to the NW and the Nowa Ruda Basin to the SE (Fig. 1b). In Permian times the main depositional area shifted to the SE, although several subsidiary basins and troughs occurred in the NW (Dziedzic 1961; Nemec 1981a).

In the northern and eastern parts of the basin, where the sequence is most complete, volcanic activity occurred in both Early and Late Carboniferous times and reached its climax in the Early Permian (Awdankiewicz 1999a, b and references therein). The activity changed from dominantly acidic, calc-alkaline in the Carboniferous, to intermediate and acidic, mildly alkaline in the Permian. The earliest volcanism was sited along the northern margin of the basin, and subsequent activity moved southeastwards, consistently with the intra-basinal depositional troughs. The activity was dominated by effusive eruptions of less-evolved magmas in the north and west, while explosive eruptions and emplacement of major subvolcanic intrusions, both of more-evolved compositions, were typical in the central and SE parts of the basin. These relationships occur in each of the three stages of volcanism, but are most clear in the early Permian event, which is recognized in a continuous outcrop across the basin. Within this outcrop a basaltic trachyandesite shield volcano and an extensive rhyolitic (low-silica) effusion have been determined in the NW and west. Near the centre, a complex of trachyandesite lava flows and domes occurs and, in the SE, a rhyolitic (high-silica) ignimbrite-related caldera with associated trachyandesitic and rhyolitic subvolcanic intrusions have been interpreted (Awdankiewicz 1998, 1999a; Awdankiewicz et al. 2003).

Subvolcanic intrusions

The earliest subvolcanic intrusions were emplaced during the early stages of the basin

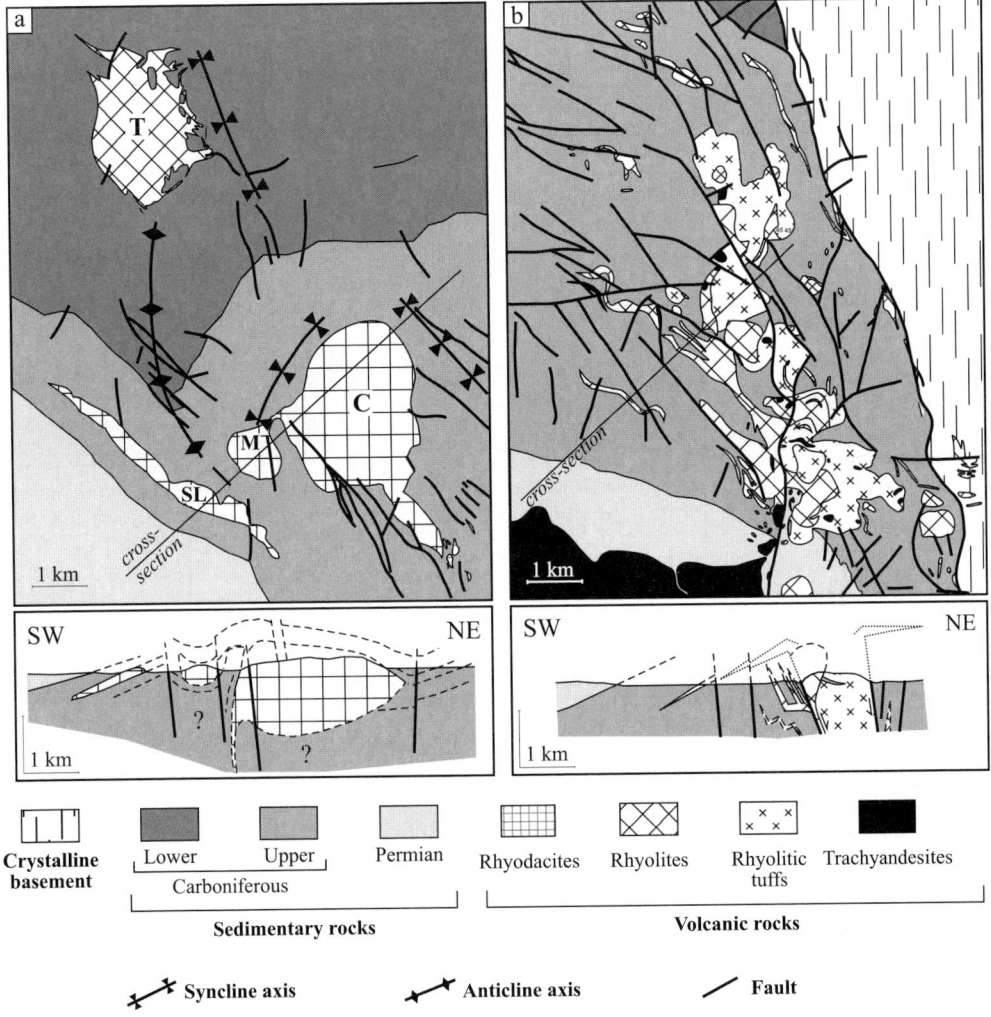

Fig. 2. Geological sketches and cross-sections of volcanic and subvolcanic complexes at the western (**a**) and eastern (**b**) margins of the Wałbrzych Trough (location of the Wałbrzych Trough is shown in Fig. 1b). Symbols in (**a**): T, Trójgarb intrusion; C, Chełmiec laccolith; M, Mniszek phacolith; SL, Stary Lesieniec lavas.

opening in Late Tournaisian or Early Viséan times (Nowakowski & Teisseyre 1971; Awdankiewicz 1999a). The intrusions, in the sedimentary rocks at the base of the sequence near the northern margin of the basin, comprise a few thin (<3.5 m) andesite sills (Fig. 4a) which can be traced for some 3 km. The ascent of the andesite magma was probably facilitated by the basement fractures which controlled the development of the basin. The andesite sills were emplaced into fresh, poorly lithified sediments, and some interpenetration of magma and sediments occurred along the intrusive contacts (load casts and flame-like structures; Nowakowski & Teisseyre 1971).

In Late Carboniferous times, two volcanic centres were active along the fault-controlled margins of the Wałbrzych Trough (Fig. 1). Along the western margin of the trough (Fig. 2a), rhyodacitic and rhyolitic magmas were both erupted, forming an extensive lava cover (Stary Lesieniec rhyodacites), and intruded, forming a few large intrusions in the folded sedimentary sequence (Awdankiewicz 1999a). The intrusions include the Chełmiec laccolith, the SE aligned Sobięcin dyke, which extends upwards from its top, and the Mniszek phacolith. In the NW part of the complex, the structural position and outcrop pattern of the Trójgarb intrusion, with its oval outcrop near the hinge of an anticline, bounded by a syncline to the east and a fault to the west, is similar to the Chełmiec intrusion, and is a laccolith rather than a plug (Dziedzic & Teisseyre 1990). Overall, the volcanic/subvolcanic complex represents a NW-aligned dome-like structure, mainly intrusive at its core, effusive lavas on its SW flanks, and an erosional unconformity at its top. The ascent of the magmas was controlled by NW-trending faults in the basement (Nemec 1979). In addition, the Late Carboniferous subvolcanic activity in that area included intrusion of thin basaltic andesite sills (known from drill cores only) further west of the Wałbrzych Basin. Distribution, geometry and other features of the sills are very similar to the Early Carboniferous andesitic sills mentioned above.

Along the eastern margin of the Wałbrzych Trough (Fig. 2b) a NNW-trending belt of possibly up to 10 maars developed in Late Carboniferous times (Nemec 1979, 1981b; Awdankiewicz 1999a). The activity commenced with explosive, phreatomagmatic eruptions of rhyolitic magmas and formation of diatremes filled with volcaniclastic rocks. Later, rhyolitic and trachyandesitic magmas intruded both into the diatreme fill and adjacent sedimentary sequence, as dykes, sills and small plugs. Transgressive sills (composite intrusive sheets of stair-and-step geometry) occur in the sedimentary sequence (Nemec 1979), while small domes and laccoliths in the diatremes possibly extend into extrusions. Emplacement of the larger intrusions deformed and fluidized the unconsolidated deposits in their contact zones (Fig. 4b). A tentative correlation of sedimentary intercalations in the diatremes suggests that the diatreme fill subsided for at least several hundred metres (Awdankiewicz 1999a).

The third major volcanic centre was a 10-km wide caldera (Fig. 3), in the SE part of the Intra-Sudetic Basin. It formed as a consequence of a voluminous (tens of cubic kilometres) ignimbrite-forming eruption (the Góry Suche Rhyolitic Tuffs; Awdankiewicz 1998, 1999a) in Early Permian times. Following the eruption, trachyandesitic and rhyolitic magmas intruded the caldera margins. The largest intrusions, in the NW and SE parts of the caldera, reflect magma supply from NW–SE faults in the basement. The intrusions were emplaced mainly into the mudstones and sandstones close to the base of the ignimbrite sheet. The c. 300 m thick and homogeneous ignimbrite sheet prevented the rise of magmas to the surface. Most of the intrusions are laccoliths and sills which intruded and marginally interdigitated with the sedimentary rocks. Locally, the peperitic and brecciated margins of the intrusions indicate fluidization of the country rocks, and, elsewhere, small-scale deformation suggests at least partial lithification. A strong silification of the intrusive contacts is found in places (Fig. 4c). However, at the SE margin of the caldera, where the ignimbrite sheet is thinner, the dome-like intrusions almost reach the surface. The final volcanic activity within the caldera, and further SE, was the emplacement of basaltic andesites and rhyolitic tuffs.

Summary

During the Permo-Carboniferous development of the Intra-Sudetic Basin there were strong links between tectonics, sedimentation and volcanic/subvolcanic activity. Tectonics controlled the location of the depositional centres, the ascent pathways of the magmas and the volcanic centres. The volcanic centres situated away from the intra-basinal depositional troughs, such as the Early Permian volcanoes in the western part of the Intra-Sudetic Basin, were characterized by few subvolcanic intrusions. In contrast, the volcanic centres located close to intra-basinal depositional troughs, such as the Late Carboniferous volcanoes near the Wałbrzych trough and Permian volcanic centres in the SE part of the Intra-Sudetic Basin, included many subvolcanic intrusions. These features suggest that magma emplacement level (lavas v. subvolcanic intrusions) was influenced by the thickness of the sedimentary basin fill. Thicker sequences within the troughs restricted significant amounts of magma below the surface in subvolcanic intrusions. The relationship reflects the role of negative density gradients between the rising magma and the country rocks as the driving force of magma ascent (e.g. Williams & McBirney 1979; Francis 1982; Francis & Walker 1987; Corry 1988). However, there are marked differences in the structure and geometry of the intrusive complexes, which result from both the magmatic and tectonic development. The emplacement of the laccoliths, and other intrusions, on the western edge of the Wałbrzych Trough was probably synchronous with folding of the sequence. On the eastern margin of the trough, and around the Permian caldera, emplacement of the intrusions was influenced by faults at the diatreme and caldera margins and by major bedding planes in the host sequence. The influence of magma composition and related physical properties on the emplacement processes and structure of the subvolcanic

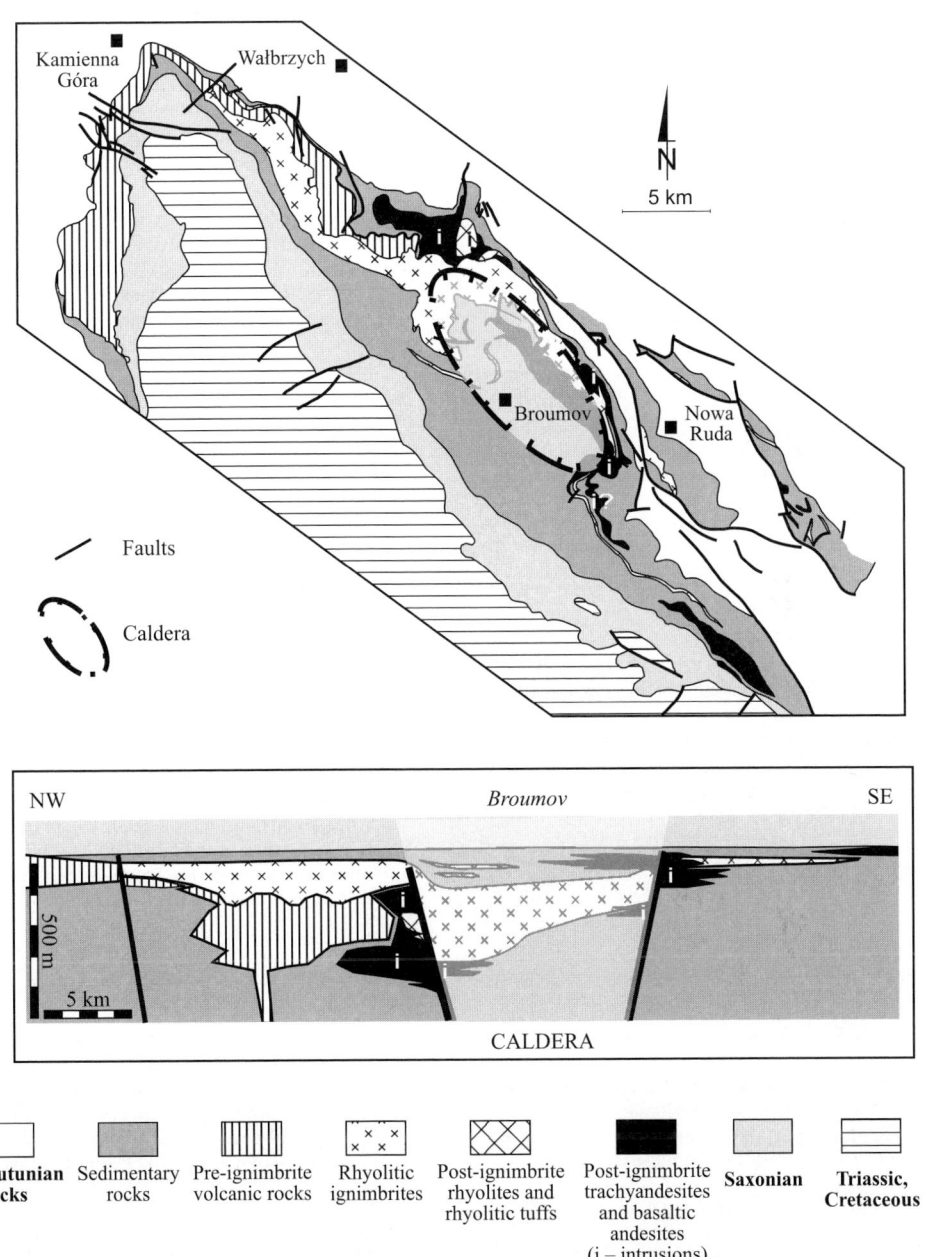

Fig. 3. Geological sketch and schematic cross-section showing the relationships between the Permian ignimbrites, related caldera, pre-ignimbrite volcanic rocks and post-ignimbrite volcanic rocks and intrusions in the Intra-Sudetic Basin.

complexes appear less distinctive, as intermediate and acidic intrusions show a very similar distribution and form (e.g. around the Permian caldera).

The subvolcanic complexes of the Intra-Sudetic Basin characterized above show numerous structural (and geochemical) similarities with those found in other Permo-Carboniferous

Fig. 4. (a) Lower Carboniferous andesite sills in conglomerates at Nagórnik village, in the northernmost part of the Intra-Sudetic Basin. (b) Late Carboniferous trachyandesitic laccolith/cryptodome in a diatreme at the eastern margin of the Wałbrzych Trough (T, trachyandesites; S, sedimentary rocks; R, rhyolitic tuffs). Emplacement of the intrusion disturbed and partly fluidized the country rocks, forming a zone of sedimentary rocks with tuff and trachyandesite blocks (S + R,T). The exposed section is c. 50 m high. (c) top of a Permian trachyandesitic laccolith (T) within mudstones (M) at Świerki, eastern part of the Intra-Sudetic Basin. The mudstones are silicified (Q) in the contact zone of the intrusion. The exposed section is c. 15 m high.

basins of Europe. For instance, felsic laccolith complexes are prominent in the Late Palaeozoic Ilfeld-, Saar–Nahe- and Saale basins (Breitkreuz & Mock 2002); diatremes with associated sills, laccoliths and domes are found in the Saar–Nahe Basin (Lorenz & Haneke 2002); and andesitic sills/laccoliths emplaced in sedimentary strata underneath a thick ignimbrite sheet are described from the Flechtingen Block (Breitkreuz et al. 2002). The analogies in the development of volcanic/subvolcanic complexes within many of the Permo-Carboniferous basins of Central Europe possibly reflect common patterns of interrelated tectonic, sedimentary and magmatic processes characteristic of the region.

The Institute of Geological Sciences, University of Wrocław, is gratefully acknowledged for the support of the study (grant 2022/W/ING/02-3). Helpful reviews of the paper, by A. Muszyński and M. Howells, are highly appreciated.

References

AWDANKIEWICZ, M. 1998. Permskie tufy ryolitowe niecki śródsudeckiej: geologia, petrologia, geochemia. *Polskie Towarzystwo Mineralogiczne, Prace Specjalne*, **11**, 51–53.

AWDANKIEWICZ, M. 1999a. Volcanism in a late Variscan intramontane trough: Carboniferous and Permian volcanic centres of the Intra-Sudetic Basin, SW Poland. *Geologia Sudetica*, **32 (1)**, 13–47.

AWDANKIEWICZ, M. 1999b. Volcanism in a late Variscan intramontane trough: the petrology and geochemistry of the Carboniferous and Permian volcanic rocks of the Intra-Sudetic Basin, SW Poland. *Geologia Sudetica*, **32(2)**, 83–111.

AWDANKIEWICZ, M., KUROWSKI, L., MASTALERZ, K. & RACZYŃSKI, P. 2003. The Intra-Sudetic basin – a record of sedimentary and volcanic processes in late- to post-orogenic tectonic setting. *GeoLines*, **16**, 165–183.

BREITKREUZ, C. & MOCK, A. 2002. Are laccolith complexes characteristic of transtensional basin systems? – Examples from Permocarboniferous Central Europe. *In*: BREITKREUZ, C., MOCK, A. & PETFORD, N. (eds) *First International Workshop: Physical Geology of Subvolcanic Systems – Laccoliths, Sills and Dykes (LASI)*. Wissenschaftliche Mitteilungen, **20**, 8–9.

BREITKREUZ, C., AWDANKIEWICZ, M. & EHLING, B.-C. 2002. Late Palaeozoic andesite sills in the Flechtingen Block, north of Magdeburg (Germany). *In*: Breitkreuz, C., Mock, A. & Petford, N. (eds) *First International Workshop: Physical Geology of Subvolcanic Systems – Laccoliths, Sills and Dykes (LASI)*. Wissenschaftliche Mitteilungen, **20**, 7–8.

BUBNOFF, S. 1924. Die tektonik am Nordostrande des Niederschlesischen Kohlenbeckens und ihr Zusammenhang mit den Kohlensäureausbrüchen in den Flözen. *Zeitschrift für das Berg-, Hütten und Salinenwesen im Preussischen Staat*, **72**, 106–138.

CORRY, C.E. 1988. Laccoliths; mechanics of emplacement and growth. *Geological Society of America Special Papers*, **220**, 110 pp.

DATHE, E. & BERG, G. 1926. *Geologische karte von Preussen und benechbarten deutschen Ländern 1:25 000. Erläuterungen zu Blatt Waldenburg. Lieferung 145*. Preußischen Geologischen Landesanstalt, Berlin, 70 pp.

DZIEDZIC, K. 1961. Utwory dolnopermskie w niecce śródsudeckiej. *Studia Geologica Polonica*, **6**, 121 pp.

DZIEDZIC, K. 1980. Subvolcanic intrusions of Permian volcanic rocks in the central Sudetes. *Zeitschrift für Geologische Wissenschaften*, **8**, 1182–1200.

DZIEDZIC, K. & TEISSEYRE, A.K. 1990. The Hercynian molasse and younger deposits of the Intra-Sudetic Depression, SW Poland. *Neues Jahrbuch für Mineralogie, Geologie und Paläontologie*, **179**, 285–305.

FRANCIS, E.H. 1982. Magma and sediment – I. Emplacement mechanism of late Carboniferous tholeiite sills in northern Britain. *Journal of the Geological Society, London*, **139**, 1–20.

FRANCIS, H. & WALKER, B.H. 1987. Emplacement of alkali dolerite sills relative to extrusive volcanism and sedimentary basins in the Carboniferous of Fife, Scotland. *Transactions of the Royal Society, Edinburgh: Earth Sciences*, **77**, 309–323.

FRANKE, W. & ŻELAŹNIEWICZ, A. 2000. The eastern termination of the Variscides: terrane correlation and kinematic evolution. *In*: FRANKE, W., HAAK, V., ONCKEN O. & TENNER D. (eds) *Orogenic Processes: Quantification and Modelling in the Variscan Belt*. Geological Society, London, Special Publications, **179**, 63–86.

GROCHOLSKI, A. 1965. Wulkanity niecki wałbrzyskiej w świetle badań strukturalnych. *Biuletyn. Instytut Geologiczny, Warszawa*, **191**. *Z badań Geologicznych na Dolnym Śląsku*, **12**, 5–68.

HOEHNE, K. 1961. Zum Alter der Porphyre im Waldenburger Bergbaugebiet (Niederschlesien). *Geologisches Jahrbuch*, **78**, 299–328.

LORENZ, V. & HANEKE, J. 2002. Relationships between diatremes, sills, laccoliths, extrusive domes, lava flows, and tephra deposits with water-saturated unconsolidated sediments in the late Hercynian intermontane Saar–Nahe Basin, SW Germany. *In*: BREITKREUZ, C., MOCK, A. & PETFORD, N. (eds) *First International Workshop: Physical Geology of Subvolcanic Systems – Laccoliths, Sills and Dykes (LASI)*. Wissenschaftliche Mitteilungen, **20**, 29–30.

LORENZ, V. & NICHOLLS, I.A. 1976. The Permocarboniferous basin and range province of Europe. An application of plate tectonics. *In*: FALKE, H. (ed.) *The Continental Permian in Central, West and South Europe*. D. Reidel Publishing Company, Dordrecht, The Netherlands, 313–342.

LORENZ, V. & NICHOLLS, I.A. 1984. Plate and intraplate processes of Hercynian Europe during the Late Paleozoic. *Tectonophysics*, **107**, 25–56.

MASTALERZ, K. 1996. Sedymentacja warstw żaclerskich (westfal) w niecce wałbrzyskiej. *In*: MASTALERZ, K. (ed.) *Z badań karbonu i permu w Sudetach*. *Acta Universitatis Wratislaviensis*, **1795**, *Prace Geologiczno–Mineralogiczne*, **52**, 21–78.

MENARD, G. & MOLNAR, P. 1988. Collapse of Hercynian Tibetan Plateau into a late Palaeozoic European Basin and Range province. *Nature*, **334**, 235–237.

NEMEC, W. 1979. *Wulkanizm późnokarboński w niecce wałbrzyskiej (synklinorium śródsudeckie)*. Ph.D. thesis, University of Wrocław, 201 pp.

NEMEC, W. 1981*a*. Tectonically controlled alluvial sedimentation in the Słupiec Formation (Lower Permian) of the Intrasudetic Basin. *International Symposium Central European Permian. Jabłonna, April 27–29, 1978*. Geological Institute, Warsaw, 294–311.

NEMEC, W. 1981*b*. Problem genezy i wieku skał wulkanoklastycznych na wschodzie niecki wałbrzyskiej. *In*: DZIEDZIC, K. (ed.) *Problemy Wulkanizmu Hercyńskiego w Sudetach Środkowych. Materiały Konferencji Terenowej. Ziemia Wałbrzyska, 30–31 Maja 1981*. Wydawnictwo Uniwersytetu Wrocławskiego, 92–105.

NOWAKOWSKI, A. & TEISSEYRE, A.K. 1971. Wulkanity karbońskie i trzeciorzędowe w północnej części niecki śródsudeckiej. *Geologia Sudetica*, **5**, 211–236.

PETRASCHECK, W.E. 1938. Zur Altersbestimmung des varistischen Vulkanismus in Schlesien. *Zeitschrift der Deutschen Geologische Gesellschaft*, **90(1)**, 20–25.

WILLIAMS, H. & MCBIRNEY, A.R. 1979. *Volcanology*. Freeman, Cooper, San Francisco 397 pp.

WOJEWODA, J. & MASTALERZ, K. 1989. Ewolucja klimatu oraz allocykliczność i autocykliczność sedymentacji na przykładzie osadów kontynentalnych górnego karbonu i permu w Sudetach. *Przegląd Geologiczny*, **4**, 173–180.

Are laccolith complexes characteristic of transtensional basin systems? Examples from the Permo-Carboniferous of Central Europe

CHRISTOPH BREITKREUZ & ALEXANDER MOCK

Institut für Geologie, TU Bergakademie Freiberg, Bernhard-von-Cotta-Str. 2, 09599 Freiberg, Germany (e-mail: mock@geo.tu-freiberg.de)

Abstract: By comparing felsic laccolith complexes prominent in the Late Palaeozoic Ilfeld-, Saar–Nahe-, and Saale basins in Germany, a characteristic pattern related to transtensional tectonics is revealed. In contrast to the central magma feeding systems recognized so far for laccolith complexes, individual units of the Late Palaeozoic Central European complexes apparently were fed synchronously by numerous feeder systems arranged laterally in a systematic pattern.

The Ilfeld Basin is a small strike-slip pull-apart basin in the SE of the Hartz Mountains cogenetic with a neighbouring rhomb horst – the Kyffhäuser. The Ilfeld Basin represents a 'frozen-in' early stage of laccolith complex evolution, with small isolated intrusions and domes emplaced within a common level at the intersections of intra-basinal Riedel shears. In the Saar–Nahe basin, numerous medium-sized felsic subvolcanic to subaerial complexes emplaced at a common level have been recognized (Donnersberg-type laccolith complex). The magmatic evolution of the Halle Volcanic Complex in the Saale Basin culminated in the more or less synchronous emplacement of voluminous porphyritic laccoliths within different levels of a thick pile of Late Carboniferous sediments (Halle-type laccolith complex). Laccoliths in the Halle area might consist of several laccoliths typical of the Saar–Nahe Basin according to outcrop pattern and host sediment distribution.

The above-mentioned three post-Variscan Central European basins are characterized by a dextral transtensional tectonic regime, leading to a model for laccolith-complex evolution:

1 Initial lithosphere-wide faulting forms pathways for magma ascent.
2 Supracrustal pull-apart leads to the formation of a transtensional basin.
3 Continued transtension gives way to decompressional melting of the mantle lithosphere, especially if fertilized by previous magmatic activity, as in the Variscan orogen. The mantle melts rise into the lower crust to differentiate, mingle or cause anatexis.
4 The melts homogenize and start crystallizing in a mid- to upper-crustal magma chamber tapped during tectonic episodes.
5 The resulting SiO_2-rich magmas ascend along major transtensional faults into thick sedimentary basin fill.

The amount of transtension and the amount of melt rising from the lithospheric mantle have a major influence on the type and size of the laccolith complex to be formed. Additionally, the presence of a mid- to upper crustal magma chamber is a prerequisite for the formation of the Donnerberg and Halle type laccolith complexes.

Based on a detailed study of one laccolith complex (i.e. the Halle Laccolith Complex, HLC) and a literature review on other laccolith complexes in Permo-Carboniferous Central Europe, we attempt to explain the tectonic controls on the formation of laccolith complexes in continental strike-slip systems. This contribution should serve as a base for discussion about the definition of new types of laccoliths beyond the classic mechanical spectrum defined by Corry (1988), which shows a 'Punch' to 'Christmas-tree' geometry.

Laccoliths and magmatism in strike slip-settings

Laccoliths are intrusive bodies with a flat lower and a curved upper contact. They are common features of intrusive mafic and felsic intracontinental magmatic provinces (Corry 1988; Friedman & Huffman 1998). Several aspects of their emplacement and geometry are not well understood. The classic approach of Gilbert (1877) in the Tertiary Henry Mountains in Utah, USA, has been developed further by Corry

From: BREITKREUZ, C. & PETFORD, N. (eds) 2004. *Physical Geology of High-Level Magmatic Systems.* Geological Society, London, Special Publications, **234**, 13–31. 0305-8719/04/$15.00
© The Geological Society of London 2004.

(1988), with finite element modelling and thorough field investigations (Johnson & Pollard 1973; see also: Jackson & Pollard 1988 and Kerr & Pollard 1998 for recent numerical modelling approaches). Corry postulated that the level of neutral buoyancy of magma and country rock is the major controlling factor for the initiation of laccolith formation. However, other parameters are important, such as the presence of fluids, the stress field in the host rock, and dynamic features of the rising magma (crystallization, viscosity, magma driving pressure, transport rate). According to analogue modelling, laccolith emplacement also requires the presence of a weak layer, e.g. a sole thrust or a less-competent lithology, near the level of emplacement (Roman et al. 1995). There, the orientation of magma flow changes from vertical to horizontal. A sill of c. 30 m thickness forms first and inflates upon reaching a critical expanse determined by the effective thickness of the overburden (Johnson & Pollard 1973; Pollard & Johnson 1973), given that the supply of magma is sufficient.

Recent studies suggest that laccolith-like mechanisms play a major part in the emplacement even of large plutonic bodies (e.g. Vigneresse et al. 1999). Furthermore, the magma chamber below many caldera complexes has a laccolithic geometry ('lacco-caldera' according to Henry et al. 1997). Depending on the surrounding geology and tectonic setting, laccolith geometries might become much more complex than the simple mechanical models suggested (Morgan et al. 1998).

Before turning to the laccoliths in the strike-slip pull-apart basins of Permo-Carboniferous of Central Europe, as a means of comparison, we would like to introduce some examples of similar tectonic environments showing different styles of magmatism. They represent different styles of deformation and magmatic activity, ranging from ancient plutonic emplacement to recent volcanic features.

In the NW corner of the Arabian plate, near the triple junction of Anatolia, Arabia and Africa, Late Cenozoic elongate volcanoes, volcanic ridges and linear clusters of adjacent volcanic vents are rooted on tension fractures, which are a kilometre or several kilometres in length and show similar development in depth. Non-volcanic tension fractures are also common (Adiyaman & Chorowicz 2002).

Late Jurassic strike-slip intra-arc basins formed along the axis of earlier Early to Middle Jurassic extensional intra-arc basins in western North America. Volcanism occurred only in releasing bends in the Late Jurassic arc, producing more episodic and localized eruptions than in the extensional arc, where volcanism was voluminous and widespread (Busby 2002).

The Ollo de Sapo domain of the northern part of the Variscan belt of Spain contains Precambrian and Ordovician metamorphic rocks intruded by the Guitiriz granite. The domain is bounded by two north–south transcurrent shear zones. Plutonism occurred in three steps:

1 development of N-S trending structural and magnetic fabrics;
2 concordant structures in granites and country rocks; and
3 development of shear zones along the eastern and western granite margins.

The proposed emplacement model involves the northwards tectonic escape of a crustal wedge – the Ollo de Sapo Domain – bounded by two shear zones acting as conjugate strike-slip zones (Aranguren et al. 1996).

The island of Vulcano is composed of four main volcanoes which date from about 120 ka to historical times. The time–space evolution of the volcanism indicates a shifting of the activity from the SE sectors towards the NW. Two main systems of NW–SE-trending right-lateral strike-slip faults affect the island. NE–SW- and north–south-trending normal faults are also present. This system of discontinuity is related to the stress field acting in the southern sector of the Aeolian Archipelago. Volcanological and geochronological data are also consistent with the opening of a pull-apart basin (Ventura 1994).

These examples show that a strike-slip tectonic environment can produce very different styles of magmatism. In the next section, we will focus on some specific examples from the Carboniferous–Permian transition in Central Europe. They will be compared to the well-investigated classic examples of the Tertiary Colorado Plateau in the Western US.

Geotectonic setting of Central Europe at the Carboniferous–Permian transition

A number of rift-related basin systems with pronounced magmatic activity developed in the terminal phases of Variscan orogenesis in the foreland and on the cratonic blocks of former Baltica (Arthaud & Matte 1977). Among these are the Oslo Graben, Norway, the Whin Sill region, Northern England, the North Sea graben systems: Central and Horngraben – the former dominated by tholeiitic flood basalts, the latter with chemically varied magmatism – and the central NE German Basin. The latter contains about 48 000 km^3 of volcanic rocks with subordinate SiO$_2$-poor lavas, but dominantly (c. 70%) SiO$_2$-rich, calc-alkaline, subaerial ignimbrites and lava domes (Fig. 1; Breitkreuz & Kennedy 1999; Benek et al. 1996).

The decaying Variscan orogen itself, on the contrary, was characterized by gravitational collapse and dextral strike-slip (Henk 1997). A number of basins developed within this tectonic framework in the later Alpine region, the Pyrenees, the Sudetic Mountains, the Thuringian Forest, SW Germany, and – focused on in this study – the Saar–Nahe, Saale, and Ilfeld basins (Fig. 1). Sedimentation in the latter basins started early (Namurian, Westphalian, and Stephanian, respectively), the onset of volcanism took place at a later stage. A long-standing volcanic evolution with a climax at around 300 Ma (as in other areas) is recorded. Also, subvolcanic SiO$_2$-rich complexes are prominent here. For these Central European systems, a complex magma genesis has been assumed, involving mantle-derived melts which experienced differentiation and mixing with anatectic crustal melts (Arz 1996; Büthe 1996; Romer et al. 2001). The intrusive complexes in the Saar–Nahe, Saale, and Ilfeld basins will be discussed in detail in the next section.

Variscan intramontane strike-slip basins with prominent laccolith complexes

The Ilfeld Basin is a small strike-slip pull-apart basin that formed cogenetically with a neighbouring rhomb horst – the Kyffhäuser – in the SE of the Harz Mountains, Germany (Fig. 2). It is characterized by subalkaline high-K magmatic rocks: pyroclastics, lavas and minor subvolcanic intrusions. The sedimentary basin fill spans the Stephanian to the Saxonian, with mainly alluvial fan deposits (conglomerates and coarse, poorly sorted partly cross-bedded sandstones), some coal-bearing fine-grained sediments, and an interlayering of clay-bearing siltstones and cross-bedded sandstones. In places silicified limestone beds occur. Compositions of the volcanic rocks cover the range: latitic–andesitic, andesitic, rhyodacitic, and rhyolitic. They formed pyroclastic flow deposits (partly ignimbrites), lavas, and ash tuffs. Radiometric dating of dykes and a rhyolite conglomerate revealed ages from 289 to 298.6 ± 1.6 Ma (Büthe 1996). The volume of intrusive/lava dome material has been estimated at about 15 km^3; the total amount of magmatic material (extrusive andesitic pyroclastics and rhyolitic lavas with a non-quantifiable amount of rhyolitic dykes) is about 24 km^3.

In the Ilfeld Basin, small isolated felsic intrusions and domes emplaced and extruded within a common stratigraphic level at the intersections of intra-basinal syn- and antithetic Riedel shears (Fig. 2, Büthe & Wachendorf 1997). Some rhyolitic domes crop out at the margin of the basin, but most magmatic centres are inferred from gravimetric and geomagnetic surveys (Büthe 1996). The emplacement character of the domes has not been shown unambiguously, but they are believed to be mainly extrusive.

The Saar–Nahe basin is filled by lacustrine, deltaic, fluvial, and alluvial fan deposits (Stollhofen & Stanistreet 1994). The predominant lithologies are grey shales with minor coal horizons, conglomerates, and sandstones interrupted by phases of volcanic activity (effusive basaltic to andesitic and rhyolitic deposits of pyroclastic flows from phreatoplinian eruptions). Four main tectonostratigraphic phases can be recognized:

1 initial proto-rift,
2 prevolcanic syn-rift,
3 volcanic syn-rift, and
4 final post-rift phases.

The intrusive activity mainly took place in the volcanic syn-rift phase of basin evolution. In the Saar–Nahe basin, around 230 km^3 of medium-sized felsic subvolcanic to subaerial complexes have been recognized (Bad Kreuznach, Donnersberg, Kuhkopf, Nohfelden, etc., Fig. 3; see Lorenz & Haneke, this volume). In the Saar–Nahe basin, there are abundant contemporaneous pyroclastic deposits related to the formation of the laccolith/dome complexes. Volcanic events are closely related to tectonic events in the basin history (Stollhofen et al. 1999).

The Donnersberg laccolith complex might represent a continuation of the Ilfeld-like scenario. Flow foliation measurements led Haneke (1987) to distinguish 15 units emplaced in lateral contact with each other within the same stratigraphic level – like balloons inflated in a box. Abundant mafic sills and dykes occur in the Donnersberg area. It is thought that the rhyolite dome intruded at a very shallow level, being exposed rapidly and depositing its own debris apron during and shortly after emplacement. A deep-seated intrusion preceded the shallow intrusion of the Donnersberg massif. The above scenario was inferred from seismic exploration of the area, and has been postulated because the pre-Donnersberg sediments are less thick in the area surrounding the Donnersberg

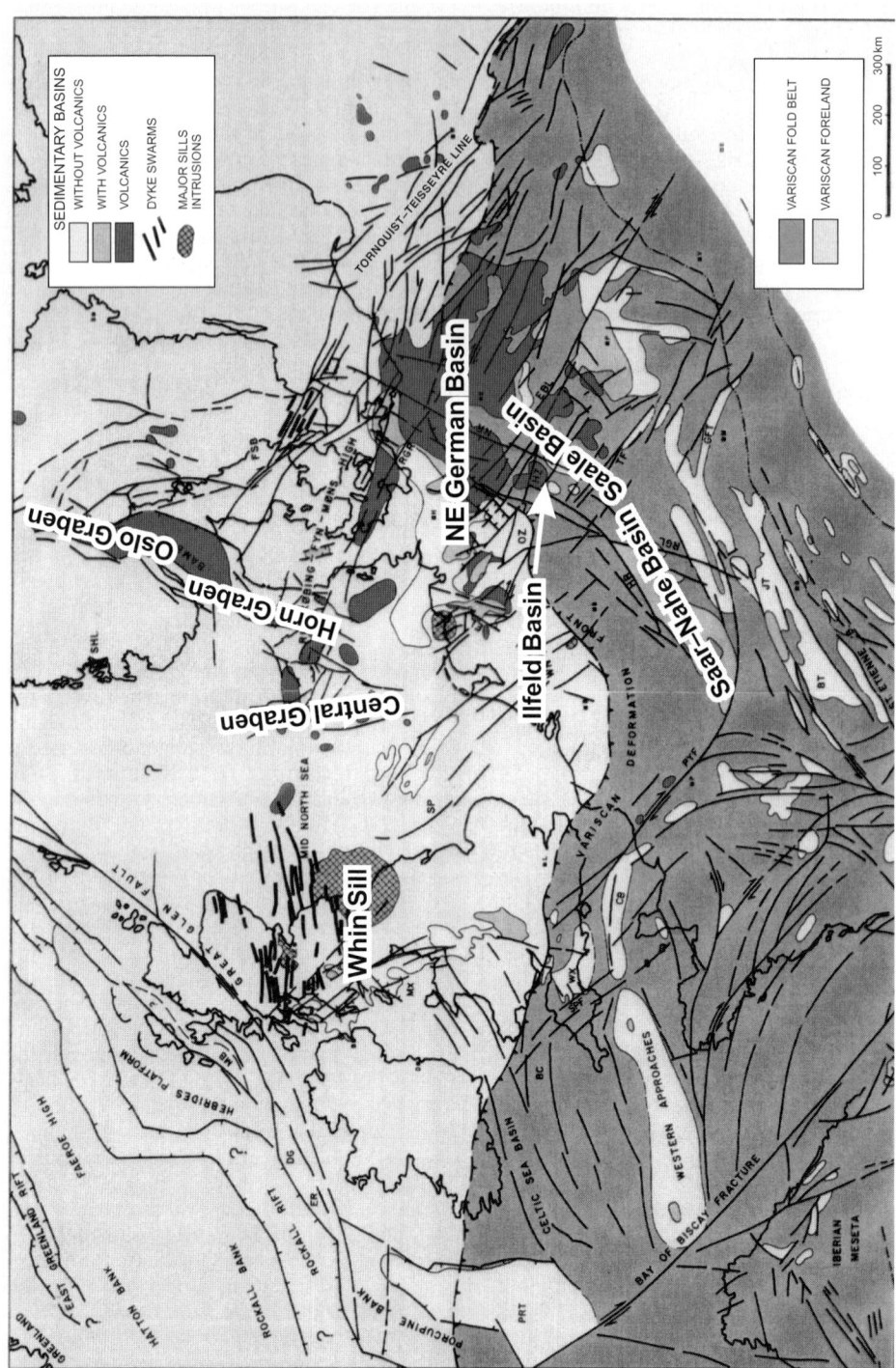

Fig. 1. Palaeogeographical map of Central Europe in Late Palaeozoic times, after Ziegler (1990). The main regions of volcano-tectonic activity are indicated, as well as the basins focused on in this study.

The Ilfeld Basin in the Harz Mts

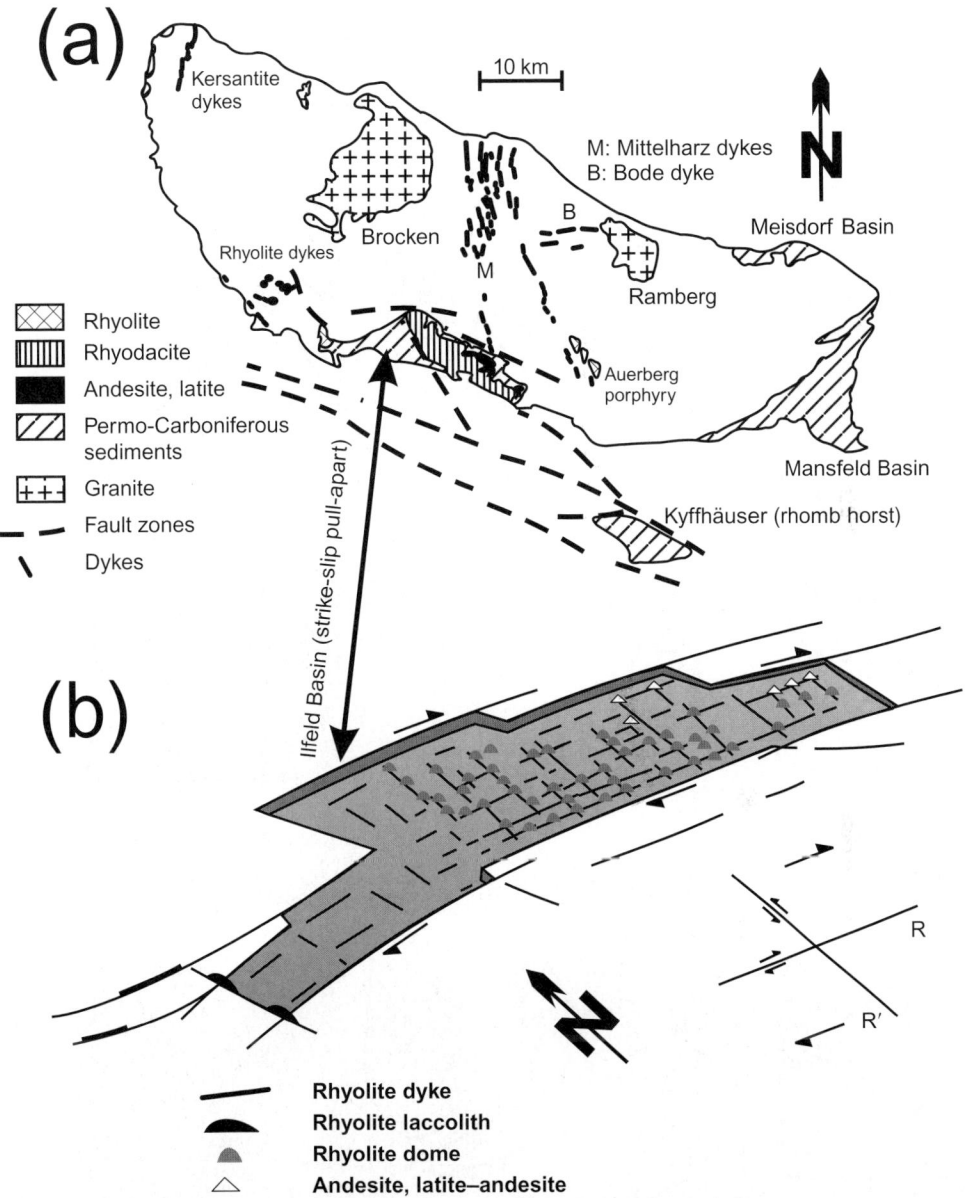

Fig. 2. (**a**) The Ilfeld Basin south of the Harz Mountains (for location, see Fig. 1) and its connection to the Kyffhäuser crystalline rise as a strike-slip pull-apart structure with a rhomb horst. (**b**) Structure and spatial pattern of intrusive bodies in the Ilfeld Basin. The correlation between points of intersecting complementary Riedel shears, and magmatic activity becomes evident (modified after Büthe 1996).

than elsewhere in the generally subsiding Saar–Nahe Basin. Rocks equivalent to the Donnersberg rhyolite have been dated with Rb–Sr to 280 Ma, and with Ar/Ar to 295–300 Ma. Arikas (1986) subdivided the Donnersberg complex into four different units according to their

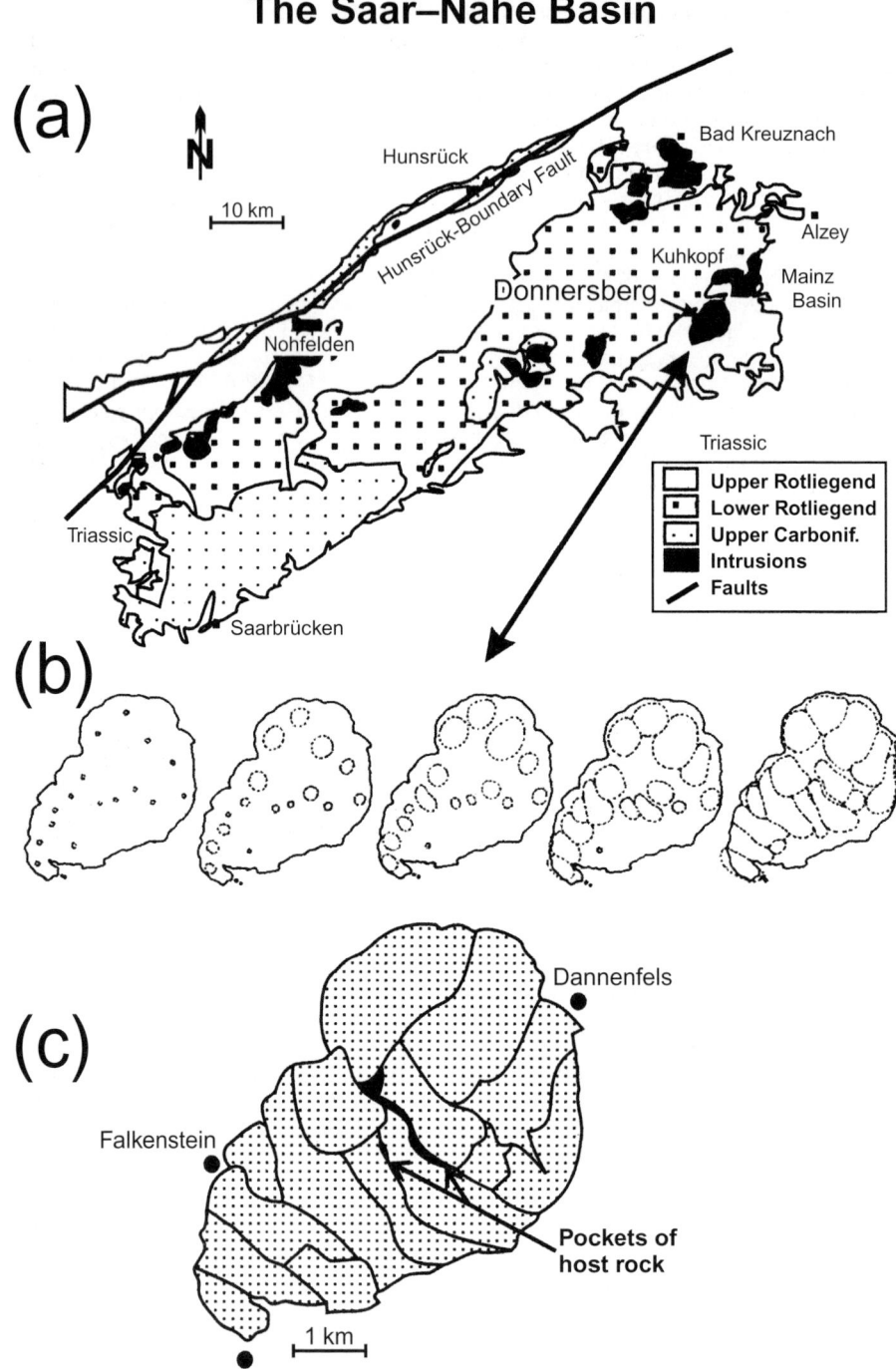

Fig. 3. (**a**) Geological map of the Saar–Nahe Basin, showing the main laccolithic intrusions (black: Arz 1996); (**b**) stepwise intrusion of the Donnersberg laccolith as envisioned by Haneke (1987); (**c**) different intrusive units of the Donnersberg laccolith with pockets of host-rock sediment squeezed in between some central units (map view: Haneke 1987).

geochemical characteristics. The volume of the Donnersberg was estimated at around 40 km^3.

The magmatic evolution of the Halle Volcanic Complex in the Saale Basin culminated in the more or less synchronous formation of >200 km^3 of porphyritic rhyolitic laccoliths which were emplaced at different levels of the thick pile of Late Carboniferous sediments (i.e. the Halle Laccolith Complex (HLC), Fig. 4; Schwab 1965; Kunert 1978; Breitkreuz & Kennedy 1999). Romer et al. (2001) reported the remarkably homogeneous composition of the laccolith complex. The main laccolith units (Wettin, Löbejün, Petersberg, Landsberg) are separated by tilted host sediments (Fig. 4; Kampe et al. 1965). These consist of a succession of grey silt- and mudstones with several fine sandstone beds and coal seams, and a fluvio-limnic succession of reddish-grey conglomerates, siltstones, clays, and sandstones with abundant volcaniclastics (Kampe & Remy 1960; Knoth et al. 1998). The large thicknesses, very thin contact aureoles and lacking or very thin chill zones within the laccoliths indicate the intrusive character of the rhyolitic intrusions (Schwab 1962; Mock et al. 1999). The coarsely porphyritic units (Löbejün, Landsberg; phenocryst content up to 35%) are thicker (>1000 m) than the finely and generally less porphyritic units (Petersberg, Wettin, <300 m, Fig. 4), and the former were emplaced at deeper levels.

Bowl-shaped flow-banding geometries have been observed in SiO$_2$-rich lavas and lava domes (e.g. Fink 1987). In contrast, intrusions develop an onion-like closed flow-banding with cupola-shapes in the upper part and bowl-shapes in the lower part (Nickel et al. 1967; Fink 1987). In the Halle laccoliths, flow structures are present only in the finely crystalline units. Flow-banding structures show complexly cupola- and bowl-shaped geometries, inferring erosion to the upper or lower part, respectively; sometimes flow-banding indicates feeder systems (Mock et al. 1999). The flow structures of the Wettin laccolith, the style of the outcrop and subcrop of the Löbejün laccolith, and the occurrence of at least four pockets of host sediment trapped during the emplacement at the top of the Löbejün laccolith suggest that individual Halle-type laccoliths may consist of several Donnersberg-type units (see Figs 4 & 6, and next section). Size distributions and the spatial arrangement of felsic phenocrysts in the Petersberg laccolith indicate that the intrusion formed by several magma batches without major cooling in between (Fig. 4 and Mock et al. 2003). Only minor pyroclastic activity associated with the intrusion of the laccoliths took place in the HLC (Büchner & Kunert 1997). Instead, the topographic heights created during laccolith intrusion led to more-pronounced erosion and subsequent exhumation of each laccolith. Abundant clasts of the porphyritic rhyolite of the laccoliths can be found in alluvial fan deposits filling the valleys between the laccolith hills.

Melt inclusion compositions in quartz phenocrysts indicate that the H$_2$O-content of the magma was originally in the range of 2–3% (Rainer Thomas, GeoForschungsZentrum, Potsdam, Germany, pers. comm.). Quartz phenocrysts in the Halle laccoliths are often broken along embayments, the fragments slightly rotated and annealed (Fig. 5). Presumably, the rising magma vesiculated upon decompression at some 3–4 km depth (Eichelberger et al. 1986); vesiculation inside embayments led to the fragmentation of quartz phenocrysts (Best & Christiansen 1997). Thus, volatile loss into the unconsolidated sediments of the Saale Basin was possible. As a result, the volatile-poor magma was emplaced as laccolith or lava, instead of erupting explosively. However, the exact conditions determining effusive (lava) or intrusive (laccolith) emplacement are still contentious.

The Saar–Nahe and Saale basins are associated with major lineaments of the late Variscan orogenesis: the Hunsrück Fault and the Northern Harz boundary fault. Early in their history, these basins showed only minor evidence of intrusive activity, but volcanism was eruptive. The Ilfeld Basin represents such an early stage of basin evolution. The inferred magmatic centres (Fig. 2) led mainly to the formation of domes and to eruptive activity. In the Saar–Nahe Basin, as in the Ilfeld Basin, a relation between eruptive centres and the pattern of strike-slip faults and Riedel shears can be shown (Stollhofen et al. 1999).

The size of these three basins has been estimated from maps or cited from the references given (Table 1). The volume of laccolithic intrusions or domes, respectively, has been estimated from the subcrop and outcrop maps (Figs 2 to 4). The areas of outcrop and subcrop have been multiplied by the exposed thickness or the true thickness prior to erosion (where indicators provide a present level of erosion: see above). The values so obtained are minimum values.

Types of laccolith complex

The types of basins and associated laccolith complexes from the previous section shall now be compared with the classic concept developed for laccoliths. Corry (1988) suggested a spectrum of laccolith types and shapes with two

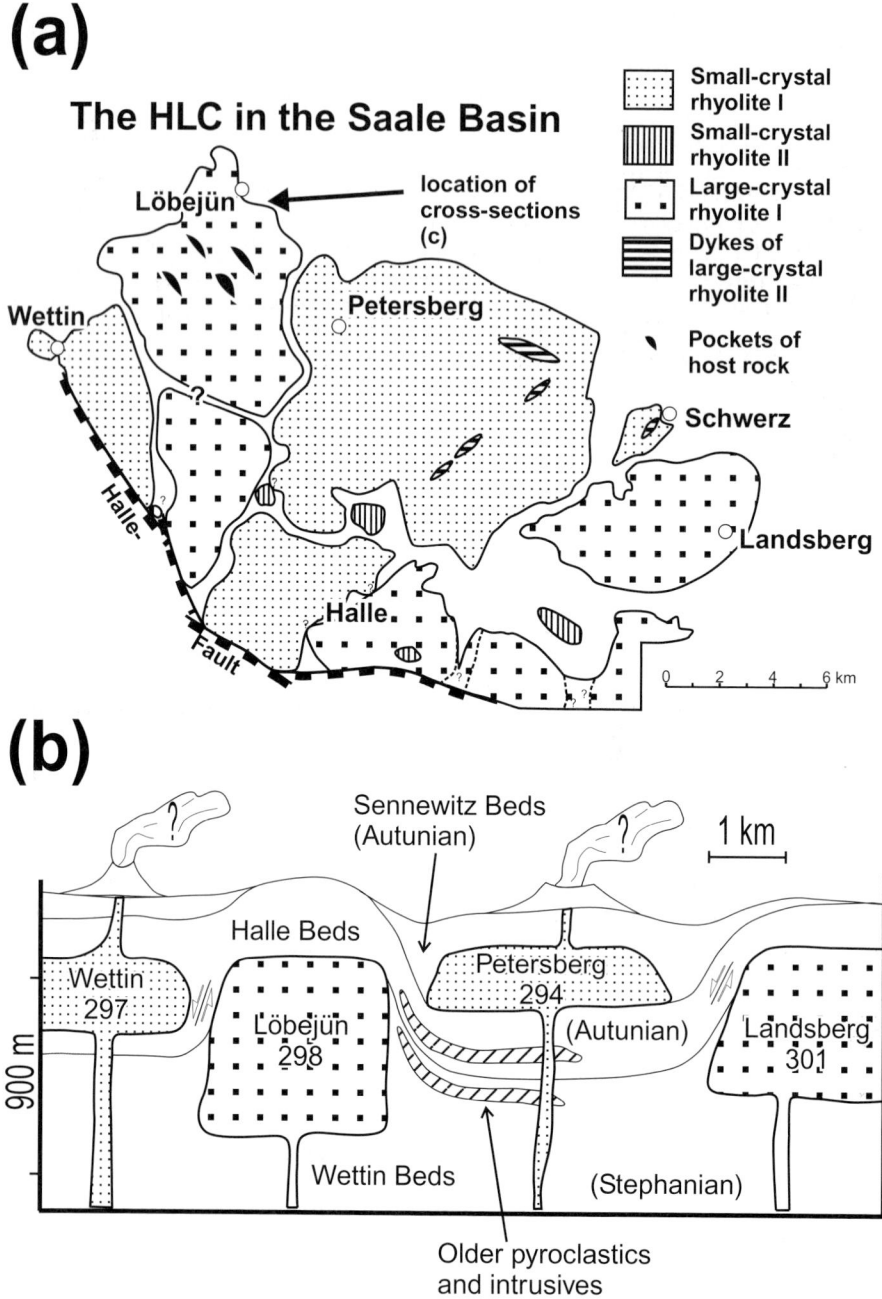

Fig. 4. (a) Map of the HLC, after B.-C. Ehling, LGBSA, Halle (pers. comm.); (b) sketch cross-section east–west through the HLC. Numbers refer to U/Pb-SHRIMP (Sensitive High-Resolution Ion Microprobe) ages from Breitkreuz & Kennedy (1999); (c) three cross-sections at the margin of the Löbejün laccolith, after Kampe *et al.* (1965) show the host-rock deformation caused by the emplacing laccolith (for location, see a); (d) plot of groundmass content v. size of K-feldspar phenocrysts for 99 samples from the porphyritic rhyolitic laccoliths of the HLC, showing the clear distinction between the large and small crystal varieties; (e) *R*-value v. depth plot of six samples from a drill core through the Petersberg laccolith, suggesting the intrusion of the laccolith by at least two batches of magma. For details on the *R*-value method, see Jerram *et al.* (1996) and Mock *et al.* (2003).

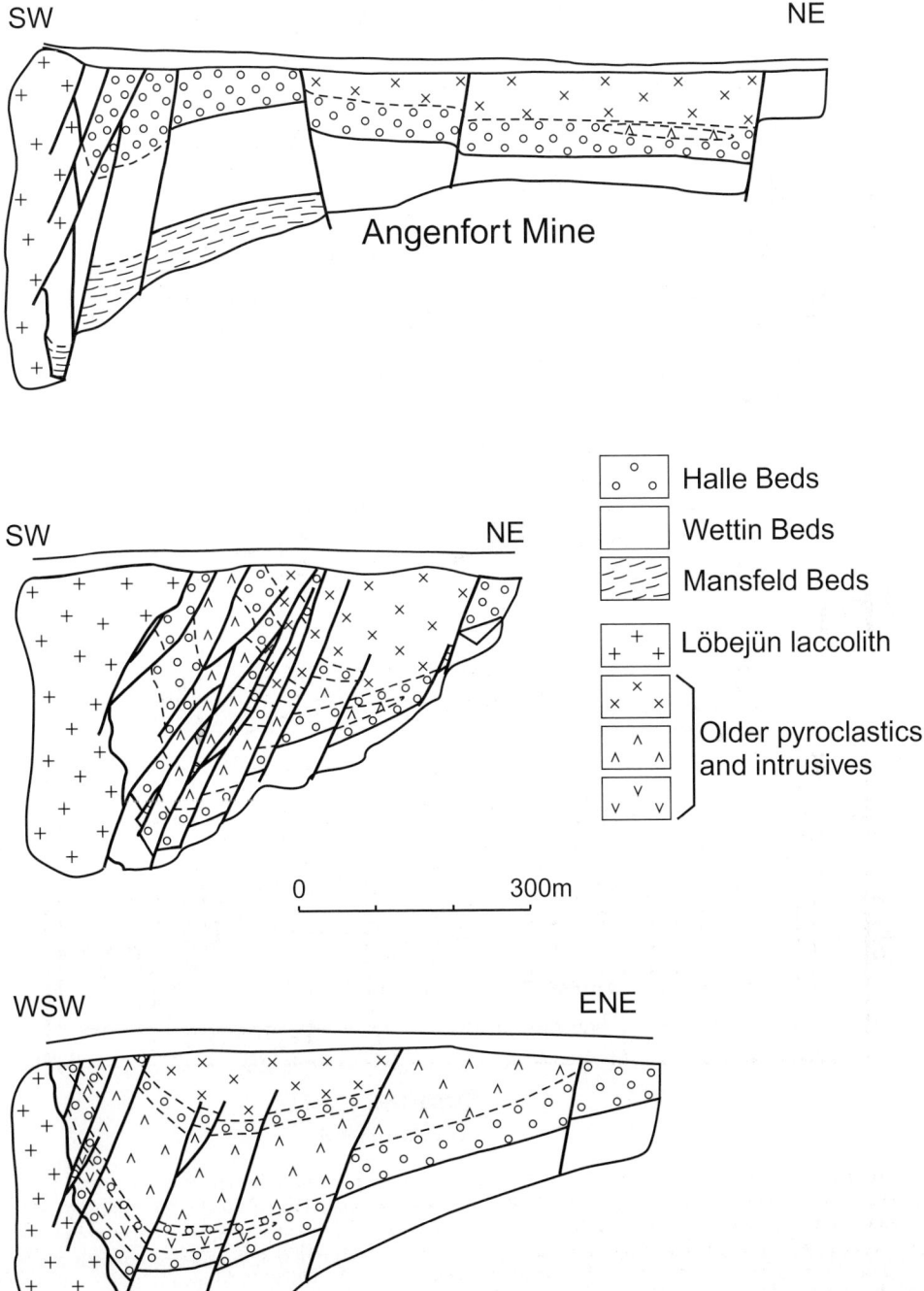

Fig. 4. *continued.*

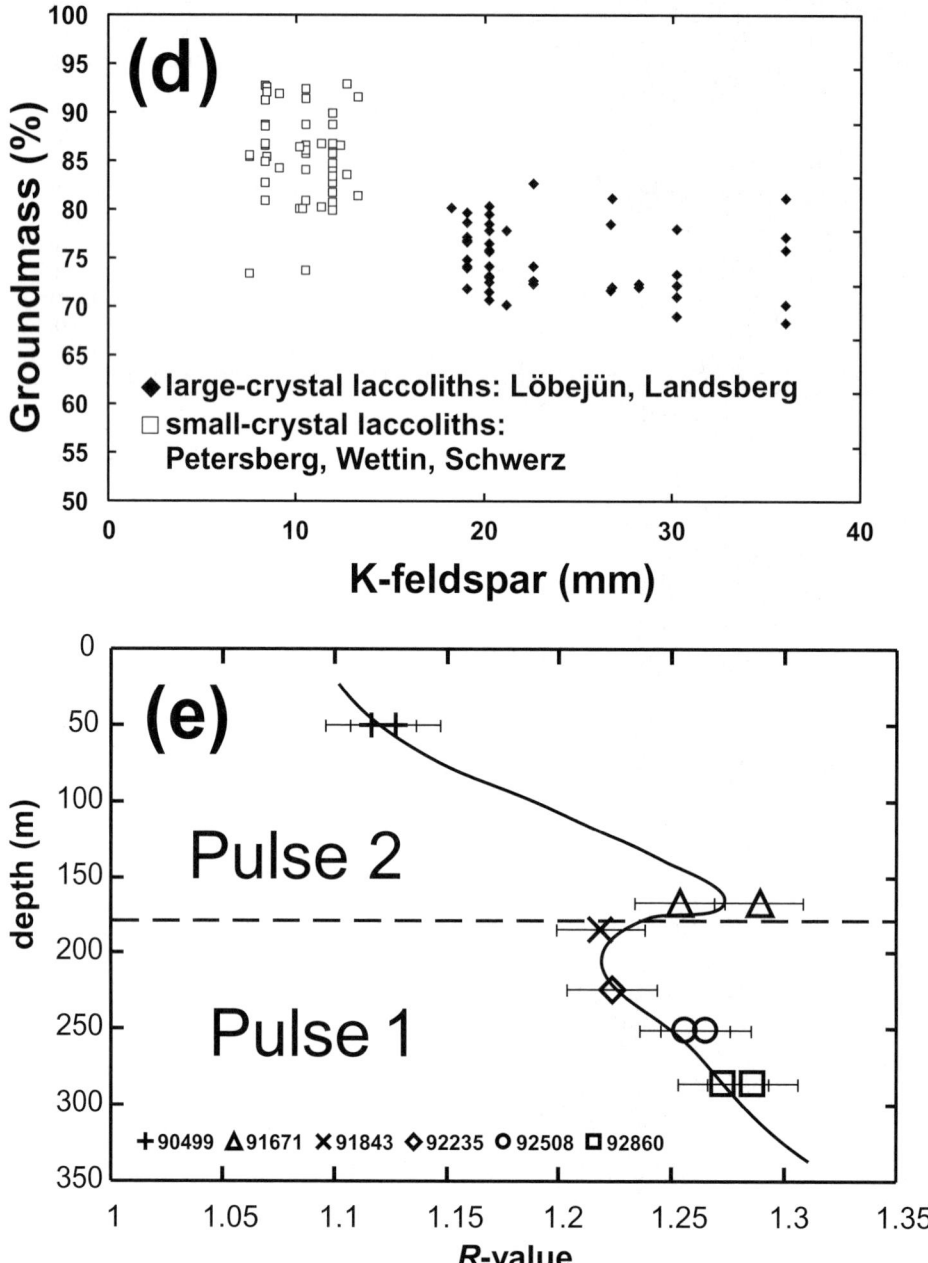

Fig. 4. *continued*.

end-member geometries: Punched and Christmas-tree. These are fed by central single-conduit plumbing systems. In contrast, the individual units of the Late Palaeozoic Central European laccolith complexes, apparently, were fed synchronously by numerous feeder systems laterally arranged in a systematic pattern (Fig. 6):

- the Donnersberg type (Saar–Nahe Basin; Haneke 1987): a group of intrusions (each about 500–1000 m in diameter) with discrete conduits, emplaced simultaneously at ± the same stratigraphic level. In the course of emplacement, the intrusive bodies come in contact with and sometimes penetrate each

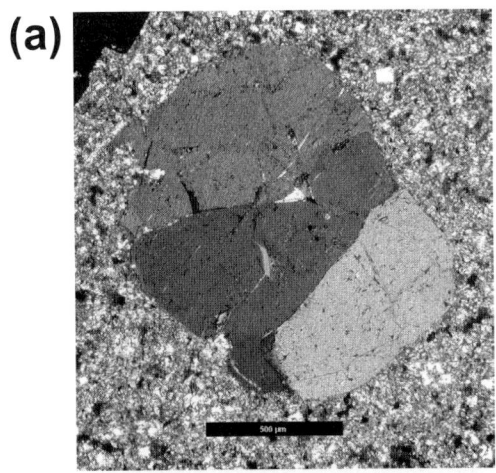
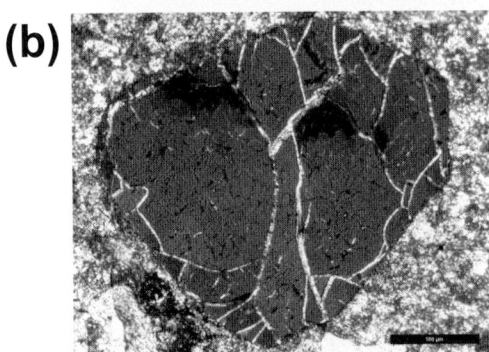

Fig. 5. Broken quartz phenocrysts from the rhyolitic laccoliths of the HLC; (**a**) sample HA 1-55 from a drill core situated in the Landsberg laccolith (Fig. 4); (**b**) sample HA 5-9 from the Schwerz Quarry near Landsberg (Fig. 4); (**c**) sample 28/8/97/4 from a quarry at the Quetzer Berg, NE of the Schwerz Quarry. For approximate locations, see Fig. 4. Scale bars are 500 μm.

other. According to Lorenz & Haneke (this volume) some of the very shallow-level intrusions breached their roofs, forming dome complexes.
- the Halle type (Saale Basin, Kunert 1978; Mock *et al.* 1999): ± synchronous emplacement of laccolith units (several km in diameter, several hundreds to (?) more than a thousand metres thick) into different stratigraphic levels by subsequent magma batches with no intermittent cooling, resulting in the laccolith units being separated by host-rock sediments. Individual Halle-type laccoliths might consist of several Donnersberg-type units (Fig. 6; see also Fig. 4 and previous section).

The above-mentioned laccolith complexes can be envisioned as being fully developed. The small strike-slip pull-apart basin of Ilfeld on the southern flank of the Harz Mountains, Germany, contains a large number of small-scale intrusive centres. They indicate the tectonic control on magma ascent and emplacement in these transtensional settings, and might even be envisaged as early stages in the magmatic evolution in these basin types (see next section and Fig. 2).

The laccolith complexes of the Halle and Donnersberg type formed in Late Palaeozoic basin systems which have been controlled by continent-scale dextral strike-slip (Arthaud & Matte 1977). The Tertiary laccolith complexes in the Paradox Basin in Utah – the classic sites of laccolith research (see above) – are the product of extensional basin-and-range tectonics (Table 2; Huffman & Taylor 1998). The complexes are slightly offset from the intersection points of major lineaments. The basin is an order of magnitude larger than the Palaeozoic basins described above.

Discussion: strike-slip control on the evolution of laccolith complexes

The laccoliths of Permo-Carboniferous Europe differ from those first described from the Western United States in a number of significant ways. Some ideas on the tectonomagmatic control of basin evolution that might account for these differences are presented in this section. As discussed below, the controlling parameters for the formation of the Central European laccolith complexes are active in different levels of the lithosphere affected by intra-continental strike-slip tectonics (Fig. 7).

Amount and rate of magmatism is related to the rate and amount of strike-slip, but also to the

Table 1. *Comparison of three Variscan intra-montane basins and the dominant laccolith complexes*

Basin, laccolith province*	Size	Volume of SiO$_2$-rich magma as laccoliths	Styles of magmatism	Tectonic setting	References
Ilfeld (900 m)	120 km²	c. 15 km³	Dominantly felsic, subalkaline: pyroclastics, lava flows and domes	Strike-slip pull-apart	Büthe (1996)
Saale Basin (c. 3000 m)	>1000 km² (cut off by faults)	c. 210 km³ (HLC only)	Dominantly felsic, calc-alkaline: laccoliths, lava flows and domes, and pyroclastics	Orogenic collapse associated with overall dextral strike-slip	Romer et al. (2001) Schneider et al. (1994) Schneider et al. (1998)
Saar–Nahe (>10 000 m)	4800–6000 km²	c. 230 km³ (Donnersberg: 40 km³)	Bimodal calc-alkaline: pyroclastic and extrusive, laccoliths becoming cryptodomes	Orogenic collapse associated with overall dextral strike-slip	Haneke (1987) Stollhofen et al. (1999)

*First column in parentheses: cumulative thickness of sedimentary basin fill.

(pre-)conditions (temperature, composition) of the mantle which undergoes partial melting. Magma generation in intra-continental (trans)tensional systems is related to decompressional melting of the asthenospheric and/or lithospheric mantle. About 10% of melt is generated on average with every GPa of lithostatic pressure released during upwelling (corresponding to an uplift of c. 35 km; Asimov 2000). In the case of melt generation in the asthenosphere, large amounts of lithospheric stretching are required (50% and more, $\beta = 1.5$, Harry & Leeman 1995), and magmatism of asthenospheric origin would occur relatively late during strike-slip basin evolution. In contrast, early magmatism, prominent during the initial stages of transtension, is characteristic of decompression of lithospheric mantle which has been fertilized with magma and fluids during previous plate-tectonic processes, such as subduction. This model has been developed for initial magmatism in the Tertiary Basin and Range Province (Harry & Leeman 1995; Hawkesworth 1995).

Applying simple models for the structure of mantle and crust under transtension (e.g. from Turcotte & Schubert 2002, p. 75), a melt percentage may be calculated considering the above mentioned relation of melt percentage to the lithostatic pressure release. The thickness of the subcontinental lithosphere after stretching is then:

$$h_{stretched} = h_{unstretched}\left(\frac{\rho_m - \rho_l}{\rho_m - \rho_s}\right)\left(1 - \frac{1}{\beta}\right) + \frac{h_{unstretched}}{\beta}, \quad (1)$$

Fig. 6. Comparison of laccolith types in Tertiary Utah and Permo-Carboniferous Central Europe (Breitkreuz & Mock 2001); (**a**) sketch cross-section of the HLC, after Breitkreuz & Kennedy (1999) showing the main laccoliths and different stratigraphic units that they are intruded into (compare with Fig. 4); (**b**) subcrop map of the porphyritic rhyolites of the HLC; (**c**) possible intrusive pattern of the Löbejün and Wettin laccoliths (enlargement from (**b**), not stippled for clarity) for comparison with the Donnersberg type; (**d**) cross-section of the Donnersberg laccolith of the Saar–Nahe Basin (Haneke 1987). It was intruded into one stratigraphic level: the Nahe Group (Saxonian/dutumian, compare with Fig. 3); (**e**) map of the Donnersberg laccolith with distinct intrusive bodies (Fig. 3; Haneke 1987); (**f**) location of the Utah laccoliths on the Colorado Plateau, showing their relation to major tectonic structures in the area. Note the different scale bars; (**g**), (**h**) for reasons of comparison, the cross-sections of two classic types of laccoliths from the La Sal and Henry Mountains Utah/USA are given. (**f**), (**g**), (**h**) after Friedman & Huffman (1998).

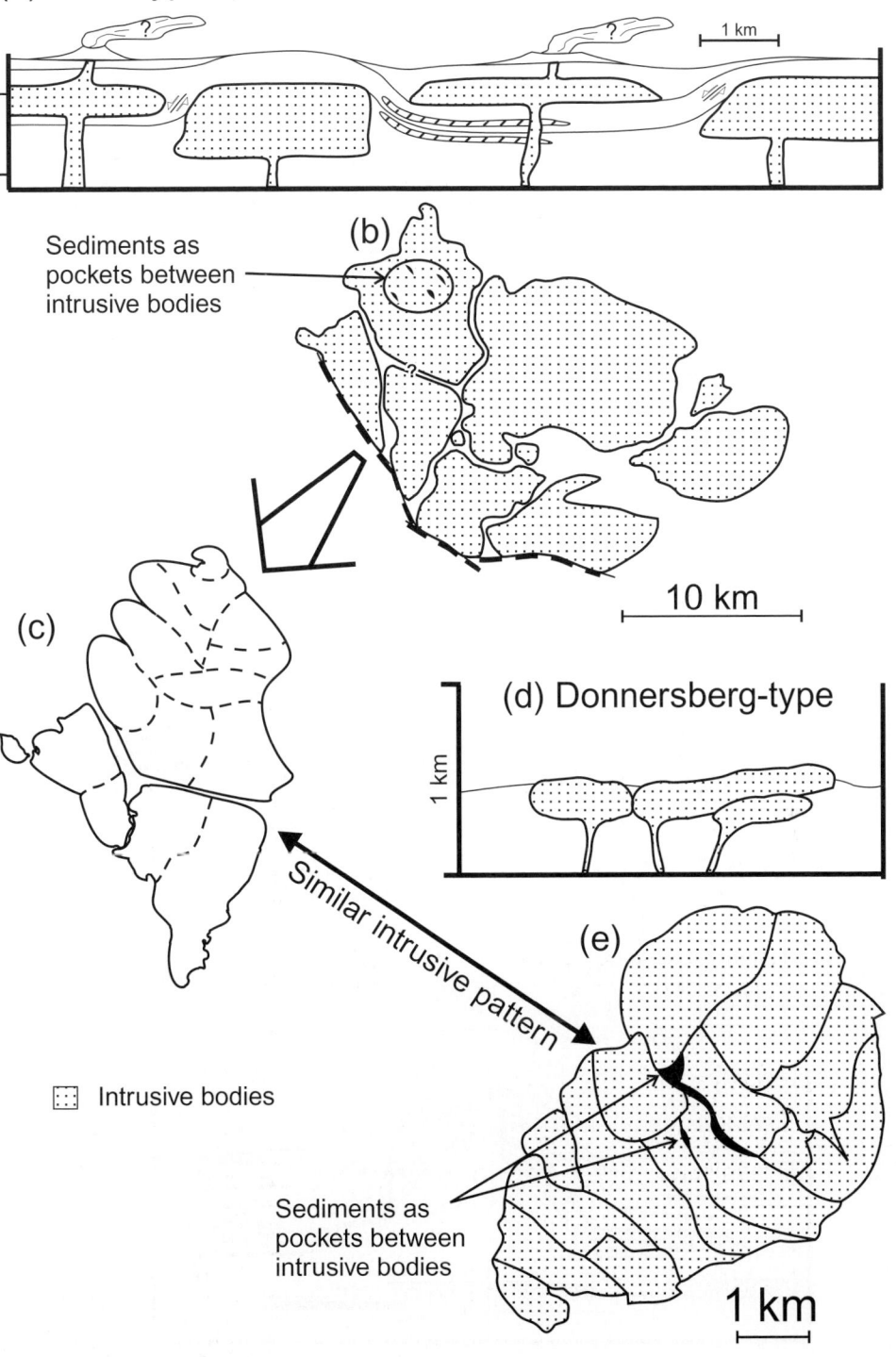

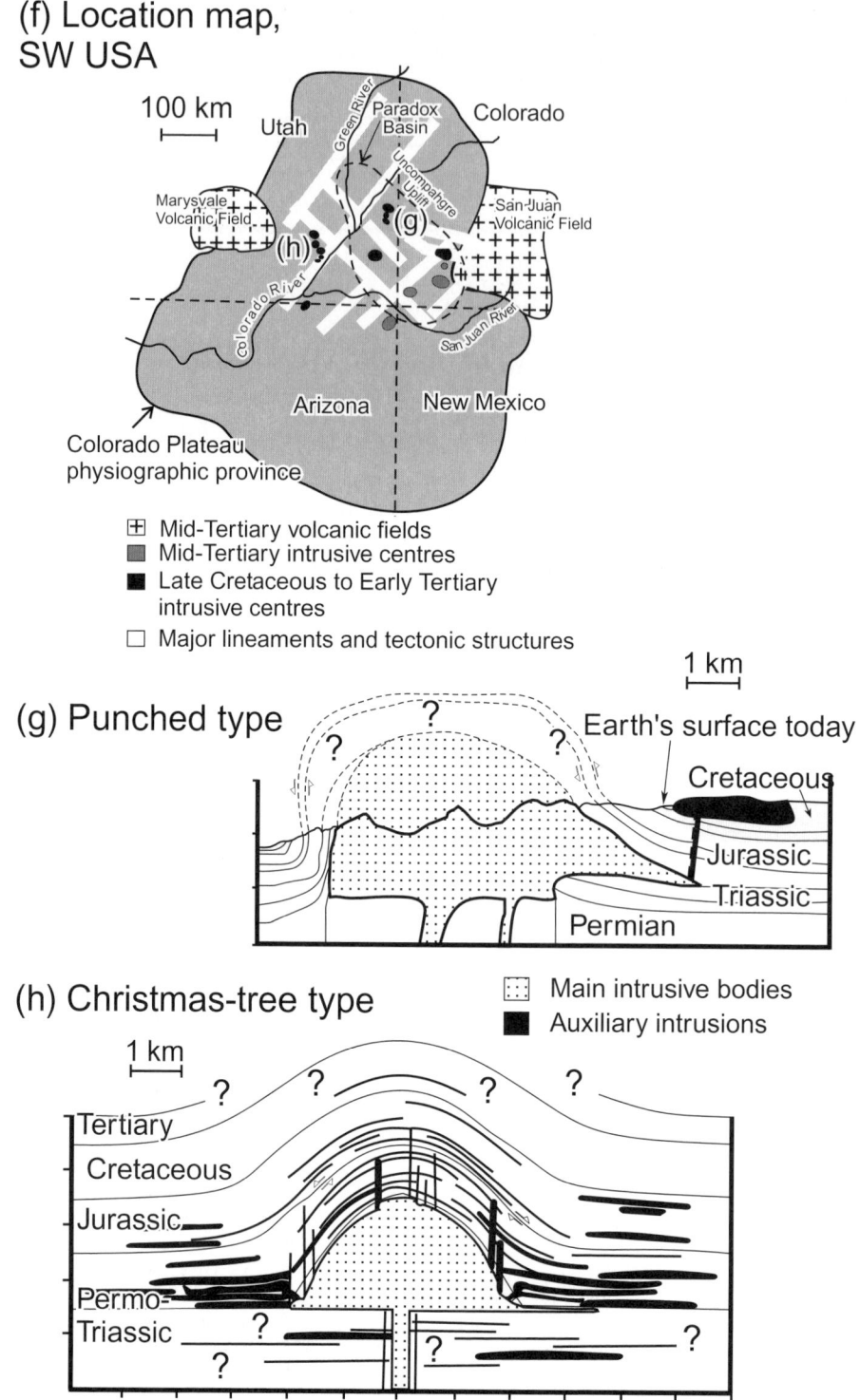

Fig. 6. *continued.*

Table 2. *Features of the laccolith complexes in the Colorado Plateau region*

Basin, laccolith province	Size	Magma volume as laccoliths	Volume of pyroclastics	Styles of magmatism	Tectonic setting	References
Colorado Plateau (e.g. Paradox Basin)	42 000 km²	139 km³	Unknown	Dominantly mafic	Subduction-related crustal extension, uplift	Friedman & Huffman (1998)

with β = stretching factor, ρ_m = density of the asthenosphere, ρ_l = density of the lithosphere, ρ_s = density of the sedimentary basin fill, and h = thickness. Assuming reasonable mean densities (ρ_m = 4500 kg m⁻³, ρ_l = 3200 kg m⁻³, ρ_s = 2200 kg m⁻³), a stretching factor β = 1.5, and an initial continental lithosphere of c. 120 km thickness (Henk 1997), the resulting thickness amounts to c. 100 km. This corresponds to a lithostatic pressure release of c. 0.6 GPa at the base of the lithosphere, and thus to a melt percentage of c. 6% (Asimov 2000). Higher temperatures in the mantle, as discussed for post-Variscan Europe (Henk 1997; Ziegler & Stampfli 2001), would result in a slightly higher percentage of melt (Harry & Leeman 1995).

Following this concept, the late start of magmatism in the Saar–Nahe, Saale, and Ilfeld basins points to a certain asthenospheric melt contribution, the amount of which has not been fully constrained as yet (see Arz 1996; Büthe 1996; Romer et al. 2001 for contributions on magma genesis). In comparison, the strong initial magmatism in the NE German Basin reflects melt generation in the lithospheric mantle (Marx et al. 1995; Benek et al. 1996; Breitkreuz & Kennedy 1999).

The formation of laccolith complexes requires evolved viscous magmas. Principally, these SiO_2-rich magmas can form by differentiation of mantle melt; by anatexis of continental crust due to magmatic underplating; and by mixing and mingling of mantle derived melts with those derived from melting of the crust. Presumably, these processes take place at the crust–mantle boundary or in the lower crust (Fig. 7). The amount of melt generated by mantle decompression and the rate of melt generation may have a strong influence on subsequent magma differentiation, anatexis and assimilation. Also, the composition of the lower crust which is affected by anatexis influences the physical characteristics of the resulting SiO_2-rich melts. For example, the H_2O content strongly influences both the degree of melting and the melt viscosity. Thus, the size and type of the laccolith complex depends on physical processes in both the mantle and the lower crust.

The presence of a pluton c. 3 km below the Donnersberg laccolith complex has been inferred from seismic sections (Haneke 1987). Similarly, the 230 km³ HLC displays a remarkably homogeneous composition (Romer et al. 2001), which could be explained best by the presence of a large magma chamber feeding the laccolith units. Thus, it appears that during evolution of the Late Palaeozoic European intra-continental strike-slip systems, mid- to upper-crustal magma chambers formed, which collected and homogenized the magma ascending from the lower crust. The size difference between the HLC on one hand and the Donnersberg and other Saar–Nahe rhyolitic complexes on the other (Table 1) is presumably related to the size of the mid-crustal magma chamber, which itself is controlled by lower-crustal and mantle processes (see above). The site of this large magma store is constrained by numerous factors, which include the level of neutral buoyancy, magma driving pressure, the structural state of the crust, stress field, the presence of fluids, and magma viscosity.

During renewed strike-slip activity, conduits open and tap the magma chamber. The timing of the tapping event relative to cooling and crystallization of the magma chamber may vary. Apparently, the Donnersberg magma chamber was tapped early, as inferred from the low phenocryst content of the laccolith units (3–9 vol. %; Haneke 1987). Tapping of the chamber beneath the HLC presumably occurred late in respect to its state of crystallization, as indicated by its high phenocryst contents (Fig. 4d). The different phenocryst contents of the Halle laccolith units might either stem from simultaneous tapping of different levels of a magma chamber zoned according to phenocryst size and content, or from differing durations of ascent and emplacement. As a consequence, the presence of a mid- to upper-crustal magma

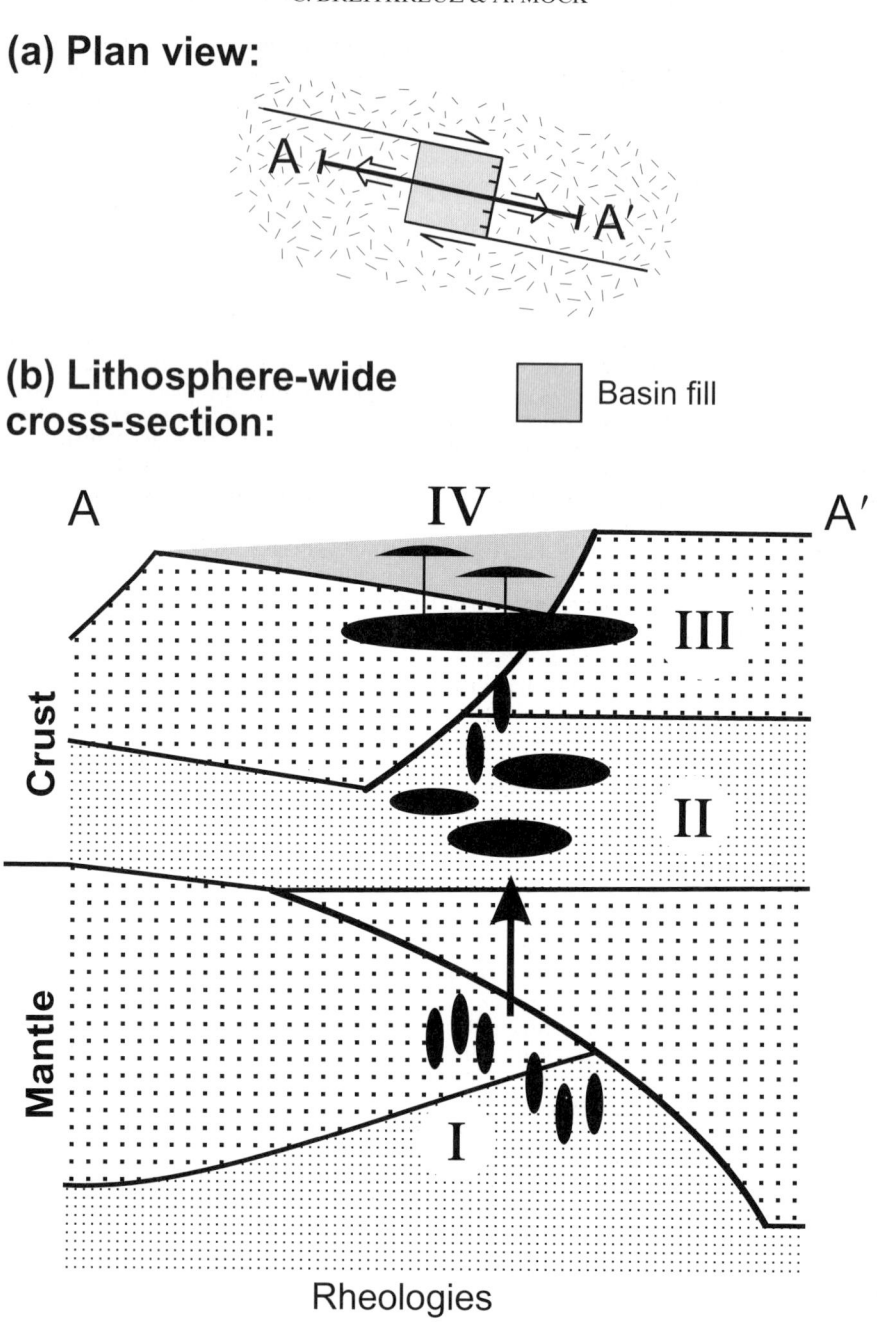

Fig. 7. Simple model of a strike-slip pull-apart basin in (**a**) plan view and (**b**) lithosphere-wide section view; different rheologies and sites of magma generation, ponding, and emplacement are shown. I, decompressional melting in the lithospheric and/or asthenospheric mantle; II, magma differentiation, anatexis, and magma mingling in the lower crust; III, magma chamber in the mid- to upper crust; IV, emplacement of laccolith complexes in the unconsolidated to semi-consolidated basin fill. See text for further explanation. Partly after Eisbacher (1996) and Reston (1990).

chamber is a prerequisite for the formation of the Donnerberg- and Halle-type laccolith complexes, with their multi-feeder systems. Laccolith-complex formation is restricted to the time lapse between the filling of the magma chamber and its advanced crystallization.

The viscous magma rises from the magma chamber through the sediments of the strike-slip basin, along conduits arranged in a spatial pattern determined by the array of intra-basinal faults. The intra-montane basins such as Saar–Nahe, Saale, and Ilfeld contained a thick pile of unconsolidated to semi-consolidated sediments when intrusive activity commenced, so that magma emplacement might have been decoupled from the regional stress regime and largely controlled by neutral buoyancy. Perhaps, in a stress-coupled environment (e.g. in the NE German basin: see above) the formation of lavas prevails, whereas in a stress-decoupled environment it is laccoliths and sills.

The volatile-rich magma vesiculates at a depth of less than 4 km, releasing H_2O and other volatiles into the unconsolidated sediments. Apparently, the rising magma batches which formed the Donnersberg laccolith complex possessed physical properties in common, since they are geochemically and texturally indistinguishable and emplaced in one stratigraphic level (Fig. 6d). In contrast, it is presumed that the emplacement of the geochemically homogeneous HLC into different levels of the sedimentary succession was controlled by different viscosities and densities, caused by different phenocryst sizes and contents. Different amounts of microlites in the groundmass would cause similar or additional effects (Cashman et al. 1999: pahoehoe–aa transition in basalts; Stevenson et al. 2001: effective viscosity as a function of microlite content in rhyolites). However, this cannot be verified in the case of the Halle samples.

Thus, it is speculated that the evolution of laccolith complexes in strike-slip systems is controlled by a number of parameters, acting at different levels of the affected continental lithosphere. The presence of melts may also have a secondary feedback effect on the strike-slip system by acting as a lubricant and, thus, enhancing tectonic activity. As such, melts form part of the tectonic system and should be described and treated like tectonic objects such as faults and folds (Vigneresse 1999).

Conclusions

On the base of the authors' own and literature studies from Permo-Carboniferous strike-slip systems in Central Europe, two types of laccolith complexes are defined, with multi-feeder systems leading to large, closely spaced laccoliths (Donnersberg and Halle type). In the USA, basin-and-range tectonics with regional uplift, on the other hand, led to widely spaced laccolith complexes with single feeder systems and punched and Christmas-tree geometries developing off the intersections of major regional tectonic lineaments. The evolution of the laccolith complexes is strongly dependent on the tectonomagmatic environment (Fig. 7). It appears that transtension in continental lithosphere controls all major phases of laccolith-complex formation:

- Initial lithosphere-wide faulting provides pathways for magma ascent.
- Supracrustal pull-apart leads to the formation of a transtensional basin with a thick pile of unconsolidated sediments.
- Continued transtension gives way to decompressional melting of the mantle lithosphere and possibly the asthenosphere, especially, as occurred in the Variscan orogen, if fertilized by previous magmatic activity. The mantle melts rise into the lower crust to differentiate, mingle, or cause anatexis.
- At mid- to upper-crustal levels, the magmas form large magma chambers that are chemically homogenized and start to crystallize to a varying degree. These chambers are tapped during episodes of tectonic activity, necessarily before complete crystallization, and
- the resulting SiO_2-rich magmas ascend along major transtensional faults into the thick sedimentary basin fill, where vesiculation and devolatilization of the magma takes place.

Conjugate intra-basinal faults (Riedel shears) provide a sieve-like system of pathways for magma ascent, in approximately the upper 4 km of the crust; however, the stress field in the emplacement level may be uncoupled from the regional stress field. The amount of transtension and the amount of melt rising from the lithospheric mantle have a major influence on the type and size of the laccolith complex to be formed, as has the formation of an upper-crustal magma chamber. Laccolith complex evolution depends on local and regional conditions in the crust and the sedimentary basin that it takes place in.

Many thanks to all the participants of the LASI workshop for an inspiring and successful meeting. N. Petford and an anonymous reviewer are thanked for their reviews. These ideas were realized by a research grant from the Deutsche Forschungsgemeinschaft to CB (Grant: Br 997/18-1,2).

References

ADIYAMAN, O. & CHOROWICZ, J. 2002. Late Cenozoic tectonics and volcanism in the northwestern corner of the Arabian Plate; a consequence of the strike-slip Dead Sea fault zone and the lateral escape of Anatolia. *Journal of Volcanology and Geothermal Research*, **117**, 327–345.

ARANGUREN, A., TUBIA, J.M., BOUCHEZ, J.L. & VIGNERESSE, J.L. 1996. The Guitiriz Granite, Variscan Belt of northern Spain; extension-controlled emplacement of magma during tectonic escape. *Earth and Planetary Science Letters*, **139**, 165–176.

ARIKAS, K. 1986. Geochemie und Petrologie der permischen Rhyolithe in Südwestdeutschland (Saar–Nahe–Pfalz–Gebiet, Odenwald, Schwarzwald) und in den Vogesen. *Pollichia-Buch*, **8**, Bad Dürkheim.

ARTHAUD, F. & MATTE, P. 1977. Late Paleozoic strike-slip faulting in southern Europe and Northern Africa: results of right-lateral shear zone between the Appalachians and the Urals. *Geological Society of America Bulletin*, **88**, 1305–1320.

ARZ, C. 1996. *Origin and petrogenesis of igneous rocks from the Saar–Nahe-Basin (SW-Germany): isotope, trace element and mineral chemistry*. Ph.D. thesis, Bayerische Julius-Maximilians-Universität Würzburg.

ASIMOV, P.D. 2000. Melting the mantle. *In*: SIGURDSSON, H., HOUGHTON, B.F., MCNUTT, S.R., RYMER, H. & STIX, J. (eds) *Encyclopedia of Volcanoes*. Academic Press, London, 55–68.

BENEK, R. *ET AL*. 1996. Permo-Carboniferous magmatism of the Northeast German Basin. *Tectonophysics*, **266**, 379–404.

BEST, M.G. & CHRISTIANSEN, E.H. 1997. Origin of broken phenocrysts in ash-flow tuffs. *Geological Society of America Bulletin*, **109**, 63–73.

BREITKREUZ, C. & KENNEDY, A. 1999. Magmatic flare-up at the Carboniferous/Permian boundary in the NE German Basin revealed by SHRIMP zircon ages. *Tectonophysics*, **302**, 307–326.

BREITKREUZ, C. & MOCK, A. 2001. Laccoliths in Cenozoic Western US and Permocarboniferous Europe: models and problems. *Tektonik & Magma, Bautzen, Exkursionsführer und Veröffentlichungen der GGW*, **212**, 30.

BÜCHNER, C. & KUNERT, R. 1997. Pyroklastische Äquivalente der intrusiven Halleschen Rhyolithe. *Mitteilungen des Geologischen Landesamtes Sachsen-Anhalt*, **3**, 37–57.

BUSBY, C.J. 2002. Climatic and tectonic controls on Jurassic intra-arc basins related to northward drift of North America. *AAPG Bulletin*, **86**, 1–197.

BÜTHE, F. 1996. Struktur und Stoffbestand des Ilfelder Beckens geodynamische Analyse einer intramontanen Rotliegend-Molasse. *Braunschweiger Geowissenschaftliche Arbeiten*, **20**, Braunschweig.

BÜTHE, F. & WACHENDORF, H. 1997. Die Rotliegend-Entwicklung des Ilfelder Beckens und des Kyffhäusers: Pull-Apart-Becken und Rhomb-Horst. *Zeitschrift für Geologische Wissenschaften*, **25**, 291–306.

CASHMAN, K.V., THORNBER, C. & KAUAHIKAUA, J.P. 1999. Cooling and crystallization of lava in open channels, and the transition of pahoehoe lava to 'a'a. *Bulletin of Volcanology*, **61**, 306–323.

CORRY, C.E. 1988. Laccoliths; mechanics of emplacement and growth. *Geological Society of America Special Papers*, **220**.

EICHELBERGER, J.C., CARRIGAN, C.R., WESTRICH, H.R. & PRICE, R.H. 1986. Non-explosive silicic volcanism. *Nature*, **323**, 598–602.

EISBACHER, G.H. 1996. *Einführung in die Tektonik*. Enke, Stuttgart.

FINK, J.H. 1987. The emplacement of silicic domes and lava flows. *Geological Society of America Special Papers*, **212**, 103–111.

FRIEDMAN, J.D. & HUFFMAN, J.A.C. 1998. Laccolith complexes of southeastern Utah; time of emplacement and tectonic setting; workshop proceedings. *US Geological Survey Bulletin*, **2158**, 292.

GILBERT, G.K. 1877. *Geology of the Henry Mountains, Utah*. US Geographical and Geological Survey of the Rocky Mountain Region.

HANEKE, J. 1987. Der Donnersberg. *Pollichia-Buch*, **10**, Bad Dürkheim.

HARRY, D.L. & LEEMAN, W.P. 1995. Partial melting of melt metasomatized subcontinental mantle and the magma source potential of the lower lithosphere. *Journal of Geophysical Research*, **100**, 10 255–10 269.

HAWKESWORTH, C., TURNER, S., GALLAGHER, K., HUNTER, A., BRADSHAW, T. & ROGERS, N. 1995. Calc-alkaline magmatism, lithospheric thinning and extension in the Basin and Range. *Journal of Geophysical Research*, **100**, 10 271–10 286.

HENK, A. 1997. Gravitational orogenic collapse vs. plate-boundary stresses: a numerical modelling approach to the Permo-Carboniferous evolution of Central Europe. *Geologische Rundschau*, **86**, 39–55.

HENRY, C.D., KUNK, M.J., MUEHLBERGER, W.R. & MCINTOSH, W.C. 1997. Igneous evolution of a complex laccolith–caldera, the Solitario, Trans-Pecos Texas; implications for calderas and subjacent plutons. *Geological Society of America Bulletin*, **109**, 1036–1054.

HUFFMAN, A.C., JR & TAYLOR, D.J. 1998. Relationship of basement faulting to laccolithic centers of southeastern Utah and vicinity. *In*: FRIEDMAN, J.D. & HUFFMAN, C.J. (eds) Laccolith complexes of southeastern Utah; time of emplacement and tectonic setting; workshop proceedings. *US Geological Survey Bulletin*, **2158**, 41–43.

JACKSON, M.D. & POLLARD, D.D. 1988. The laccolith–stock controversy; new results from the southern Henry Mountains, Utah. *Geological Society of America Bulletin*, **100**, 117–139.

JERRAM, D.A., CHEADLE, M.J., HUNTER, R.H. & ELLIOTT, M.T. 1996. The spatial distribution of grains and crystals in rocks. *Contributions to Mineralogy and Petrology*, **125**, 60–74.

JOHNSON, A.M. & POLLARD, D.D. 1973. Mechanics of growth of some laccolithic intrusions in the Henry Mountains, Utah; I, Field observations, Gilbert's model, physical properties and flow of the magma. *Tectonophysics*, **18**, 261–309.

KAMPE, A. & REMY, W. 1960. Mitteilungen zur Stratigraphie im Raume des Petersberges bei Halle. *Montanwissenschaftliche Berichte – Deutsche Akademie der Wissenschaften, Berlin*, **2**, 364–374.

KAMPE, A., LUGE, J. & SCHWAB, M. 1965. Die Lagerungsverhältnisse in der nördlichen Umrandung des Löbejüner Porphyrs bei Halle (Saale). *Geologie*, **14**, 26–46.

KERR, A.D. & POLLARD, D.D. 1998. Toward more realistic formulations for the analysis of laccoliths. *Journal of Structural Geology*, **20**, 1783–1793.

KNOTH, W., KRIEBEL, U., RADZINSKI, K.-H. & THOMAE, M. 1998. Die geologischen Verhältnisse von Halle und Umgebung. *Hallesches Jahrbuch für Geowissenschaften B*, **Beiheft 4**, 7–34.

KUNERT, R. 1978. Zur Platznahme rhyolithischer Laven. *Zeitschrift für Geologische Wissenschaften*, **6**, 1145–1160.

MARX, J., HUEBSCHER, H.-D., HOTH, K., KORICH, D. & KRAMER, W. 1995. Vulkanostratigraphie und Geochemie der Eruptivkomplexe. In: PLEIN, E. (ed.) *Stratigraphie von Deutschland I – Norddeutsches Rotliegendbecken*. Courier Forschungshefte Senckenberg, Frankfurt, 54–83.

MOCK, A., EXNER, M., LANGE, D., BREITKREUZ, C., SCHWAB, M. & EHLING, B.-C. 1999. Räumliche Erfassung des Fließgefüges der kleinporphyrischen Lakkolithe im Halle-Vulkanit-Komplex. *Mitteilungen des Geologischen Landesamtes Sachsen-Anhalt*, **5**, 169–175.

MOCK, A., JERRAM, D.A. & BREITKREUZ, C. 2003. Using quantitative textural analysis to understand the emplacement of shallow level rhyolitic laccoliths – a case study from the Halle volcanic complex, Germany. *Journal of Petrology*, **44**, 833–849.

MORGAN, S.S., LAW, R.D. & NYMAN, M.W. 1998. Laccolith-like emplacement model for the Papoose Flat Pluton based on porphyroblast-matrix analysis. *Geological Society of America Bulletin*, **110**, 96–110.

NICKEL, E., KOCK, H. & NUNGÄSSER, W. 1967. Modellversuche zur Fliessregelung in Graniten. *Schweizerische Mineralogische und Petrographische Mitteilungen*, **47**, 399–497.

POLLARD, D.D. & JOHNSON, A.M. 1973. Mechanics of growth of some laccolithic intrusions in the Henry Mountains, Utah; II, Bending and failure of overburden layers and sill formation. *Tectonophysics*, **18**, 311–354.

RESTON, T.J. 1990. Shear in the lower crust during extension; not so pure and simple. *Tectonophysics*, **173**, 175–183.

ROMAN, B.T., GAPAIS, D. & BRUN, J.P. 1995. Analogue models of laccolith formation. *Journal of Structural Geology*, **17**, 1337–1346.

ROMER, R., FÖRSTER, H.-J. & BREITKREUZ, C. 2001. Intracontinental extensional magmatism with a subduction fingerprint: the late Carboniferous Halle Volcanic Complex (Germany). *Contributions to Mineralogy and Petrology*, **141**, 201–221.

SCHNEIDER, J.W., RÖSSLER, R. & GAITZSCH, B. 1994. Time lines of the Late Variscan volcanism – a holostratigraphic synthesist, *Zentralblatt für Geologie und Paläontologie*, **5/6**, 477–490.

SCHNEIDER, J.W., SCHRETZENMAYR, S. & GAITZSCH, B. 1998. Excursion Guide Rotliegend Reservoirs at the Margin of the Southern Permian Basin, *Leipziger Geowissenschaften*, **7**, 15–44.

SCHWAB, M. 1962. Über die Inkohlung der Steinkohlen im Nördlichen Saaletrog bei Halle. *Geologie*, **11**, 917–942.

SCHWAB, M. 1965. Tektonische Untersuchungen im Permokarbon nördlich von Halle/Saale. *Freiberger Forschungshefte*, **C139**, 1–109.

STEVENSON, R.J., DINGWELL, D.B., BAGDASSAROV, N.S. & MANLEY, C.R. 2001. Measurement and implication of 'effective' viscosity for rhyolite flow emplacement. *Bulletin of Volcanology*, **63**, 227–237.

STOLLHOFEN, H. & STANISTREET, I.G. 1994. Interaction between bimodal volcanism, fluvial sedimentation and basin development in the Permo-Carboniferous Saar–Nahe Basin (south-west Germany). *Basin Research*, **6**, 245–267.

STOLLHOFEN, H., FROMMHERZ, B. & STANISTREET, I.G. 1999. Volcanic rocks as discriminants in evaluating tectonic versus climatic control on depositional sequences, Permo-Carboniferous continental Saar–Nahe Basin. In: PEDLEY, M. & FROSTICK, L. (eds) Unravelling tectonic and climatic signals in sedimentary successions. *Geological Society, London, Special Publications*, 801–808.

TURCOTTE, D.L. & SCHUBERT, G. 2002. *Geodynamics*. Cambridge University Press, Cambridge.

VENTURA, G. 1994. Tectonics, structural evolution and caldera formation on Vulcano Island (Aeolian Archipelago, southern Tyrrhenian Sea). *Journal of Volcanology and Geothermal Research*, **60**, 207–224.

VIGNERESSE, J.L. 1999. Should felsic magmas be considered as tectonic objects, just like faults or folds? *Journal of Structural Geology*, **21**, 1125–1130.

VIGNERESSE, J.L., TIKOFF, B. & AMEGLIO, L. 1999. Modification of the regional stress field by magma intrusion and formation of tabular granitic plutons. *Tectonophysics*, **302**, 203–224.

ZIEGLER, P. 1990. *Geological Atlas of Western and Central Europe*. Shell Internationale Petroleum Maatschappij, Den Haag.

ZIEGLER, P.A. & STAMPFLI, G.M. 2001. Late Palaeozoic–Early Mesozoic plate boundary reorganization: collapse of the Variscan orogen and opening of Neotethys. In: CASSINIS, G. (ed.) *Permian Continental Deposits of Europe and Other Areas. Regional and Correlations*. Museo Civico di Scienze Naturali di Brescia, Brescia, Monografie di Natura Bresciana, **25**, 17–34.

Peperitic lava lake-fed sills at Ság-hegy, western Hungary: A complex interaction of a wet tephra ring and lava

ULRIKE MARTIN [1] & KÁROLY NÉMETH [2,3]

[1] TU-Bergakademie, Institut für Geologie, Freiberg, D-09596, Germany
(e-mail: uli.martin@hotmail.com)
[2] Geological Institute of Hungary, Department of Mapping, 14 Stefánia út, Budapest, H-1143, Hungary (e-mail: nemeth_karoly@hotmail.com)
[3] Eötvös University, Department of Regional Geology, 14 Stefánia út, Budapest, H-1143, Hungary

Abstract: Ság-hegy is the remnant of a complex volcano consisting of several phreatomagmatic pyroclastic sequences preserved in immediate contact with a thick (c. 50 m) coherent lava body. Due to the intensive quarrying, the inner part of the lava has been removed, leaving behind a castle-like architecture of pyroclastic rocks. The outcrop walls thus demonstrate the irregular morphology of the lava, which was emplaced in a NW–SE-trending ellipsoidal vent zone in a phreatomagmatic volcano. Pyroclastic beds in the quarry wall are cross-cut by dykes and sills, inferred to have been fed from a central magma zone. Thin (<10 cm) strongly chilled, black, angularly jointed aphanitic basaltic lava mantles the pyroclastic sequence, and has a corrugated margin as a consequence of sudden chilling against the cold and wet phreatomagmatic tephra in the inner wall of the tuff ring crater. These corrugated zones are inferred to be a characteristic textural feature, indicating extensive mixing of lava and host tephra which led to peperite formation along the outer rim lava lake. A spectrum of peperite formed along the lake margin, and fluid oscillation, due to fluidization of the wet tephra, disrupted a steam envelope formed around the lava, causing basaltic magma to invade and mix with the phreatomagmatic tephra. The presence of unconformities in the tephra ring facilitated the formation of sills fed from the central lava body.

In the shallow subsurface and in surface levels, non-explosive magma–water interaction may occur at the margin of intrusions, lava lakes or feeder dykes, leading to fluidal and brittle fragmentation and variable intermixing of the host sediment and magma respectively (Kokelaar 1982; Kokelaar 1986; Wohletz 1986; Busby-Spera & White 1987). These processes often include fluidization and intrusion of the host sediment by the magmatic body. Peperite, a common consequence of this interaction, is a rock formed essentially *in situ* by disintegration of the intruding magma and mingling with unconsolidated, typically wet, surrounding sediment (White *et al.* 2000; Skilling *et al.* 2002). Phreatomagmatic volcanoes, such as tuff cones, tuff rings and maars, form when magma interacts explosively with surface and/or groundwater (Lorenz 1986; White 1991), and are common settings where intruding magma may encounter a large volume of wet volcaniclastic sediment and form shallow intrusions with peperitic margins. Once a crater is established, water access becomes restricted, (White & Houghton 2000), and a lava lake may develop. With a continuous magma supply, the frequency of magma–water interaction is probably a function of the rate of water-recharge in the country rock (Lorenz 1984, 1985; Aranda-Gomez & Luhr 1996; Carn 2000; Németh *et al.* 2001). If the recharge rate is low, then the fragmentation style may became magmatic, comprising Strombolian and/or Hawaiian explosive eruptions or lava effusion (Lorenz 1986). Peperites are only observed in volcanic fields where erosion has exhumed the diatremes of such volcanoes (Lorenz *et al.* 2002). Peperitic buds on dykes are also rarely well exposed in maar/diatreme fields (Hooten & Ort 2002). In contrast, lava lakes are common and often well preserved in association with old intra-continental monogenetic volcanic fields. Because it commonly had a phreatomagmatic origin, the tephra of pyroclastic edifices was moist or wet and was unconsolidated during the eruption. Therefore, pyroclastic edifices are good study areas for describing the processes of interaction between the host pyroclastic edifices and magma in lava lakes. Unfortunately, good exposures of contact zones between the lava lake and the host phreatomagmatic pyroclastic edifice are rare. However, outcrops such as the Little Hungarian Plain and the Bakony–Balaton

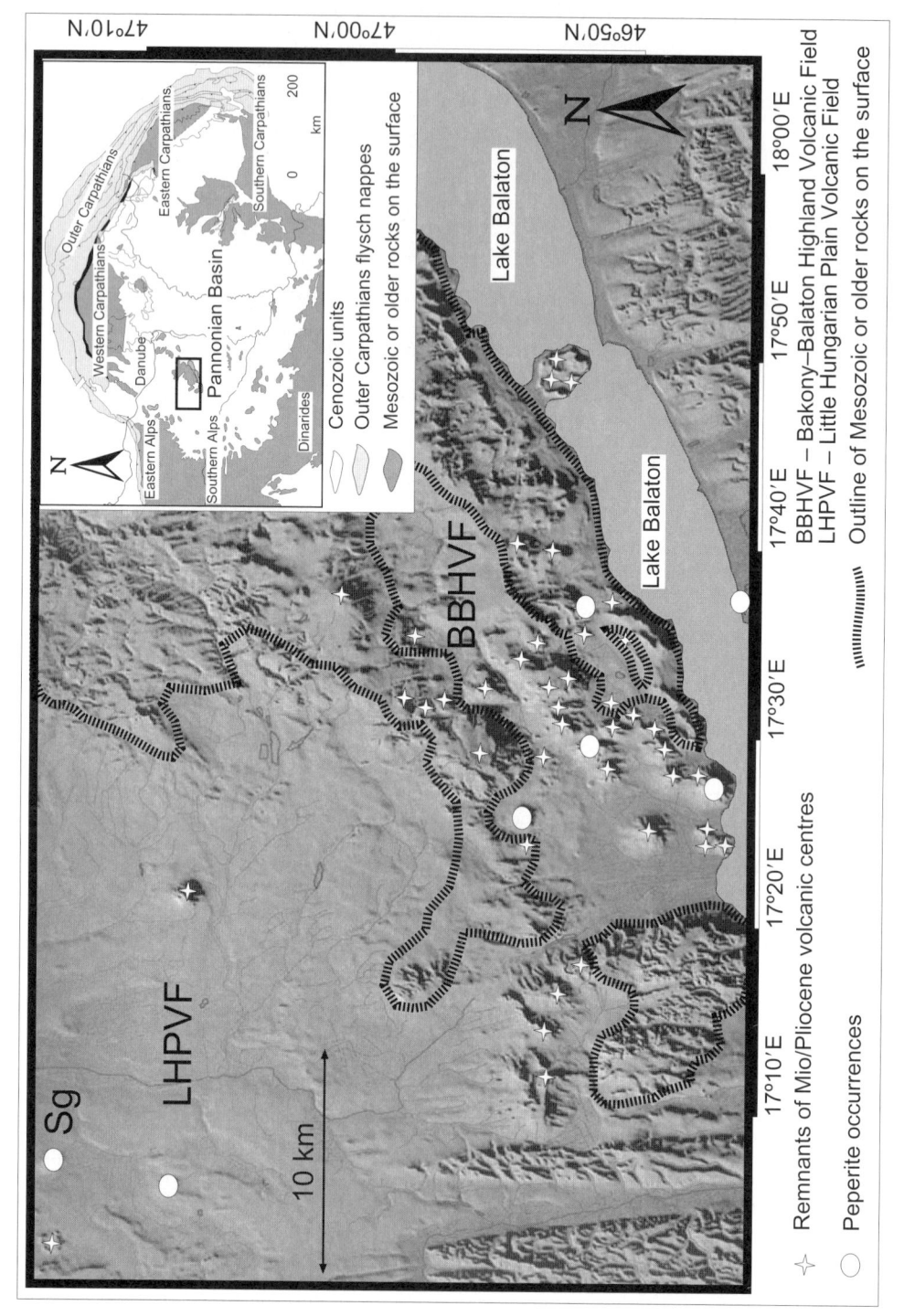

Fig. 1. Sketch map of volcanic fields in western Hungary. Note the distribution of peperite occurrences.

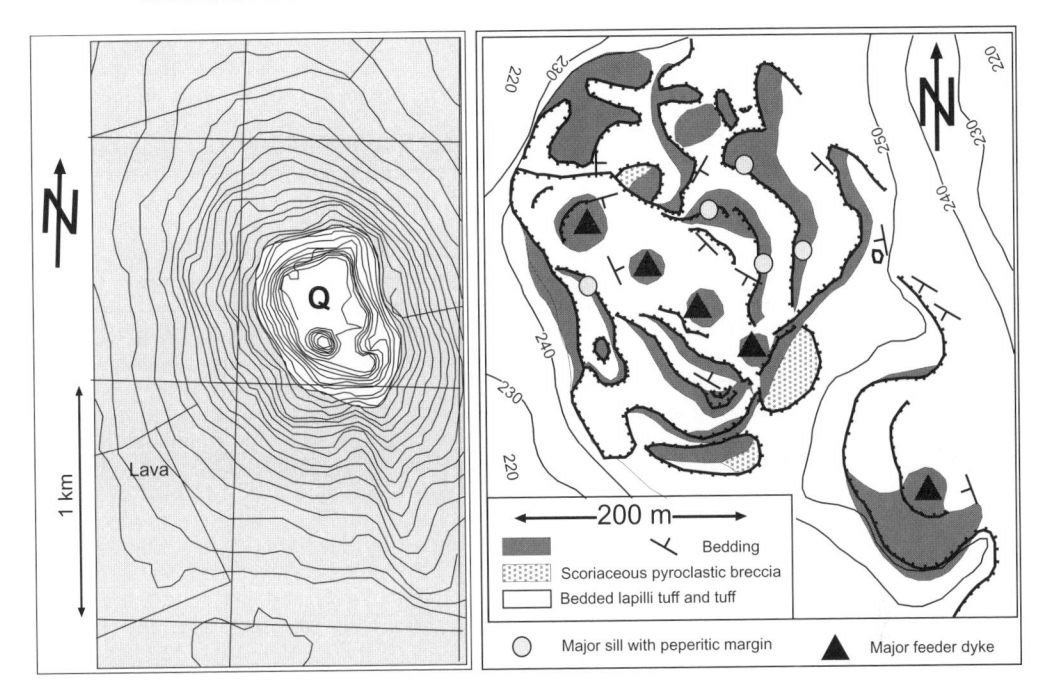

Fig. 2. Topography and simplified geology of the area around Ság-hegy. Note the complex quarry (Q) morphology largely mimic a steep-walled vent zone of a phreatomagmatic volcano.

Highland Volcanic Fields (LHPVF, BBHVF – Pannonian Basin, western Hungary) (Fig. 1) include several Miocene/Pliocene volcaniclastic successions that were penetrated by numerous shallow mafic intrusions. It has been recognized only recently that the contact zones between lava lake remnants and host phreatomagmatic units are well preserved in those outcrops and they show contact features that have not yet been described in detail (Martin & Németh 2002).

Ság-hegy is a complex phreatomagmatic volcano located in the central part of the LHPVF (Fig. 2), where such contact zones are well exposed and preserved. This paper describes such a contact zone in detail. The relationships observed demonstrate intrusive, shallow sill emplacement, presumably fed by a lava lake. Due to excellent outcrop access, the Ság-hegy deposits permit an almost three-dimensional picture to be drawn for the interface between a dynamic lava lake that was feeding shallow intrusions into the host (pyroclastic) deposits, and a tephra ring.

Geological setting

Intra-plate volcanism in the Pannonian Basin began with trachyandesitic to trachytic volcanism, which was followed by alkaline basaltic volcanism: the latter becoming more prominent from Late Miocene to Pleistocene times (Stegena et al. 1975). The volcanic fields consist of eroded remnants of Mio/Pliocene scoria cones, tuff rings and maars (Németh & Martin 1999).

The Little Hungarian Plain, where Ság-hegy is located, is a Neogene sedimentary basin filled by Miocene to Pliocene, predominantly siliciclastic sediments, up to 6000 m thick (Tari et al. 1992; Horváth 1993). The basement rocks of this basin consist of crystalline units belonging to the Upper and Lower Austro-Alpine terrain (Horváth 1993). Formation of the basin occurred during the uplift of Penninic metamorphic core complexes and coincident development of an extensional basin system bounded by low-angle normal faults (Tari et al. 1992; Horváth 1993). Several seismic, magnetotelluric and magmatic studies suggest that a supracrustal fault and an asthenospheric dome may be present in the axis of the Little Hungarian Plain (Horváth 1993), and that both may have significance in the development of the volcanic field. In the Neogene, just shortly before the volcanism started, a large lake known as the Pannonian Lake occupied the Pannonian Basin (Kázmér

1990). As a consequence, lacustrine sand-(stones), mud(stones), and marls are widespread in the Pannonian Basin (Kázmér 1990), with the Little Hungarian Plain among the largest sub-basins where these deposits accumulated. Recent studies based on facies analyses of Late Miocene fluvio-lacustrine sediments of the Pannonian Lake system, together with high-resolution seismic profiles and palaeontological evidence suggest that lacustrine sedimentation related to the Pannonian Lake ceased about 9 to 8 Ma ago in western Hungary (Magyar et al. 1999). A southward-prograding delta system gradually filled up the Pannonian Lake, and the region had become an alluvial plain by the Pliocene (Jámbor 1989; Juhász 1994; Juhász et al. 1997; Magyar et al. 1999). Large, shallow lakes may have developed in the region, especially during wet seasons. Consequently, volcanism occurred in subaerial settings, along fluvial valleys which probably contained swamps, small streams or shallow lakes – all providing substantial surface water and groundwater to allow phreatomagmatic volcanism. Water-saturated sediments (mainly mud) played an important role in the magma–water interaction (Németh & Martin 1999). Ság-hegy is a complex phreatomagmatic volcano on a major NW–SE-trending fault zone (Jugovics 1971). Based on published K/Ar radiometric ages, the Ság-hegy volcano is interpreted to be approximately 5–6 Ma old (Balogh et al. 1986).

Phreatomagmatic pyroclastic units

The basal pyroclastic series of the former tephra ring at Ság-hegy comprises bedded, unsorted and ungraded to normal graded, alternating tuff and lapilli tuff beds (Fig. 3a). Soft-sediment deformation, cross-bedding, undulating bedding, accretionary lapilli and deep, plastically deformed impact sags (Fig. 3a) are common and consistent with phreatomagmatic explosive eruptions similar to those that formed a tuff ring or tuff cone at Ság-hegy. Juvenile clasts (the size of fine ash to fine lapilli) are predominantly angular sideromelane glass shards with none-to-high vesicularities (Fig. 3b) suggestive of interaction of magma and a variable amount of water. The vesicular sideromelane shards tend to be stretched and slightly fluidal, and are intensely palagonitized. Pyroxene microliths, as well as occasional olivine microphenocrysts, are included in the glass shards. Tachylite shards are present but less common than sideromelane. Juvenile lithic clasts are rare, and are predominantly microgabbroid rock fragments up to coarse lapilli size. Accidental lithic clasts (<5 cm in diameter, up to 50 vol.% by visual estimation) were predominantly derived from the Late Miocene fluvio-lacustrine units immediately beneath the volcanic sequence. They often form clot-like, plastically deformed fragments or single crystals. Large (centimetre-to-decimetre-scale) mica-rich irregularly shaped clots are common, especially in medium-bedded lapilli tuffs in the lower pyroclastic sequence. Similar mica-enriched clots are also common in the fine matrix-supported lapilli tuffs and fine tuffs. Large bed-parallel platy blocks of intact sand-, silt- and mudstone chunks are prominent in the lapilli tuff beds. In some beds, these siliciclastic clasts show the effects of intense heat alteration, such as hematite enrichment and small radial joints resembling mud-cracks. The deepest exposed stratigraphic level is formed by thickly bedded, structureless or weakly stratified lapilli tuffs and/or tuff breccias rich in accidental lithic clasts. Fifty metres above the basal exposure, thinly bedded tuff and lapilli tuff are more prominent. They are unsorted and rich in accretionary lapilli and/or armoured lapilli, and contain numerous mineral groups derived from the underlying Neogene sediments. The overlying beds have numerous bomb sags (Fig. 3), scour-fill structures and vesiculated tuff layers (the latter term coined by Lorenz 1974), and also show soft-sediment slump, dish structures and undulating bedding contacts.

These bedding features, shard morphology and varied vesicularities, and abundant accidental clasts suggest that the pyroclastic units at Ság-hegy formed from phreatomagmatic eruptions during interaction of rising basaltic magma and water-saturated unconsolidated sediments. The pyroclastic units are inferred to have been deposited by alternating base surges and fall-out, which gradually built an initial tephra ring around the erupting vent(s). The large amount of accidental lithic clasts and especially the abundant broken mineral grains derived from the Neogene fluvio-lacustrine units indicate that the magma interacted predominantly with groundwater. The scarcely of coherent blocks of country rocks and the abundance of mineral grains suggest the eruptions took place within soft sediment ('soft rock' environment; Lorenz 2000, 2003a, 2003b). Accidental lithic fragments, which were derived from pre-Neogene sedimentary units are rare, indicating a relatively shallow locus for the phreatomagmatic explosions, which were presumably driven by surface- or near-surface water in unconsolidated and wet sediments. The lower massive beds resemble deposits from high-concentration, laminar mass flows, such as volcanic debris flow. The angular,

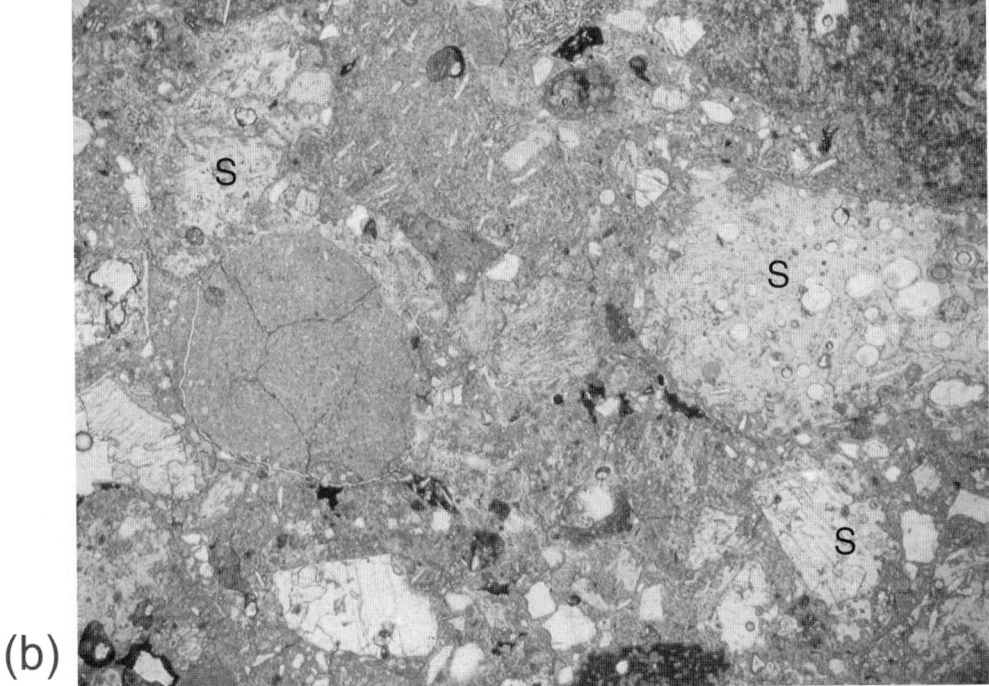

Fig. 3. (a) Ballistic bombs (arrowed) in accretionary lapilli bed from the phreatomagmaticpyroclastic units of Ság-hegy. Coin is 2 cm in diameter. (b) Microphotograph of a lapilli tuff from the phreatomagmatic pyroclastic units. Note the angular, weakly vesicular sideromelane glass shard (s) and the bright angular quartz derived from Neogene fluvio-lacustrine units (arrowed). Plane-parallel light; the short side of the image is 2 mm.

Fig. 4. View into the quarry pit, looking SE. The pit largely represents the shape of the lava rocks (coherent lava, CL) removed by quarrying. Thus the shape of the quarry resembles the shape of the original phreatomagmatic vent. Locally on the quarry wall, there is a thin, shark-skin-like lava crust preserved. S, scoriaceous capping units; lines represent the inferred margin of the former lava lake. Quarry wall (Q) is about 20 m high. Lines represent inferred margin of former lava lake and phreatomagmatic units (PYX).

ragged and irregular shape of the juvenile pyroclasts, as well as the presence of chilled glassy pyroclasts such as volcanic glass and/or glassy juvenile fragments with low to moderate vesicularity indicate a primary, eruption-fed origin for these deposits. The subhorizontal bedding characteristics and the abundance of coarse lapilli and block-size juvenile fragments are suggestive of deposition from pyroclastic density currents in a near-vent setting. The abrupt change from thick massive beds to thin- and well-bedded ones, of the pyroclastic units in the upper and lower part may indicate changes in the eruptive environment from shallow subaqueous to subaerial.

Corrugation zones along the lava lake margin

Due to intensive quarrying in past decades, the inner part of a coherent lava mantling the centre of the Ság-hegy phreatomagmatic volcanic complex has been completely removed, leaving behind a castle-like architecture of pyroclastic rocks. The outcrop walls thus represent more or less the original irregular basal surface morphology of the lava lake emplaced in the crater of a phreatomagmatic volcano (Fig. 4). The quarry walls, constructed of pyroclastic beds, are truncated by coherent lava layers (Fig. 5), which seem to be connected with a central magma zone (presumably to the already quarried lava) through narrow (centimetre-to-decimetre-scale) lava necks. There are large areas (tens of m^2), where a thin (<10 cm) screen of strongly chilled, black, angularly jointed aphanitic basaltic lava adheres to the preserved pyroclastic sequence (Fig. 6) – a relationship interpreted to be the irregular contact zone of the inner crater wall of the tuff ring and the lava lake. These chilled contact zones in places seem to cover entire quarry walls, giving an impression that they are lava flows intersected with the pyroclastic sequence. The lava at its boundary with the pyroclastic rocks is undulatory and resembles a 'shark skin' or 'elephant hide' texture. The contact shows mm-to-cm-scale undulations or ridges in the lava, with a centimetre-to-decimetre-scale wavelength. There is no systematic distribution of these, or any systematic size variations. Similar features have been reported from the Peninsula Tuff

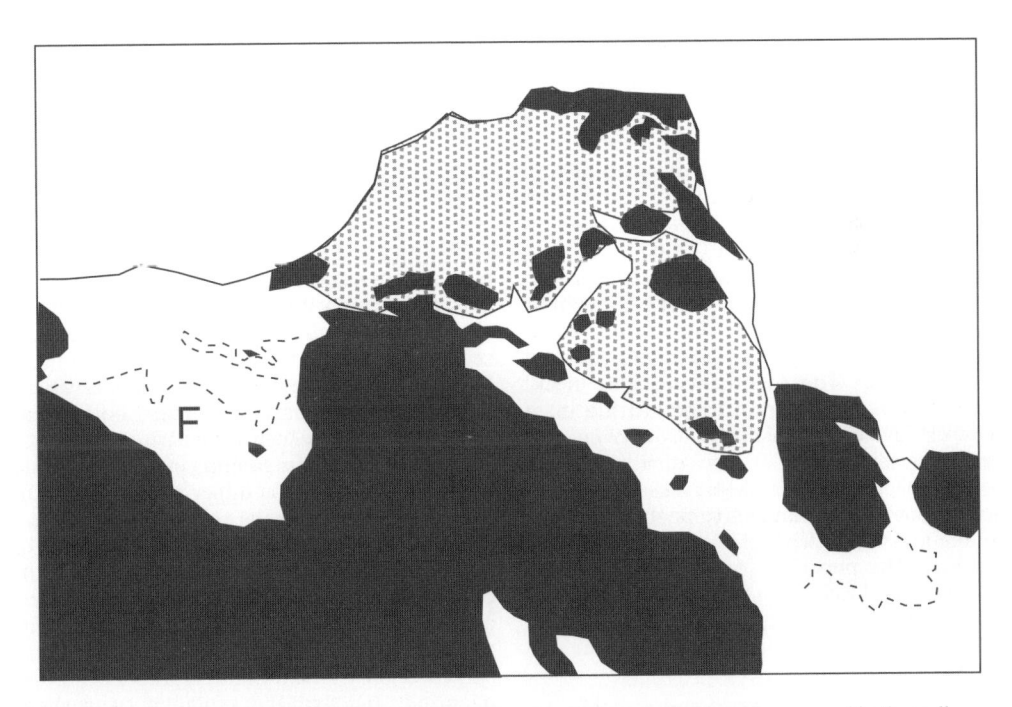

Fig. 5. View of the quarry wall. Note the chaotic appearance of the pyroclastic rocks exposed in the wall. There are large irregularly shaped areas of coherent lava, mimicking interbedded lava layers in the pyroclastic units. Also note that the coherent lava is always surrounded by a light-coloured, strongly fluidized, now homogenized sediment halo (F). The lava masses are interpreted as lava protrusions (on a metre scale) and sills that invaded the tephra. Due to fluidization, large, dispersed, pillow-shaped lava masses are often detached from the main lava-lake body (see also Fig. 13).

Fig. 6. Corrugation features at the contact of the main lava-lake body with the host pyroclastic units. Note the thin chilled crust covering the pyroclastic rock wall. These dark lava crusts are easy to misinterpret as large coherent lavas interfingering with the pale pyroclastic units; however, they clearly form a thin crust on the former phreatomagmatic inner crater wall. Hammer is about 40 cm long.

Cone, California (Lavine and Aalto 2002); however, the preservation of the Californian locality is far poorer, and does not show corrugated lava and pyroclastic units in contact together. At Ság-hegy, these corrugated zones often feed centimetre-to-metre-scale, straight or slightly twisted protrusions (Fig. 7). In general, the edges of the protrusions are usually rounded to subangular. The (centimetre-scale) 'neck' zones connecting the protrusions to the main coherent lava body (which is quarried away) are clearly preserved (Fig. 7). The small undulating contacts against the pyroclastic rocks are prominent. Some of the sills are fragmented and have centimetre- to decimetre-thick globular peperitic margins. The peperitic zones have a random distribution. Microphenocrysts of plagioclase and pyroxene are aligned mainly parallel to the sill margin, and are less abundant along the margin.

Large (few metres-scale) peperitic sills fed from the lava lake

Feeder dykes also intruded the phreatomagmatic deposits, and often display peperitic margins. The peperitic margins are more prominent in the lower parts of the section, suggesting that the sediments were water-saturated and unconsolidated (Fig. 8). The sills in the lower part of the sequence have jagged and blocky brecciated margins and they were intruded along or close to unconformities in the tuff ring sequence (Fig. 9). The centre of the sill(s) and larger lava bodies are aphanitic and non-vesicular, whereas the margins of these bodies and smaller clasts show high vesicularities.

In contrast, reverse faulting of the pyroclastic units is common in the upper section (Fig. 10), suggesting that there the rocks were drier and more brittle during intrusion, although thin peperitic margins are also present. In the fault planes, the margins of the intrusions are generally sharp, but show centimetre-scale undulations and boudinage-like structures. They are surrounded by a halo of finely dispersed, homogenized, and fluidized sediment (with slight orientation of the grains), a halo which may be focused along reverse fault surfaces extending away (on a scale of metres) from the tip of the intrusion. The common relationship of the intrusive bodies with thrust-faulted and differentially rotated pyroclastic units close to each other in

Fig. 7. View perpendicular to the lava crust, showing small-scale (decimetre-scale) protrusions of lava invading the host pyroclastic units. Dashed line – zone of disturbed tephra around the main intrusive body. Arrow points to thick line showing a detached dm-scale sill intruded into the host tephra.

the exposure levels of the exposure indicate that thrusting was initiated by the intrusions rather than the intrusions having invaded the existing thrust planes. This peperitic zone in the upper exposure level is thinner and sharper than peperitic sill and/or dyke margins at lower stratigraphic levels, and is interpreted to be the result of exhaled magmatic gases and vapour that in combination may have promoted the formation of these peperites.

The centre of the sill(s) and larger lava bodies are aphanitic and non-vesicular, whereas the margins of these bodies and smaller clasts show high vesicularities.

Fluidization halo around the lava lake margin and lava lake-fed intrusions

All the coherent lava bodies at Ság-hegy, especially those which were involved in the peperite forming process, are enclosed by a centimetre-to-metre-scale strongly fluidized zone of now homogenized whitish, fine-enriched sediment (Fig. 11). This halo is generally easy to distinguish from the undisturbed lapilli tuff and tuff by its lighter colour and lack of sedimentary structures. This halo is often narrow (centimetre-scale) in places where globular peperite has developed, whereas it is much wider (metre-wide) in areas where blocky peperite is present, indicating that the peperitic zones may have acted as pathways for dispersed lava clasts deep into the host tephra via clastic dykes (Fig. 8). These zones may be areas where a steam envelope was disrupted and direct contact between water-rich sediment and the magma lead to local disruption of the sill along its margins (Fig. 12). The enrichment in fine sediment in the peperites and along sill margins is suggestive of fluidization processes enhanced by the heat of the emplacing magma. Fluidization of host tephra (Fig. 13) occurred, forming fines-enriched zones at the centimetre to metre scale along the margin of sills fed from the lava lake, and drove the excess pore fluid pressure concentration at the advancing magma head by heating the pore fluids in the host tephra. Similar fines-enriched zones are developed around the main lava-lake unit as a result of the heat of the lava occupying the crater and the increased pressure effect on the underlying host sediments at the base of the lava lake. The lava and sills have glassy margins and grids of baked mud. Mica flakes in the fluidized and

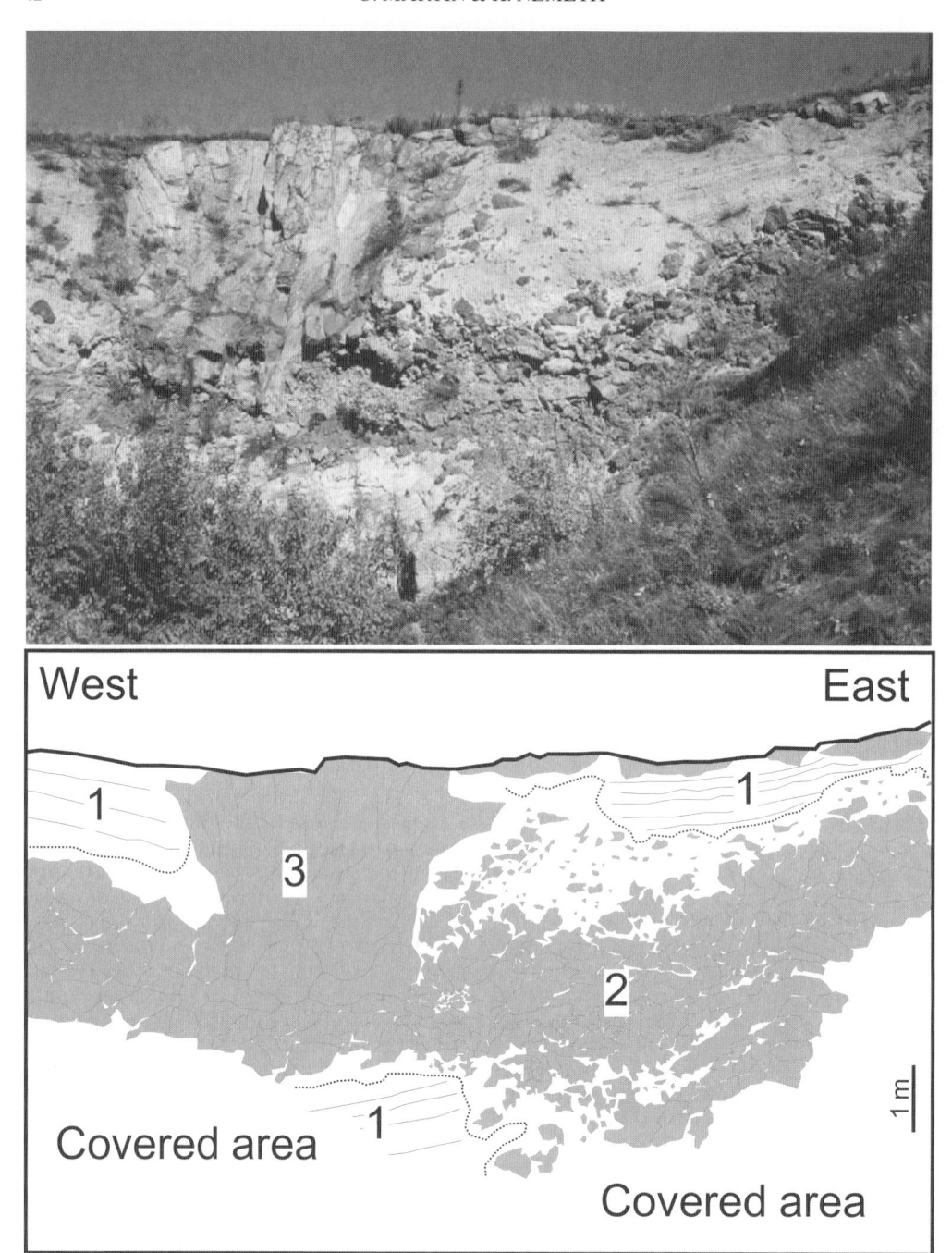

Fig. 8. Major sill intrusion from the lava lake into the pyroclastic unit. The coherent lava body is sandwiched between lapilli tuff beds (1), and resembles a lava flow (2). The intrusive origin of this lava is supported by the presence of highly irregular peperitic lower and upper margins and a pale-coloured strongly fluidized now completely homogenized, fines-enriched halo surrounding the lava (2). Note the oblique dyke (3) with a sharp contact with both, the sill (2) and the pyroclastic units (1), indicating late-stage emplacement, presumably after the tephra dried out significantly. Numbers represent the time sequence of these events.

Fig. 9. Oblique view of the major sill shown on Figure 8. Note the pale-coloured fluidized, now homogenized halo around the sill (H). The sill clearly intruded into an unconformity (arrowed) in the phreatomagmatic unit.

now homogenized zones are commonly oriented parallel to margin contacts. Small-scale (centimetre-to-decimetre) subvertical zones initiating from sill terminations have been recognized on the basis of uniformly oriented mica flakes as well as the presence of dispersed peperite.

Intrusion along unconformities in the tuff-ring sequence may have enhanced sill formation fed from the central, large-volume lava lake unit that gradually filled the vent zone due to decreased stress, which allowed an easier emplacement. The lava lake-fed sills have jagged and brecciated margins, and intrusion occurred preferentially along unconformities of any type in the tuff-ring sequence.

Discussion

At Ság-hegy, there is a clear transition from margins of sills or dykes (globular peperite) to disrupted-texture (blocky peperite) in the same locality. The presence of mixed globular and blocky peperite at the same locations indicates a change in the fragmentation and mixing mechanism of host sediment and intruding magma during wet sediment–magma interaction. Intrusion along unconformities in the tephra-ring sequence may have enhanced the formation of sills, due to decreased shear strength, which allowed an easier emplacement. The initial magma fragmentation and mixing with sediment is interpreted to have been the result of tearing apart of magma and shaping of the magma–sediment interface into globular, pillow-shaped bodies by contact–surface interaction. During a second stage, blocky peperite developed along the margin of sills and the lava-lake margin, due to phreatomagmatic reactions, including the breakdown of insulating vapour films at the sediment–magma interface during strong oscillations of pore-water. The presence of peperitic zones and associated fluidization haloes along the lava and lake-derived sill margins indicates that pore-water was easily remobilized from the host tephra, due to the heat of the lava. It would imply that the temperature difference was relatively high along the entire lava body and the host sediment was heavily water saturated as well as loosely packed. The textural gradation between corrugation zones and large-scale undulating lava-lake margins, and between small irregularly shaped sills and larger, strongly fragmented sills

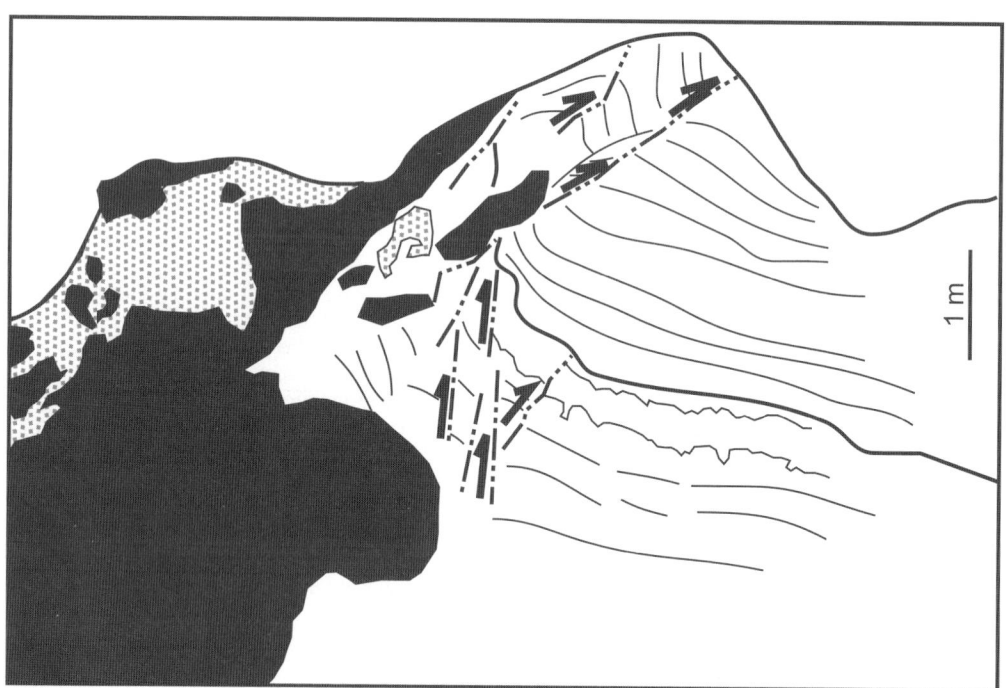

Fig. 10. Reverse faulting of the upper phreatomagmatic pyroclastic units, faulting probably attributed to displacement due to the expanding lava lake. Note the lava intruded in the fault plane. Stippled pattern represents brecciated, homogenized tuff breccia along the intrusion.

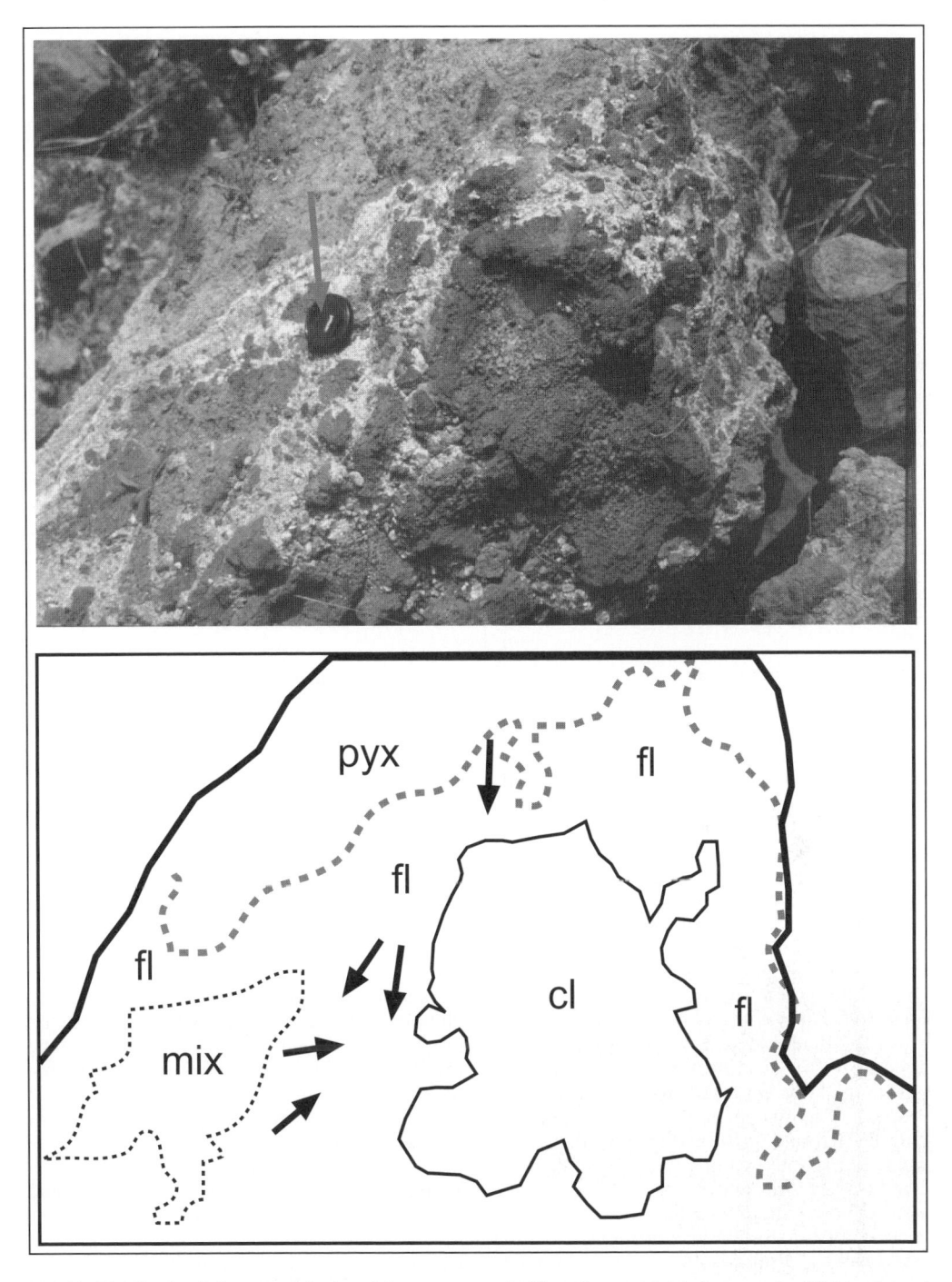

Fig. 11. Fluidization halo around the lava lake at meso-scale. Note the completely homogenized light-coloured halo (fl) containing highly vesicular lava clasts (cl). Outside the halo, the original bedding of the pyroclastic unit (pyx) is still preserved. The arrow in the upper photo points to a lens cap with a 5 cm diameter. Arrows in the lower photo show slight alignment of detached coherent lava clasts from the feeding lava. Also note that the coarse-grained lapilli tuff at the margin of the intrusion consists of mixed fragments (mix) from the pyroclastic host and the coherent lava. The lens cap is about 5 cm in diameter.

Fig. 12. Fluidization channel initiated from the tip of a sill. Note that the fluidization pipe created a homogenized zone, rich in dispersed coherent lava clasts. Hammer is about 40 cm long.

with widely dispersed peperitic zones suggests a similar gradation in the process responsible. The widespread distribution of the peperitic sill margins, apparently interpretative of stratigraphic position, indicates that the weight of the lava lake was not enough to suppress effectively the pore-fluid oscillation at the base of the lake. There was no clear morphological trend from features characteristic of Raleigh–Taylor instability (such as elongated, streamlined, flute-shaped lobes) to fully dispersed peperitic margins relative to stratigraphic position. However, the lower part of the lava lake is not exposed, which is where flute-shaped undulations might be expected. The lack of a stratigraphic relationship between these features may be a result of the elongated, fissure-like shape of the vent system at Ság-hegy (Fig. 14). The lava lake that fills a fissure-like vent has a steep contact with the wet tephra. The tephra is chaotically displaced by the intruding lava. The steep contact forms an optimal channel for vertical fluid movement that creates a continuous pore-water oscillation disrupting the steam envelope along the emplacing lava lake (Fig. 14). Conditions were favourable for peperitic rather than explosive phreatomagmatic disruption, because:

1 the low water content of the tephra was insufficient, and
2 the magma discharge rate may have been relatively high.

The latter caused relatively large magmatic pressures on the inner crater wall: enough to suppress explosive disruption. The common presence of brittle fractures in the upper pyroclastic units indicates drier conditions, but the occurrence of fluidization haloes there, as well as globular peperite in the small sills, indicate that the deposit was still wet, and soft enough to be involved in fluidization as well as coarse mixing of lava and tephra. The different mechanical responses of the pyroclastic pile may reflect the time that it took to:

1 fill the fissure-like vent/crater zone by lava and
2 reach the uppermost, partially dried tephra.

Phreatomagmatic tephras are likely to dry out relatively quickly (days to weeks), and, because there is no indication that the lava lake was emplaced in stages, it is plausible that the lava lake formed a single event, and in a shorter time than it took for the phreatomagmatic tephra to dry out. However, tephra deep in the volcanic

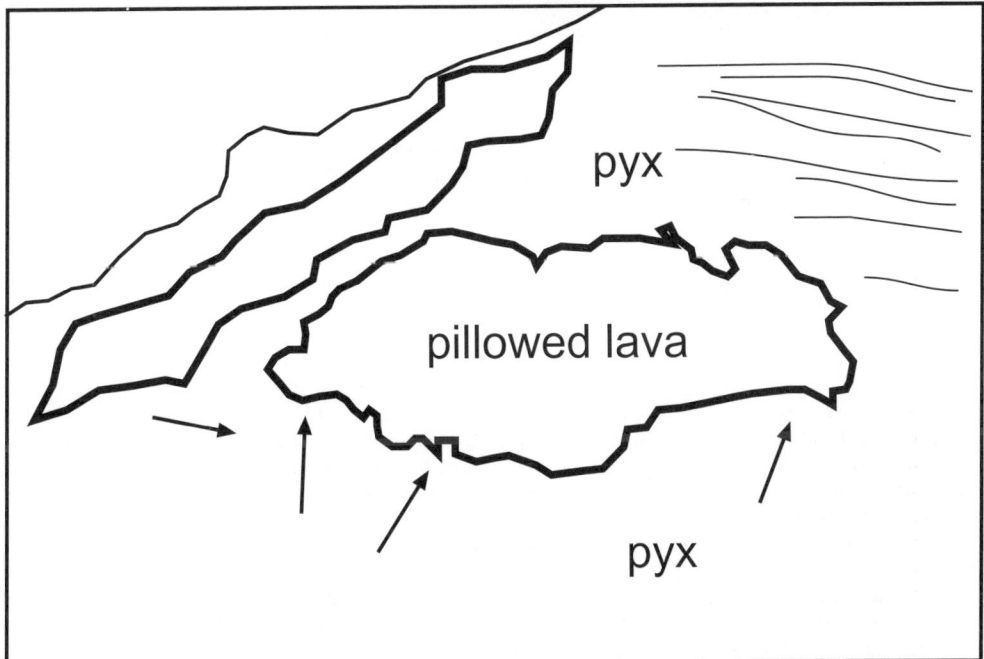

Fig. 13. Detached megapillow in an initially fluidized and now homogenized (arrows) tephra host (pyx). Note that the original bedding is still preserved away from the fluidized zone (lines on line drawing). However, due to the coarse-grained texture of the host rock, it is not easy to recognize. Hammer is about 40 cm long.

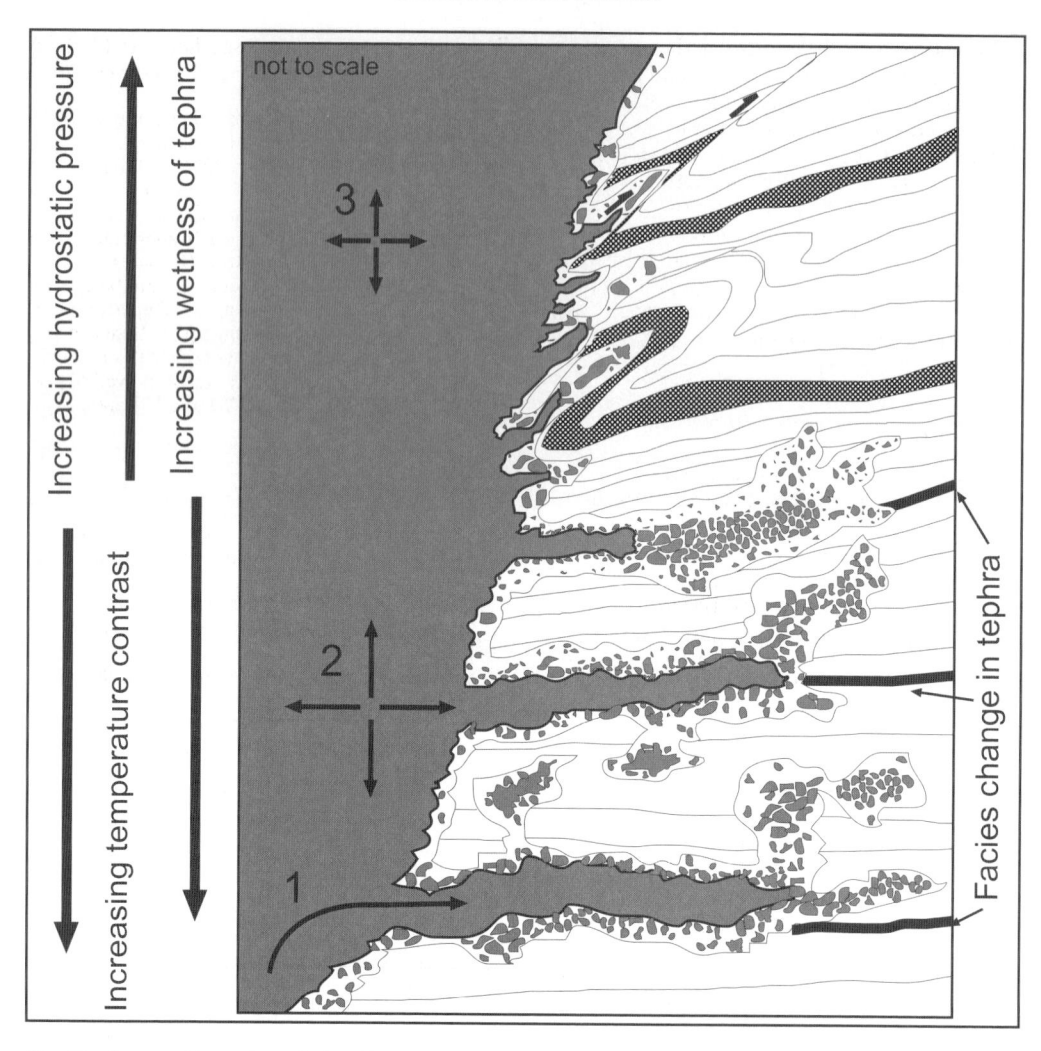

Fig. 14. Cartoon illustrating relationships and features of lava-lake-fed intrusions at Ság-hegy. The steep-walled, fissure-like phreatomagmatic vent structure might be a reason for the mixed distribution of contact features which do not show a significant stratigraphic relationship. However, brittle deformation is more prominent up-section, indicating that the higher levels of the pyroclastic core were drier prior to the appearance of the lava lake. No. 1 on the lava lake represents movement of meld into host (arrow) due to the great pressure of the lava lake on the bottom of the lake. No. 2 and 3 represent smaller pressure (arrows) caused by the weight of the lava lake and as a consequence the variability of the intrusion styles along the lava lake and the host sediment.

edifice would have remained wet for a longer period, particularly if the volcano formed in a shallow lake. The drier conditions of the upper stratigraphic levels could also be a result of the water in the tephra seeping downward and making the underlying tephra more water saturated, which they (not all beds) should not have been immediately after deposition.

The peperitic margins of the intrusive bodies and their field relationship to large coherent lavas, e.g. the lava lake at Ság-hegy, record the complex eruption history of an emplacing magma body into a wet vent zone of a phreatomagmatic volcano. The presence of peperitic intrusions also highlights the slight time-delay (hours to days) between formation of the phreatomagmatic tephra ring/mound and the emplacement of basaltic intrusions.

Lava-lake-fed sills with peperitic margins are very likely to be present in other similar volcanic settings. Lava lakes have also been described in western Hungary, but in less detail than for the Ság-hegy outcrop (Martin & Németh 2002). Similarly, there are irregular contact zones of lava lakes emplaced in the crater of Plio/Pleistocene phreatomagmatic volcanoes in southern Slovakia (Konecny & Lexa 2000) and along intrusive bodies emplaced into maar-crater-filling lacustrine units reported from phreatomagmatic volcanoes in the Eger rift (Suhr & Goth 1996).

Conclusions

Ság-hegy is a phreatomagmatic volcano with a complex pyroclastic sequence. Interaction occurred between water-saturated mud and vent-filling pyroclastic deposits such as base-surge, phreatomagmatic air-fall and interbedded volcanic debris flow deposits. Based on the field relationships, it is suggested that coherent lava layers in the phreatomagmatic pyroclastic units are intrusive and were connected to a central lava unit that was emplaced as a lake within the Ság-hegy crater.

The presence and form of peperites at the margin of the lava lake and associated lava-lake-fed intrusions are related to a complex interaction of magma intruding into a wet vent-to-crater zone in a phreatomagmatic volcano. The peperitic lava lake and sill margins support the interpretation that the tephra was still unconsolidated, wet and easily incorporated with the intruding melts. However, the variable development of peperitic margins at similar stratigraphic positions indicates that the compaction and dewatering of the tephra was spatially inhomogeneous. The common brittle deformation of the upper pyroclastic units by the intruding lava-lake-fed sills indicates that these pyroclastic units were close to the surface and dried out more quickly than the deeper tephra units. The presence of reverse-faulted pyroclastic blocks also suggests that the tephra units were easily displaced: consistent with a position in the original tuff ring close to the surface. The continuous connection between small-scale marginal corrugations, large lava pillows and blocks, and fully developed peperitic sill margins, indicates a cogenetic relationship. Lava lakes associated with wet phreatomagmatic volcanoes similar to Ság-hegy are common in fluvio-lacustrine basins with abundant surface and shallow subsurface water (like the Pannonian Basin in the Late Miocene to Pleistocene). The features described at Ság-hegy are likely to be common in volcanoes in these settings.

Partial funding for this research by the DAAD, German-Hungarian Academic Exchange Program 2002 (DAAD-MÖB 4616–2001), DIG grant MA244011 to U.M., the OTKA (Hungarian Science Foundation) F 043346 Grant and the Magyary Zoltán Post-doctoral Fellowship are greatly appreciated. The State of Saxony Travel Grant to K.N. made it possible to attend the LASI Conference in Freiberg. The positive attitude of the Mapping Department of the Geological Institute of Hungary (T. Budai and G. Csillag) significantly helped to complete this research. The final version of this paper was improved by the constructive reviews by J. Smellie and V. Lorenz.

References

ARANDA-GOMEZ, J.J. & LUHR, J.F. 1996. Origin of the Joya Honda maar, San Luis Potosi, Mexico. *Journal of Volcanology and Geothermal Research*, **74**, 1–18.

BALOGH, K., ARVA-SOS, E., PECSKAY, Z. & RAVASZ-BARANYAI, L. 1986. K/Ar dating of post-Sarmatian alkali basaltic rocks in Hungary. *Acta Mineralogica et Petrographica, Szeged*, **28**, 75–94.

BUSBY-SPERA, C.J. & WHITE, J.D.L. 1987. Variation in peperite textures associated with differing host-sediment properties. *Bulletin of Volcanology*, **49**, 765–775.

CARN, S.A. 2000. The Lamongan volcanic field, East Java, Indonesia: physical volcanology, historic activity and hazards. *Journal of Volcanology and Geothermal Research*, **95**, 81–108.

HOOTEN, J.A. & ORT, M.H. 2002. Peperite as a record of early stage phreatomagmatic fragmentation processes: an example from the Hopi Buttes volcanic field, Navajo Nation, Arizona, USA. *Journal of Volcanology and Geothermal Research*, **114**, 95–106.

HORVÁTH, F. 1993. Towards a mechanical model for the formation of the Pannonian basin. *Tectonophysics*, **226**, 333–357.

JÁMBOR, A. 1989. Review of the geology of the s.l. Pannonian formations of Hungary. *Acta Geologica Hungarica*, **32**, 269–324.

JUGOVICS, L. 1971. A Kisalföld bazalt és bazalttufa elöfordulásai [Basalts and basalt tuffs of the Little Hungarian Plain]. *MÁFI Évi Jelentés 1970-röl*, 79–101.

JUHÁSZ, E., KOVÁCS, L., MÜLLER, P., TÓTH-MAKK, A., PHILLIPS, L. & LANTOS, M. 1997. Climatically driven sedimentary cycles in the Late Miocene sediments of the Pannonian Basin, Hungary. *Tectonophysics*, **282**, 257–276.

KÁZMÉR, M. 1990. Birth, life, and death of the Pannonian Lake. *Palaeogeography, Palaeoclimatology, Palaeoecology*, **79**, 171–188.

KOKELAAR, B.P. 1982. Fluidisation of wet sediments during the emplacement and cooling of various igneous bodies. *Journal of the Geological Society of London*, **139**, 21–33.

KOKELAAR, P. 1986. Magma–water interactions in sub-aqueous and emergent basaltic volcanism. *Bulletin of Volcanology*, **48**, 275–289.

KONECNY, V. & LEXA, J. 2000. Pliocene to Pleistocene alkali basalt diatremes and maars of Southern Slovakia: a common model for their evolution. *Terra Nostra*, **6**, 220–232.

LAVINE, A. & AALTO, K.R. 2002. Morphology of a crater-filling lava lake margin, The Peninsula tuff cone, Tule Lake National Wildlife Refuge, California: implications for formation of peperite textures. *Journal of Volcanology and Geothermal Research*, **114**, 147–163.

LORENZ, V. 1974. Vesiculated tuffs and associated features. *Sedimentology*, **21**, 273–291.

LORENZ, V. 1984. Explosive volcanism of the West Eifel volcanic field, Germany. *In*: KORNPROBST, J. (ed.) *Kimberlites I.: Kimberlites and Related Rocks*. Elsevier, Amsterdam, 299–307.

LORENZ, V. 1985. Maars and diatremes of phreatomagmatic origin: a review. *Transactions of the Geological Society of South Africa*, **88**, 459–470.

LORENZ, V. 1986. On the growth of maars and diatremes and its relevance to the formation of tuff rings. *Bulletin of Volcanology*, **48**, 265–274.

LORENZ, V. 2000. Formation of maar–diatreme volcanoes. *Terra Nostra*, **2000/6**, 284–291.

LORENZ, V. 2003a. Maar-diatreme volcanoes, their formation, and their setting in hard-rock or soft-rock environments. *GeoLines – Papers in Earth Sciences, Prague, Czech Republic*, **15** [Hibsch 2002 Symposium Volume], 72–83.

LORENZ, V. 2003b. Syn- and post-eruptive processes of maar-diatreme volcanoes and their relevance to the accumulation of post-eruptive maar crater sediments. *Földtani Kutatás, Budapest, Hungary*, **XL** (1–2), [Proceeding of the 7th International Alginite Symposium], 13–22.

MAGYAR, I., GEARY, D. & MÜLLER, P. 1999. Paleogeographic evolution of the Late Miocene Lake Pannon in Central Europe. *Palaeogeography, Palaeoclimatology, Palaeoecology*, **147**, 151–167.

MARTIN, U. & NÉMETH, K. 2002. Interaction between lava lakes and pyroclastic sequences in phreatomagmatic volcanoes: Haláp and Badacsony, western Hungary. *Geologica Carpathica*, **53**, Special Issue – Proceeding for the XVIIth Congress of Carpathian–Balkan Geological Association, Bratislava, CD-version [ISSN 1335-0552].

NÉMETH, K. & MARTIN, U. 1999. Large hydrovolcanic field in the Pannonian Basin: general characteristics of the Bakony–Balaton Highland Volcanic Field, Hungary. *Acta Vulcanologica*, **11**, 271–282.

NÉMETH, K., MARTIN, U. & HARANGI, S. 2001. Miocene phreatomagmatic volcanism at Tihany (Pannonian Basin, Hungary). *Journal of Volcanology and Geothermal Research*, **111**, 111–135.

SKILLING, I.P., WHITE, J.D.L. & MCPHIE, J. 2002. Peperite: a review of magma–sediment mingling. *Journal of Volcanology and Geothermal Research*, **114**, 1–17.

STEGENA, L., GÉCZY, B. & HORVÁTH, F. 1975. Late Cenozoic evolution of the Pannonian Basin. *Tectonophysics*, **26**, 71–90.

SUHR, P. & GOTH, K. 1996. Erster Nachweis tertiärer Maare in Sachsen. *Zentralblatt für Geologie und Paläontologie Teil I [Stuttgart]*, **1995 1/2**, 363–374.

TARI, G., HORVÁTH, F. & CRUMPLER, J. 1992. Styles of extension in the Pannonian Basin. *Tectonophysics*, **208**, 203–219.

WHITE, J.D.L. 1991. Maar–diatreme phreatomagmatism at Hopi Buttes, Navajo Nation (Arizona), USA. *Bulletin of Volcanology*, **53**, 239–258.

WHITE, J.D.L. & HOUGHTON, B.F. 2000. Surtseyan and related eruptions. *In*: SIGURDSSON, H., HOUGHTON, B., MCNUTT, S., RYMER, H. & STIX, J. (eds) *Encyclopedia of Volcanoes*. Academic Press, New York, 495–512.

WHITE, J.D.L., MCPHIE, J. & SKILLING, I. 2000. Peperite: a useful genetic term. *Bulletin of Volcanology*, **62**, 65–66.

WOHLETZ, K.H. 1986. Explosive magma–water interactions: thermodynamics, explosion mechanisms, and field studies. *Bulletin of Volcanology*, **48**, 245–264.

Emplacement textures in Late Palaeozoic andesite sills of the Flechtingen–Roßlau Block, north of Magdeburg (Germany)

MAREK AWDANKIEWICZ[1], CHRISTOPH BREITKREUZ[2] & BODO-CARLO EHLING[3]

[1]*University of Wrocław, Institute of Geological Sciences, Department of Mineralogy and Petrology, ul. Cybulskiego 30, 50–205 Wrocław, Poland*
(e-mail: mawdan@ing.uni.wroc.pl)
[2]*TU Bergakademie Freiberg, Institut für Geologie/Paläontologie, Bernhard-von-Cotta-Strasse 2, D-09599 Freiberg, Germany*
[3]*Landesamt für Geologie und Bergwesen, Sachsen-Anhalt, Köthener Str. 34, 06118 Halle, Germany*

Abstract: During Late Palaeozoic times, andesite magmas intruded a 100-m thick sequence of Late Carboniferous lacustrine to alluvial siliciclastic rocks, sandwiched between folded Namurian sediments at the base and a thick, partly welded rhyolitic ignimbrite sheet at the top, in the Flechtingen area. An intrusive complex, comprising of two main sills up to 200 m thick and over 20 km in lateral extent, was formed (the Flechtingen Sill Complex, or the 'lower andesites', previously interpreted as lava flows). In the supposed feeder area the andesitic magmas locally pierced the ignimbrite seal, forming isolated pipes and domes (the 'upper andesites'). Thickness variations of the sills suggest ponding of the andesite magma within former depositional troughs and syn-emplacement deformation of the host sequence with the formation of swells, basins and fault-bounded grabens at the top of the sills. Locally, thin 'failed' sills are present. In places, sill margins show domains of flattened and aligned vesicles and planar, sharp contacts to the host sediments. However, in many outcrops and drill cores the sill margins consist of variable andesite breccias and peperites. These fragmental rocks reflect auto- to quench-clastic brecciation of chilled andesite magma (*in situ* breccias and perlite), variable magma–sediment interactions and later brecciation by hydrothermal fluids.

Basic to intermediate sill complexes are a prominent feature of the Permo-Carboniferous volcanic zones in Europe. Sundvoll and Larsen (1993) provide an overview about the knowledge of basic sill–dyke complexes in the Oslo Graben and its vicinity. Large and voluminous sills (up to hundreds of cubic kilometres) were emplaced in Northern England (Whin Sill, etc; Francis 1988). A number of large lava–sill–dyke complexes are known from the intra-montane basins of the decaying Variscan orogen: the Saar–Nahe Basin in SW Germany (Jung 1967; Lorenz & Haneke, this volume), the Thuringian Forest in central Germany (Katzung & Obst 1996), and the Sudetic basins in SW Poland (Awdankiewicz 1999; Awdankiewicz, this volume).

In this paper, combining sections from quarries and drill cores, detailed texture descriptions of igneous/sediment contacts are given for a Late Palaeozoic andesitic sill complex and associated minor stocks and lavas in Eastern Germany, near Flechtingen (Fig. 1). The intrusive complex described, defined here as the Flechtingen Sill Complex (FSC), has been previously interpreted as a lava succession (Benek *et al.* 1973). The FSC crops out on the Flechtingen–Roßlau Block, uplifted during Mesozoic compressional tectonics, and forms a part of a large volcanic province: the NE German Volcanic Province (NEGVP) in the Central European Basin. At least 50 000 km^3 of predominantly calc-alkaline, SiO$_2$-rich volcanics (Benek *et al.* 1996) developed in a relatively short time (302–297 ± 3 Ma; Breitkreuz & Kennedy 1999) in the NEGVP.

The Late Palaeozoic Flechtingen volcano-sedimentary succession was used in previous studies as a lithostratigraphic key section for the subdivision of much of the NEGVP, which, except for the Flechtingen area, is covered with younger sediments and only exposed by hundreds of drill holes, in places down to 8 km depth (Hoth *et al.* 1993a, b). Thus, the recognition of the intrusive nature of the FSC, and the texture descriptions presented here, carry important regional-stratigraphic implications. These findings also provide a tool for the identification of

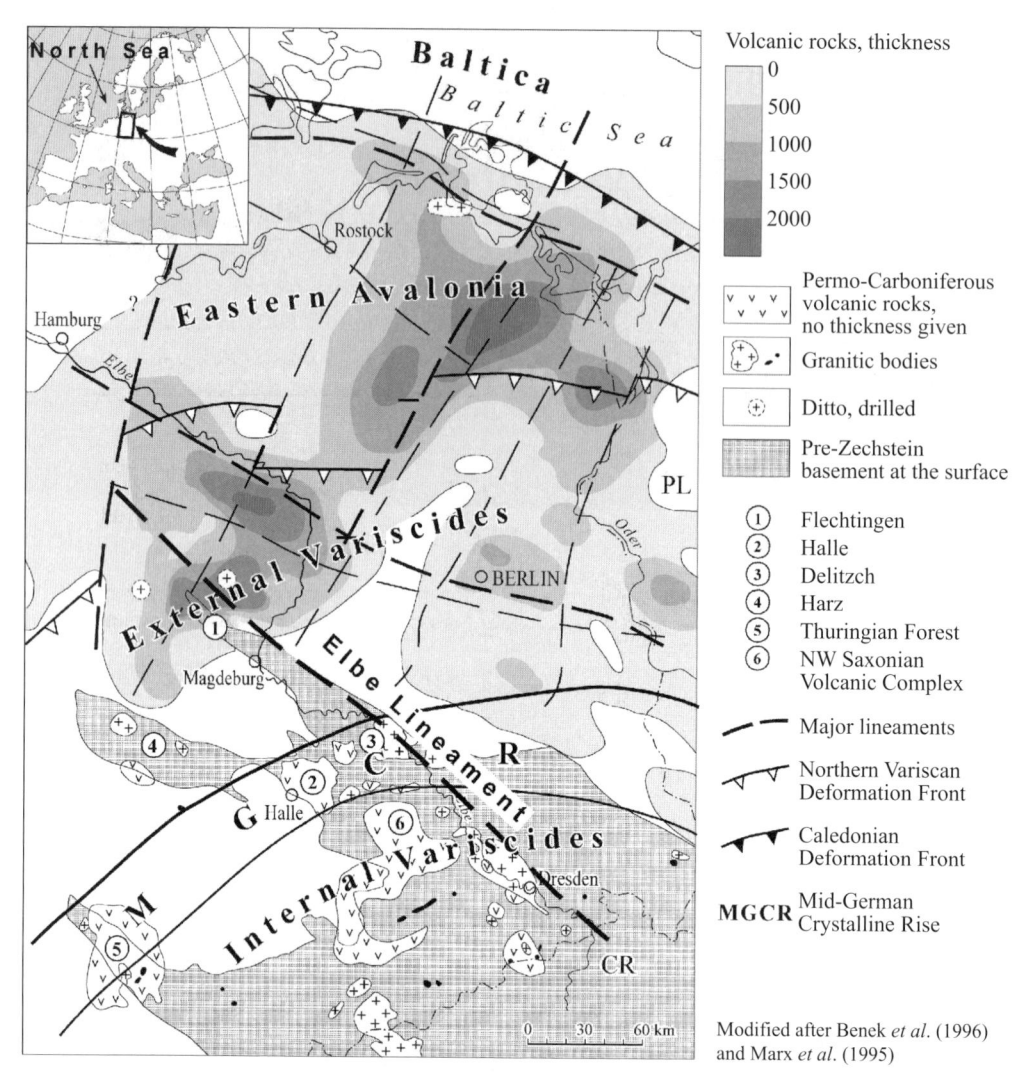

Fig. 1. Geological sketch map of eastern Germany depicting, in the lower part, outcrops of pre-Zechstein rocks, Late Palaeozoic volcanic and intrusive complexes (1 to 6), and, in the upper part, subcrop distribution and thickness of Late Palaeozoic volcanic rocks.

mafic sill complexes in other parts of the Central European Basin and beyond.

The Late Palaeozoic volcano-sedimentary succession of the Flechtingen–Roßlau Block

The c. 1400-m thick Stephanian–Lower Rotliegend succession of the Flechtingen–Roßlau Block rests unconformably on folded Namurian greywackes, and dips southwestwards at low to moderate angles (Fig. 2). According to previous studies, the succession starts with two or three thick sheets of andesitic lava ('Older or Lower Andesitoids'; Benek et al. 1973; Benek & Paech 1974), which are underlain, intercalated and overlain by, thin lacustrine to alluvial sediments of the Süplingen-, Bodendorf- and Eiche beds (Burchardt & Eisenächer 1970; Benek et al. 1973). Detailed sedimentological investigation, the results of which will be published elsewhere (see Egenhoff & Breitkreuz 2001), and the results presented here, indicate that the Süplingen-, Bodendorf- and Eiche beds had been separated by andesite intrusions, and

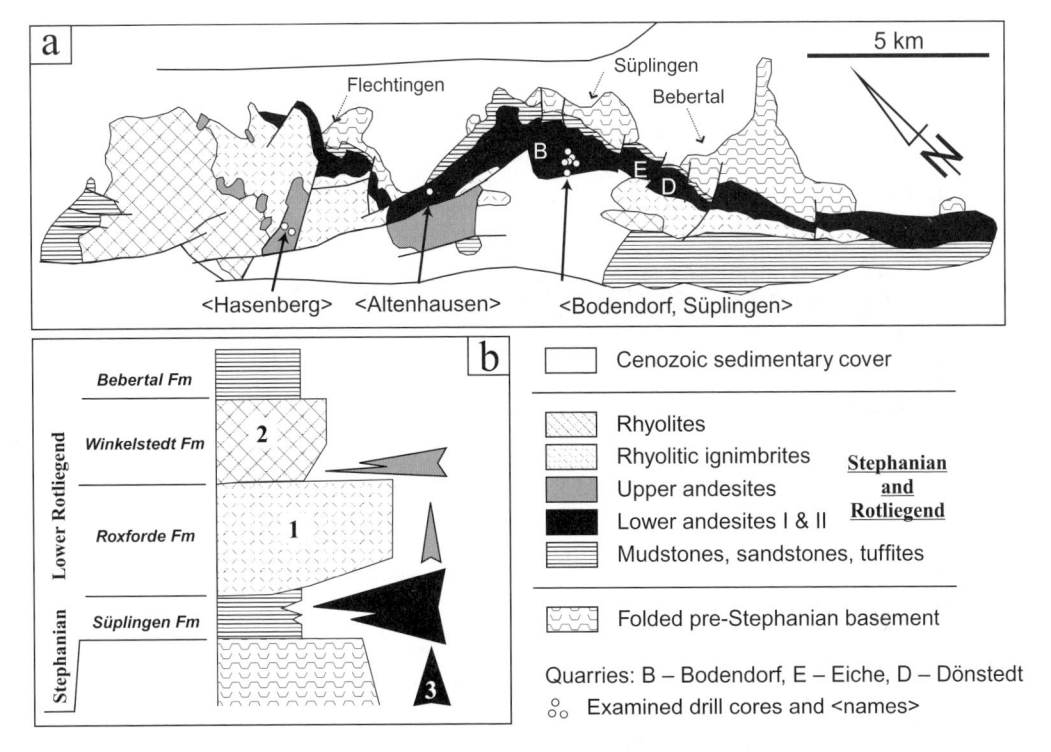

Fig. 2. Geological map (**a**), (after Benek *et al.* 1973 #1224) and revised lithostratigraphic log (**b**), (after Plein, 1995 #813, and German Stratigraphic Commission 2002) of the Late Palaeozoic volcano-sedimentary succession of the Flechtingen–Roßlau Block; Numbers 1, 2 and 3 represent the succession of formation/emplacement presumed in this contribution; for location see Figure 1.

belong to a single lacustrine to alluvial sedimentary unit of Stephanian age, named the Süplingen Formation. Deposition of this formation was accompanied by SiO_2-rich explosive volcanism, as indicated by the presence of pumice and lava fragments at some levels. Soft-sediment deformation indicates the non-lithified status of the sediments during intrusion. The andesite intrusions hosted in the Süplingen Formation are grouped into the Flechtingen Sill Complex (FSC).

A major ignimbrite sheet, belonging to the Röxforde Formation, covers the Süplingen Formation (Benek *et al.* 1973). The partially welded and up to 650-m thick rhyodacitic pyroclastic flow deposits comprise a lower lithic-rich unit, and an upper lithic-rich, garnet-bearing unit (the Steinkuhlenberg and Holzmühlenthal members, respectively), the latter dated at 302 ± 3 Ma (Breitkreuz & Kennedy 1999). On top of the ignimbrite, rhyolitic subaerial lava was emplaced (Winkelstedt Formation, Fig. 2). In the NW part of the outcrop area, isolated andesitic stocks and domes were emplaced within and on top of the silica-rich volcanic units. These 'Younger or Upper Andesitoids' have been already considered as intrusions by Benek *et al.* (1973), and we discuss below their cogenetic emplacement in the framework of the FSC. The Flechtingen volcanic succession is covered with Permian sediments (the Bebertal Formation).

The andesites of the Flechtingen area show a strong hydrothermal alteration and a partial to total replacement of the primary igneous phases with various secondary minerals (Ehling *et al.* 1995). However, the lower and upper andesites differ in their phenocryst content and size (Burchardt & Eisenächer 1970; Benek *et al.* 1973). The lower andesites are aphyric to weakly porphyritic, with the largest phenocrysts *c.* 2 mm long. The upper andesites are porphyritic, with up to *c.* 30% phenocrysts, up to *c.* 4 mm long. The phenocrysts are albite (± calcite, kaolinite and sericite) pseudomorphs after plagioclase and chlorite (± opaques and calcite) pseudomorphs after pyroxene. The main groundmass components are lath-shaped plagioclase and variable amounts of interstitial quartz, alkali

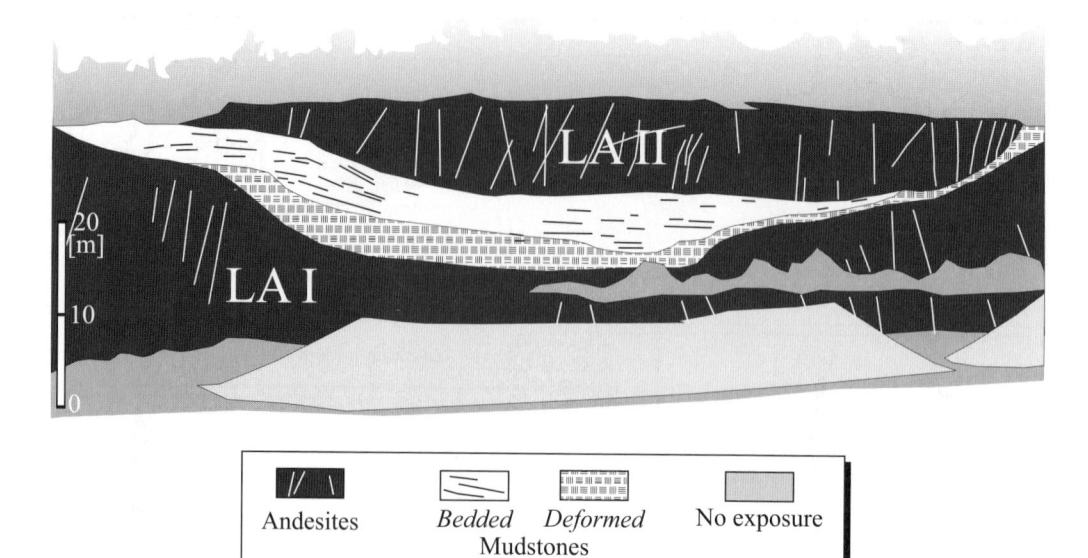

Fig. 3. Sketch of a photo from a wall in the Bodendorf Quarry, depicting the lower andesites I and II (LA I and II) with an intercalation of deformed and bedded sediments (Süplingen Formation).

feldspar, calcite, chlorite and hematite. Some main- and trace-element analyses of both units are discussed by Benek et al. (1973); more complete element analyses are present in diagrams published by Benek et al. (1996) and Marx et al. (1995).

Structure, lithology and emplacement processes of the andesite complexes

The Flechtingen Sill Complex

The FSC consists of thick, essentially conformable, massive andesite sheets, hosted within and intercalated with fine-grained sedimentary rocks (Fig. 2). Two major successive sheets of the complex are distinguished as the lower andesites I (LA I) and the lower andesites II (LA II): the latter characterized by a more coarse-grained groundmass and by quartz xenocryst content of up to 20% (Benek et al. 1973). The lower andesites I extend for c. 14 km along the NW and central parts of the Flechtingen Block, and are up to 100 m thick. At the Alte Mühle near Süplingen, the sharp basal contact with sediments of the Süplingen Formation is exposed. The lower andesites II crop out in the central part of the area and further southwestwards for at least 20 km, and are up to 200 m thick. However, field and drill-core data suggest that the main andesite sheets locally show thin offshoots or split into thinner component units separated by sedimentary pockets or lenses. The basal contact of the lower andesite II was exposed in 1995 at the base of the Dönstedt Quarry (Fig. 6a).

Well-exposed sections in active quarries reveal various deformation structures within the andesite–sedimentary complex. Large-scale structures include shallow synclines as well as grabens and half-grabens (Fig. 3). In the Eiche–Dönstedt quarries, five west–east-oriented synclines and grabens, up to c. 100 m wide, can be identified along a c. 1.5 km section of the lower andesite II top (Fig. 4). The covering ignimbrite sheet was also affected by the graben formation.

On a smaller scale, sedimentary rocks adjacent to the andesite sheets, both in quarries and drill cores, show bending, folding and fracturing, and locally the andesites are discordant relative to the deformed bedding of the sedimentary rocks (Figs 5 & 6d). All the deformation structures mentioned tend to be asymmetrical, but no preferred directions have been identified so far. In places, the sedimentary rocks penetrate the andesites as clastic wedges and clastic dykes, up to c. 3 m long. The andesites also host domains of the sedimentary rocks, both in the margins of the sheets and their

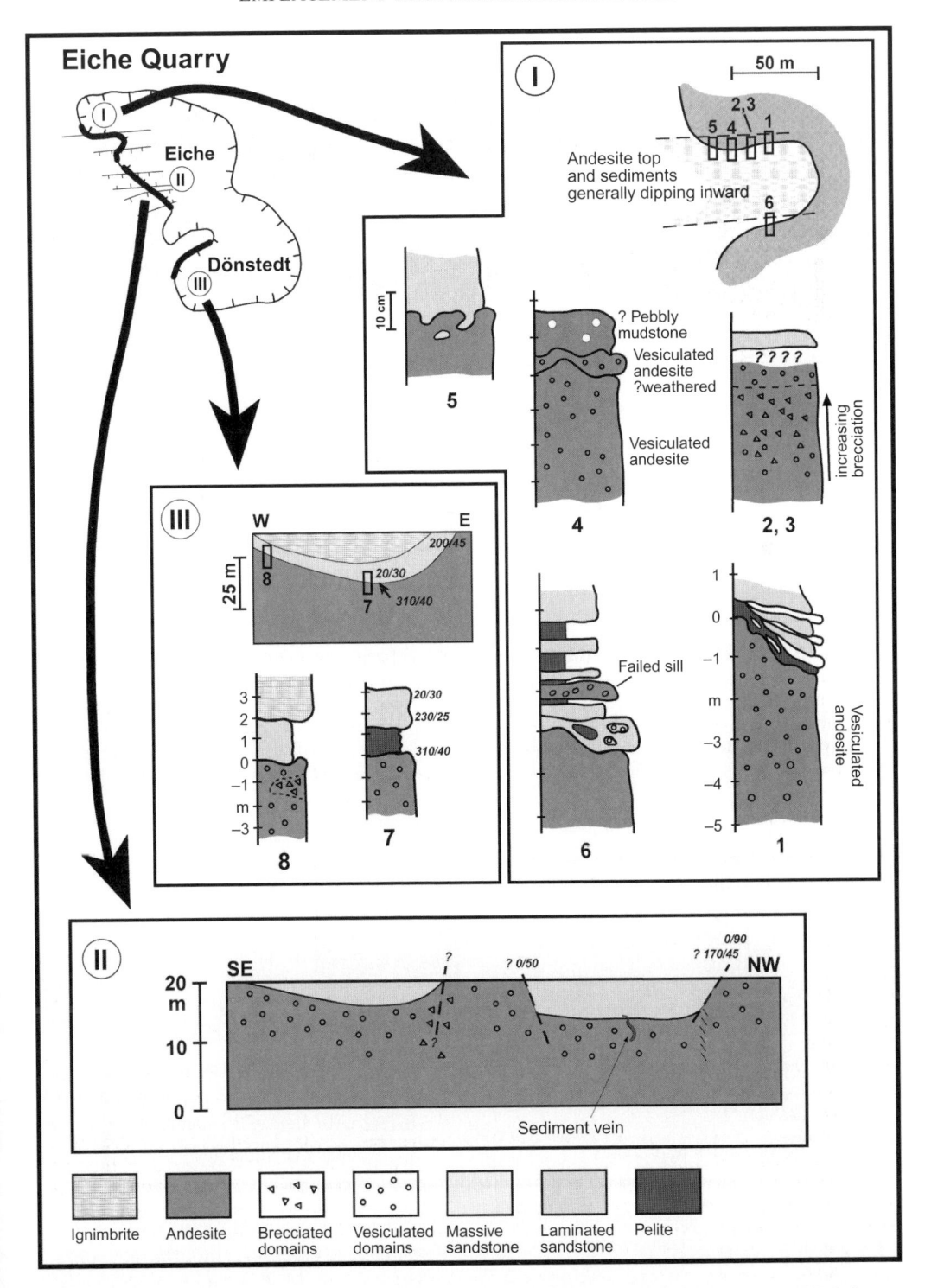

Fig. 4. Sketch of the Eiche–Dönstedt Quarry NW of Bebertal, depicting detailed sections I, II and III; I shows a map view of the quarry with the position of profiles 1 through 6; II represents a sketch from a photo, showing the graben and horst structure at the top of LA II.

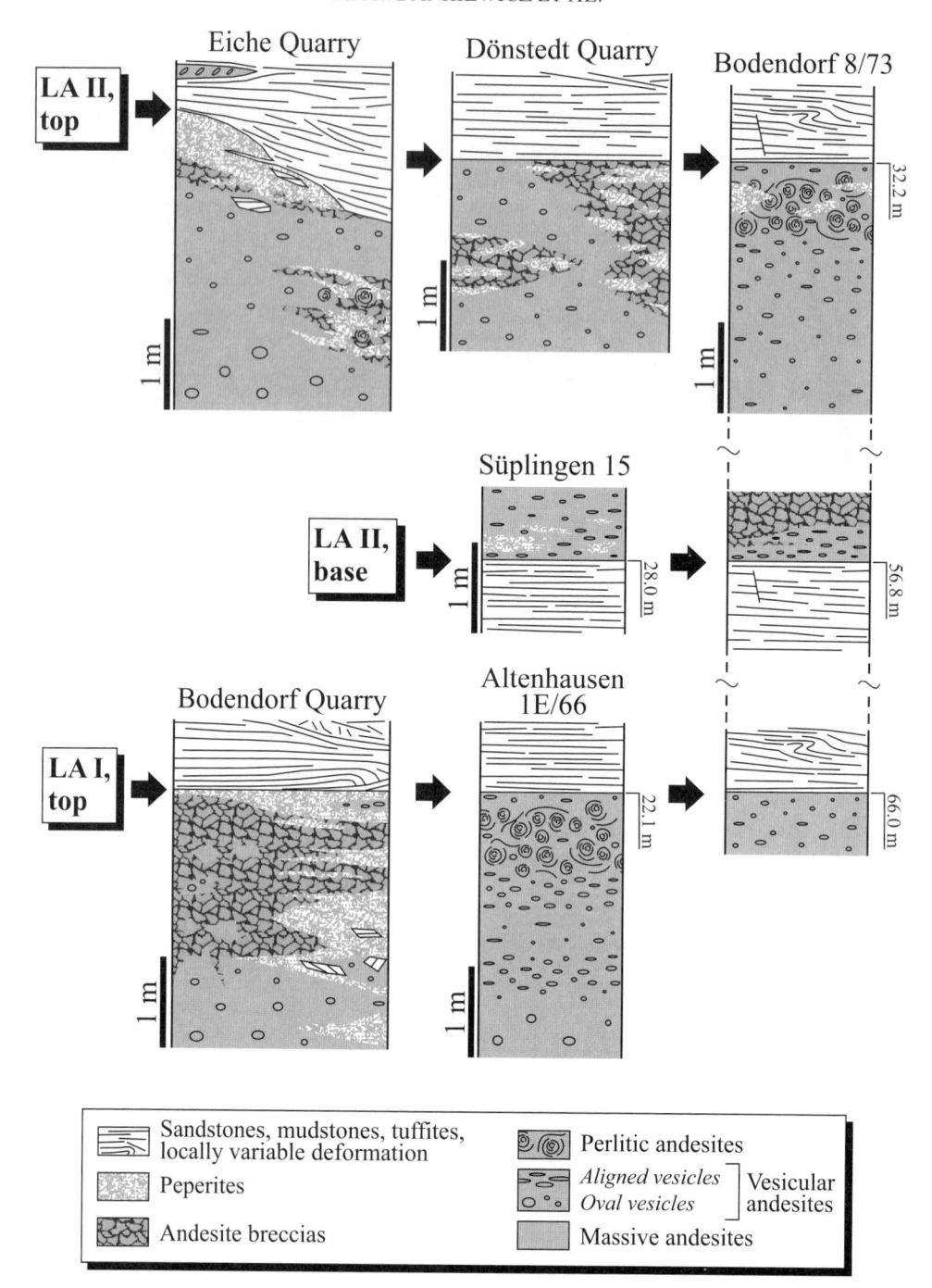

Fig. 5. Schematic logs of quarry walls and drill cores featuring different textures in the contact zones between the andesitic sills and host sediments; for location see Figure 2.

interiors. The largest sedimentary rafts observed in the SE part of the Bodendorf Quarry are up to 1 m thick and over 5 m long.

The andesite sheets generally show a monotonous internal structure and consist largely of massive rocks with predominant subvertical joints, and prismatic to irregular joints near the margins. Veins of calcite and quartz, as well as

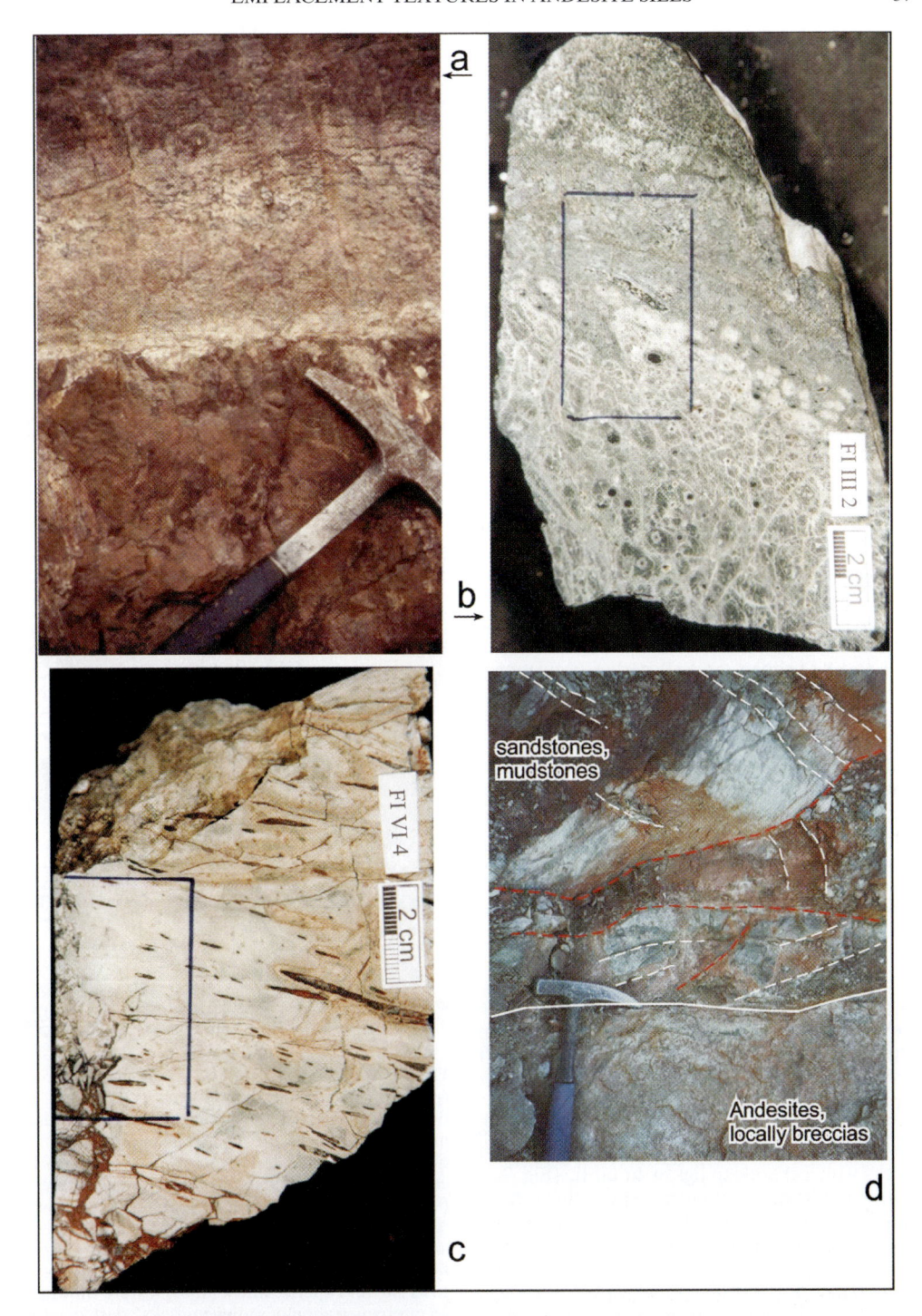

Fig. 6. (a) Basal contact of the lower andesite II in Dönstedt Quarry. In 1995, a 10-m long sharp contact to structureless fine sediments was exposed; note the vesicles in the andesite aligned parallel to the margin; the light colour is carbonate alteration. (b) Perlitic texture at the top contact, drill core Altenhausen 1E/66, 22.8 m depth. (c) Andesite with flattened vesicles aligned parallel to the margin, drill core Bodendorf 8/73, 57.0 m depth. (d) LA I top overlain by tilted and faulted sediments, Bodendorf Quarry (location: centre of Fig. 3). White line, top of andesite; white broken lines, bedding in sedimentary rocks:; red broken lines, thrust planes.

breccias of massive andesite blocks cemented with these minerals, are found locally. The upper, c. 10–20-m thick zones of the sheets are characterized by large, sparsely distributed vesicles (up to nearly 10 cm in diameter, partly filled with silica minerals, carbonates and hematite). However, the marginal zones of the sheets, 1 to 3 m thick, show a strong lithological variation and consist of massive to vesicular andesites, perlitic andesites, andesite breccias and peperites.

Representative sections of the marginal zones of the andesite sheets are shown in Figs 4 & 5. The vesicular andesites in the contact zones are usually characterized by a strong flattening and alignment of vesicles (Fig. 6c); a c. 30-cm thick andesite offshoot in the Eiche quarry (Section 6 in Fig. 4) shows imbrication of flattened vesicles. The vesicles are usually filled with quartz, carbonates, chlorite or clay minerals. Well-developed perlitic textures (Fig. 6b) can be seen in the cores: Altenhausen 1E and Bodendorf 8, and relics of perlites are also found in the Eiche Quarry in some breccia blocks. The vesicle alignment and perlitic textures are most characteristic of non- to weakly brecciated margins of the andesite sheets, and all perlite occurrences are found in the top contact zones of the andesites (however, only two basal contacts were available for detailed study). Most of the contact zones show at least a weak brecciation of the andesites and some deformation of adjacent sedimentary rocks. In sections with significant brecciation of andesites, the sedimentary rocks are also more strongly deformed. The brecciated margins are characterized by an interdigitation of various lithologies, especially well seen in the quarries, where the contact zones can be traced laterally for several metres. The andesite breccias and peperites occur both as small patches (centimetres to decimetres wide) in massive to vesicular andesites, as well as larger, 1–2 m thick, discontinuous layers along the contact zones. The boundaries of the various lithologies are variable, sharp to gradational.

The andesite breccias are monolithological, relatively fine-grained, usually matrix-supported rocks with a chaotic to jigsaw-fit fabric (Fig. 7a). These rocks consist of petrographically uniform andesite clasts, usually less than 5 mm in size, with the largest up to c. 20 mm long. The clasts are isometric to elongate, and angular to subrounded in shape. They are uniformly dispersed within an aphanitic matrix of microcrystalline quartz with abundant hematite staining, with some broken plagioclase crystals and chlorite pseudomorphs after ferromagnesian minerals. The matrix of the breccias compares well to the

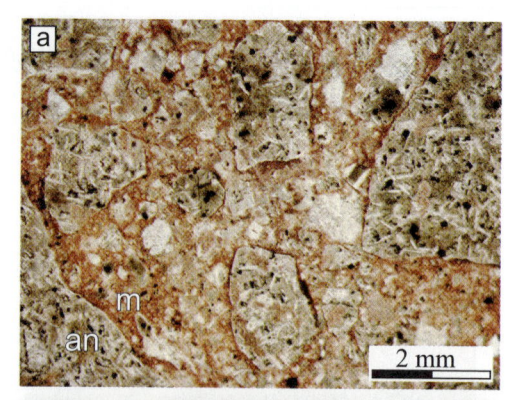

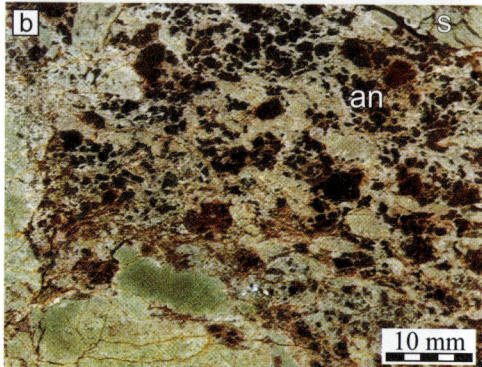

Fig. 7. (a) Photomicrograph of the andesite breccia (plane-polarized light). Andesite clasts ('an') and broken crystals derived from the andesite are set in a matrix ('m') of hematite-stained microcrystalline quartz. Bodendorf Quarry, top of LA I sill (location: centre of Fig. 3). (b) Peperite with a clastic texture ('an', dark-brown to reddish-coloured andesite clasts; 's', yellowish to green sedimentary matrix). Most andesite clasts are angular to subangular and jigsaw fabric is seen in many places. Eiche Quarry, top of LA I sill. (c) Peperite with a fluidal texture showing an interpenetration of sedimentary ('s', dark greenish to grey) and igneous components ('an', yellowish to greenish) and lobate interfaces between them. Bodendorf Quarry, top of LA I sill (location: centre of Fig. 3).

interstitial components of non-brecciated andesite (hematite-stained quartz with acicular plagioclase microliths), which most likely represents a devitrified glass.

The peperites represent texturally variable mixtures of two components:

1 andesitic and
2 sedimentary siliciclastic.

The latter corresponds to sandstones and mudstones, often with a substantial tuffaceous admixture (devitrified glass shards). Most of the peperites show clastic textures and jigsaw to chaotic fabrics (Fig. 7b). Such rocks consist of andesite clasts in a sedimentary matrix or, more rarely, sedimentary clasts in the andesitic matrix. The clasts are up to 10 cm in size and angular to lobate in shape. Both the andesite and sedimentary clasts may show some petrographic variation, in contrast to the monotonous appearance of the andesite breccias described above. Some andesite clasts show evidence of chilling and granulation of their margins. However, some of the peperites are apparently not clastic rocks. Rather, they represent intimate mixtures of fluidally interpenetrating andesitic and sedimentary components, with irregular to lobate inclusions of one in the other (Fig. 7c).

Lithological gradations between the andesite breccias and peperites are common. The transitional rocks are andesite breccias, which host small patches and streaks of sedimentary rocks as well as peperitic domains, and show a stronger textural and petrographic variation compared to the sediment-free andesite breccias.

In addition, quartz, calcite, chlorite, clay minerals and hematite patches and veins of variable size (from microscopic to several cm wide) are common in andesites and breccias of the contact zones. Locally, cracked rock cut with a network of these mineral veins grades into breccias with a matrix of quartz, calcite, etc. These mineral accumulations often occur concentrated close to the boundaries of various lithologies (e.g. andesite–sedimentary rocks, andesite–andesite breccia, andesite breccia–peperite) and crosscut the lithological boundaries.

Interpretation

The lower andesites, considered as lava flows so far, are re-interpreted in this study as a subvolcanic intrusive complex: the Flechtingen Sill Complex, that was emplaced into fine-grained Stephanian sedimentary rocks sandwiched between folded basement rocks and the thick ignimbrite cover. The intrusive nature of the andesites is evidenced by the structures and lithologies in their contact zones, including:

1 local andesite offshoots,
2 sedimentary xenoliths,
3 peperites and
4 sediment clastic dykes.

In particular, the two latter features demonstrate that magma–wet sediment interactions must have occurred along both the basal and top margins of the andesite sheets during their emplacement, apparently as intrusions (cf. Martin & Nemeth 2002). Furthermore, the overall structure of the andesite sheets and some characteristics of their host sedimentary rocks are inconsistent with the emplacement of andesites as hundreds of metres thick lava flows. Lava of intermediate composition is typically developed as block lava with a thick mantle of relatively coarse-grained autoclastic breccias, and the invasion of such thick flows into a fluviatile/lacustrine environment would leave a clear record in the contemporaneous deposits, including, for example, the abundant supply of coarse-grained andesitic detritus or strong hyaloclastite development and the creation of morphological scarps and strong facies changes. Neither thick autoclastic deposits, nor volcanism-related variation are observed in the studied succession, and thus effusive andesitic volcanism in the Flechtingen area is considered to have been unlikely. However, tuffaceous components observed in the sedimentary rocks (altered glass shards, quartz and biotite phenocrysts) testify to contemporaneous explosive eruptions of intermediate to acidic magmas in adjacent areas.

It is further considered that the andesite intrusions were emplaced into the host sedimentary rock after the deposition of the rhyolitic ignimbrites. The thickness of sedimentary intercalations between the component andesite sheets and the ignimbrites does not exceed 20 m, which is an order of magnitude less than the thickness of the intrusions. It seems likely that such a thin sedimentary mantle would easily be destroyed and pierced by the intruding magma if the thick, massive ignimbrite cover was not present. This would eventually lead to a subaerial andesitic volcanism, and the evidence against such activity was discussed above. Effusive eruption of the andesites followed by plunging of the flows into wet, unlithified fluvio-lacustrine deposits (Rawlings et al. 1999), which represents yet another interpretation to be considered, can also be excluded for the reasons mentioned.

From these constraints it follows that the andesitic magmas rising through the heterogeneous near-surface section of the Flechtingen

Block (composed of the rigid, folded basement overlain by fine-grained sedimentary strata and a thick, relatively massive ignimbrite sheet) preferentially intruded into the sedimentary rocks, which apparently represented a weak horizon within this succession. The ignimbritic overburden presumably caused an overpressure in the wet unconsolidated sediments. The emplacement depth of the andesitic intrusions is thus constrained by the thickness of the ignimbrite, which reaches c. 650 m to the NW and decreases southeastwards (Fig. 2). Although the amounts of possible compaction or erosion of the ignimbrite during its post-depositional history are unknown, the intrusion depth of the andesite magma can roughly be estimated as several hundred metres (down to 600–800 m below the Permian palaeosurface).

The location of the FSC conduit system is not known. Reconstruction of the palaeoflow direction using magnetic measurements (AMS) is hampered by the strong thermal overprint that the rocks experienced. Vesicle alignment at the sill margins could be determined at a few places within the quarries; however, palaeoflow indicators would be local, and the interpretation may remain ambiguous. The position of the plugs and flows of the upper andesites, which are considered as cogenetic to the FSC, point to a conduit system located west of the town of Flechtingen (Fig. 2).

As described above, the lower andesites generally represent a complex of two stacked sheet-like intrusions. The structure of the complex and the geometry of the component sheets are gradational between those typical of laccoliths and sills. The presence of the stacked sheets resembles the Christmas-tree laccolith of Corry (1988), and the aspect ratio (thickness/length ratio) of the component sheets (c. 0.007 for LA I and 0.01 or less for LA II) is intermediate between that of classical laccoliths (often over 0.1) and extensive doleritic sills (typically 0.001 or less).

An unusual feature of the intrusive complex at Flechtingen is the uneven morphology of the upper surfaces of the intrusive sheets, with the synclines and grabens (Figs 3 & 4). These features can be generally considered as intrusion-related deformation structures. A syn-intrusive formation of the synclines and grabens might have been induced by some rheological heterogeneity of the intruded country rocks, with more rigid parts (e.g. more lithified) forming obstacles against the intruding magma. A syn- to post-intrusive local sagging of roof-rocks into the solidifying intrusions might also have played a role. However, these structures may also be related to a specific growth mode of the intrusions. Corry (1988) discussed the problem of laccolith growth, and concluded that laccoliths initially spread laterally as thin sills and then inflate to their final thickness. Although Corry rejected other models, the synclines and grabens discussed here may represent specific deformation structures predicted by other models of laccolith emplacement (Davis 1925; Hunt et al. 1953; Corry 1988), in particular the model of lateral growth of a thick sheet, or that of a simultaneous lateral growth and thickening. A verification of these hypotheses would require a separate, detailed structural study.

The structures and lithologies of the contact zones of andesites and their host rocks shed light on the processes of magma–country rock interactions during the emplacement of the intrusions. The contact zones of the FSC may be classified into two types:

1. non-brecciated to weakly brecciated, with weak deformation of country rocks, and
2. strongly brecciated, with significant deformation in the country rocks and local discordance of the intrusion margins.

The former contact type is not common, and the latter type is most widespread. No relationship is found between the contact type and lithology of the country rocks (bedded mudstones and sandstones in both cases), and lateral gradations between the two types – along the same contact zone – are observed (e.g. along the top of lower andesites I at Bodendorf Quarry). These features suggest that the development and final structure of the contact zones depended on a complex interplay of various factors, possibly including the variable degree of lithification and pore-water content of the sedimentary rock as well as variable flow rate of magma or changes of magma rheology (due to cooling and degassing) during the emplacement. Another important factor possibly involved was the conformable v. discordant character of the magma–host rocks interface. At even a slightly discordant margin, the intruding magma cross-cuts bedding planes and is thus more efficiently exposed against pore fluids in the host. In consequence, processes such as quenching of magma or fluidization of sediments and peperite formation are enchanced, resulting both in a stronger brecciation of the intrusion margin and a stronger deformation of the country rocks. Such interpretation is consistent with the relationships between the intensity of brecciation at sill margins and the deformation of adjacent sedimentary rocks in the FSC, and similar links (stronger fragmentation at the discordant margins of intrusions) can also be inferred from the descriptions of other contact zones (e.g. Walker & Francis 1987; Goto 1997).

The dominant processes affecting intrusion margins at the conformable, non-brecciated contacts in the FSC were cooling and vesiculation of the intruding magma. These processes led to the formation of perlitic and vesicular andesites with stretched bubbles. The perlites are rare in the FSC, a fact consistent with the subvolcanic nature of the magmatic bodies, and the presence of perlites along only the top contacts of the sills may further suggest that cooling rates at the bottom contacts were lower that at the top, too – low for the perlite formation.

Fragmentation and deformation of both the cooling magma and host rocks, as well as interpenetration and mixing of the igneous and sedimentary components, affected the brecciated, partly discordant margins of the intrusions in the FSC. The andesite breccias probably formed due to quenching-induced fragmentation, with some influence of flow-related brecciation. The latter process might have been of minor importance, as no aligned fabrics can be seen in the andesite breccias (however, the largely isometric clasts of the breccias are not good kinematic indicators). Although the andesite breccias can be genetically considered a type of hyaloclastite, they are characterized by relatively large proportion of hypocrystalline to crystalline, non-glassy clasts. Another process involved at the brecciated margins – magma–sediment interaction – resulted in the formation of the sedimentary xenoliths and peperites. The relative scarcity of peperites in the FSC compared to other localities (in particular subvolcanic intrusions or cryptodomes emplaced into fresh sediments at the sea-floor, e.g. Howells et al. 1991; Allen 1992) indicates the host sediments were partly consolidated (and/or pore-water poor?) at the time of andesite intrusion, but contained also wet, less consolidated domains. The textural heterogeneity of the peperites is similar to that observed in other occurrences worldwide (reviewed recently by Skilling et al. 2002) and suggests that a range of processes were involved in their formation. Some of the described peperites are similar to the blocky type, while others resemble the globular peperites of Busby-Spera and White (1987), and their formation possibly involved quench-related fragmentation of magma, fluidization of wet sediments and fuel-coolant type interactions (FCI). Other peperites found at Felchtingen texturally resemble mixtures of two liquids with different viscosities and their formation possibly involved hydrodynamic mingling between the andesitic magma and the liquefied sediments, aided by explosive phreatomagmatic processes, as suggested by Zimanowski & Büttner (2002).

The textural heterogeneity of the peperites points to their multi-stage formation, with variation of the dominant processes with time. The final stages of intrusive margin development also involved brecciation due to the hydraulic fracturing mechanism, as indicated by jigsaw-fit breccias with calcite or silicate cement and abundant hydrothermal veins cross-cutting the andesite breccias and peperites.

The upper andesites

The upper andesites comprise seven small, oval to aligned, poorly exposed outcrops, most of which straddle the rhyolite/ignimbrite boundary west of Flechtingen. These outcrops range from c. 0.2 to nearly 3 km in length, and the two largest are partly fault-bounded. The second largest outcrop has been penetrated with two boreholes (Fig. 2): Hasenberg 5/66 and Hasenberg 8/66: 70 m and 98 m deep, respectively. Archival samples and unpublished core descriptions constrain the structure of this igneous body and its relationship to the underlying rocks (Fig. 8).

The Hasenberg 5/66 core, located near the NW margin of the andesite outcrop, passed through the andesites into sedimentary rocks below. The following lithologies, in upward succession, are found there:

- c. 13 m of sandstones overlain by volcanogenic sandstones, pebbly sandstones and mudstones;
- c. 22 m of sandy andesite breccias with intercalations of massive andesites;
- c. 35 m of massive and vesicular andesites.

The sandstones in the lowermost part of the section are massive, with indistinct planar alignment of their components. Bedding, if present, is thicker than the largest core samples preserved for study (c. 10 cm long). The sandstones are composed of quartz, chlorite and clay minerals.

The volcanogenic deposits found above show well-defined subhorizontal and wavy lamination and load casts. Finer-grained laminae are composed of quartz, feldspars and glass shards replaced by chlorite and clay minerals with variable hematite staining. Coarser-grained laminae consist of devitrified hypocrystalline andesite clasts, showing pseudomorphs after phenocrysts of plagioclase and ferromagnesian minerals in a groundmass of microcrystalline quartz, clay minerals and chlorite. Pink andesite clasts are less common (see below).

The sandy andesite breccias in the middle part of the section are characterized by a lack of bedding, chaotic structure and compact to open framework. The clasts are subangular

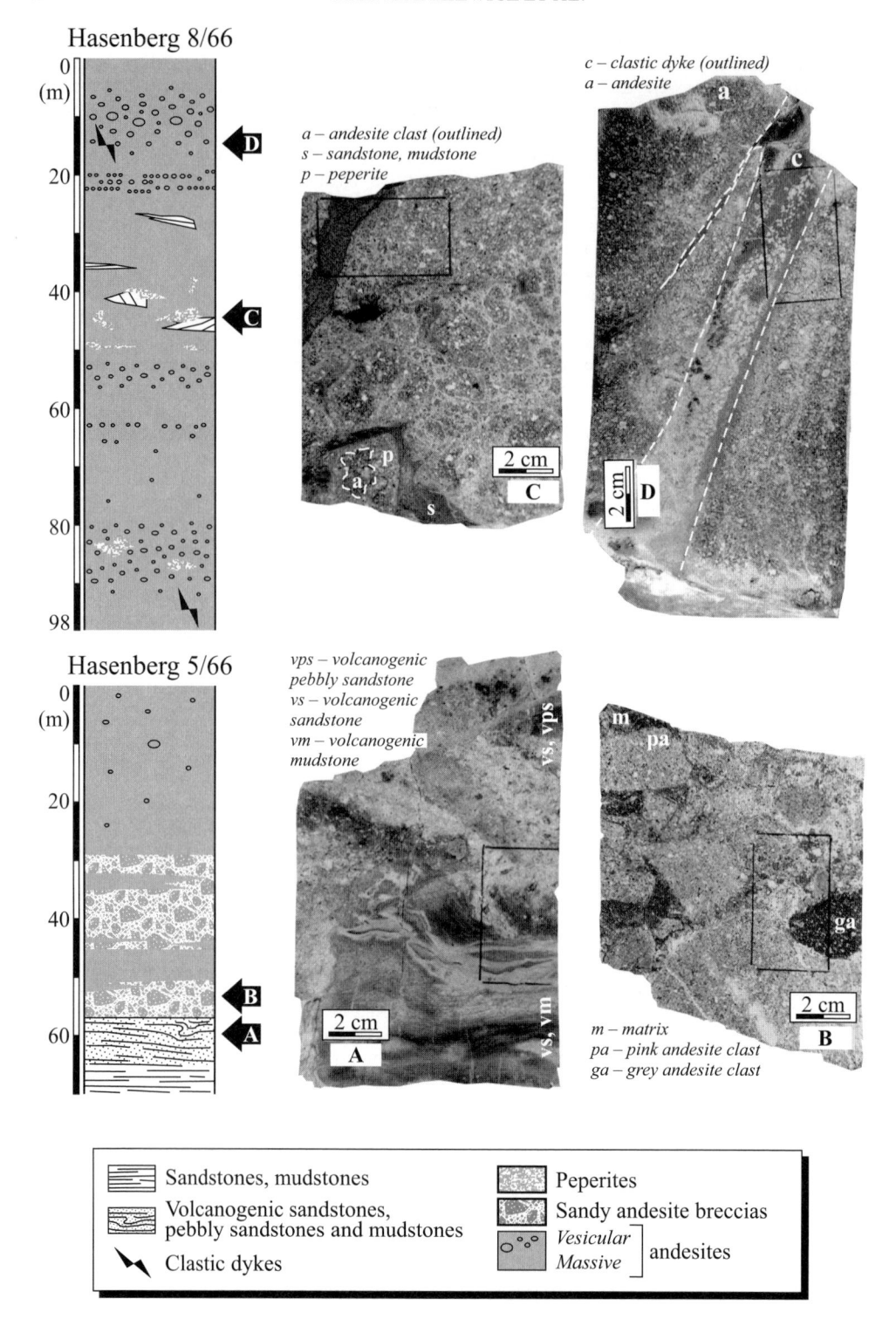

Fig. 8. Schematic logs of the Hasenberg 5/66 and 8/66 drill cores, exposing basal and internal textures, respectively, of upper andesite lava plugs and flows. Sample photos: (**a**) sandstones and pebbly sandstones; (**b**) sandy andesite breccia; (**c**) peperite, (**d**) clastic dyke. For the location of the drill holes see Figure 2.

to subrounded in shape, and the largest are up to 10 cm in size. The dominant components of the breccias are pink, porphyritic andesite clasts. Less abundant, and more typical of the finer matrix fractions, are grey to greenish andesite fragments. Both types of andesite clasts contain up to *c.* 30% phenocrysts (plagioclase replaced by albite and white mica, and chlorite pseudomorphs after, possibly, pyroxene) in a groundmass of small plagioclase laths. The interstitial groundmass components of the andesites are hematite-stained alkali feldspar and quartz (in the pink clasts) or chlorite and clay minerals (in the grey to greenish clasts). The matrix of the breccias also contains quartz grains.

The Hasenberg 8/66 core penetrated the central part of the andesite outcrop and represents the internal part of the andesite body. The core predominantly consists of massive and vesicular to amygdaloidal andesites, with the amygdales filled with quartz, chlorite and carbonates. Flow-banding is locally defined by variable vesicle and/or phenocryst content. The andesites are porphyritic, with *c.* 20–30% phenocrysts. The phenocrysts are less than 2 mm long and comprise plagioclase (strongly replaced by albite, calcite and white mica) and chlorite pseudomorphs, most likely after pyroxene. The microcrystalline groundmass consists of acicular and anhedral feldspars, quartz, small opaque grains, chlorites, clay minerals and carbonates. The acicular feldspars are usually arranged into fan-like aggregates and spherulites. Around a depth of 75 m the andesite is inhomogeneous and consists of alternating dark- and light-coloured bands and patches. The light and dark andesites differ in their groundmass interstitial components, which are anhedral feldspars or chlorites, respectively.

Lumps of sedimentary clastic rocks (up to several centimetres in size) are found in the middle and lower parts of the section. They are typically several centimetres in size and range from aligned clasts to irregular patches and streaks, which grade into a network of thin, irregular veins penetrating the host andesites. Locally, patches of peperites consist of irregular, lobate andesite clasts, many of which show chilled margins. The sedimentary matrix of the peperites is finer grained (mudstones) compared to the dominant xenoliths (sandstones). In addition, thin, steeply inclined clastic dykes (<2 cm and *c.* 70°, respectively) cut the andesites at depths of 14.6 m and 95 m. The sedimentary rocks in xenoliths, peperites and clastic dykes represent tuffaceous sandstones and mudstones. The tuffaceous sandstones are composed of devitrified glass shards (replaced with microcrystalline quartz and/or chlorite) and variable amounts of quartz, feldspars, carbonates and opaque minerals. The mudstones consist of small quartz and feldspar grains in a microcrystalline matrix of quartz, chlorites and clay minerals and calcite.

Interpretation

The upper andesites are interpreted in this study as plugs and lava domes or short flows (Benek *et al.* 1973). The plugs were preferentially emplaced along the boundary of contrasting lithologies: rhyolitic lava and ignimbrites, and locally pierced their host rocks, resulting in lava effusion. Data from the drill cores discussed above point to a complex structure of the lava flows, which document both intrusive and effusive stages of their development. The sedimentary xenoliths, peperites and clastic dykes found in the interior of the flow (e.g. core Hasenberg 8/66) reflect some incorporation of wet, unconsolidated country rocks (i.e. sediments of the Süplingen Formation) by the rising andesite magma, and variable, possibly multistage magma–wet sediment interactions (Skilling *et al.* 2002). Local fluidization of the unconsolidated deposits caused veining of the cooling andesite magma with sediments adjacent to the xenoliths. Also, peperite formation occurred due to chilling and fragmentation of the andesite melt and mixing of the resulting clasts with the fluidized sedimentary component. The peperites may be classified as the globular type (Busby-Spera & White, 1987), characteristic of magma interaction with fine-grained deposits of low permeability consistent with their finer grained matrix compared to the associated xenoliths.

The rock sequence in the Hasenberg 5/66 core is interpreted as a basal part of a lava flow overlying contemporaneous, partly volcanogenic deposits. This sequence reflects autoclastic brecciation processes and increasing supply of volcanogenic detritus into the depositional environment in front of an advancing lava flow. The autoclastic origin of the sandy andesite breccias is documented by their interdigitation with massive lavas as well as structural and textural features. Small amounts of detrital quartz found in the matrix of the breccias were possibly derived from disintegrated sedimentary xenoliths or from the basement of the lava.

Emplacement scenario of andesitic magmas in the Flechtingen Block

During the Variscan orogeny, the Flechtingen area formed part of the Rhenohercynian

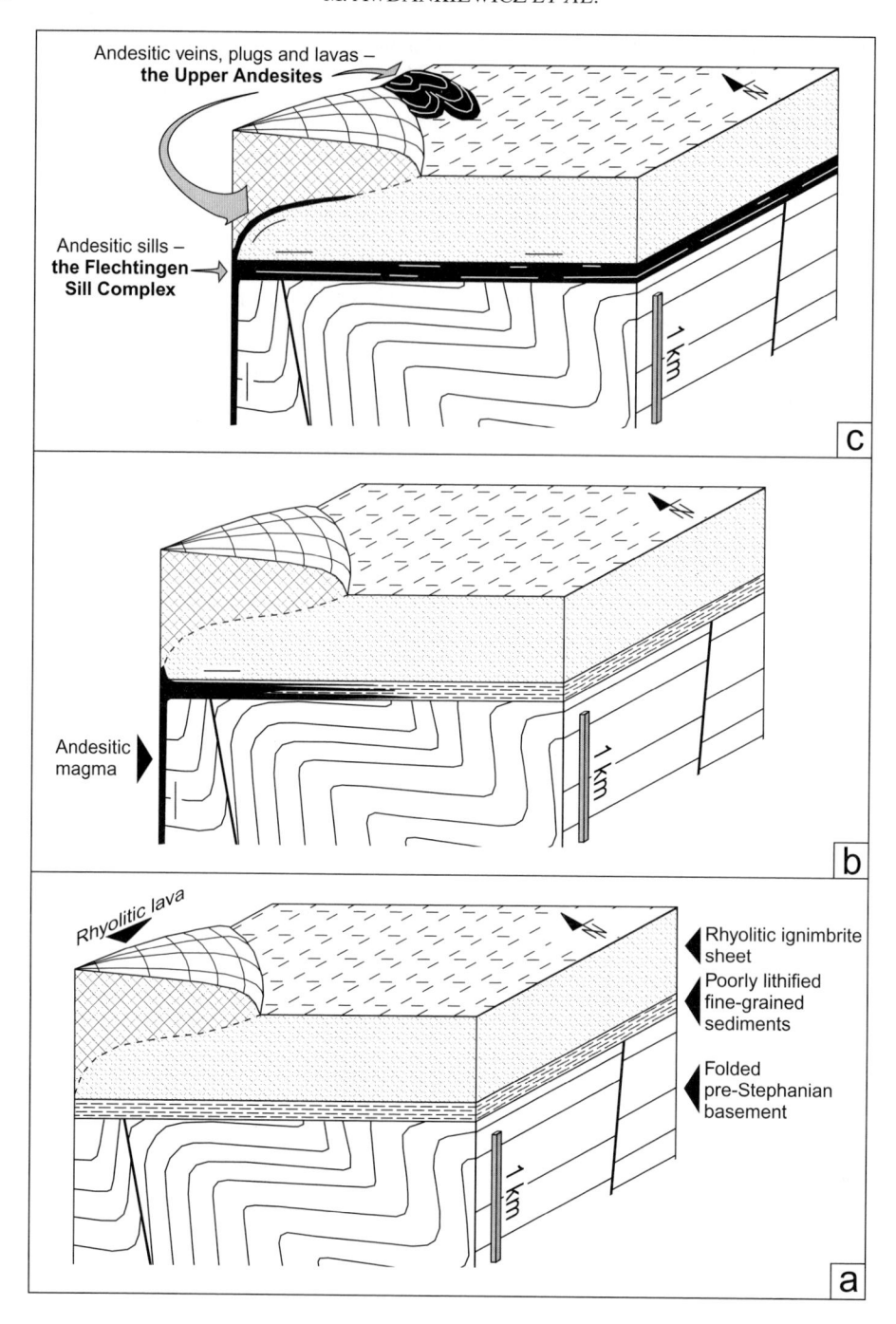

Fig. 9. A simplified model of the andesite magma emplacement in the Flechtingen Block. (**a**) Pre-andesitic stage. Folded pre-Stephanian basement is overlain by a Stephanian–Lower Rotliegend volcano-sedimentary sequence. Poorly lithified siliciclastic sediments are sandwiched between the rigid basement and a thick, massive ignimbrite sheet. (**b**) Andesitic magma rises through the basement along faults and intrudes laterally, forming sills within the weak sedimentary horizon. (**c**) The Flechtingen Sill Complex develops as magma is supplied from depth. Locally, the andesitic magma is emplaced along the rhyolite/ignimbrite boundary in the sequence above, and also extrudes at the surface, forming the so-called Upper Andesites.

fold-and-thrust belt. Intensive post-orogenic magmatism took place at the Carboniferous–Permian transition in/on a consolidated block which experienced only weak Late Carboniferous basin development with a thin sediment fill (Eigenfeld & Schwab 1974). Thus, the supracrustal setting for magma ascent and eruption/emplacement of the early rhyolitic (pyroclastics and extrusives) and successive andesitic phases (sill complex, with subordinate plugs and flows, featured here) was that of solid blocks under a dextral transtensional regime (Arthaud & Matte 1977). This is in contrast to other late- to post-Variscan basins in Central Europe (e.g. Saar–Nahe Basin, Saale Basin), where magmatism took place on and in a thick pile of basin sediments. In the Flechtingen area, apparently, the unconsolidated sediments of the Süplingen Formation, sandwiched between the Variscan basement and the thick welded ignimbrite (Fig. 9a & 9b) were the first weak zone that the andesitic melts encountered during their ascent (see Einsele 1982).

The FSC was emplaced into partly unconsolidated sediments. Thus, the time lapse was 'geologically' short between the sedimentation of the Süplingen Formation, the deposition of the thick ignimbrite and the FSC emplacement. This time constraint makes it likely that a portion of the very mantle melt, which initiated the formation of the SiO_2-rich, garnet-bearing ignimbrite (presumably via anatexis and magma mingling in the lower crust), rose to the upper crust, 'shortly' after on giving way to the FSC emplacement and the formation of the upper andesites. The high content of xenocrystic quartz in some andesite units (Benek et al. 1973), underlines the hybrid nature of the andesitic magma: mantle melts mixed with crustal anatectic melts.

The FSC melts penetrated the Süplingen Formation in the form of a number of sills (Fig. 9b). Two of these inflated to great thicknesses (LA I and LA II), enclosing sediment rafts of metres to hundreds of metres in size. Thin failed sills are preserved locally (Goto 1997). The textural characteristics of intrusion margins of the FSC demonstrate that the contact zones of sills in partly consolidated to unconsolidated fine-grained siliciclastic rocks represent a highly dynamic setting during the magma emplacement. The interplay of both intrusion-related and host-rock-related factors sets up different local conditions and enhances different processes from place to place. In consequence, a strong lateral variation of lithologies and textures along the intrusive margins develops (cf. Skilling et al. 2002). The final stage of the andesitic magmatism included the formation of intrusive veins and plugs along the rhyolite/ignimbrite boundary and local effusion of short lava flows, where the ignimbrite seal was pierced by the andesite magma (Fig. 9c).

We would like to thank the employees of the Haniel Baustoff-Industrie for allowing us access to the quarries. A one-month fellowship given to MA by the German Academic Exchange Service (DAAD) and the financial support of the study by the Institute of Geological Sciences, University of Wrocław (grant 2022/W/ING/02-3 to MA) are highly acknowledged. Funds given to CB by the German Research Foundation (DFG grants Br 997/10 and 11) are also acknowledged. The authors wish to thank D. A. Jerram and an anonymous reviewer for their helpful comments and suggestions on the paper.

References

ALLEN, R.L. 1992. Reconstruction of the tectonic, volcanic, and sedimentary setting of strongly deformed Zn–Cu massive sulphide deposits at Benambra, Victoria. *Economic Geology*, **87**, 825–854.

ARTHAUD, F. & MATTE, P. 1977. Late Palaeozoic strike-slip faulting in southern Europe and Northern Africa: results of right-lateral shear zone between the Appalachians and the Urals. *Geological Society of America Bulletin*, **88**, 1305–1320.

AWDANKIEWICZ, M. 1999. Volcanism in a late Variscan intramontane trough: Carboniferous and Permian volcanic centres of the Intra-Sudetic Basin, SW Poland. *Geologia Sudetica*, **32**, 13–47.

BENEK, R., ET AL. 1996. Permo-Carboniferous magmatism of the Northeast German Basin. *Tectonophysics*, **266**, 379–404.

BENEK, R. & PAECH, H.-J. 1974. Zur Paläotektonik des Permosiles im Gebiet der Flechtinger Scholle (Bezirk Magdeburg). *Zeitschrift für Geologische Wissenschaften*, **2**, 1143–1155.

BENEK, R., PAECH, H.J. & SCHIRMER, B. 1973. Zur Gliederung der permosilesischen Vulkanite der Flechtinger Scholle. *Zeitschrift für Geologische Wissenschaften*, **1**, 867–878.

BREITKREUZ, C. & KENNEDY, A. 1999. Magmatic flare-up at the Carboniferous/Permian boundary in the NE German basin revealed by SHRIMP zircon ages. *Tectonophysics*, **302**, 307–326.

BURCHARDT, I. & EISENÄCHER, L. 1970. Neue Ergebnisse zur Gliederung der Vulkanitserie im Gebiet des Flechtinger Höhenzuges (Subherzyne Scholle). *Geologie*, **19**, 813–825.

BUSBY-SPERA, C.J. & WHITE, J.D.L. 1987. Variation in peperite textures associated with differing host-sediment properties. *Bulletin of Volcanology*, **49**, 765–775.

CORRY, C.E. 1988. Laccoliths; mechanics of emplacement and growth. *Geological Society of America Special Papers*, **220**, 110 pp.

EGENHOFF, S.O. & BREITKREUZ, C. 2001. Fazielle Entwicklung und stratigraphische Revision oberkarbonischer Sedimente im Flechtinger

Höhenzug (nördlich Magdeburg). *Schriftenreihe der Deutschen Geologischen Gesellschaft, Heft* **13**, p. 33.

EHLING, B.C., KOCH, M.M., MATHEIS, G. & STEDINGK, K. 1995. Polystage alteration of Permosilesian magmatic rocks and its significance for mineralising processes in the NE-Rhenoherzynian Belt (Germany). *Zentralblatt für Geologie und Paläontologie Teil 1*, **5/6**, 561–565.

EIGENFELD, F. & SCHWAB, M. 1974. Zur geotektonischen Stellung des permosilesischen subsequenten Vulkanismus in Mitteleuropa. *Zeitschrift für Geologische Wissenschaften*, **2**, 115–137.

EINSELE, G. 1982. Mechanism of sill intrusion into soft sediment and expulsion of pore water. *Scientific Results, Deep Sea Drilling Project*, **64**, 1169–1176.

FRANCIS, E.H. 1988. Mid-Devonian to early Permian volcanism: Old World. *Geological Society, London, Special Publications*, **38**, 573–584.

GERMAN STRATIGRAPHIC COMMISSION (ed.) 2002. *Stratigraphic Table of Germany 2002*. Geo-Forschungs Zentrum, Potsdam.

GOTO, Y. 1997. Interlayered sill–sediment structure at the base of a Miocene basaltic andesite sheet intrusion, Rebun Island, Hokkaido, Japan. *Journal of Mineralogy, Petrology and Economic Geology*, **92**, 509–520.

HOTH, K., HUEBSCHER, H.-D., KORICH, D., GABRIEL, W. & ENDERLEIN, F. 1993a. Die Lithostratigraphie der permokarbonischen Effusiva im Zentralabschnitt der Mitteleuropäischen Senke – der permokarbone Vulkanismus im Zentralabschnitt der Mitteleuropäischen Senke. *Geologisches Jahrbuch, Reihe A*, **131**, 179–196.

HOTH, K., RUSBÜLT, J., ZAGORA, K., BEER, H. & HARTMANN, O. 1993b. Die tiefen Bohrungen im Zentralabschnitt der mitteleuropäischen Senke – Dokumentation für den Zeitabschnitt 1962–1990. *Schriftenreihe für Geowissenschaften*, **2**, 7–145.

HOWELLS, M., REEDMAN, A. & CAMPBELL, D.G. 1991. *Ordovician (Caradoc) Marginal Basin Volcanism in Snowdonia (North-West Wales)*. London, HMSO for the British Geological Survey, London, 191 pp.

JUNG, D. 1967. Die Mineralassoziationen der Palatinite und ihrer Aplite. *Ann. Univ. Sarav., Reihe Mathematisch.-Naturwissenschaftliche Fakultät, Saarbrücken*, **5**, 1–130.

KATZUNG, G. & OBST, K. 1996. Spätvariszischer basischer Magmatismus – der Höhenberg-Sill im Thüringer Wald. *Zeitschrift der Deutschen Geologischen Gesellschaft*, **147**, 11–38.

MARX, J., HUEBSCHER, H.-D., HOTH, K., KORICH, D. & KRAMER, W. 1995. Vulkanostratigraphie und Geochemie der Eruptivkomplexe. Stratigraphie von Deutschland I – Norddeutsches Rotliegendbecken. *Courier Forschungsinstitut Senckenberg, Frankfurt*, **183**, 54–83.

PLEIN, E. (ed.) 1995. Stratigraphie von Deutschland I: Norddeutsches Rotliegendbecken – Rotliegend-Monographie Teil II. *Courier Forschungs-Institut Senkenberg*, **183**.

RAWLINGS, D.J., WATKEYS, M.K. & SWEENEY, R.J. 1999. Peperitic upper margin of an invasive flow, Karoo flood basalt province, northern Lebombo. *South African Journal of Geology*, **102**, 377–383.

SKILLING, I.P., WHITE, D.L. & McPHIE, J. 2002. Peperite: a review of magma–sediment mingling. *Journal of Volcanology and Geothermal Research*, **114**, 1–17.

SUNDVOLL, B. & LARSEN, B.T. 1993. Rb–Sr and Sm–Nd relationships in dyke and sill intrusions in the Oslo Rift and related areas. *Norges Geologiske Undersøkelse*, **425**, 25–42.

WALKER, B.H. & FRANCIS, E.H. 1987. High-level emplacement of an alkali dolerite sill into Namurian sediments near Cardenden, Fife. *Transactions of the Royal Society, Edinburgh: Earth Sciences*, **77**, 295–307.

ZIMANOWSKI, B. & BÜTTNER, R. 2002. Dynamic mingling of magma and liquified sediments. *Journal of Volcanology and Geothermal Research*, **114**, 37–44.

High-level volcanic–granodioritic intrusions from Zelezniak Hill (Kaczawa Mountains, Sudetes, SW Poland)

KATARZYNA MACHOWIAK[1], ANDRZEJ MUSZYŃSKI[1] & RICHARD ARMSTRONG[2]

[1]*Institute of Geology, Adam Mickiewicz University, Maków Polnych 16, 61-606 Poznań, Poland (e-mail: anmu@amu.edu.pl)*
[2]*Research School of Earth Sciences, The Australian National University, Canberra ACT 0200, Australia*

Abstract: New petrological and geochemical data on high level ($c.$ 2 km^2) silicic lava domes and laccoliths from the Kaczawa Mountains, Sudetes, SW Poland, are presented. The system comprises a carapace facies of exposed ignimbrites and spherlulitic rhyolites. Recovered core (drilled to 55 m) includes volcanic rocks ranging in composition from andesite to rhyodacite, and a plutonic facies of microgranite and granodiorite. Country rocks (greenschist-facies metavolcanogenic rocks) are contact metamorphosed to hornfels and cut by kersantite veins and a pipe breccia of diatremic origin. New ^{206}Pb–^{238}U zircon mineral ages from the volcanic and granitic rocks yield ages of 315 to 316 Ma, making the Zelezniak Hill complex the oldest magmatic rocks so far dated in Avalonia.

We present new data on high-level volcanic and plutonic rocks from Zelezniak Hill, situated in the SE region of the Kaczawa Mountains (Sudetes), in the NE part of the Bohemian Massif (Fig. 1). Magmatism in the study area appears relatively localized and is characterized by a few limited outcrops, so that field data were acquired mostly from the study of loose blocks and heads. Numerous radiating kersantite dykes cut the complex and extend outwards to a radius of 7 km (Zimmermann & Berg 1932; Manecki 1966; Baranowski *et al.* 1990, 1998). Polymetallic ore deposits from a disused mine at Stara Gora (former Altenberg) are a product of post-magmatic hydrothermal activity.

In this short paper, we report new data concerning the Late Carboniferous hypabyssal magmatic body of Zelezniak Hill, Poland. Petrological data combined with observations from physical volcanology are used to help to constrain the subsurface geology of the intrusive body.

Geological setting

The plutonic rocks comprising Zelezniak Hill were intruded into Lower Palaeozoic schists along a fault system bounding the Swierzawa (S) and Bolkow (B) tectonic units (Baranowski *et al.* 1990 and Fig. 1). Other volcanic rocks crop out northwards in the North Sudetic Depression (NSD) and southwards at the Inner Sudetic Depression (ISD). During the final stages of the Variscan orogeny, both depressions formed an intra-montane basin, with volcanic activity interspersed with molasse-type sedimentation. The age of volcanic activity in the NSD is Lower Permian (Autunian) and in the ISD is Upper Carboniferous and Lower Permian. The volcanism is bimodal in character, ranging from trachyandesites and andesites to rhyodacites and rhyolites (Awdankiewicz 1999).

In the north and NE of the study area, within the Fore Sudetic Block (Strzegom, Strzelin, Niemcza) massifs of granodiorite are exposed. Similar rocks occur in drill holes at the Odra Fault Zone, while granitoids of the Karkonosze (German Riesengebirge) pluton crop out south of Zelezniak Hill. All of the granitoid rocks are Early to Late Carboniferous in age. Zimmermann & Berg (1932), the authors of a geological map of the area, concluded that the volcanic activity was accompanied by doming that resulted in the formation of radial fractures around the intrusive core and contact aureole (Fig. 2). Subsequently, Majerowicz & Skurzewski (1987) have shown that the intrusive rocks (granites and rhyolites) comprise a subvolcanic laccolith.

Petrography

The Zelezniak Hill complex is a diversified magmatic body composed of rhyolite, rhyodacite, dacite, trachyandesite and microgranite. The latter are fine-grained, porphyritic rocks.

From: BREITKREUZ, C. & PETFORD, N. (eds) 2004. *Physical Geology of High-Level Magmatic Systems.*
Geological Society, London, Special Publications, **234**, 67–74. 0305-8719/04/$15.00
© The Geological Society of London 2004.

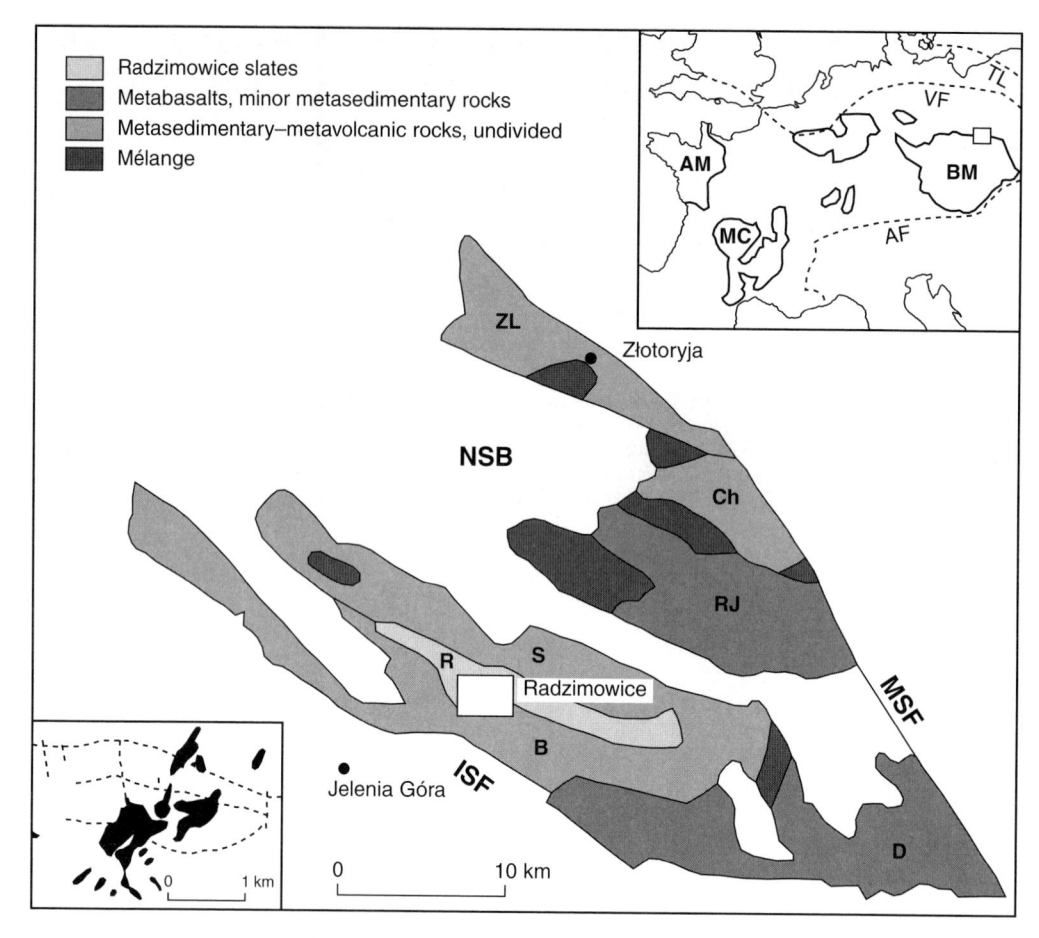

Fig. 1. Geological sketch map of the Kaczawa Mountains and their location within the Variscan belt (based on Baranowski *et al.* 1998; modified by the authors). Main tectonic units are: B, Bolkow; Ch, Chelmiec; D, Dobromierz; R, Radzimowice; RJ, Rzeszowek–Jakuszowa; S, Swierzawa; ZL, Złotoryja–Luboradz; NSB, North Sudetic Basin; ISF, Intra-Sudetic Fault; MSF, Marginal Sudetic Fault. Inset map: AF, Alpine Front; AM, Armorican Massif; BM, Bohemian Massif; MC, Massif Central; VF, Variscan Front; TL, Teisseyre–Tornquist Line.

Equigranular and coarse-grained (0.5–1.5 cm) granodiorites have also been drilled on the western slope of Zelezniak Hill at a depth of 55 m (Fig. 2). This series contains massive rocks identified as 'core' facies, as well as many volcanic rock varieties of different groundmass texture, e.g. spherulitic, micropoikilitic, and granophyric. Such groundmass textures are the products of volcanic glass recrystallization, indicative of an outer, quickly cooled 'carapace' facies, probably in contact with the surface (Cas & Wright 1987). The volcanic rocks differ in phenocryst content (ranging from 5 to 35%) and in mineral composition. The acidic rocks consist of plagioclase quartz and biotite phenocrysts, while the more basic rocks are rich in amphibole.

There is a zone around the nearby polymetallic lode of the Stara Gora mine which shows extensive hydrothermal alteration, including sericitization of feldspars, chloritization of biotite, muscovitization of chlorite, and pyritization. Late K-feldspar metasomatism has apparently overprinted all mineral alteration. A single ignimbrite sample containing pseudomorphic replacement of shards with a curvilinear surface (blocky shards) was found, along with samples displaying flow textures and containing recrystallized rock fragments (Fig. 3).

Generally, the granitic rocks are typically porphyritic microgranites and intermediate-composition porphyries (Fig. 3B–D). Such rock varieties are indicative of a hypabyssal setting.

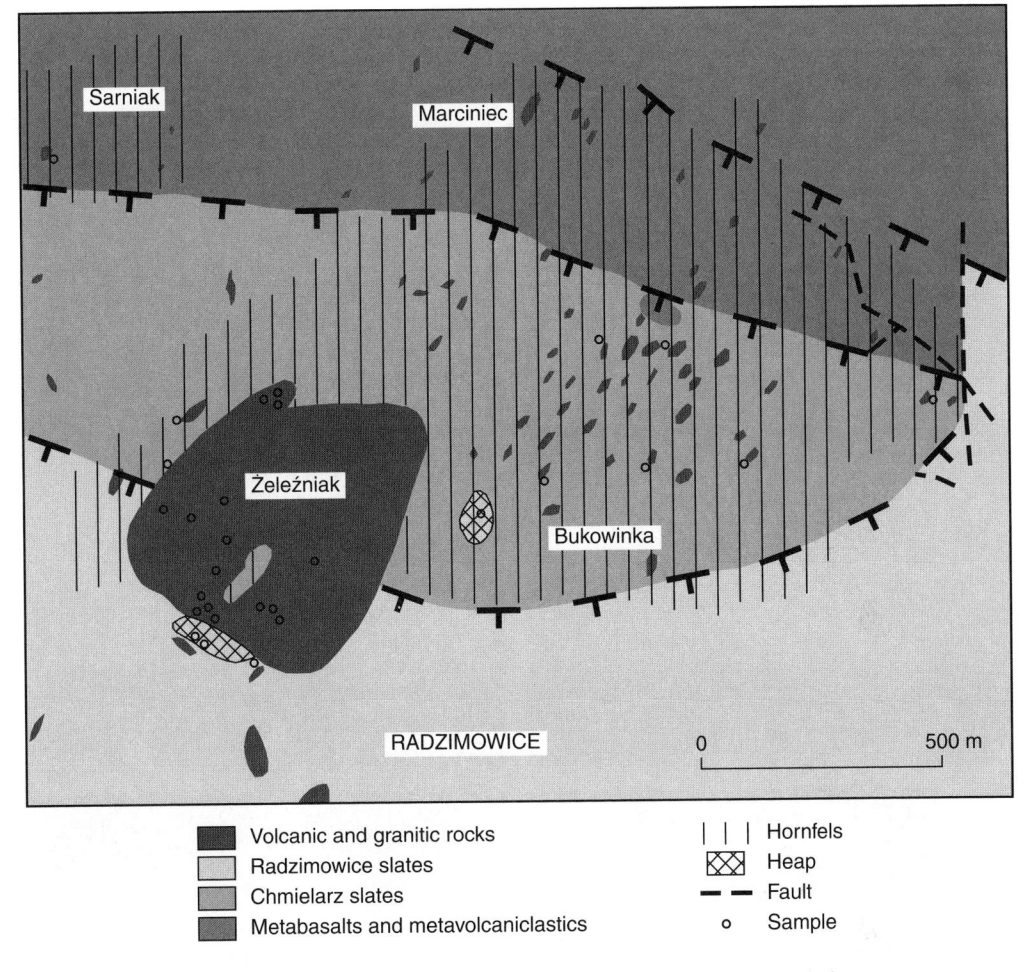

Fig. 2. Geological map of the Zelezniak Hill area: 1:10 000. Sample sites are shown as circles.

The mineralogy of the equigranular granodiorites differs from other plutonic rocks of the complex, in that they contain plagioclase, minor K-feldspar and biotite flakes, as well as abundant amphibole, accessory titanite and ore minerals (Fig. 3F). These volcano-granitic rocks are accompanied by kersantite and breccia occurrences. Fine-grained and porphyric textural varieties of lamprophyre were found. The porphyritic variety contains pseudomorphoses after olivines and glomeroporphyritic clinopyroxene clusters. All kersantite samples enclose numerous plagioclases and phlogopites; moreover, they comprise xenoliths of an unknown origin that are completely overprinted by carbonates.

Breccias from Zelezniak Hill comprise various clast types, including volcanic and adjacent metamorphic rocks, and basic rocks with abundant clinopyroxene, magnetite and dark mica (Fig. 3H). The breccia cement contains plagioclase (oligoclase) composition feldspar, and diopside (Fig. 3G). The breccia is probably of diatremal origin.

Geochemical features and age

Chemical analyses were done at the Activation Laboratories Ltd in Ontario, Canada. All research procedures and standard information which was applied to the 4-Lithores package are available at the ACTLABS homepage: (http://www.infomine.com/index/suppliers/activation_laboratories_ltd.html).

More than 90% of the granitoid intrusive rocks of Zelezniak Hill are weakly peraluminous to peraluminous S-types (Chappell & White

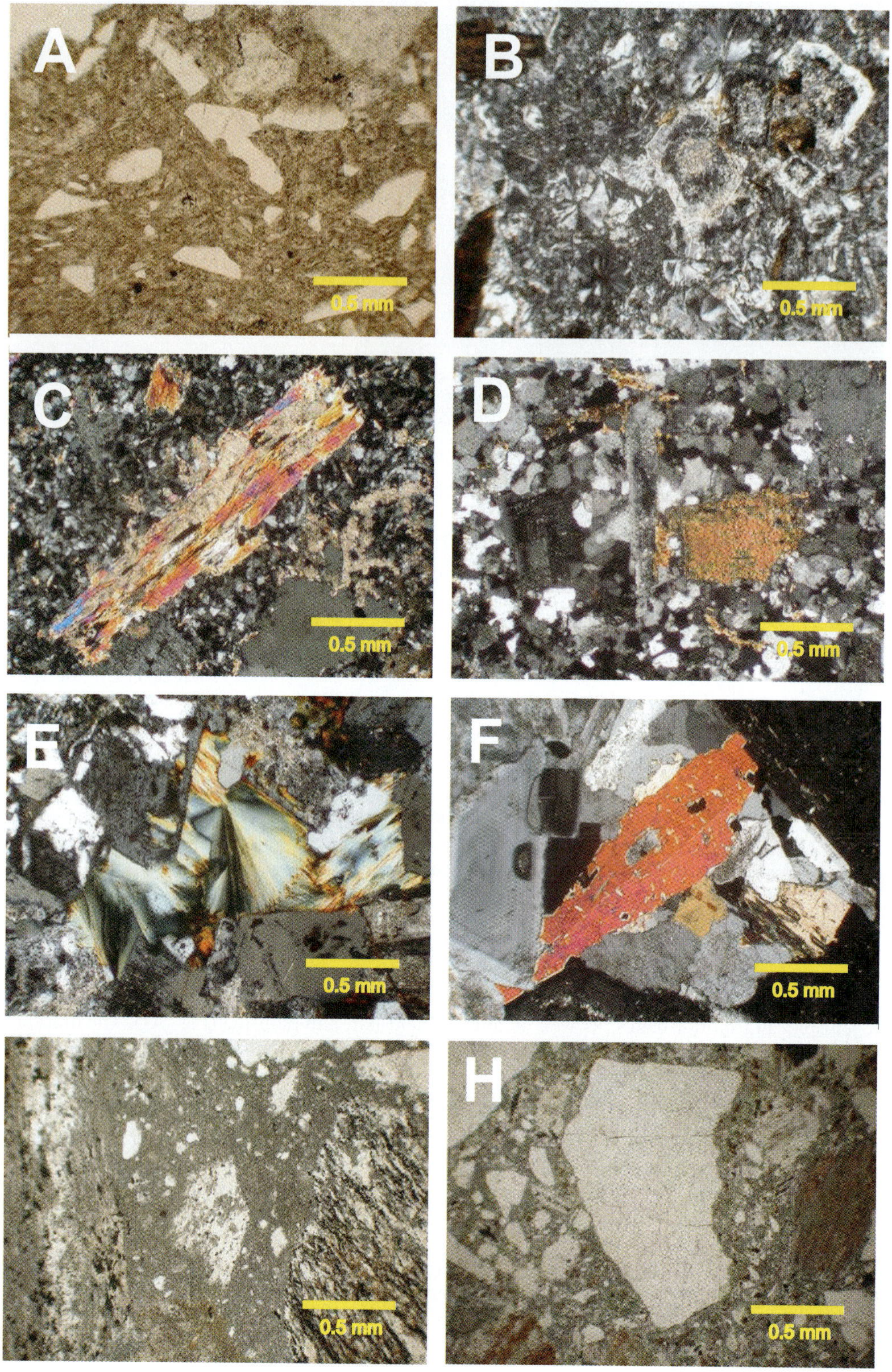
A
0.5 mm
B
0.5 mm
C
0.5 mm
D
0.5 mm
E
0.5 mm
F
0.5 mm
0.5 mm
H
0.5 mm

1977). On an Ab–An–Or diagram they plot in the granite, adamellite, trondhjemite and granodiorite fields. The volcanic rocks plot on the Winchester and Floyd diagram within the fields of rhyolite, dacite, rhyodacite and trachyandesite (Machowiak 2002). The chemical compositions of the analysed rocks are variable. The SiO_2 contents in the granitoids and volcanic rocks range from 66 to 77 wt %. There are also noticeable variations in K_2O (0.55–8.95%), Na_2O (0.18–5.33%), CaO (0.03–3.52%) and MgO (0.16–3.07%). Trace-element ratios of Th/U (2.8–8.45), Rb/Sr (0.04–1.9), Ce/Pb (0.97–12.0) and Rb/Cs (19.0–81.2) show a wide degree of chemical diversity. Trace-element concentrations and ranges, including Sr (92–555 ppm), Cs (1.5–6.7 ppm), Rb (20–182 ppm) and Nb (8.9–14.9 ppm), combined with revelant Harker plots, indicate that the composition of the intrusive rocks from Zelezniak Hill reflect a combination of source (partial-melting) processes and fractional crystallization. One candidate protolith for the majority of these magmas is recycled lower crustal material of tonalite–trondhjemite composition. Limited $^{87}Sr/^{86}Sr$ isotope data for the magmatic rocks show typical continental crustal values, ranging from $0.711\ 985 \pm 12$ to $0.717\ 670 \pm 10$ for the volcanites and $0.708\ 563 \pm 13$ to $0.728\ 809 \pm 14$ for the granitoids.

Zircon U–Pb age dating

An age study of the magmatic rocks from Zelezniak Hill was undertaken on zircon grains from two samples: a rhyodacite and a granite. This work was done at the Research School of Earth Sciences ANU, Canberra, using the SHRIMP II ion microprobe. All data were reduced following Williams (1998), using the SQUID Excel Macro of Ludwig (2000). Results are shown in Table 1. The $^{206}Pb-^{238}U$ age of granite crystallization is 316.7 ± 1.2 Ma. The calculated rhyodacite age is 315 ± 1.8 Ma. Some zircons have an old core of Proterozoic age: 2598.2 ± 4.6 and 2063 ± 13 Ma (Machowiak 2002).

Geometry of the intrusive complex

The shape and dimensions of the Zelezniak Hill volcanic–plutonic complex are difficult to establish, due to the lack of noticeable relationships between the different rock types. However, it is certain that a considerable amount of the upper part of the rock cover has been removed by erosion. The presence of the diatremal breccias suggests they are the remnants of a chimney pipe. We propose that the magmatic rocks from Zelezniak Hill may represent a multi-pulse injection lava dome, from which short-lived volcanic activity resulted in limited venting of material to the surface (Fig. 4).

Conclusions

Volcanic rocks from Zelezniak Hill can be divided into two facies types, comprising a core and a carapace. The volcanic rocks are accompanied by small bodies of ignimbrites and diatremic breccias. Fine-grained microgranites represent the exposed rocks of the core facies, while drilling has revealed the presence at deeper levels of more mafic granodiorites. The presence of granitoids at depth is supported by the presence of a contact aureole extending up to 2 km around the body, with the typical contact mineral assemblages of diopside–andradite–andalusite–corundum–biotite–hornblende. The protolith of the magmatic rocks may have been rejuvenated Precambrian lower continental crust of tonalitic–trondhjemitic composition – possibly with partial melts of upper-crustal metasedimentary rocks. The U–Pb dating of zircon grains suggest the main stage of the magmatic activity took place between 315 and 316 Ma ago. These ages from Zelezniak Hill are the oldest documented ages of Variscan volcanic activity within the Sudetes Mountains and also within Avalonia.

While the shape and dimensions of the intrusive complex are uncertain at present, we suggest that the magmatic rocks from Zelezniak Hill comprise a multi-pulse injection lava dome that occasionally vented to the surface.

Fig. 3. Selected photomicrographs of the rock textures from the core and carapace facies of the Zelezniak Hill complex. (**A**) Ignimbrite, broken quartz phenocrysts in groundmass containing numerous pseudomorphs after fine-grained blocky-shards. (**B**) Rhyolite with spherulites in groundmass (carapace facies). (**C**) Trachyandesite with amphibole and K-feldspar phenocrysts in the microgranular groundmass of plagioclase composition (core facies). (**D**) Porphyraceous microgranite with biotite and zonal plagioclases in fine-grained quartz–feldspar mosaic. (**E**) Fine-grained granite with hydrothermal chlorites. (**F**) Granodiorite from drill borehole (Zelezniak, 55 m) with zonal plagioclase, amphibole and titanite. (**G**) Pipe diatremic breccia composed of irregular metamorphic and minor volcanic clasts within a plagioclase–diopside groundmass. (**H**) Pipe diatremic breccia with abundant irregular volcanic clasts; groundmass of plagioclase–diopside composition.

Table 1. Summary of SHRIMP U-Pb zircon data for sample 14. The data for the Phanerozoic zircons are shown in the first part of this table, and the older components in the second

Grain spots	(2) %²⁰⁶Pb_c	U (ppm)	Th (ppm)	²³²Th/²³⁸U	²⁰⁶Pb* (ppm)	Total ²³⁸U/²⁰⁶Pb ± %	Total ²⁰⁷Pb/²⁰⁶Pb ± %	(1) ²⁰⁷Pb*/²⁰⁶Pb* ± %	(1) ²⁰⁶Pb*/²³⁸U ± %	(1) ²⁰⁷Pb*/²³⁵U ± %	err corr	(2) ²⁰⁶Pb/²³⁸U age
1.1	–	542	311	0.59	23.6	19.75 1.5	0.05268 0.62	0.05086 1.8	0.05052 0.62	0.3543 1.9	.335	318.4 ± 2.0
2.1	0.12	547	274	0.52	23.6	19.93 1.4	0.05368 0.64	0.05131 1.9	0.05002 0.64	0.3539 2.0	.315	315.2 ± 2.0
3.1	0.07	562	379	0.70	24.5	19.66 1.4	0.05337 0.64	0.05177 1.8	0.05075 0.64	0.3623 1.9	.343	319.5 ± 2.0
5.1	0.31	313	140	0.46	13.2	20.37 1.9	0.0550 0.76	0.0517 2.5	0.04889 0.77	0.3482 2.6	.294	308.0 ± 2.4
6.1	–	512	385	0.78	22.2	19.80 1.5	0.05240 0.65	0.05240 1.5	0.05050 0.65	0.3649 1.6	.405	317.7 ± 2.1
7.1	0.02	309	124	0.41	13.2	20.07 2.0	0.0528 0.85	0.0528 2.0	0.04982 0.85	0.3629 2.1	.394	313.4 ± 2.6
8.1	0.12	655	474	0.75	27.8	20.19 1.3	0.05355 0.61	0.0511 2.3	0.04937 0.62	0.3479 2.4	.258	311.2 ± 1.9
9.1	0.22	580	309	0.55	25.2	19.76 1.3	0.05449 0.62	0.0524 2.5	0.05047 0.64	0.3648 2.6	.248	317.5 ± 2.0
10.1	2.01	628	325	0.53	24.3	22.21 1.3	0.06794 0.70	0.0582 4.5	0.04448 0.77	0.357 4.6	.168	278.4 ± 2.0
11.1	–	309	130	0.44	13.3	19.90 1.8	0.05258 0.76	0.0503 3.0	0.05011 0.77	0.347 3.1	.245	316.1 ± 2.4
12.1	0.80	403	184	0.47	17.6	19.67 1.5	0.05913 0.90	0.0514 5.6	0.05036 0.96	0.357 5.7	.169	317.3 ± 2.8
13.1	0.05	603	275	0.47	25.8	20.03 1.3	0.05305 0.64	0.0496 2.5	0.04970 0.65	0.3399 2.6	.251	313.9 ± 2.0
14.1	0.03	195	92	0.49	8.48	19.71 2.4	0.0530 0.89	0.0509 3.2	0.05060 0.90	0.355 3.3	.271	318.9 ± 2.8
15.1	0.02	447	233	0.54	19.4	19.75 1.6	0.05290 0.68	0.0511 4.0	0.05052 0.72	0.356 4.1	.177	318.4 ± 2.2
16.1	0.28	274	216	0.82	11.7	20.05 2.0	0.0549 0.81	0.0549 2.0	0.04989 0.81	0.3774 2.1	.381	313.0 ± 2.5
17.1	0.02	501	292	0.60	21.7	19.86 1.4	0.05291 0.91	0.05125 1.9	0.05026 0.92	0.3551 2.1	.434	316.7 ± 2.9

Errors are 1–sigma; Pb_c and Pb* indicate the common and radiogenic portions, respectively.
Error in standard calibration was 0.18% (not included in above errors but required when comparing data from different mounths)
(1) Common Pb corrected using measured ²⁰⁴Pb
(2) Common Pb corrected by assuming ²⁰⁶Pb/²³⁸U – ²⁰⁷Pb/²³⁵U age-concordance

Grain spots	(1) %²⁰⁶Pb_c	U (ppm)	Th (ppm)	²³²Th/²³⁸U	²⁰⁶Pb* (ppm)	Total ²³⁸U/²⁰⁶Pb ± %	Total ²⁰⁷Pb/²⁰⁶Pb ± %	(1) ²⁰⁷Pb*/²⁰⁶Pb* ± %	(1) ²⁰⁶Pb*/²³⁸U ± %	(1) ²⁰⁷Pb*/²³⁵U ± %	²⁰⁶Pb/²³⁸U age	²⁰⁷Pb/²⁰⁶Pb age	err corr
4.1	1.09	298	219	0.76	14.5	17.64 1.1	0.0623 2.5	0.0572 2.5	0.05632 1.1	0.444 2.7	353.2 ± 3.7	500 ± 55	.402
4.2	1.30	448	216	0.50	28.6	13.440 0.64	0.06665 3.3	0.0611 3.3	0.07389 0.68	0.622 3.3	459.5 ± 3.0	642 ± 70	.204
11.2	2.94	642	927	1.49	248	2.223 0.51	0.17516 0.27	0.1742 0.3	0.4493 0.51	10.790 0.58	2 392 ± 10	2 598.2 ± 4.6	.883

Errors are 1–sigma; Pb_c and Pb* indicate the common and radiogenic portions, respectively.
Error in standard calibration was 0.18% (not included in above errors but required when comparing data from different mounths)
(1) Common Pb corrected using measured ²⁰⁴Pb

Table 2. Summary of SHRIMP U-Pb zircon data for sample 26. The data for the Phanerozoic zircons are shown in the first part of this table, and the older components in the second

Grain spots	(2) % 206Pb_c	U (ppm)	Th (ppm)	232Th/238U	206Pb* (ppm)	Total 238U/206Pb ± %	Total 207Pb/206Pb ± %	(1) 207Pb*/206Pb* ± %	(1) 206Pb*/238U ± %	(1) 207Pb/238U ± %	err corr	(2) 206Pb/238U age
1.1	0.29	147	57	0.40	6.20	20.35 1.0	0.0548 2.7	0.0427 10	0.285 10	0.04840 1.1	.111	308.3 ± 3.1
2.1	12.78	1092	539	0.51	45.5	20.63 0.53	0.15405 0.61	0.0512 12	0.298 12	0.04227 0.92	.076	267.0 ± 1.4
3.1	6.20	583	293	0.52	22.2	22.53 0.62	0.1012 1.2	0.0527 11	0.303 11	0.04170 0.95	.085	263.0 ± 1.7
4.1	1.91	358	177	0.51	15.1	20.38 0.71	0.0677 1.5	0.0485 8.4	0.320 8.4	0.04789 0.86	.102	303.0 ± 2.1
5.1	14.42	1248	613	0.51	54.0	19.84 0.51	0.1673 0.77	0.0432 15	0.254 15	0.04268 0.93	.062	272.2 ± 1.5
6.1	0.38	101	38	0.39	4.35	20.03 1.2	0.0556 3.1	0.0495 5.1	0.338 5.3	0.04955 1.2	.233	312.9 ± 3.8
7.1	0.05	265	324	1.26	11.4	20.01 1.0	0.0530 1.9	0.0472 3.2	0.323 3.4	0.04960 1.1	.315	314.1 ± 3.3
8.1	3.39	1141	440	0.40	43.5	22.53 0.53	0.07886 0.86	0.0512 5.7	0.302 5.7	0.04286 0.64	.111	270.7 ± 1.4
9.1	0.19	577	422	0.76	25.0	19.85 0.62	0.05424 1.3	0.0494 3.2	0.341 3.3	0.05008 0.65	.197	316.3 ± 2.0
10.1	41.99	1717	1356	0.82	86.2	17.120 0.48	0.3861 0.75	0.023 60	0.102 60	0.03275 1.6	.027	214.8 ± 1.7
11.1	19.61	1106	520	0.49	55.0	17.271 0.51	0.2097 1.1	0.0549 16	0.354 16	0.04670 1.2	.072	293.3 ± 1.8
11.2	0.09	384	100	0.27	25.9	12.755 0.66	0.05756 1.2	0.05438 1.6	0.585 1.8	0.07807 0.66	.378	486.0 ± 3.2
12.1	–	586	426	0.75	25.0	20.18 0.61	0.05100 1.3	0.05115 1.3	0.3496 1.5	0.04957 0.61	.420	312.4 ± 1.9
13.1	0.14	240	136	0.58	10.3	19.94 1.2	0.0538 2.0	0.0564 3.5	0.391 3.7	0.05032 1.2	.336	315.1 ± 3.8
15.1	2.02	591	266	0.47	24.3	20.84 0.63	0.0685 2.2	0.0514 6.5	0.333 6.5	0.04696 0.73	.111	296.1 ± 1.9
16.1	0.07	565	313	0.57	24.2	20.05 0.79	0.05320 1.4	0.0502 2.4	0.3440 2.5	0.04970 0.80	.316	313.6 ± 2.5
17.1	0.21	675	550	0.84	29.4	19.72 0.60	0.05446 1.2	0.0564 2.0	0.3949 2.1	0.05082 0.61	.287	318.1 ± 1.9

Errors are 1–sigma; Pb_c, and Pb* indicate the common and radiogenic portions, respectively.
Error in standard calibration was 0.18% (not included in above errors but required when comparing data from different mounths)
(1) Common Pb corrected using measured 204Pb
(2) Common Pb corrected by assuming 206Pb/238U – 207Pb/235U age-concordance

Grain spots	(1) % 206Pb_c	U (ppm)	Th (ppm)	232Th/238U	206Pb* (ppm)	Total 238U/206Pb ± %	Total 207Pb/206Pb ± %	(1) 207Pb*/206Pb* ± %	(1) 206Pb*/238U ± %	(1) 207Pb*/238U ± %	206Pb/238U age	207Pb/206Pb age	err corr
14.1	1.16	143	74	0.53	43.9	2.806 0.81	0.12947 0.75	0.1274 0.75	0.3555 0.81	6.247 1.1	1,961 ± 14	2,063 ± 13	.733

Errors are 1–sigma; Pb_c, and Pb* indicate the common and radiogenic portions, respectively.
Error in standard calibration was 0.18% (not included in above errors but required when comparing data from different mounths)
(1) Common Pb corrected using measured 204Pb

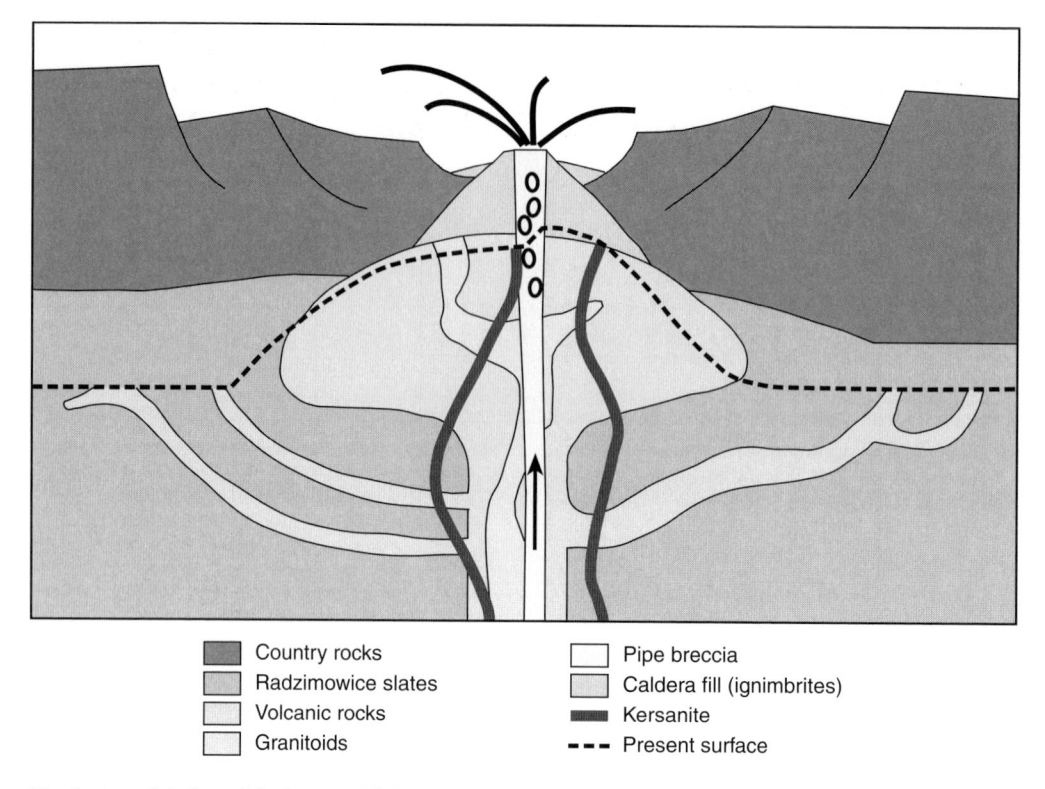

Fig. 4. A model of possible shape and lithotype relationships at the Zelezniak Hill complex (not to scale).

Legend:
- Country rocks
- Radzimowice slates
- Volcanic rocks
- Granitoids
- Pipe breccia
- Caldera fill (ignimbrites)
- Kersanite
- --- Present surface

This study was supported by State Committee grants for Scientific Research (KBN) 6PO4D 03414 and 6PO4D 005 20. N. Petford is thanked for help with editing an earlier draft of the manuscript.

References

AWDANKIEWICZ, M. 1999. Volcanism in the late Variscan intramontane trough: Carboniferous and Permian volcanic centers of the Intra-Sudetic Basin, SW Poland. *Geologia Sudetica*, **32(1)**, 13–47.

BARANOWSKI, Z., HAYDUKIEWICZ, A., KRYZA, R., LORENC, S., MUSZYŃSKI, A. & URBANEK, Z. 1990. Outline of the geology of the Gory Kaczawskie (Sudetes, Poland). *Neues Jahrbuch für Geologie und Paläontologie Abhandlungen*, **179(2–3)**, 223–257.

BARANOWSKI, Z., HAYDUKIEWICZ, A., KRYZA, R., LORENC, S., MUSZYŃSKI, A. & URBANEK, Z. 1998. The lithology and origin of the metasedimentary and metavolcanic rocks of the Chelmiec Unit (Gory Kaczawskie, Sudetes). *Geologia Sudetica*, **31**, 33–59 (English summary).

CAS, R.A.F. & WRIGHT, J.V. 1987. *Volcanic Succession. Modern and Ancient.* Allen and Unwin, London, 528 pp.

CHAPPELL, B.W. & WHITE, A.J.R. 1977. Two contrasting granite types. *Pacific Geology*, **8**, 173–174.

LUDWIG, K.R. 2000. *SQUID 1.00, A User's Manual.* Berkeley Geochronology Center, Special Publication, **2**, 17 pp.

MACHOWIAK, K. 2002. *Petrology and age of igneous rocks from Zelezniak Hill (Kaczawa Mts.).* Unpublished Ph.D. thesis (in Polish only).

MAJEROWICZ, A. & SKURZEWSKI, A. 1987. Granites from Wojcieszow area in the Gory Kaczawskie. *Acta Universitatis Wratislaviensis. Prace Geologiczno-Mineralogiczne*, **X**, 265–274 (in Polish only).

MANECKI, A. 1965. Mineralogical and petrographical study of the polymetallic veins from the area of Wojcieszow (Lower Silesia). *Prace Mineralogiczne Komitetu Nauk Mineralogicznych PAN, Oddzial w Krakowie*, **2**, 7–65 (in Polish only).

WILLIAMS, I.S. 1998. U–Th–Pb geochronology by ion microprobe. *In*: MCKIBBEN, M.A., SHANKS III, W.C. & RIDLEY, W.I. (eds) Applications of microanalytical techniques to understanding mineralizing processes. *Reviews in Economic Geology*, **7**, 1–35.

ZIMMERMANN, E. & BERG, G. 1932. *Geologische Karte von Preussen und Benachbarten Deutschen Landern, 1 : 25 000.* Blatt Kauffung, Berlin.

Relationship between diatremes, dykes, sills, laccoliths, intrusive–extrusive domes, lava flows, and tephra deposits with unconsolidated water-saturated sediments in the late Variscan intermontane Saar–Nahe Basin, SW Germany

VOLKER LORENZ[1] & JOST HANEKE[2]

[1,]*Institut für Geologie, Universität Würzburg, Pleicherwall 1, D-97070 Würzburg, Germany (e-mail: vlorenz@geologie.uni-wuerzburg.de)

[2]Landesamt für Geologie und Bergbau Rheinland-Pfalz, Emy-Röder-Str. 5, D-55133 Mainz, Germany (e-mail: jost.haneke@lgb-rlp.de)

Abstract: The late Variscan intermontane Saar–Nahe Basin underwent an intensive episode of synsedimentary intra-basinal magmatism, with magmas ranging from tholeiitic basalts to rhyolites. Volcanism began late in the sedimentary history of the basin, after accumulation of about 5000–5500 m of continental sediments. Basic to silicic maar–diatremes formed mostly on hydraulically active faults or fault intersections. Basic to intermediate sills were emplaced at depths between about 2500 m and almost the original surface. Some sills inflated considerably in thickness. Silicic laccoliths intruded in the same depth range. Ongoing volume inflation of some laccoliths led to huge intrusive–extrusive domes, rock falls and probably block-and-ash flows and even to extensive thick lava extrusions. Some domes are composite or show evidence for magma mingling. In the Baumholder–Idar–Oberstein area, lava flows reach a cumulative thickness of 800–1000 m. Outside this thick lava pile, flows are concentrated in several thinner series. Basic to intermediate lava flows were frequently inflated to a thickness of up to 40 metres and were emplaced like thick flood basalts. Silicic tephra deposits are widespread and mostly phreatomagmatic in origin.

The specific formation of maar–diatremes, sills, laccoliths and most tephra deposits is related to the uppermost 1500 to about 2500 m of the continental sediments of the basin fill. During volcanism and subvolcanism, these sediments were largely unconsolidated and water-saturated, and thus this soft sediment environment influenced very specifically the emplacement of the magmas in the basin. Inflation of laccoliths in this environment caused slumping and washing away of the updomed unconsolidated roof sediments. Consequently, the effective initial overburden decreased with time by this particular process of unroofing, and, upon further inflation, larger inflating intrusive domes became extrusive.

Diatremes, dykes, sills and laccoliths are found in many geological environments in the world, and, especially the basic to intermediate bodies, can be frequent in sediment basins characterized by synsedimentary volcanism. Such basins and their volcanism occur in different tectonic environments: continental rift zones as, for example, in the West European Continental Rift Zone (Wimmenauer 1974; Keller et al. 1990), the Oslo Rift (Neumann et al. 1992), the East African Rift Zone (Baker 1987), and their successor rifts which finally evolve into volcanic passive continental margins (Storey et al. 1992; Planke et al. 2002). In addition intra-basinal volcanism characterizes late-orogenic intermontane basins as, for example, the late Variscan intermontane basins of Europe (Lorenz & Nicholls 1976, 1984; Lorenz 1992a, b; Hoth et al. 1993; Beneck et al. 1996; Mock et al. 2002a, b), the late Caledonian intermontane basins (Andersen 1998; Osmundsen et al. 1998), basin-and-range provinces (e.g. the Basin-and-Range Province of the western USA and northern Mexico (Scholz et al. 1971; Armstrong & Ward 1991; Fitton et al. 1991; Lipman & Glazner 1991; Liu 2001; Bennett et al. 2003), and, finally, some foreland basins, e.g., the Molasse Basin north of the Alps in the Hegau area (e.g. Keller et al. 1990), or the Western Interior Seaway in Canada with the newly discovered Cretaceous to Early Tertiary kimberlite diatremes (Field & Scott Smith 1999). Except for the foreland basins, most of the above basin types are related to pronounced crustal extension (Dewey 1988; Lipman & Glazner 1991; Metcalf & Smith 1995).

Many subvolcanic bodies and diatremes are

---

* To whom correspondence should be addressed

relevant with regard to exploration for hydrocarbons (Planke et al. 2002), a variety of other commodities (ore deposits, diamonds, road metal, dimension stones), or geothermal energy (Wohletz & Heiken 1992), and thus volcanological expertise is required in the respective exploration studies. As a consequence of the various interests in these subvolcanic features, the intrusion of subvolcanic bodies such as sills and laccoliths frequently raises the question: why did the magmas reach such high levels in the crust but did not break through to the Earth's surface and erupt subaerially? A similar question has been asked with respect to the formation of maar–diatreme volcanoes: why did ultrabasic, basic and intermediate magmas form these maar–diatremes at their respective localities, and why did they not form ordinary scoria cones and lava flows, and, similarly, why did intermediate to silicic magmas form diatremes at specific localities but did not form ordinary pumice cones or extrusive domes and lava flows?

It is the purpose of this paper to elucidate the emplacement of diatremes, dykes, sills, laccoliths, intrusive–extrusive domes, lava flows and tephra deposits in the late Variscan intermontane Carboniferous–Permian Saar–Nahe Basin in SW Germany. The basin fill shows many excellent exposures and is characterized by a multitude of such subvolcanic and volcanic bodies, and, via its well-known lithostratigraphy and many large-scale published maps, it offers the possibility for an integrated study of these subvolcanic and volcanic bodies. In part, this paper is intended to review the large existing knowledge, published over many years mostly in regional journals, diploma and Dr.rer.nat. theses, and in a number of geological survey maps. We especially refer to the geological maps compiled by Konzan et al. (1981) for the SW part of the Saar–Nahe Basin (Saarland) and Dreyer et al. (1983) for the main part of the Saar–Nahe Basin (in Rheinland–Pfalz). In addition, however, this paper is intended to integrate:

1 aspects of the continental sediments deposited in the Saar–Nahe Basin prior and during synsedimentary volcanism, and
2 new volcanological aspects of the various subvolcanic and volcanic bodies, especially on the interaction of the rising magmas with unconsolidated water-saturated sediments.

Thus, this paper aims at presenting genetic emplacement models of the subvolcanic and volcanic systems that might be useful for the understanding of similar volcano-sedimentary systems worldwide and for modelling of the emplacement of magma in such environments.

Late orogenic compression and extension in the Variscan collision belt and the formation of the late Variscan intermontane basins

The Saar–Nahe Basin is one of about 70 late-orogenic intermontane basins of the Variscan continental collision belt (Lorenz & Nicholls 1976, 1984; Arthaud & Matte 1977; Franke 1989; Franke & Oncken 1990; Ziegler 1990; Henk 1997, 1998, 1999). During the final phase of the compressive tectonics of the Variscan orogenic belt in SE, central, southern, and SW Europe as well as in NW Africa, a phase of crustal extension started in the central part of the orogenic belt in Upper Viséan times (Lorenz & Nicholls 1984; Lorenz 1992a). At the former surface, this crustal extension resulted in intermontane basins with purely continental sediments and intensive widespread volcanic activity. The zone of crustal extension was enlarged in Namurian and Early Westphalian times. Despite this enlargement at right angles to the strike of the belt, it was still flanked on both sides by receding zones of flysch troughs (on continental crust) and their respective successor molasse troughs and associated compressional deformation in the respective inner trough walls (Lorenz & Nicholls 1976, 1984). In Late Westphalian, Stephanian and Early Permian times, i.e. in the time period after the compressive deformation had ended on both flanks of the orogenic belt, the number of these late-orogenic basins dramatically increased, as did volcanicity inside the basins, but, to some extent, also in between the basins. In Stephanian, but especially in Early Permian times, crustal extension even spread across the northern Variscan front into northern Germany (Lorenz & Nicholls 1976, 1984; Breitkreuz & Kennedy 1999), Britain, the area of the present North Sea, and part of Poland. All of these late Variscan intermontane basins contain continental sediments, and most of them also contain volcanic rocks. Crustal extension, brittle in the upper and ductile in the lower crust, was responsible for reducing the thickened orogenic crust of the Variscan collision belt to more or less its present-day crustal thickness (Lorenz & Nicholls 1984; Henk 1993a–c, 1997, 1998, 1999) – not taking into account Mesozoic and Cenozoic crustal thickening in the Alpine orogenic belts and crustal thinning in the West European Continental Rift Zone. Lithospheric extension was responsible for decompressional melting in the upper-mantle and widespread associated melting in the lower crust as well as in the elevated heat flow in late Variscan times in the orogenic belt (Lorenz 1992a, b; Henk 1997, 1998, 1999). In the late-orogenic basins and foreland basins, the elevated heat flow (Buntebarth 1983) resulted in 'high-temperature' diagenetic, respectively anchimetamorphic processes, including the high coalification ranks of the Westphalian, Stephanian, and Lower Permian coals in the paralic foreland and limnic intermontane basins. The elevated heat flow also resulted in the high degree of alteration of the volcanic and subvolcanic rocks (Lorenz & Nicholls 1976; Teichmüller et al. 1983).

Similar late-orogenic basins characterized by similar sediments and volcanics and subvolcanics include the Devonian Old Red basins of Norway, Great Britain and East Greenland, i.e. the late Caledonian intermontane basins (Lorenz & Nicholls 1984; Dewey 1988; Ziegler 1990; Dewey et al. 1993). Another zone of widespread crustal extension with many basins characterized by synsedimentary tholeiitic to rhyolitic volcanism is the Basin-and-Range Province in the western USA and in northern Mexico (Scholz et al. 1971; Christiansen & Lipman 1972; Lorenz & Nicholls 1976, 1984; Coward et al. 1987; Eddington et al. 1987; Harry et al. 1993). The Basin-and-Range Province, however, is not related to a late-orogenic tectonic environment.

The intermontane Saar–Nahe Basin

Geological maps of the Saar–Nahe Basin (Fig. 1) were published by Konzan et al. (1981) for the SW part (Saarland) and by Dreyer et al. (1983) for the NE part of the Saar–Nahe Basin. The basin is one of the largest and best-exposed late-orogenic intermontane basins of the European Variscan belt. The basin is located on and immediately to the SE of the boundary between the central crystalline zones of the orogenic belt (Saxothuringian Zone and Moldanubian Zone) and the former shelf of Northern Europe–North America (Baltica–Laurentia–Avalonia), i.e. at the boundary of the Saxothuringian Zone in the SE and the Rhenohercynian Zone in the NW. This boundary zone represents the suture between the southern Europe and northern Europe continental plates in middle to late Variscan times (Lorenz 1976; Lorenz & Nicholls 1976, 1984; Franke 1989; Franke & Oncken 1990). Located below the NW Saxothuringian crust, the presence of a slice of Rhenohercynian lower crust may be responsible for a double Moho (Oncken 1997). At the northern margin of the Saar–Nahe Basin, a pronounced fault zone, the Southern Hunsrück Boundary Fault (SHBF on Fig. 1) represents a large southward-dipping detachment fault, and is responsible for the Saar–Nahe Basin being a half-graben (Henk 1990, 1992, 1993a–c). The Saar–Nahe Basin is more than 150 km long (NE–SW) and up to 80 km wide (NW–SE). Because of onlapping Upper Permian coastal Zechstein deposits (Dittrich 1996) and Lower Triassic sediments (Bunter Sandstone) in the south and west and Tertiary sediments in the Rheingraben in the NE, the exposed part of the Saar–Nahe Basin extends for only 130 km NE–SW and for up to 40 km NW–SE (Fig. 1). Because of the productive coal measures, many drill holes have penetrated the coal-bearing and overlying sediments in the SW of the Saar–Nahe Basin, i.e. in the Saarland. A few hydrocarbon exploration drill holes have also analysed the sediments of the Saarland and the NE of the Saar–Nahe Basin (Habicht 1966). Since 1987 a number of research drill holes have penetrated almost the whole volcano-sedimentary sequence of the Rotliegend in the NE half of the basin (Haneke & Lorenz 1990; Haneke 1991, 1997, 1998; Haneke & Stollhofen 1994).

Subsidence of the late-orogenic basin started in Westphalian times at c. 315 Ma and lasted for c. 45 Ma until about 270 Ma (Lippolt & Hess 1983, 1989; Hess & Lippolt 1986, 1988; Lippolt et al. 1989; Henk 1992, 1993a–c). The lithostratigraphy of the continental sediments (shales, black shales, sandstones, arkoses, conglomerates, breccias, limestones, and coal deposits, deposited in respective lacustrine, fluviatile, deltaic, and alluvial fan environments) is given in Figure 2. Thus the Saar–Nahe Basin, like all other late Variscan intermontane basins, did not experience seawater influx and shallow marine sedimentation, as did, in contrast, e.g. the Tertiary Upper Rhine Graben of the European Continental Rift Zone.

The sediments which accumulated in the Saar–Nahe Basin during pre-rift, non-volcanic syn-rift, volcanic syn-rift and post-rift thermal subsidence phases add up to about 6.5 km thickness, with the depocentre migrating from Late Westphalian to Early Permian times from the SW towards the NE (Teichmüller et al. 1983; Henk 1990, 1993a–c; Stollhofen 1991, 1998; Henk & Stollhofen 1994; Stollhofen & Stanistreet 1994). During the long period of crustal extension (Hess & Lippolt 1988) and thus of extended subsidence of the basin, movements along the Southern Hunsrück Boundary Fault (SHBF), the detachment fault at depth and along NW–SE-trending transfer faults were

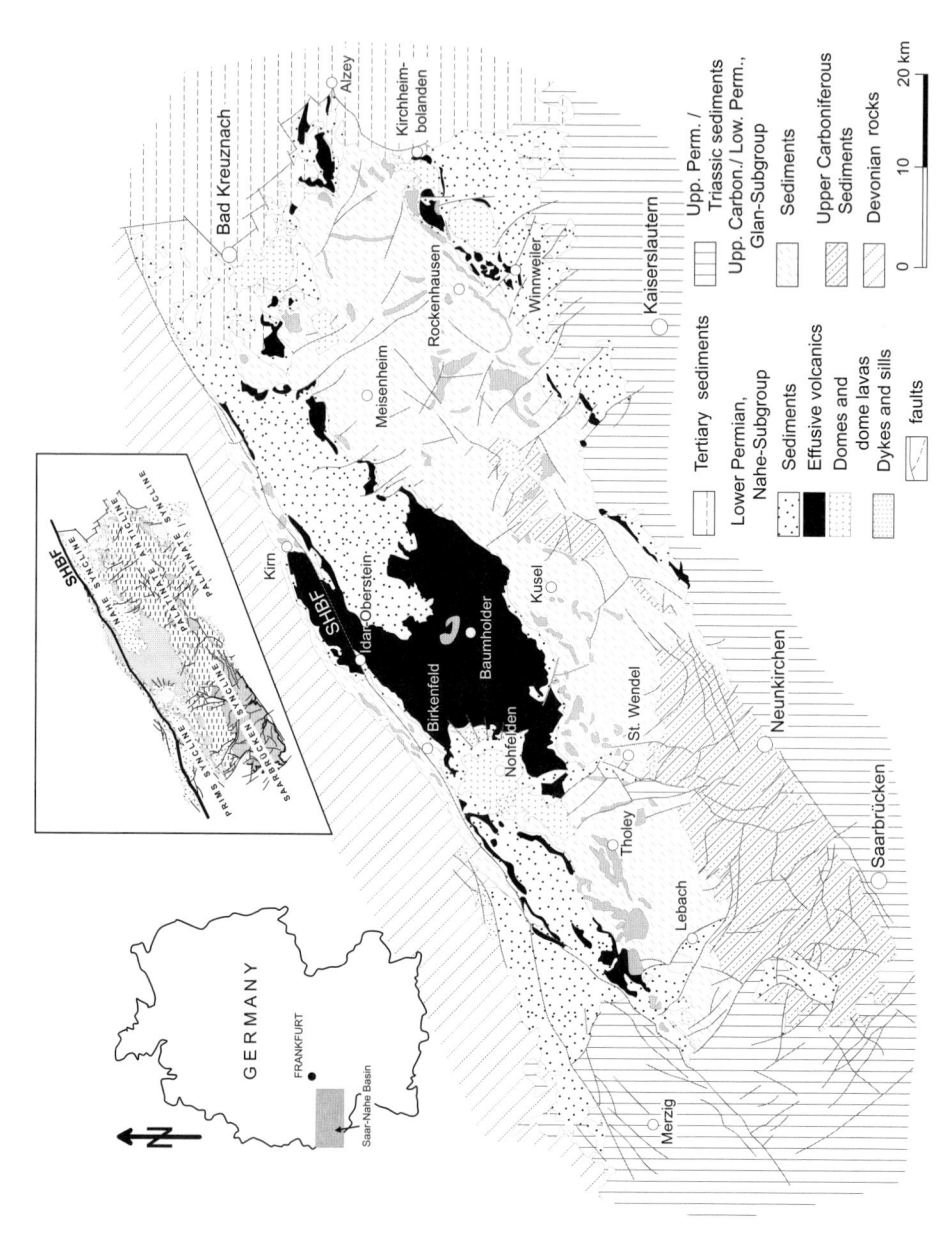

Fig. 1. Schematic map of the exposed part of the late Variscan intermontane Saar–Nahe Basin, showing major areas of effusive volcanics, domes, and dykes and sills. SHBF represents the Southern Hunsrück Boundary Fault. Base map after Dreyer et al. (1983). Left inset shows the location of the Saar–Nahe Basin within Germany. Right inset shows the Saar–Nahe Basin with its major anticlines and synclines.

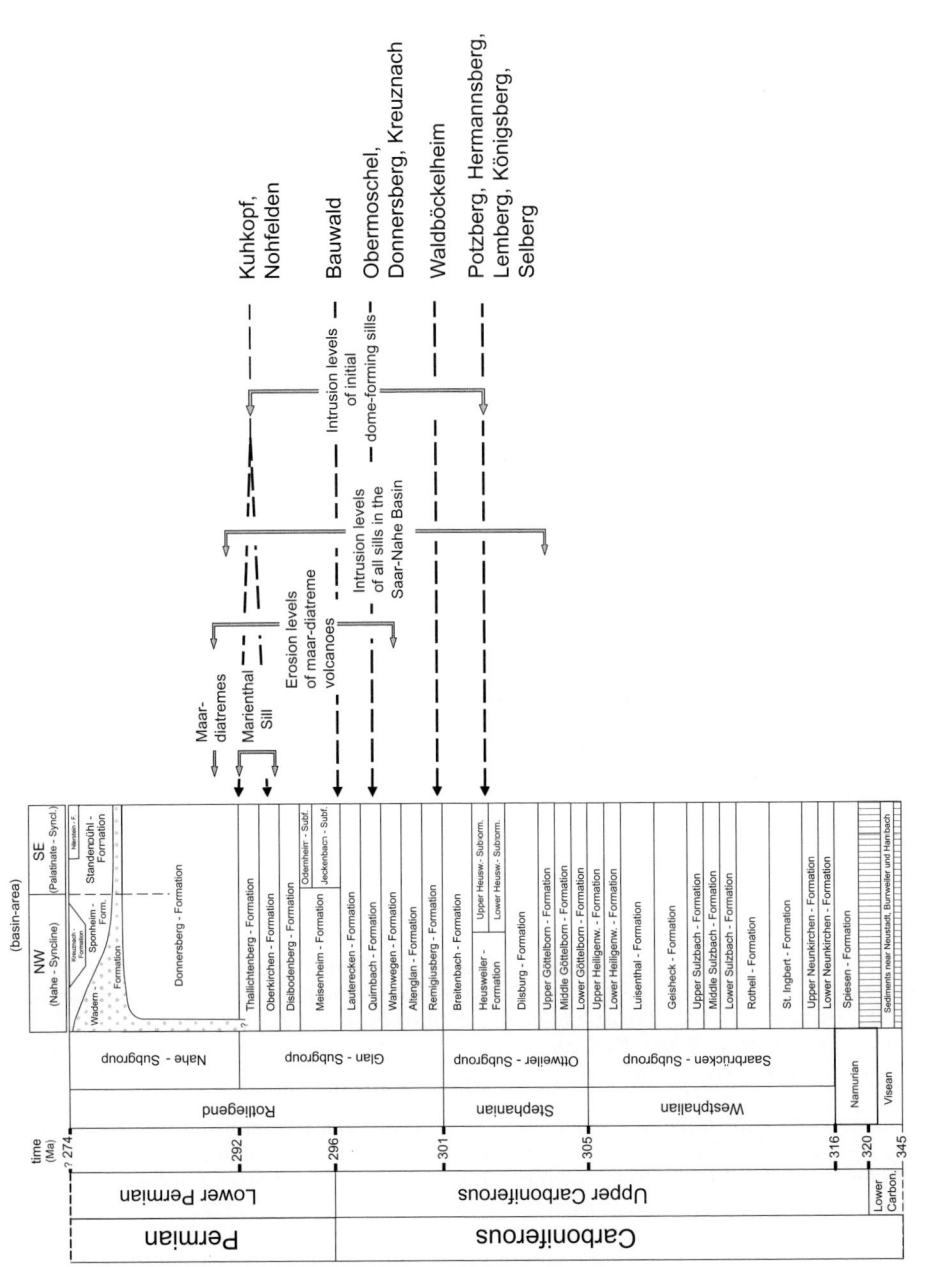

Fig. 2. Lithostratigraphic table of the Upper Carboniferous and Lower Permian sediments of the Saar–Nahe Basin and the stratigraphic level of the surface level contemporaneous with the eruption of a number of maar–diatreme volcanoes; the present range of erosion levels of maar-diatreme volcanoes, the range of intrusion levels of exposed sills; and levels of the initial sill-type intrusion of the various domes.

repeatedly active and compartmentalized the basin floor and fill (Haneke et al. 1979; Henk 1990, 1993a–c; Stollhofen 1991, 1993, 1994a–d, 1998; Henk & Stollhofen 1994; Stollhofen & Stanistreet 1994). As a result, there exist tectonically controlled thickness and facies variations not only across the SHBF but also across large open folds and many transfer faults (Haneke et al. 1979; Stollhofen 1994a–d, 1998; Stollhofen & Stanistreet 1994).

During the long-lived crustal extension and consequent subsidence of upper-crustal rocks along the detachment fault beneath the Saar–Nahe Basin, the basin fill was, as stated above already, deformed into large open folds (Henk 1990, 1993a–c): the SW–NE-trending Saarbrücken Anticline and its NE continuation, the Palatinate Anticline, in the central part of the exposed part of the basin, the Prims Syncline and its NE continuation, the Nahe Syncline, both just to the SE of the SHBF, respectively NW of the Saarbrücken and Palatinate anticlines. To the SE of the anticlines, the Palatinate Syncline follows. However, the Palatinate Syncline is mostly covered by Upper Permian littoral (Dittrich 1996) and Lower Triassic fluvial sediments of the Palatinate Forest (Pfälzer Wald). Both axes of the Palatinate Anticline and neighbouring Nahe Syncline plunge at a few degrees towards the NE. Because of the large open folds, the various sediments and volcanic rocks are exposed along the flanks of the anticlines and synclines and, therefore, linear exposures of specific horizons along their strike are characteristic, with the dips of the respective horizons being mostly between 15° and 35° (Fig. 1). Along the SHBF, subsidence of the basin fill in Carboniferous to Permian times was inverted later on, probably in Late Mesozoic to Early Tertiary times, and the basin fill was thrust towards the NW, reaching – close to the thrust – SE dips of up to 70–80°.

During deposition of the Westphalian, Stephanian and lower part of the Rotliegend (Glan-Subgroup, formerly called Lower Rotliegend), volcanism was active only outside the basin, in all probability in the Moldanubian zone of the Variscan Belt (Königer et al. 1995, 2002; Königer & Lorenz 2002, 2003). Voluminous plinian and/or phreatoplinian eruptions caused deposition of many thin distal ash layers in the Saar–Nahe Basin and many other Late Variscan basins (Königer & Lorenz 2002, 2003; Königer et al. 2002). It was almost exclusively during the upper half of the Rotliegend, especially during the Donnersberg-Formation which represents the lower half of the Nahe-Subgroup (the latter formerly called Upper Rotliegend) that intensive volcanicity affected the Saar–Nahe Basin itself (Figs 1, 2 & 3) (Lippolt & Raczek 1979, 1981; Lippolt & Hess 1983, 1989; Lippolt et al. 1989). Konrad & Schwab (1970), Theuerjahr (1971), Schwab (1971a), and Haneke (1987) assume some intrusive activity already during sedimentation of the Glan-Subgroup. Intra-basinal volcanism expressing itself in lava flows or thick pyroclastic beds, however, is unknown from the Glan-Subgroup.

Inside the Saar–Nahe Basin the intensive volcanism was caused by silica-saturated (quartz-normative) magmas covering the full range between tholeiitic basalts and rhyolites (Jung & Vinx 1973; Lorenz 1987, 1992b; Arz 1996; Arz & Lorenz 1994a–b, 1995a–b, 1996; von Seckendorff et al. 2004). Volcanism expressed itself in the Saar–Nahe Basin by maar–diatreme volcanoes, dykes, sills, laccoliths, intrusive–extrusive domes, lava flows, and various tephra deposits. On the basin floor, sediments and volcanics as well as tephra deposits are interbedded, thus volcanism was synsedimentary in character. To the NW, outside the Saar–Nahe Basin, the neighbouring Hunsrück was one of the source regions for the sediments of the basin fill, and consequently must have been a mountain range with deeply incised valleys. In these Hunsrück mountains, volcanism was restricted to three probably also Lower Permian rhyolite dykes (Dreyer et al. 1983), their potential surface volcanics, and an ignimbrite vent, 1 km in diameter, with an intrusive rhyolite plug (Minning & Lorenz 1983; Negendank 1983) which represents the source of ignimbrites in the Prims and Nahe synclines (see below).

In the following sections, various subvolcanic and volcanic features are described from the Saar–Nahe Basin and interpreted in their physical volcanology and in their relationship with the sediments deposited previously and simultaneously.

Maar–diatreme volcanoes

Maar–diatreme volcanoes form when rising magma interacts explosively with groundwater. Formation of maar–diatreme volcanoes by such phreatomagmatic respectively thermohydraulic explosions has been investigated intensively using world-wide field data and their interpretation as well as many experiments (Lorenz 1971a–c, 1972, 1973, 1974, 1976, 1985, 1986, 1992a, b, 1998, 2003a, 2003b; Cas & Wright 1987; Zimanowski et al. 1997a–c; Zimanowski 1998; Büttner & Zimanowski 1998; Schmincke 2000).

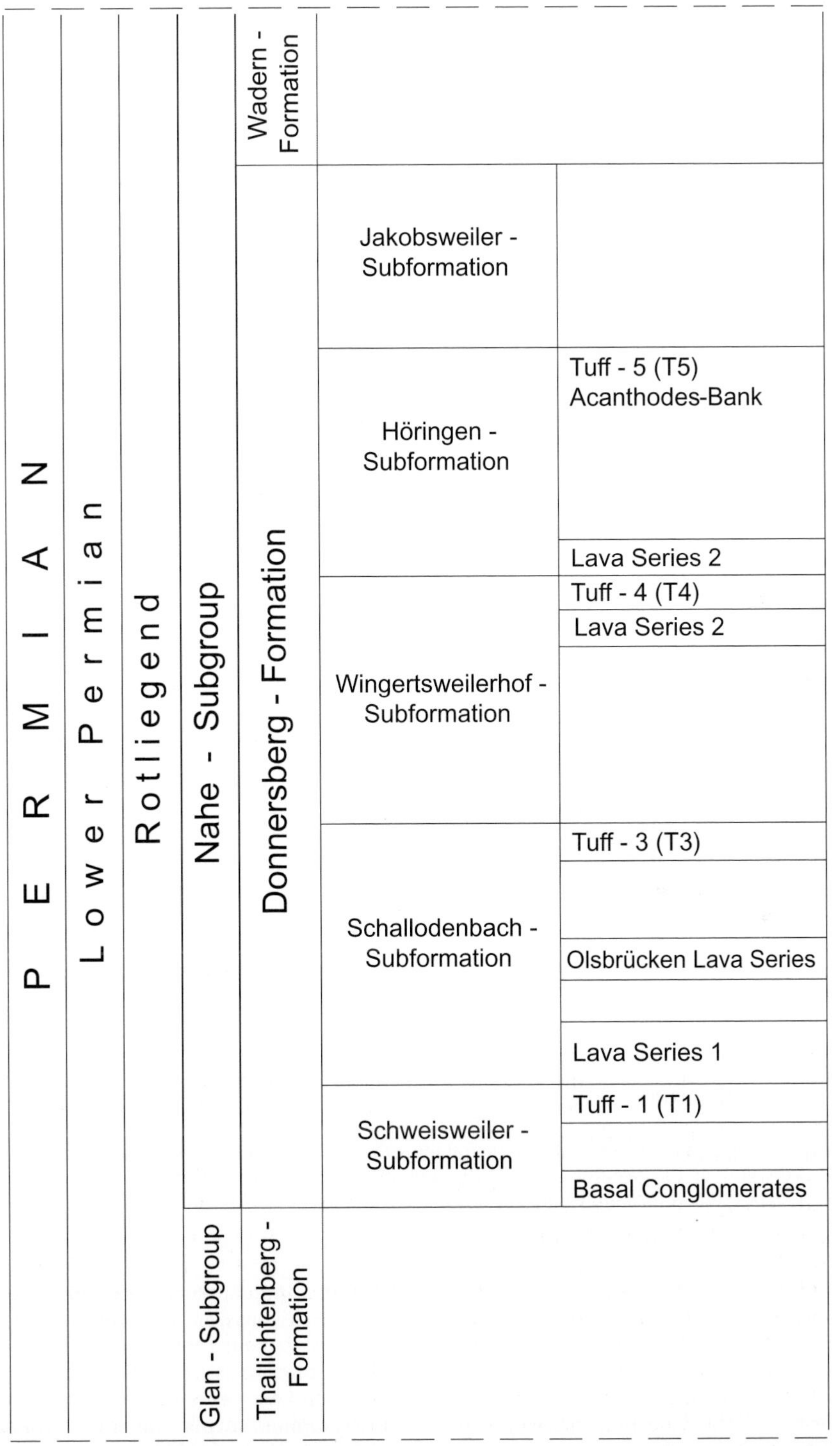

Fig. 3. Lithostratigraphic sequence of continental sediments, major tephra horizons, and major lava flows of the Donnersberg-Formation in the Nahe-Subgroup of the Rotliegend. The Nahe-Subgroup represents the former Upper Rotliegend in the Saar–Nahe Basin.

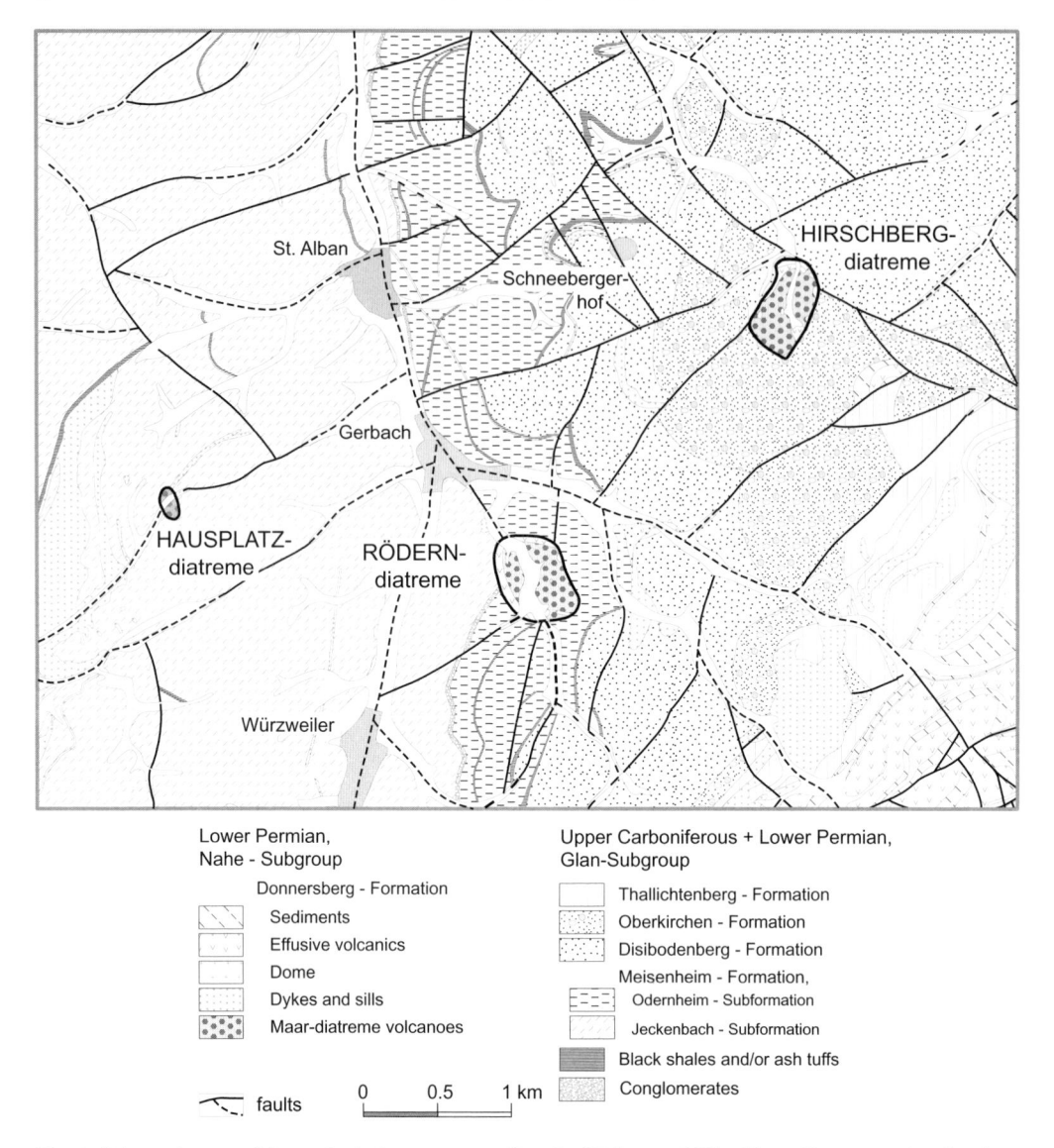

Fig. 4. Schematic map of the geological area surrounding the Rödern and Hirschberg diatremes, also showing the relationship between the location of diatremes and faults. The faults are assumed to have been repeatedly active during deposition of the Rotliegend sediments.

Magmas that erupted and formed maar–diatreme volcanoes in the Saar–Nahe Basin cover the whole range from tholeiitic basalts to rhyolites. About 26 maar–diatreme volcanoes have so far been recognized in the sedimentary fill of the basin. The volcanoes are exposed at different levels below their respective original surface, depending on the particular age, localization, and amount of erosion of the diatremes and sedimentary surrounding rocks (Figs 2, 4 & 5). Only the diatremes are preserved. Late Palaeozoic erosion has removed the maar craters and maar tephra-rings. However, some distal tephra beds from several maar–diatreme volcanoes have been preserved in the respective stratigraphic levels of the Donnersberg-Formation (Lorenz 1971a).

During the volcanism of the Donnersberg-Formation the maar–diatremes erupted at different times. The oldest diatremes known to

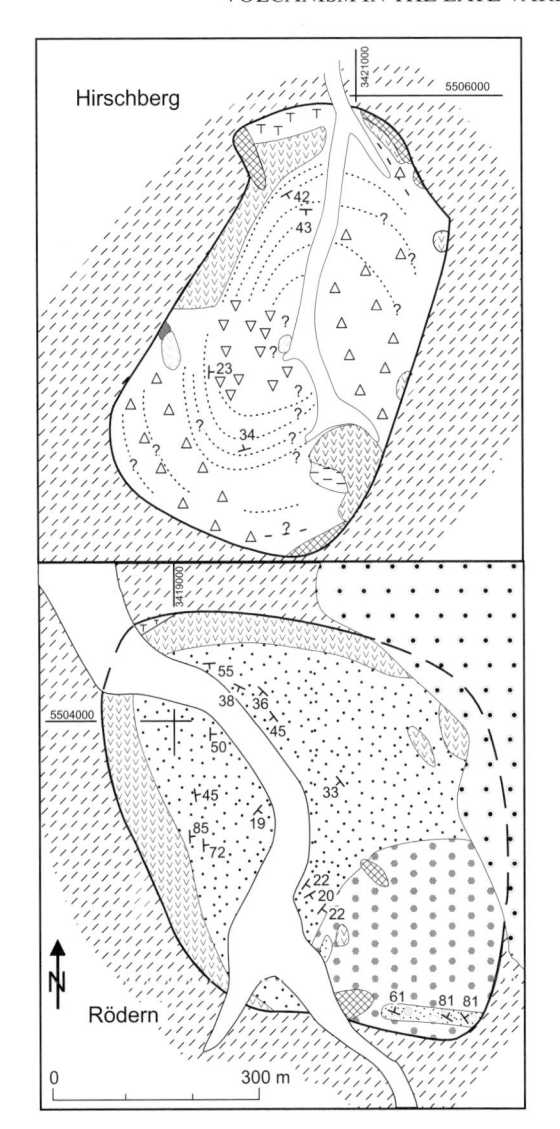

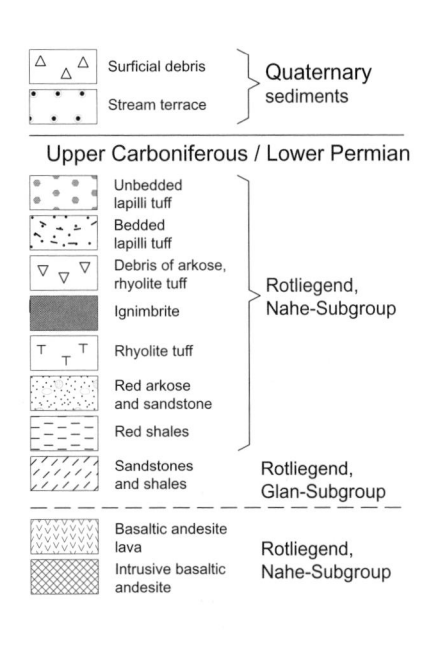

Fig. 5. Geological maps of the two diatremes: Rödern and Hirschberg, in the Saar–Nahe Basin, NE of the Donnersberg dome.

have erupted are the Rödern, Hirschberg, and Mörsfeld diatremes that were erupted during deposition of the Lava Series 1 (Figs 2, 4 & 5), whereas the Wolfsgalgen, Spitzenberg and Waldböckelheim diatremes are younger and were erupted after the Lava Series 2 (Fig. 3). The age of most other diatremes relative to the lithostratigraphy of the Donnersberg-Formation is not known. The upper diameter of the cone-shaped diatremes varies because of the level of exposure and the length of the eruptive activity of the individual volcano i.e. the larger the volcano, the longer it is active (Lorenz 1985, 1986, 1998). The largest diatreme in the Saar–Nahe Basin, the Falkenstein diatreme, has a maximum diameter of 1520 m (Lorenz 1971b) and the smallest diatremes in the basin are the rhyolitic diatremes that are several tens to hundreds of metres in diameter and located above the Waldböckelheim rhyolite laccolith (Geib 1956, 1972, 1975).

Inside the diatremes there usually exist bedded and unbedded lapilli tuffs (Figs 5 & 7). The bedding of the tuffs indicates that tephra

Fig. 6. Sample of indurated tephra from the Hirschberg diatreme. The juvenile ash grains and lapilli are reddish and surrounded by a pale alteration rim. The dark ash grains and lapilli consist of chlorite pseudomorphs after olivine and/or pyroxenes. The larger lapilli represent fragments from the basaltic andesite of the Lava Series 1, which formed the Earth's surface at the time that the Hirschberg maar–diatreme volcano erupted. In the matrix there are many individual grains of quartz, feldspar, and mica (not visible in hand specimen), derived from the surrounding sediments which were unconsolidated at the time that the maar–diatreme volcano was formed.

and reworked tephra must have been laid down successively on the respective maar crater floor, and, consequently, because of ongoing eruptions, subsided within the diatremes for several hundred metres up to more than 1000 m (Lorenz 1985, 1986). The bedded and unbedded tuffs contain only the mineral constituents of the surrounding presently indurated sediments, i.e. quartz, feldspar, mica, clay minerals, and their pebbles (milky quartz, quartzite, black chert, rhyolite) (Figs 6 & 9). However, they lack medium- to coarse-grained sedimentary rock clasts (like sandstone, conglomeratic sandstone, or conglomeratic arkose clasts) from these surrounding indurated sediments. Within the Rödern diatreme, however, there occur some clasts of black bituminous shale, which point to brittle to plastic behaviour of these fine-grained sediments during explosive activity.

From the relationship given above, a very important interpretation results for the emplacement of the maar–diatreme volcanoes, and this relationship has already been pointed out by Lorenz (1986, 1998, 2003b) for the Saar–Nahe Basin: the medium- to coarse-grained sediments surrounding the diatremes at the time of diatreme formation, and from which the mineral grains and pebbles were derived, consequently must still have been unconsolidated and in this state down to those depths to which the final diatreme root zones have penetrated downwards (Lorenz 2003b). Since synvolcanic sediment

Fig. 7. Part of a sandstone block which subsided within the Rödern diatreme and was injected by tephra dykelets and stringers. The serrated contact relationship between the sandstone and the tephra proves that at the time of the eruption of the diatreme the sandstone was unconsolidated and thus a soft rock (see also Fig. 8).

deposition occurred in the Saar–Nahe Basin in a mostly fluviatile and lacustrine environment – less frequently in an alluvial environment – the groundwater table at the time of volcanism must have been rather close to the surface and intermittently even at the surface. Consequently, the underlying unconsolidated sediments must have been water-saturated basin-wide at the time of volcanism in the Donnersberg-Formation. The medium- to coarse-grained sediments must have acted as rather permeable pore aquifers. This unconsolidated water-saturated state can be expected when volcanism occurs during sedimentation within an active graben structure (Lorenz 2003b). The medium- to coarser-grained sedimentary rocks presently occurring in this state, therefore, must have become diagenetically indurated only after the diatremes had formed, as did the originally unconsolidated tephra inside the diatremes. In addition to the sediment components, a number of diatremes contain volcanic rock clasts derived from solidified lava flows (Fig. 6) and sills. Supportive evidence for the interpretation given above comes from the fact that in hard-rock environments, i.e. within indurated sediments, or within volcanic, plutonic or metamorphic rocks, maar–diatreme volcanoes always contain clasts of these hard rocks of sandstone, basalt, granite, quartzite, or gneiss. Even an impacting meteorite does not fragment completely all involved hard rocks, like indurated sandstone or granite, into their mineral grains, neither do maar–diatremes fragment hard rocks of any rock type completely into their individual minerals (Lorenz 1985, 2003b).

In the larger diatremes there exist subsided blocks of older volcanic or sedimentary rocks derived from higher stratigraphic levels. The sediment and volcanic blocks in the Rödern diatreme subsided for about 500–550 m along the margin of the growing diatreme (Lorenz 1971a). Within the reddish sandstone blocks there occur dykelets and swirly stringers of tephra with irregular boundaries (Figs 5, 7 & 8). Also, individual juvenile ash grains or lapilli are found

Fig. 8. Part of a sandstone block which subsided within the Rödern diatreme and was injected by tephra dykelets and stringers. The relationship between the sandstone and the tephra dykelets and stringers proves that the sandstone at the time of the eruption of the diatreme was unconsolidated and thus subsided inside the diatreme as a block of soft rock.

within and surrounded by sandstone. These relationships point to mixing of sand and tephra and thus also to the unconsolidated state of the presently indurated sandstone at the time of diatreme formation and later joint diagenetic induration (Lorenz 2003b).

What is also worth realizing about the diatremes in the Saar–Nahe Basin is that wherever a diatreme was surrounded by a red sediment horizon, the bedded tephra, despite having subsided for several hundreds of metres from the crater floor, became oxidized at the level of the red beds. Where drab-coloured beds surround the subsided tephra, however, the subsided tephra beds were reduced and turned green (Fig. 9). Thus groundwater with either oxidizing or reducing capacity must have penetrated from the surrounding sediments into the diatremes, a fact also pointing to the sediments in the vicinity of the diatremes having been permeable at the time of diatreme formation (Lorenz 1972). It seems quite conceivable that the presently red surrounding sediments became oxidized during their diagenesis only after formation of the diatremes.

The depth of exposure of the most deeply exposed diatremes below their probable syn-eruptive surface (Lava Series 1 or even higher up: Lava Series 2 or 3) is 1000 m to possibly as much as 1500 m, as in the case of the Niedermoschel diatreme. This diatreme is cut by erosion in sediments of the Wahnwegen-Formation. Even this large diatreme, which must extend to an even greater depth, shows the same lack of country rock clasts in its lapilli tuffs. Consequently at least 1000–1500 m, possibly even 2000–2500 m of sediments below the syn-eruptive surface, were unconsolidated (at least the medium- to coarse-grained sediments) and consequently water-saturated when the diatremes formed explosively.

In the NE of the Saar–Nahe Basin, a number of diatremes, as, for example, the diatremes Rödern, Hirschberg and Hausplatz on sheet 6313 Dannenfels 1 : 25,000 (Haneke & Lorenz 2000), are localized on faults and particularly on intersections of faults (Fig. 4). From the study of the sedimentary horizons and their thickness and facies variation across faults, it was realized that most faults had been active synsedimentarily

Fig. 9. Bedded tephra from the Rödern diatreme which was originally deposited on the crater floor tens to possibly in excess of 100 m below the original surface and which then subsided syn- and post-eruptively to a depth of about 600 m below the original surface. The thinly bedded tephra sample on the left shows imbrication of clasts within a number of thin beds, which points to the emplacement of the tephra by a base surge. The sample on the left is reddish, whereas the sample on the right is greenish, both due to post-eruptive diagenetic processes (oxidation or reduction, respectively) affecting the subsided tephra.

(Haneke *et al.* 1979; Stollhofen 1994*a–c*; Haneke & Lorenz 2000). Thus it can be assumed that the diatremes erupted mostly at sites where hydraulically active faults or even hydraulically active fault intersections provided sufficient groundwater for the phreatomagmatic explosions (Fig. 4).

Near the end of their activity, magma intruded many of the diatremes non-explosively and formed dykes, sills or plugs in these diatremes. Thus the groundwater supply required for the explosive eruptions seems to have been exhausted during this late intrusive activity. In some diatremes (Rödern and Hirschberg) the late intrusions occupy only small dykes, with some of them representing some of the least-altered volcanic rocks in the Saar–Nahe Basin (Lorenz 1971*a*; Nicholls & Lorenz 1973; Göpel 1977; Arz 1996; Arz & Lorenz 1996; von Seckendorff *et al.* 2004). However, at the present erosion surface of the Falkenstein rhyodacite diatreme, about one-third of the diatreme is occupied by late rhyodacite intrusions (Lorenz 1971*b*).

The same relationship between the tephra colours of diatremes and the surrounding formerly unconsolidated sediments and their colours occurs in the Carboniferous Midland Valley, Scotland. In the diatreme at the Parade (at Dunbar), on the south side of the Firth of Forth, red Carboniferous sediments surround red-bedded tephra. In contrast, in the diatremes between Elie and St Monance on the north side of the Firth of Forth, greenish bedded tephras are surrounded by drab-coloured Carboniferous sediments (Francis 1962, 1970; Lorenz 1972, 1985).

In this chapter on maar–diatreme volcanoes in the Saar–Nahe Basin, it has been stated that during maar–diatreme activity in the Donnersberg-Formation the uppermost 1500 to possibly 2500 m of sediments were to a large extent unconsolidated and, consequently, they must have been water-saturated at that particular period of time. During the same time period of maar–diatreme formation, all other volcanic and subvolcanic bodies, as well as pyroclastic deposits, were also emplaced in the Saar–Nahe Basin. Thus the respective magmas must have partly or fully traversed the same unconsolidated water-saturated sediments and become emplaced either intrusively within these unconsolidated sedimentary sequence or managed somehow to reach the respective surface. In the following chapters the diverse subvolcanic and volcanic features and pyroclastic deposits will be described and analysed in their relationship with the sediments of the Saar–Nahe Basin.

Dykes

Dykes are the main feeder structures for near-surface emplacement of sills and also for volcanism at the Earth's surface (Halls & Fahrig 1987; Parker et al. 1990). In an area characterized by intensive volcanism, they should be frequent and cut the subvolcanic sediments or other rock types. Dykes, however, are rare in the Saar–Nahe Basin and absent over large areas and levels of the exposed part of the basin. Most of the dykes occur in the SW part of the basin, in a restricted region between the Saarbrücken Anticline and the Prims and Nahe Synclines, between the St Wendel Graben and the Herrmannsberg (Fig. 1). In this region, 14 basalt to basaltic andesite dykes have been mapped with most of them extending for several kilometres along a NNW–SSE trend (Konzan et al. 1981; Dreyer et al. 1983). The longest dyke has a length of eight kilometres. The dykes are a few metres to up to about ten metres thick and probably attained their thickness via inflation in thickness during emplacement, as it is widely known from dyke emplacement (Wright & Fiske 1971; Delaney & Pollard 1981; Wadge 1981). They occur south of the thick pile of lava flows of the Baumholder area and may, in fact, have been feeders of some of the lava flows from the thick lava pile of Baumholder-Idar-Oberstein, which is up to 800 or even 1000 m thick. Many more dykes may exist underneath this thick lava pile. One dyke north of Kusel trends NE–SW and extends at its NE end to the base of the lava pile. Another dyke worth mentioning is located west of St Wendel and follows the western fault of the St Wendel Graben for three km, which indicates that the fault had been active already prior to the emplacement of this dyke. Schwab (1971b) describes a number of short dykes from the basal lava complex of the thick lava pile of the Baumholder area. Farther to the NE in the Saar–Nahe Basin, there are even fewer dykes than in the area south of the thick lava pile described above. This correlates with the reduced total thickness of the lava flows in the NE. Nevertheless, there are a number of sills in the NE, which in all probability require nearby feeder dykes.

In the NE of the Saar–Nahe Basin, Diehl (1981, 1982) investigated a few dyke–sill systems and realized a cogenetic emplacement across the separating faults (Fig. 11).

All of the basic to intermediate dykes in the Saar–Nahe Basin have neither been investigated volcanologically in any detail nor in respect to intrusion direction applying the study of the anisotropy of magnetic susceptibility (AMS). Their relationship with the unconsolidated water-saturated sediments has not been investigated either. On the other hand, the magmas involved in the phreatomagmatic formation of the maar–diatreme volcanoes must have traversed via dykes the deeper sedimentary levels of the Saar–Nahe Basin.

Silicic dykes have not yet been discovered cutting in a normal fashion through the sediments of the Saar–Nahe Basin. Rhyolitic dykes occur, however, in the updomed sediments above the laccoliths and inside the intrusive–extrusive domes, and are related to their emplacement (see below). Silicic dykes that have cut through Westphalian and Stephanian and in some cases even through the lowermost Rotliegend strata must exist at depth and have fed the laccoliths and intrusive–extrusive domes either centrally or possibly tangentially (Hyndman & Alt 1987). These feeder dykes are not exposed because, with one exception (Kuhkopf W, see below), the base of these laccolithic and extrusive–extrusive bodies is not exposed.

Sills

The Carboniferous–Lower Permian sediments of the Saar–Nahe Basin contain many sills (Fig. 1) (Ammon & Reis 1903, 1910; Schuster 1914, 1933; Reis 1922; Konzan et al. 1981; Dreyer et al. 1983), and because of this fact the Saar–Nahe Basin represents the best area to study sills and their emplacement processes in Central Europe. Most of the sills in the basin are tholeiitic basalts and basaltic andesites, with the majority being basaltic andesites (Schwab 1968, 1981; von Seckendorff et al. 2004). There are a few intermediate sills, and the so-called kuselites, which represent particularly altered intrusive bodies and probably are, as intermediate in composition (Schuster & Schwager 1911; Koch 1938; Bederke 1959; Jung 1970; Vetter & Jung 1971; Köhler 1987). With the possible exception of a ?sill SW of the Wilzenberg Dome SW of Idar-Oberstein, silicic sills do not occur in the stratigraphic levels exposed in the Saar–Nahe Basin. Evidence for the above volcanic bodies being sills is given by field relationships. In many good outcrops, the volcanic bodies are in part concordant with the under- and overlying sediments; in part they are slightly or even distinctly discordant or may even extend some apophyses into the overlying sediments. In addition, sediments both below and above the volcanic rock became contact-metamorphosed for 10 cm to about 1–2 m, depending on the thickness of the respective sill, its dispensable heat content, and the thermal conductivity of the surrounding water-bearing sediments. Depending on the depth of intrusion, sills are non-vesicular at greater depth, or display a very thin zone of tiny vesicles at both contacts at intermediate depth (several hundred metres below the original surface), or are even rather vesicular like a lava flow very close to the upper contact if the sill intruded very close, i.e. several tens of metres, to the original surface (see below). Even these vesicular sills show contact-metamorphosed overlying sediments.

Because of the large-scale open folds (Saarbrücken and Palatinate anticlines and the adjacent Prims and Nahe synclines in the NW and the Palatinate Syncline in the SE), the sills in the Saar–Nahe Basin are exposed along-strike only. Therefore, part of the original extent has been eliminated by erosion. Without drilling and/or geophysical investigations, downdip extent (see, for example, Cruden & McCaffrey 2002; McCaffrey & Cruden 2002; Planke et al. 2002; Thomson 2002) cannot be established for the Saar–Nahe sills.

Most of the sills intruded into the sediments of the Glan-Subgroup. Few sills are present in the Stephanian or in the basal sediments of the Nahe-Subgroup. The sills mostly intruded into black, laminated pelites or slightly coarser shales. A few intruded into sandstones or arkoses (e.g. the kuselite sill of Ruppertsecken, NW of the Donnersberg rhyolite in the NE of the Saar–Nahe Basin, Fig. 10). Some sills are thin and short. Many sills, however, extend laterally for up to 12 km and may be in excess of 200 m thick. The classic type locality of the original tholeiite, the Schaumberg tholeiite near the town of Tholey (which originally was spelled Tholei, Steininger 1819, p. 119; Daly 1952), is a sill up to 230 m thick, 9 km long along strike, and composed of an altered basaltic andesite, differentiated *in situ* (Jung 1958). Some sills intruded individually, and some intruded in groups with the individual sills stacked vertically above each other, with the same geochemistry and almost the same petrography. If intruded in groups, one of the sills usually has a much more pronounced thickness (Konzan et al. 1981; Dreyer et al. 1983). The maximum thickness is usually developed in the central part of the sill and decreases gradually towards the margins (e.g. the Dielkirchen sill). Thick sills, in general, must be considered to have inflated in thickness during ongoing intrusion in order to attain their final thickness and to uplift the overlying rock sequence largely intact. If, in contrast, from the very beginning they were intruded laterally with a thickness of several ten or even several 100 m, the overlying rocks would have had to go through a migrating monocline of that scale at the front of the laterally expanding sill, and would certainly have become distorted, intensively disrupted, and even intruded by the magma from the advancing sill.

During lateral intrusion a number of sills came up against a synsedimentary fault and did not intrude across it, probably because of a facies barrier. The Ruppertsecken Sill NW of the Donnersberg, for example, abutted against a fault and then inflated in thickness, finally reaching 120 m (Fig. 10).

In the Saar–Nahe Basin the sills intruded at various depths. Since volcanism in the Saar–Nahe Basin is assumed to have been active almost only during the deposition of the Donnersberg-Formation of the Nahe-Subgroup (former Upper Rotliegend) the intrusion depth of a sill is considered to be at least the depth below the base of the first lava flows of the Lava Series 1 in the Donnersberg-Formation – or even deeper when the syn-intrusion Earth's surface is assumed at even higher levels of the volcanic stratigraphy (lavas and tephra) in the Donnersberg-Formation (Fig. 2). At several laccoliths (Königsberg, Herrmannsberg, Lemberg) the deepest updomed sills intruded Stephanian sediments, and at the intrusive–extrusive Donnersberg and Kreuznach domes, the deepest updomed sills had intruded sediments of the Quirnbach- and the overlying Lauterecken-Formation respectively. Many sills intruded intermediate levels of the Glan-Subgroup (former Lower Rotliegend) (Figs 1 & 2). Thus, judging from the known thickness of sediments and the assumption given above, the sills intruded at a maximum depth of about 2000 m, possibly even at 2500 m. In the Donnersberg region in the NE of the Saar–Nahe Basin, the sill intruded at the highest stratigraphic level is the Marienthal Sill. Initial intrusion was at a depth of about 200 m below the surface. Southwestwards from its feeder plug at the Platte (Haneke & Lorenz 2000), over 6 km, the sill did gradually cut upward across the stratigraphic sequence

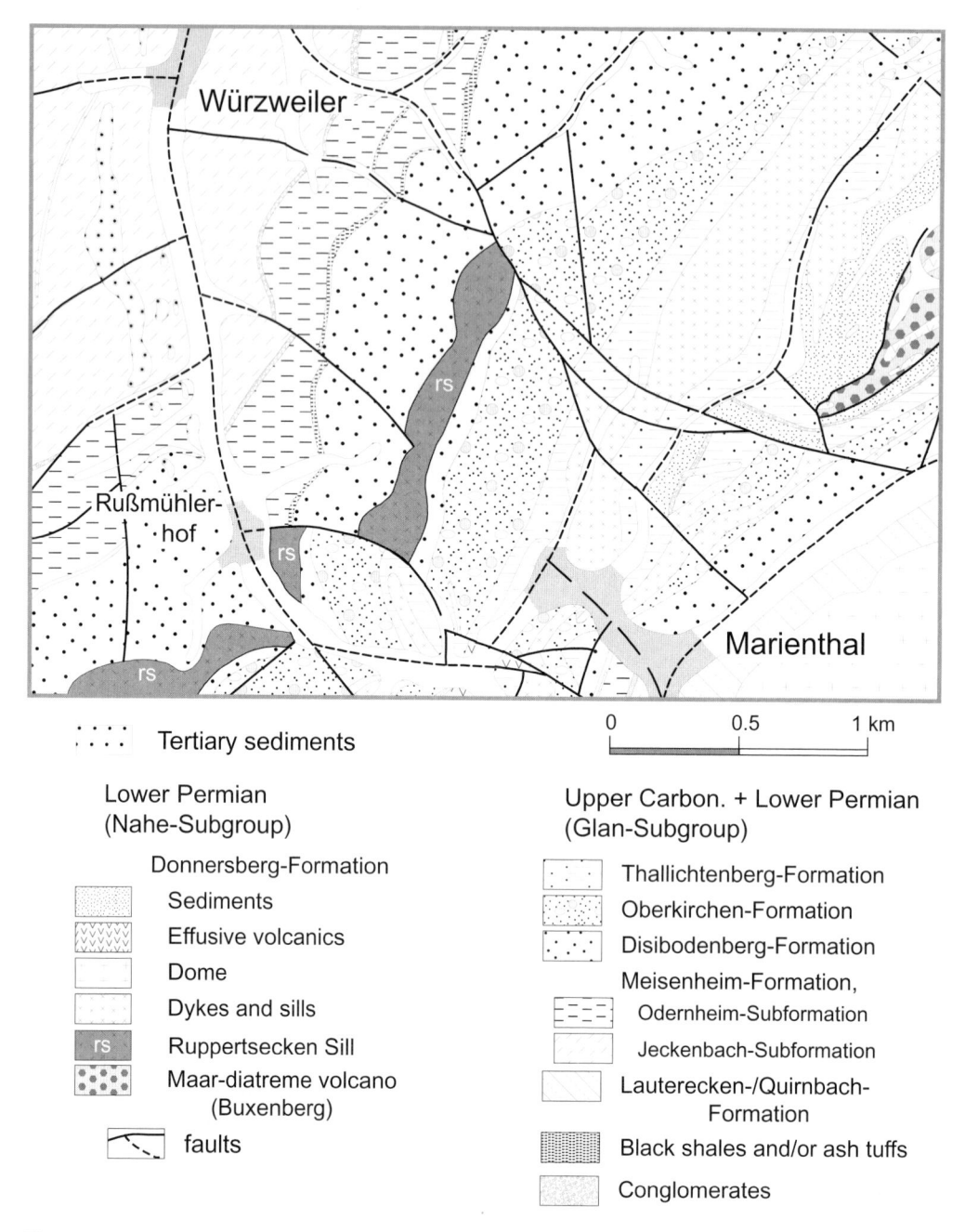

Fig. 10. Geological map of the NE part of the Ruppertsecken Sill and its surrounding geology, with the NE margin of the sill abutting against a NW–SE-trending fault.

and, at its end, it reached a few metres below the first lava flow of the Lava Series 1 (Donnersberg Grenzlager). Judging from its petrography and geochemistry the sill, in all probability, fed this lava flow (Schwab 1965, 1967, 1968, 1981). The sills that intruded the closest to the original surface are frequently rather vesicular near their upper surface and thus resemble lava flows, as, for example, the sills NE of the Platte plug and the Mannbühl kuselite sill NE of the

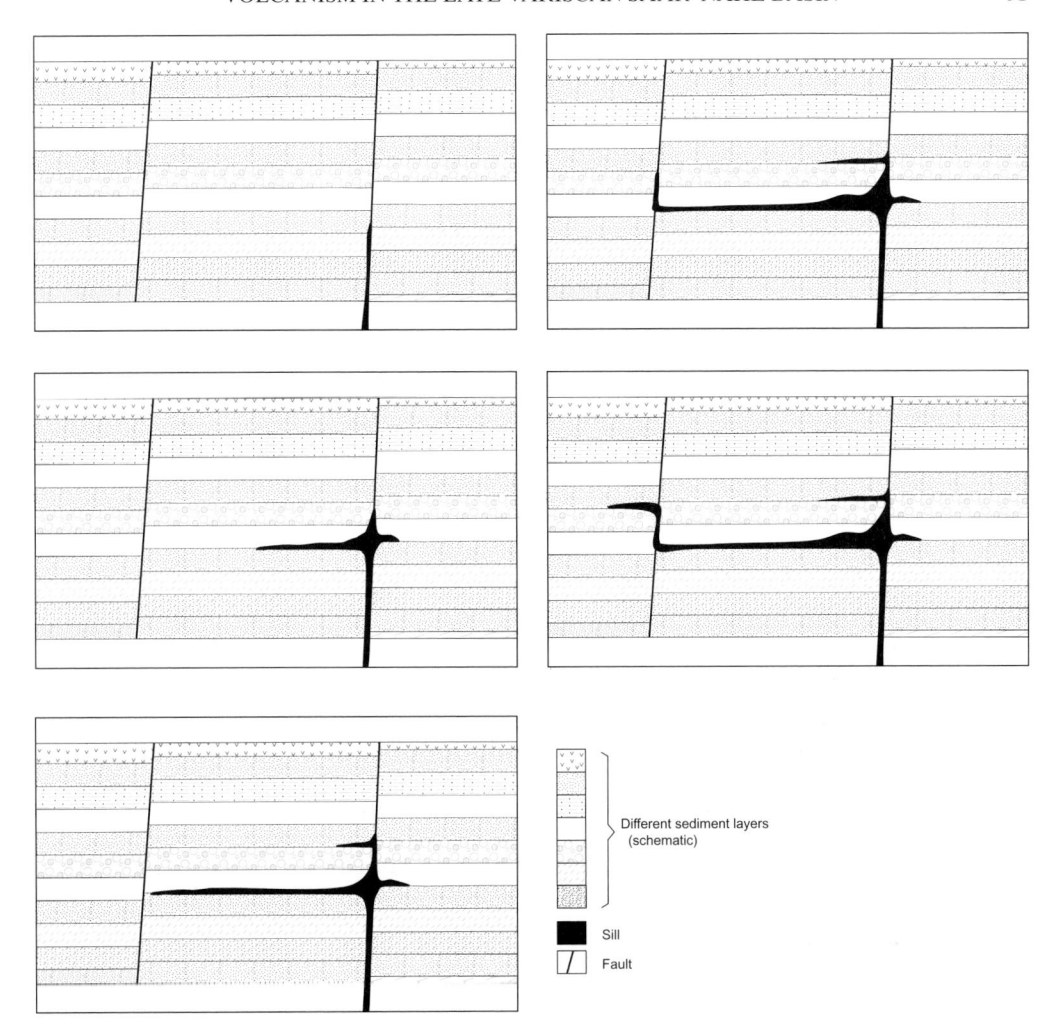

Fig. 11. Sequence of schematic diagrams of a dyke–sill system in the Saar–Nahe Basin intruding successively along faults and bedding planes within a sediment sequence which had been faulted synsedimentarily.

Donnersberg. Their overlying contact-metamorphosed sediments, however, prove that the underlying volcanic rock had its origin in a sill. Thus, the uppermost levels, sills in the Saar–Nahe Basin were able to intrude into, i.e. several metres to several tens of metres below the original surface, does not conform to the model of Mudge (1968), who assumed a depth range of sill intrusions in flat-lying sediments between about 900 m and about 2300 m, a depth range already challenged by Pollard & Johnson (1973).

For a number of sills in the NE part of the Saar–Nahe Basin, Reis (1921) and Diehl (1981, 1982) showed that they intruded as fingered sheets (Pollard *et al.* 1975) and are elongated in a NE–SW direction (30°). The polarity of the intrusion direction has not been worked out yet. Diehl (1981, 1982) also showed that a number of dykes and sills were emplaced cogenetically, forming dyke–sill systems (Fig. 11).

As discussed above the Saar–Nahe maar–diatremes point to unconsolidated water-saturated sediments in the uppermost 1500 to possibly 2500 m of the sediment fill of the basin at the time of their volcanic activity in the Donnersberg-Formation. Since the sills intruded within the same sedimentary sequence and within the same depth range, the sediments within which they intruded, therefore, must have been the same water-saturated sediments. Thus it may well be that the preference of the sills for the sediments of the Upper Stephanian and Glan-Subgroup may indicate that the even deeper sediments in the basin had been already indurated to such an extent that their yield strength did not allow the emplacement of magma as sills.

The obvious preference of the sills to intrude black, laminated pelites, however, has not been investigated in the Saar–Nahe Basin. In contrast to the coarser sediments, the black pelites – despite their porosity and large amounts of pore fluids – must have been rather impermeable for groundwater flow. It is also possible that diagenesis of the pelitic sediments had already become more advanced when compared with that of the coarser sediments. Hence the pelitic sediments could have been partially indurated at even shallower depth than the coarser sediments. The diagenetic state of the different sediments at the various depths of intrusion of the sills at the time of their intrusion, however, has not yet been investigated.

Other areas comparable with the Saar–Nahe Basin, where sills intruded unconsolidated water-saturated sediments, are some other Carboniferous–Permian intermontane basins of the Variscan orogenic belt (Breitkreuz & Mock 2002; Breitkreuz et al. 2002a, b; Wilson et al. 2004), as, for example, the Briançonnais/France, Flechtingen Block/Germany (Breitkreuz et al. 2002), the Carboniferous Midland Valley, and adjacent areas in Scotland (Francis 1982, 1991; Liss et al. 2002), the Karoo of southern Africa (Cox 1992; Marsh et al. 1997; Planke et al. 2002) and of Antarctica (Elliot 1992), and volcanic continental margins as, for example, along the Atlantic Ocean: Namibia (Gerschütz 1996), Brazil (White 1992), Greenland (Larsen & Marcussen 1992), and Norway (Skogseid et al. 1992; Planke et al. 2002; Thomson 2002).

Laccoliths, cupolas and intrusive–extrusive domes

In the Saar–Nahe Basin there exist large volumes of silicic rocks (rhyolites and the less-frequent alkali-feldspar trachytes or alkali-feldspar quartz trachytes). Their geochemistry, isotope geochemistry, and petrology have been discussed by von Seckendorff et al. (2004). The silicic magmas were either emplaced at the surface as tephras interbedded with sediments and lava flows (see below) or they were emplaced in the sedimentary fill as laccoliths, intrusive–extrusive domes, associated dykes, ?sills, lava flows, and even as diatremes. The intrusive–extrusive domes started as laccoliths, and, upon further inflation, some laccoliths even breached the surface and thus, at their higher levels, became extrusive (Figs 12 & 13). Therefore, laccoliths and intrusive–extrusive domes are treated jointly in this chapter. Because of Late Palaeozoic and younger erosion, several laccolithic bodies are exposed, but some laccoliths are still covered by updomed roof sediments. In the Saar–Nahe Basin, the concealed and some exposed laccoliths have been called cupolas. For the Frankelbach, Hohlborner Hof and Stahlberg cupolas (Figs 1 & 12), it is not known if they are really underlain by a laccolith. Firstly the cupolas and laccoliths and then the extrusive–intrusive domes will be dealt with each group – from the NE to the SW.

Laccoliths and cupolas

Waldböckelheim cupola The very distinctive Waldböckelheim cupola in the NE of the Nahe Syncline is about 10 km × 6 km in size (Fig. 12). It is assumed to be underlain and updomed by a rhyolite laccolith, because the hydrocarbon exploration drill-hole Waldböckelheim 1 penetrated rhyolite below the sediments of the Remigiusberg-Formation, at a depth of about 2500 m below the ground surface (Habicht 1966). In addition, at the NE margin of the cupola there are five rhyolite diatremes filled by tephra and small late-intrusive rhyolite bodies (Geib 1956, 1972, 1975). The cone-shaped dacitic intrusive body of the Welschberg may form a protrusion derived from a central body of the then composite laccolith underneath. The youngest rocks clearly involved in the updoming of the cupola belong to the lava flows of the Lava Series 3 and possibly the basal sediments of the overlying alluvial fan deposits of the Wadern-Formation. Thus, the laccolith intrusion must be at least slightly younger than the Lava Series 3 and must have been intruded at a depth below the original surface of at least 2500 m (Habicht 1966).

Bauwald laccolith The Bauwald laccolith, about 1 km south of the Lemberg and about 5 km SE of the Waldböckelheim cupola (Fig. 12) consists of a dacite (Vinx 1974; Atzbach 1983; Göpel 1977; Stollhofen 1986). It has an irregular shape: 4.5 km in maximum diameter. Atzbach (1983) considered the Bauwald to represent a lava flow, but steeply tilted sediments and some basic sills at its margins reveal that according to these contact relationships the Bauwald appears to be a laccolith which updomed its roof sediments and enclosed sills. The present-day almost horizontal top formed during Tertiary marine littoral abrasion. Locally, on the flat top, the

Fig. 12. Schematic map of the exposed part of the late Variscan intermontane Saar–Nahe Basin, showing the location of cupolas, laccoliths and intrusive–extrusive domes. 1, Kreuznach intrusive–extrusive dome; 2, Waldböckelheim cupola; 3, Bauwald Laccolith and Lemberg intrusive–extrusive dome; 4, Obermoschel cupola; 5, Kuhkopf laccolith and intrusive–extrusive dome; 6, Donnersberg intrusive–extrusive dome; 7, Stahlberg cupola; 8, Nußbach cupola; 9, Hohlborner Hof cupola; 10, Königsberg laccolith; 11, Frankelbach cupola; 12, Selberg cupola; 13, Herrmannsberg cupola; 14, Potzberg cupola; 15, Wilzenberg intrusive dome; 16, Nohfelden intrusive–extrusive dome; 17, Horst intrusive dome; 18, Himmelberg intrusive dome. SHBF represents the Southern Hunsrück Boundary Fault. Base map after Dreyer et al. (1983).

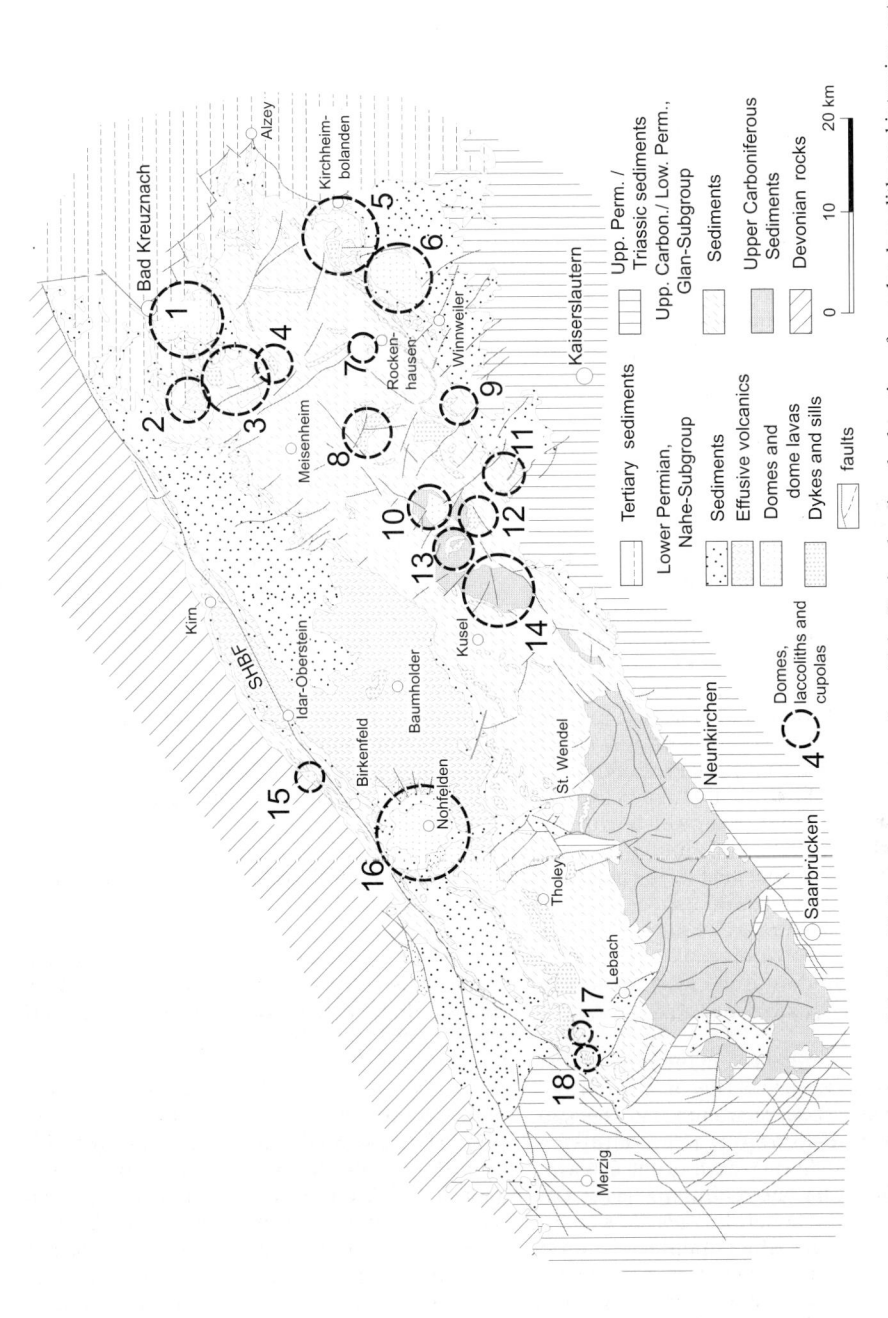

volcanic rock is slightly vesicular (Stollhofen 1986). The marginal tilted rocks consist of sediments of the upper part of the Glan-Subgroup (Meisenheim-Formation to Thallichtenberg-Formation) and basal levels of the Donnersberg-Formation of the Nahe-Subgroup. The northern indentation of the laccolith is an area of updomed sediments of the Meisenheim-Formation and is penetrated by a few thick dykes of the same dacite as forming the laccolith. Thus, this area is assumed to be underlain by a deeper part of the laccolith. In the SE of the Bauwald, the laccolith interfingers a little with the uplifted sediments, showing differential intrusion of the laccolith magma into its own roof sediments. Under this assumption, the laccolith at depth would occupy a more regular, almost subrounded rectangular shape. It is unknown whether the intrusive body had access to the original surface and became extrusive. In many outcrops the Bauwald dacite shows enclaves of a more basic rock, which points to magma mingling at depth prior to emplacement of the laccolith (Göpel 1977; von Seckendorff et al. 2004).

Obermoschel cupola The Obermoschel cupola, also located in the NE part of the Saar–Nahe Basin (Fig. 12), is about 7.5 km in NE–SW diameter and about 3.5 km m in NW–SE diameter. The central area consists of updomed sediments of the Quirnbach-Formation and overlying Lauterecken-Formation and three basaltic intrusive bodies as well as three diatremes (Krupp 1981). It is not certain, but it may well be that the cupola is underlain by a laccolith. It is even possible that this laccolith may be a composite laccolith with a central basaltic andesite body, which might have penetrated its silicic envelope and the overlying roof sediments – a possible evolution similar to that of the Herrmannsberg laccolith, which is described below.

Palatinate Anticline cupolas In the central area of the Palatinate Anticline (Fig. 12), from the NE towards the SW, there are four cupolas and laccoliths, i.e. the Königsberg laccolith, the Selberg cupola just to the south of the Königsberg, the Herrmannsberg cupola and the Potzberg cupola (Dreyer et al. 1983). The group of laccoliths and cupolas updomed Upper Carboniferous sediments (Stephanian and lower Glan-Subgroup) with the latter being deformed into synclines between the individual dome structures.

Königsberg laccolith The impressive Königsberg Hill (Fig. 12) exposes the central part of a laccolith, which was responsible for the updoming of its former roof sediments and of two basic pre-laccolith sills (Dreyer 1970; Dreyer et al. 1983). The updomed area has a diameter of 5 km and the amount of updoming is about 600 m or even more. The Königsberg rhyolite itself has a diameter of 2.5 to 3 km and rises from the neighbouring Lauter Valley (186 m a.s.l.) to 567 m a.s.l. Locally, the rhyolite shows columnar jointing at its contact with the surrounding Stephanian sediments and also flow-banding (Negendank 1965; Dreyer 1970; Atzbach et al. 1974). Towards the SE, two rhyolite dykes (or ?sills) accompany the laccolith. The youngest clearly updomed sediments belong to the Quirnbach-Formation, and the basic sills of the Donnersberg-Formation indicate that the surface was part of the Donnersberg-Formation. Towards the west the rhyolite is not in contact with Stephanian sediments but with the younger sediments of the Glan-Subgroup (Remigiusberg-Formation). Either the rhyolite punched through the Stephanian sediments in this area, or a pre-laccolith fault allowed the rhyolite to get into this contact relationship. Because of erosion it is unknown whether the Königsberg rhyolite breached the surface and became extrusive, or if it even erupted tephra or lava.

Selberg cupola Just to the south of the Königsberg there is the lower hill of the Selberg (Fig. 12). Its updomed area of Stephanian and lower Glan-Subgroup sediments has a diameter of 3 km, and the amount of updoming was about 500 m. In the eastern part of the cupola there are the above-mentioned three rhyolite dykes (Dreyer et al. 1983), which certainly must either have protruded from the underlying laccolith towards the original surface along extensional fissures or, likewise, from the Königsberg laccolith. The large intermediate to basic intrusive body in the SW may also indicate a protrusion from depth, possibly from an intermediate to basic core of the then composite laccolith.

Herrmannsberg cupola The Herrmannsberg (Fig. 12) to the SW of the Königsberg in its central area also exposes updomed Stephanian and lower Glan-Subgroup sediments – and two basaltic andesite sills (Dreyer et al. 1983). The youngest updomed sediments belong to the Lauterecken-Formation. The amount of updoming in the central area of the cupola in comparison with the least-affected sediments at the margins of the cupola is about 600 m. Doming of the sediments resulting in the cupola is assumed to have been caused by an underlying laccolith. In the central area of the cupola there are two arcuate rhyolite dykes resembling ring dykes or cone sheets. The Herrmannsberg has been mapped in some detail (Grimm & Stapf 1991) but it is still unknown whether the flow-banding of the rhyolite dykes is inclined towards the centre of the cupola (making them cone sheets) or whether it is inclined almost vertically or even outward (making them ring dykes). In the central area of the cupola there is also a thick, rather medium- to coarse-grained basaltic andesite (a so-called palatinite; Jung 1967) trending NW–SE. According to Dreyer *et al.* (1983) and Grimm & Stapf (1991) the NW end of this basaltic andesite is in contact with the eastern rhyolite dyke. In all probability the rhyolite dykes and the central basaltic andesite dyke protruded from an underlying composite laccolith – similar to the Kuhkopf composite laccolith described later.

In the NE and locally in the NW, an intermediate to basic sill has been updomed with the surrounding sediments of the Altenglan-Formation. This indicates that even the basal sediments and volcanic rocks of the Donnersberg-Formation had been originally involved in the updoming of the laccolith roof – similarly to the situation at the neighbouring Potzberg. The cupola extends towards the NW, which may indicate a subterranean extension of the laccolith towards the NW.

Potzberg cupola The Potzberg cupola (Fig. 12), SW of the Herrmannsberg, has a maximum diameter of about 9 km NNE/SSW and a minimum diameter of about 7.5 km WNW/ESE. The cupola does not expose any silicic or basic volcanic rocks in its central area of updomed Stephanian and lower Glan-Subgroup sediments of the Rotliegend. In the NW, the lowermost Glan-Subgroup sediments contain one interbedded basic to intermediate sill, which predates the updoming event. The youngest updomed sediments belong to the lowermost Meisenheim-Formation, whereas the updomed sill indicates that emplacement of the laccolith occurred during the Donnersberg-Formation. Therefore, the original surface at the time of laccolith emplacement was in all probability a surface forming part of the Donnersberg-Formation. The roof was updomed for about 500–600 m.

At the level of exposures all four cupolas and laccoliths during their updoming affected sediments reaching from the Stephanian to the Meisenheim-Formation of the Glan-Subgroup. Since sills intruded into Stephanian and Glan-Subgroup sediments during the time of deposition of the sediments and lava flows of the Donnersberg-Formation, but since they themselves were also affected by the updoming, the intrusion of the laccoliths must have occurred during the Donnersberg-Formation. Quite conceivably it caused also updoming of the originally overlying sediments of the Donnersberg-Formation. At three of the four cupolas rhyolitic dykes intruded the updomed roof sediments either above the laccoliths or marginally in respect to the laccoliths.

Wilzenberg dome Southwest of Idar–Oberstein, at the southern margin of the exposed Hunsrück Mountains but NW of the Southern Hunsrück Boundary Fault (SHBF), there exists the Wilzenberg intrusive rhyolite body (Fig. 12) within undifferentiated sediments of the Glan-Subgroup. The rhyolite is rather altered and has a diameter of 1.6 km (NE–SW) by 1 km (NW–SE). It is in contact with the Lower Devonian Hunsrück sediments (Müller 1982; Dreyer *et al.* 1983). This rhyolite is accompanied towards the SW by a smaller dyke- or sill-like rhyolite. Because of a lack of informative outcrops it is not clear what relationships exist between the rhyolite dome and its surrounding sediments and also between the rhyolite dome and the dyke- or sill-like rhyolite in the SW, and thus an intrusion model has not yet been worked out.

Himmelberg dome The silicic dome of the Himmelberg (Fig. 12) at the villages of Schmelz and Außen is about 1.1 km in diameter and located in the SW of the Saar–Nahe Basin, 16 km SW of the present margin of the Nohfelden rhyolite lava flow (Konzan *et al.* 1981, 1989; von Seckendorff 1990). The geology and petrology of the Himmelberg dome have been described by von Seckendorff (1990) and von Seckendorff *et al.* (2004). The volcanic rock consists of a K-rich garnetiferous quartz–alkali-feldspar trachyte and contains five lati-andesite plugs and three dykes. Locally a convex–concave contact was exposed between the trachyte and lati-andesite, pointing to intrusion of the plugs and dykes at a time when the trachyte was still molten, i.e. still in a ductile state. Von Seckendorff (1990) assumed an intrusion of the lati-andesite plugs and dykes at a time following the emplacement of the trachyte. At two contacts, older sediments of the Glan-Subgroup (Meisenheim-Formation and ?Oberkirchen-Formation) were seen to underlie the volcanic Himmelberg rock with a shallow dip towards its centre, i.e. in an unclear contact relationship with it (von Seckendorff 1990). At the immediate contact, the sediments are fragmented and thus possibly volcanotectonically deformed. Sediments overlying the quartz–alkali-feldspar trachyte of the Himmelberg seem to be younger and to have been deposited on top of the volcanic rock. A few vesicles (about 0.3 mm in size) occur in the marginal facies of the Himmelberg and, slightly more frequently, in the Himmelberg volcanic rock clasts found at a distance of 1 km in a reworked position in the alluvial fan deposits of the Wadern-Formation. The vesicles point to near-surface or even surface emplacement. According to von Seckendorff (1990) the Himmelberg quartz–alkali-feldspar trachyte was probably emplaced extrusively. The intrusive contacts with earlier sediments; the rather round shape of the Himmelberg Dome; and the reworked clasts found in younger Lower Permian sediments all suggest that the Himmelberg Dome may have been emplaced as an intrusive body which, upon further inflation, might have become an extrusive dome with a short lava flow (von Seckendorff 1990).

Von Seckendorff (1990) briefly describes three bodies (100–900 m in maximum diameter) of alkali-feldspar trachyte from the margins of the Horst lati-andesite intrusion (5 × 1.5 km in area; located 500 m to the NE of the Himmelberg). No contacts are exposed, and thus the emplacement of these silicic bodies remains unclear. At the Wehlenberg–Weltersberg near Düppenweiler, 5 km SW of the Himmelberg Dome, K-rich garnetiferous quartz–alkali-feldspar trachyte was emplaced and formed the subvolcanic intrusion near Düppenweiler (von Seckendorff 1990).

Intrusive–extrusive domes

Donnersberg intrusive–extrusive dome The Donnersberg rhyolite is an intrusive–extrusive dome, which is located in the NE of the Saar–Nahe Basin on the SE flank of the Palatinate Anticline (Fig. 12). The dome structure and surrounding geology have been a matter of interest since the early studies in the Saar–Nahe Basin (von Gümbel 1846, 1848; Schuster 1914; Schmidt & Reis 1915; Reis 1921, 1922; Schwab 1967; Lorenz 1971b, c; Haneke et al. 1983; Haneke & Lorenz 1987). Thus this intrusive–extrusive dome will be dealt with first. The Donnersberg dome reaches a height of 687 m a.s.l. It occupies an area of about 19.5 km² and has a diameter of 6.5 km NE/SW and is 5 km across. The rhyolite contains about 3–9.5% phenocrysts (mostly 1–2 mm in size) of quartz, alkali feldspar, plagioclase, the hydrous phase biotite, and opaques in a felsitic groundmass of quartz–feldspar intergrowth (Theuerjahr 1973, 1986; Arikas 1986; Haneke 1987). The rhyolite is flow-banded and non-vesicular. The flow-banding has been studied intensively and 15 individual dome units have been realized, with some of them separated by narrow septa of the former roof sediments (Lorenz & Haneke 1981; Haneke 1987; Lorenz et al. 1987). The septa between the individual dome units in principle are similar to but in a more evolved state than the synclines between the four cupolas and laccoliths of the Königsberg–Potzberg group (see also Breitkreuz & Mock, this volume). The sediments and volcanic rocks (intrusive and effusive: Lava Series 1) surrounding the rhyolite dome in the NW and north have been updomed and tilted away from the rhyolite. The updomed sediments belong to the sequence from the Quirnbach-Formation up to the lower part of the Donnersberg-Formation (Lorenz & Haneke 1981; Haneke 1987; Lorenz et al. 1987; Haneke & Lorenz 2000). Thus, initial intrusion and inflation of the rhyolite magma into a laccolith occurred in the sediments of the Quirnbach-Formation. The sediments of the Glan-Subgroup contain two basic to intermediate sills. The main sill, which had been involved in the updoming, is the Marienthal sill (in the NW and north) that intruded into the sediments at the Glan-Subgroup/Nahe-Subgroup boundary, whereas small relicts of the second sill occur in the deepest updomed sediments (Quirnbach-Formation) close to the contact with the rhyolite in the NW and north. It is worth noting that within the updomed sedimentary sequence there is a gap between the Lauterecken-Formation and the Meisenheim-Formation, which can be localized at a roughly concentric fault. Thus about 650 m of sediments do not occur in the updomed series surrounding the north and west of the Donnersberg dome at the present surface, probably because they had been overthrust by the older sediments during uplift and lateral growth of the dome (Fig. 13).

Because a thick rhyolite breccia (proximal)

Fig. 13. Sequence of schematic diagrams indicating the evolution of the emplacement of an inflating silicic laccolith in the Saar–Nahe Basin (intruding originally as a sill and fed from a deeper ?laccolithic intrusion) and formation, via further inflation, of an intrusive–extrusive dome. When the dome became extrusive, rock falls and probably also block-and-ash flows formed. The evolution of the Donnersberg intrusive–extrusive dome, thus deduced, has been used as an example for these schematic diagrams.

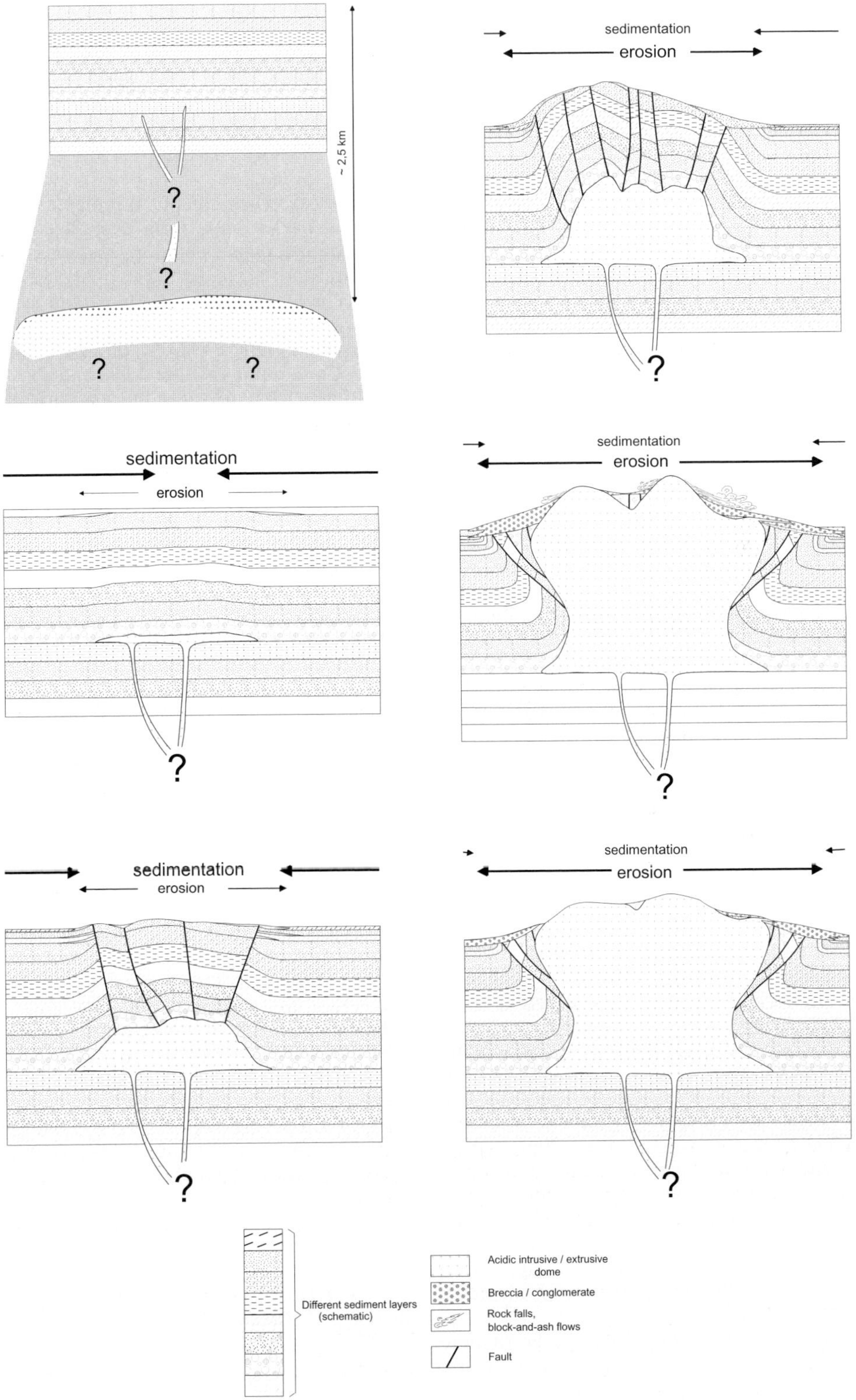
~ 2,5 km
?
?
?
?
sedimentation
erosion
?
sedimentation
erosion
?
sedimentation
erosion
?
sedimentation
erosion
?
sedimentation
erosion
?
Different sediment layers
(schematic)
Acidic intrusive / extrusive
dome
Breccia / conglomerate
Rock falls,
block-and-ash flows
Fault

and conglomerate (distal) in the Donnersberg-Formation are derived from the dome and surround it in the north, east and SW, there must have been ongoing inflation of the original laccolith, with consequent updoming of its overlying sediments to such an extent that the rhyolite finally breached the surface and formed an extrusive dome (Fig. 13). The whole intrusive–extrusive dome with the associated rhyolite breccia/conglomerate has an estimated volume of 23 km^3. The extrusive dome formed a volcanic mountain rising above the almost horizontal graben-floor up to a height of possibly 1000 m. Compared with the extrusion rate of recent domes, the extrusion of the Donnersberg dome lasted maybe about 380–1700 years (Haneke 1987; Lorenz et al. 1987).

Predominantly on the SE slope of the Donnersberg dome, there are exposed about 70 rhyolite dykes (Haneke 1987). They are from only a few centimetres to in excess of 50 cm thick and can be traced for up to several tens of metres. Obviously the cooling carapace of the extruding and inflating rhyolite dome started to rupture, thus allowing magma from the interior of the dome to be squeezed up along the fractures.

The badly exposed contact between the Donnersberg rhyolite and its surrounding updomed sediments shows little thermal contact effects, as has been pointed out for a number of laccoliths by Corry (1988). Reis (1921) mentioned tourmaline at one outcrop (Seedell). Hornfelsed black shales are found for several 100 m along the SE contact, at Reich Geschiebe. The septum of conglomeratic sediments in the Spendel valley displays metre-sized blocks of contact-quartzitic conglomeratic sandstone (Lorenz & Haneke 1981; Haneke 1987).

Extrusion of the large rhyolite dome requires erosion of its updomed former roof sediments prior to the deposition of the widespread rhyolite breccia. The lack of recognizable reworked sedimentary rocks, i.e. the lack of sedimentary rock clasts derived from the updomed roof of the intrusive and inflating Donnersberg laccolith had been an intriguing problem and was originally not understood. If the sediments, as is assumed here, were unconsolidated and water-saturated, their updoming would have led to slumping (as suggested for such sediments by Corry 1988), and their flowing and being washed away during heavy rainfalls, and consequently to their mixing with the regional sediment influx (see also below).

In an initial depositional phase, the rhyolite breccias with their non-vesicular clasts (with a maximum size of 1–2 m close to the rhyolite) extended laterally for at least a distance of 8 km towards the SW and 7 km to the NE (Haneke 1987; Haneke & Lorenz 1987; Bahr 1989; Brod 1989; Stollhofen & Haneke 1993). Towards the NW and west rhyolite breccia and conglomerate were certainly deposited but have been eroded since. In the north, east and SW and close to the rhyolite dome, the rhyolite breccia unconformably overlies steeply dipping sediments uplifted by the dome from deep stratigraphic levels of the Glan-Subgroup. Over the above given transport distance, the angular clasts of the breccia increasingly give way to pebbles because of reworking on alluvial fans and in fluvial environments (Brod 1989; Stollhofen & Haneke 1993). Close to the rhyolite dome and above the basal far-reaching rhyolite breccia horizon, the overlying stratigraphic sequence is in part interbedded with the proximal breccia, thus rhyolite clasts continued to be shed from the rhyolite dome. In part, most obvious in the north, the respective stratigraphic sequence (from the level of the Tuff 3 to the Quartzite Conglomerate) is thinned. The thinning, with partial wedging out of particular horizons towards the rhyolite, indicates that the rhyolite magma was still rising slowly and continued its extrusion and thus continued with some dragging upward and uplifting the sediments and lava flows in its close vicinity. Thus the extrusive activity was rather long-lived and is equivalent to the regional deposition of 600 m of sediments, lava flows, and tephra beds of the Donnersberg-Formation (Haneke 1987). During the whole period of this ongoing extrusion, rhyolite breccia was shed and deposited in normally or inversely graded beds of a thickness of several tens of centimetres to several metres. With the exception of the beginning (see above), the breccia was shed to radial distances of about 1.5 km. Breccia deposition thus peaked during initial, probably rapid extrusion, and then decreased with time. Only once more, was a second but lower maximum in rhyolite breccia production reached and breccia shed to a distance of at least 4 km to the NE: also showing evidence for having been reworked by debris flows and fluvial transport. With increasing distance from the source, siliciclastic material is interbedded with the basal and the second peak rhyolite breccias, and the rhyolite clasts show increasing effects of rounding, due to reworking as already stated above. Locally, rhyolite clasts even occur in the 1-m thick lacustrine *Acanthodes* limestone bed (a regional marker bed) close to the rhyolite dome, indicating that at that particular time the still-active dome was rising out of a widespread but shallow lake. In the rhyolite breccia, small amounts, and, with increasing distance from the dome, increasing amounts of sediment-derived minerals and pebbles occur between the rhyolite clasts. Close to the rhyolite dome, in all probability, they were derived from the updomed sediments. Farther away the siliciclastic material may in part represent the regional influx. Where the breccias are clast-supported, a former matrix could have been lost by water washing through the deposits or they may represent rock-fall deposits lacking fines. It is highly probable that, during the extrusion of the rhyolite dome, the breccias developed from the certainly blocky carapace of the dome (still lacking abundant small clasts), and slope instability and consequent slope failure of the dynamically evolving rhyolite dome. Therefore, they represent rock falls or even block-and-ash flows (Fig. 13). During the dynamic transfer of the clastic material from the dome downslope, fragmentation of larger clasts gave rise to the formation of the many small and even very small clasts, and thus to the poor sorting. Once deposited on the slopes and on the surrounding plain, debris flows and some fluvial activity reworked these mass deposits and was responsible for the final deposition. Interbedded with the distal rhyolite conglomerate beds, towards the SW of the dome, some ash tuff beds may represent ash clouds derived from the rock falls and block-and-ash flows.

With ongoing but reduced later extrusion of the extrusive dome, rock falls and possibly block-and-ash flows may have occurred less frequently and thus given rise to less frequent and thus proximal deposits only. It is evident from an analysis of the Tertiary to Recent history of the Donnersberg area and rhyolite erosion products formed during this time period that from Oligocene to Recent times very little erosion has occurred at the Donnersberg rhyolite dome – which has been a hill, rising up to 400 m above the surrounding area, during this whole period of time. This lack of an appreciable amount of erosion over the last 40 million years makes rock falls or block-and-ash flows generated by active dome growth and associated slope failure in the lowermost Permian also very likely.

Seismic investigations suggest that a plutonic body underlies the Donnersberg intrusive–extrusive laccolith, with its surface located at a depth of about 2700 to 3000 m (Haneke 1987). Intrusion of this plutonic body is assumed to have caused slight doming of its roof sediments and consequent thinning of the sediments of Oberkirchen-Formation and Thallichtenberg-Formation, i.e. prior to the intrusion of the initial Donnersberg sill (Haneke 1987). If the thinning of these sediments is in fact related to the deep intrusion, then the deep intrusion might represent a laccolith-shaped body intruded already near the end of the Glan-Subgroup.

Kuhkopf laccolith and intrusive–extrusive dome
The Kuhkopf dome structure (Fig. 12), located just 2 km NE of the Donnersberg, is in plan view an irregularly shaped dome complex (Fig. 12). At the present surface it consists in part of alkali-feldspar trachyte only; in part it consists of a composite laccolith with an internal basaltic andesite facies and a marginal envelope of alkali feldspar trachyte (Jung 1967; Hinzmann 1987; Lorenz *et al.* 1987; Bargenda 1988; Carle 1988; von Seckendorff 1990; von Seckendorff *et al.* 2004). Bargenda (1988) and Carle (1988) both mapped the dome complex, and, in evaluating the flow-banding, worked out that the dome complex consists of five individual intrusive units – each emplaced at a slightly different level and/or inflated to different volumes. In the area south of the village of Orbis, the northernmost unit is concealed beneath contact-metamorphosed sandstones of the Disibodenberg-Formation (Haneke & Lorenz 2000). The next three units farther south and SW intruded initially more or less into the same stratigraphic level. Via pronounced inflation they either punched locally through younger sediments and volcanics of the Lava Series 1 – or they uplifted their roof sediments and volcanics along some pre-existing synsedimentarily active faults, and, consequently, at their margins, came into contact with sediments and volcanics otherwise occurring above them (Haneke & Lorenz 2000: Geological Sheet 6313 Dannenfels). These dome units finally breached the overlying sediments of the youngest Glan-Subgroup (Oberkirchen-Formation and Thallichtenberg-Formation) and sediments, pyroclastic beds and lava flows (Lava Series 1) of the basal Donnersberg-Formation of the Nahe-Subgroup and, at least in part, reached the surface. As the present surface is an erosion surface (Donnersberg rhyolite breccia is overlying these Kuhkopf dome units), it is not known if these three extrusive Kuhkopf units formed a high relief resulting in rock falls and/or block-and-ash flows and even in a lava flow. In the locally vesicular dome unit 3, alkali-feldspar trachyte magma rose along fractures into the cooling carapace of the inflating unit, forming some dykes (Bargenda 1988; Carle 1988). Towards the west, unit 4 underlies in part up-domed sediments of the Oberkirchen, Thallichtenberg and basal Donnersberg formations, the latter containing lavas of the Lava Series 1 and tephra (T1 and T2) (Haneke & Lorenz 2000).

Because of later tilting, the westernmost unit (5), a laccolith, shows both the lower and the upper laccolith/sediment contacts. The stratigraphic rock units below and above the laccolith belong to the basal sediment levels of the Donnersberg-Formation, and thus make this western Kuhkopf laccolith the stratigraphically highest laccolith in the Saar–Nahe Basin, i.e. the laccolith intruded just above the base of the Donnersberg-Formation in the Nahe-Subgroup, i.e. approximately 450 m below the original surface. It has to be stated that the basal contact of this laccolith unit cuts across a few previously active faults and thus in an oblique angle cuts across several stratigraphic horizons.

As indicated above, this westernmost Kuhkopf laccolith unit is composite. There are no breccias along the contact between the marginal alkali-feldspar trachyte and the basaltic andesite in the interior. At several localities along the upper internal contact, a convex–concave boundary between the two rocks had been exposed in a quarry, indicating, as does the lack of breccias between the two rock facies, that the two magmas intruded simultaneously or the basaltic andesite magma intruded into the alkali-feldspar magma at a time when the latter was still in a partially molten or at least in a ductile state. At the lower and upper zone of the trachyte envelope, the flow-banding of the alkali-feldspar trachyte is oriented parallel to the basal and top contact with the sediments of the Donnersberg Formation, and it is also parallel to the contact with the central basaltic andesite. The basaltic andesite locally at its lower contact also shows flow-banding parallel to the lower contact. For about 1 km along its lower contact, the basaltic andesite displays a fine-grained chilled margin (up to about 10 m thick) pointing to a higher temperature of the basaltic andesite magma in comparison with the alkali-feldspar trachyte magma (or still ductile hot rock). The interior of the basaltic andesite, exposed in several large quarries, is medium-grained, massive and in the past was known as a palatinite (Schuster 1914; Reis 1921; Jung 1967). At the upper contact the basaltic andesite is only slightly finer grained than in the interior facies. The top part of the alkali-feldspar trachyte is about 100 m thick.

Taking the field aspects together, the following model is proposed: for this Kuhkopf laccolith unit: at first the alkali-feldspar trachyte magma intruded in a sill-like manner into the basal fine-grained sediments of the Donnersberg Formation, 450 m below the original surface, and then it started to inflate. Simultaneously or slightly later, basaltic andesite magma, probably rising within the same feeder dyke, intruded the alkali-feldspar trachyte magma, which was still behaving/reacting in a ductile fashion. The basaltic andesite magma, which was at a more elevated temperature, became chilled against the cooler underlying trachytic magma but, during its own inflation against the dynamically evolving upper contact with the overlying trachyte magma, it hardly became chilled at all at this upper interface (Lorenz et al. 1987). Because later on this composite laccolith became tilted and partially eroded, it is not known whether the two units later protruded through the updomed roof sediments (possibly similar to the situation at the Herrmannsberg, see above) and whether they even reached the original surface and formed lava flows and tephra.

Kreuznach intrusive–extrusive dome The Kreuznach intrusive–extrusive dome is located north of the Donnersberg Rhyolite Dome, also at the NE end of the exposed Saar–Nahe Basin (Fig. 12). It is 8 × 6–7 km in size and occupies a volume of possibly 30–40 km^3 (Lorenz 1973). Thus, it is one of the largest rhyolite domes in the Saar–Nahe Basin. In the north and NE it is overlain by Tertiary and Quaternary sediments and thus conceals its contact relationships with the Carboniferous–Permian sediments. The rhyolite contains up to 20–30% of phenocrysts of quartz, K-feldspar, plagioclase, the hydrous phase biotite and pseudomorphs after amphibole and clinopyroxene set in a quartz–feldspar groundmass (Theuerjahr 1973, 1986; Arikas 1986). In several areas, especially in a quarry near the village of Traisen, there are occurrences of basaltic andesite enclaves (Arikas 1986; Wörrlein 2002; von Seckendorff et al. 2004). The individual enclaves show spherical to ellipsoidal and cauliflower shapes. Many enclaves contain xenocrysts of K-feldspar which originally grew as phenocrysts in the host rhyolite magma and then only had been mechanically transferred from the host magma into the magma of the basaltic andesite enclaves (Waight et al. 2000; Wörrlein 2002). Along the 200-m high Rotenfels cliff and the Nahe and Alsenz valleys the rhyolite frequently displays flow-banding and locally also vesicles. The largest vesicles are found in the Traisen Quarry, where a few reach 5 cm in diameter. In this quarry, even a few basaltic andesite enclaves show vesicles, indicating vesiculation due to decompression when the enclaves were transported upwards within the rising rhyolite magma while they were still in the liquid state.

The rhyolite dome started as a laccolith and

updomed sediments from a depth of about 2 km (Quirnbach-Formation) up to the Donnersberg-Formation, including a few sills, such as the Norheim palatinate sill, and the basaltic andesite lava flows of both the Lava Series 1 and 2. Within this thick series of sediments and volcanics, sediments between the Lauterecken-Formation and the basal Donnersberg-Formation (roughly 1500 m thick) are absent from the uplifted series along the southern margin of the rhyolite (in the Altenbamberg area) because they became suppressed by overthrusting of older sediments (Quirnbach and Lauterecken formations) on to the younger sediments and volcanic rocks of the Donnersberg-Formation (Lorenz 1973). This outward thrusting is assumed to be the result of near-surface radial growth, i.e. inflation, of the dome. Via continued inflation in volume, the originally laccolithic intrusive dome finally breached its roof and became extrusive, forming not only an extrusive dome but also an extensive lava flow. Rock falls and probably even block-and-ash flows (see the section on the Donnersberg Rhyolite Dome) indicate slope failure of the growing and laterally enlarging dome. These rock falls and block-and-ash flow deposits and their more distal reworked facies can be traced for about 17 km towards the west into the Sobernheimer Wald (Bock 1986). There is no evidence of a breccia underneath the rhyolite breccia consisting of rock clasts from those sedimentary rocks, which previously had been updomed and breached by the Kreuznach intrusive–extrusive dome. As at the Donnersberg dome, this is believed to be the result of the respective sediments having still been largely unconsolidated and water-saturated, as is discussed below. Because of younger erosion it is uncertain to what height the Kreuznach rhyolite extruded, and how long it might have been actively extruding and forming rock falls and possibly block-and-ash flows.

Lava from the extrusive dome flowed towards the east and SE of the dome on to the rather horizontal Saar–Nahe Basin floor, and formed a thick and widespread lava unit (Schopp 1894; Lorenz 1973; Geib & Lorenz 1974; Dörhöfer 1989). This lava flow is still preserved over 7 km towards the east, displaying columnar jointing near the base, flow-banding, some pebble-like intraclasts, a few flow folds, and autoclastic breccias – the latter especially near its eastern flow front (Lorenz 1973; Dörhöfer 1989). Ramp structures are not evident. Within the eastern margin of the flow, a phenocryst-poor facies must have occurred as a lens either inside or as a slice on top of the flow as is indicated by a Tertiary coastal pebble beach containing pebbles and boulders derived from a rhyolite coastal cliff. In addition, this Tertiary beach deposit contains large blocks of autoclastic rhyolite breccia, with the clasts consisting predominantly of the normal phenocryst-rich rhyolite, but, to a small degree, also of the phenocryst-poor rhyolite (Schopp 1894; Lorenz 1973). Due to younger erosion, especially in Tertiary times, the original top of the rhyolite lava flow has not been preserved. Towards the SE, the rhyolite flowed over a distance of 9 km into the area of the present village of Kriegsfeld, where a $c.$ 150 $\times$ 150 m large block of this rhyolite lava had subsided during the eruptive activity of the Wolfsgalgen maar–diatreme volcano and had, therefore, escaped erosion (Lorenz 1973; Gehring 1998; Gehring & Lorenz 1998).

At the southern margin (in the Altenbamberg area) intrusion tectonics was studied by Lorenz (1973) and Krawinkel (1988). During its inflation, the rhyolite updomed, overturned and finally thrusted sideways (i.e. southwards) the deepest uplifted sediments (Quirnbach and Lauterecken formations) and enclosed sills. In the uplifted and partially brecciated black shales close to the rhyolite and above the thrust surface in the village of Altenbamberg there exist many slickensides, thus pointing to a sufficient amount of diagenetic induration of the shales prior to the inflationary emplacement process of the rhyolite dome.

Along the southern margin of the rhyolite there also exists an autoclastic friction breccia (Lorenz 1973; Krawinkel 1988), up to 50–70 m thick. Some rhyolite clasts carry hornfelsed clasts of greenish and reddish shales baked on to their surface. This points to elevated temperatures of the marginal rhyolite facies during its brittle deformation and mixing with the sediments during formation of the autoclastic breccia. Near its southern contact, in the Altenbamberg area, some intra-rhyolite autoclastic breccias also occur (Lorenz 1973).

Lemberg intrusive–extrusive dome The Lemberg Dome started as a laccolith in the NE of the Saar–Nahe Basin SW of the Kreuznach Dome (Fig. 12). It is a composite dome consisting of rhyolite in its centre and a rhyodacite along its margin. Its geology, petrography and geochemistry, including its isotopic composition, were studied by Vinx (1974), Stollhofen (1986), Arz & Lorenz (1994c), Arz (1996) and von Seckendorff *et al.* (2004). The dome is up to 2 km in diameter and rises from the deepest outcrop level at the Nahe River, 121 m a.s.l., over a height difference of 301 m to a height of 422 m a.s.l. During its intrusion and inflation the dome updomed Stephanian and Glan-Subgroup sediments (Remigiusberg-Formation to Thallichtenberg-Formation) and basal levels of the Donnersberg-Formation, including lava flows of the Lava Series 1 and 2 as well as several basic sills and at least one rhyodacite sill interbedded with the Glan-Subgroup sediments. Towards the west the effect of this updoming of the roof sediments reaches to a maximum of 1500 m radially away from the contact of the laccolith.

It is highly probable that the dome fed some of the intermediate lava flows of the Lava Series 3 (Nickel 1979, 1981) and then it must, at least in part, have become extrusive. However, a detailed comparative petrographic and geochemical investigation has not yet been performed. The Leisberg rhyolite lava flow in the north and the Norheimer Wald rhyolite lava flows in the NE of the Lemberg may have been fed by the central rhyolite body of the composite Lemberg Dome. If so, the Lemberg Dome must have evolved into an intrusive–extrusive dome.

Nohfelden intrusive–extrusive dome The large Nohfelden intrusive–extrusive dome also forms part of the group of large rhyolite domes in the Saar–Nahe Basin. It occurs in the SW of the basin (Fig. 12) and is located at the (apparent offset) transition of the Nahe Syncline to the Prims Syncline. It started as a laccolith and, as is seen at its NE margin, sediments of the Oberkirchen-Formation or possibly even of the basal Donnersberg-Formation (Konzan *et al.* 1981, 1989) were updomed by the laccolith. It breached the overlying sediments and extruded forming an extrusive dome, 7 × 5–6 km in size, and lava flow/s (Müller 1982) and rock falls or even block-and-ash flows. The lava flow extends up to 8 km to the SSW whereas the rhyolite breccia of the rock falls and block-and-ash flows (and their reworked facies) extends for about 8 km to the SW. Upon reaching the surface, the dome must have risen to some height, and, during prolonged extrusion, its slopes must have become repeatedly unstable and collapsed, forming these rock falls or even block-and-ash flows. See also Müller (1982).

Lava flows

Lava flows are a very characteristic feature of the Saar–Nahe Basin. They are, as pointed out above, very widespread but restricted to the Donnersberg-Formation (Figs 1 & 3). Their composition varies between tholeiitic basalts and rhyolites (von Seckendorff *et al.* 2004). In the region of Baumholder and Idar-Oberstein, a maximum thickness of about 800 m to possibly 1000 m is present (Bambauer 1960). Towards the NE and SE, the lava flows interfinger with sediments and tephra beds and several distinct stratigraphic levels (Figs 1 & 3) have been termed Lava Series 1 (= Donnersberg-Grenzlager), the Olsbrücken Lava Series (= Olsbrücken-Lager), the Lava Series 2 (= Winnweiler Lager) and the Lava Series 3 (Lorenz 1973; Haneke & Lorenz 2000). In the Prims Syncline (Fig. 1) two unnamed lava series exist (Minning & Lorenz 1983). The thick lava pile in the region of Baumholder and Idar-Oberstein might have represented some kind of shield volcano. Towards the NE the shield-volcano lavas clearly extend into lavas of the Lava Series 2. It is not clear, however, if some basal lava flows from the shield volcano originally extended across the area of the Palatinate Anticline and formed some of the lava flows of the Lava Series 1 and of the Olsbrücken Lava Series.

For most of the lava flows, the respective vents or feeder dykes are unknown. Scoria cones are also unknown. It has been pointed out above that the dykes in the SW of the Saar–Nahe Basin could have been feeder dykes for some flows. It seems conceivable also that in the Donnersberg region the basal basaltic andesite lava flows of the Lava Series 1 could have been fed by the same plug at the Platte which also fed the Marienthal Sill (Schwab 1967, 1968). Many of the basic to silicic lava flows can be traced for kilometres and, in the case of basalts and basaltic andesites, have a maximum thickness of up to 15 m to in excess of 40 metres (Schwab 1968, and see below) whereas intermediate to silicic flows (andesites to rhyolites) have a thickness of up to 200 m or even more. Thick tholeiitic basalt and the more frequent thick basaltic andesite lava flows display a massive and medium-grained core facies texture, and thus qualify for the grain-size-dependent term 'dolerite'. In the relatively thin lower crust of some lava flows, pipe vesicles have been recognized. The rather thick upper lava crust contains a number of vesicle zones parallel to the lava surface. Early Permian weathering has destroyed the delicate original surface textures, but the surface geometry was horizontal to locally slightly convex/concave. A pahoehoe surface has been recognized only once so far, i.e. in a cliff just east of the bridge across the Nahe River south of the town of Sobernheim. In the sediments directly overlying several lava flows, clasts eroded from the respective flows have been incorporated, also proving emplacement of the magma as lava flows. Rarely, some very crude columnar jointing can be seen. Typical columnar jointing, however, has not been recognized, with the exception of small short columns just beneath the upper surface, as, for example, at the lava flows just south of the Nahe river, opposite Sobernheim. Typical lower and upper aa lava crusts have never been recognized. Despite the fact that intermittently some sediments and tephra of the Donnersberg-Formation were deposited in lacustrine environments (Stollhofen 1991, 1998), subaqueous emplacement of lava flows has not yet been recognized. Thus, these tholeiitic basalt and basaltic andesite lava flows are assumed to have been subaerially emplaced pahoehoe lava flows.

On the more-or-less horizontal surface of the graben structure (with a number of small fault scarps at the surface: Haneke *et al.* 1979; Stollhofen 1991, 1993, 1994*a–d*, 1997, 1998, 2000; Stollhofen & Lorenz 1992, 1993*a, b*, 1994; Stollhofen & Henk 1994; Stollhofen & Stanistreet 1994; Stollhofen *et al.* 1999, but a lack of pronounced topographic gradients) the subaerial lavas of basaltic and basaltic andesite composition spread widely. On the horizontal basin floor, many of the flows, despite their low viscosity and large lateral extent, finally attained a thickness of up to 15 or even in excess of 40 m: for example, lava 1 in Lava Series 2 in research drill hole Jakobsweiler (Haneke & Lorenz 1990); several of the basaltic andesite flows of the Lava Series 1 NW of the Donnersberg; the Olsbrücken Lava Series near Olsbrücken; the Lava Series 2 on both sides of the Palatinate Anticline (Schwab 1965, 1967, 1981; Häfner 1977; Haneke & Lorenz 1990; Grill 1992); lava flow 2 in Lava Series 2 SW of the Donnersberg Dome; and lava 1 of Lava Series 2 at Altenbamberg (Lorenz 1973).

The intermediate to silicic lava flows are usually very thick, up to in excess of 100 m, probably because of the volumes extruded, their high viscosity and because they had to spread on an almost horizontal surface. Flow-banding and autoclastic breccias at the base and/or at the top (block lava) are a typical feature of some flows: e.g. of the thick Hochstein Lava near Hochstein, the thick basal lava of the Lava Series 2 in the Langenthal quarry near Monzingen, in several lavas of the Lava Series 3 in the Sobernheimer Wald, of the Gangelsberg near Duchroth, at the lava cliffs along the Nahe River between Bahnhof Boos and Oberhausen, and again along the Nahe River between Norheim and Niederhausen. In the latter area, ramp structures can be recognized locally (Nickel 1979, 1981). The rhyolite lava at the Leisberg, NW of the Lemberg Dome and probably fed by this dome (see below), shows a number of flow folds with almost horizontal fold axes. The rhyolite lava of Bad Kreuznach (in the area of Fürfeld and Wendelsheim) is in excess of 200 m thick, almost non-vesicular, flow-banded, and locally displays columnar jointing. Autoclastic breccias occur at the base, locally in the interior of the flow but especially at the flow front at the Steiger Berg (Lorenz 1973; Dörhöfer 1989). The silicic lavas do not show the distinct variation in vesicle distribution as described by Fink (1981); Manley & Fink (1987) and Manley (1995), and neither has a clastic texture related to annealing of reheated fragmental debris (Manley 1996) been recognized so far. As has been already pointed out above, the magma of the silicic lava flows reached the surface only via the formation of intrusive–extrusive domes (Kreuznach, Lemberg and Nohfelden domes).

Tephra deposits

As has been stated already above, many thin and very fine-grained ash tuffs are interbedded with mostly fine-grained lacustrine sediments of the Upper Carboniferous and lowermost Permian (Glan-Subgroup). They represent distal silicic ash deposits, and are assumed to have been derived from powerful and voluminous plinian and/or phreatoplinian eruptions outside the Saar–Nahe Basin, in all probability from eruptions in the Moldanubian zone of the Variscan belt (Königer 1999, 2000; Königer & Stollhofen 2001; Königer & Lorenz 2002, 2003; Königer *et al.* 2002). These distal ash beds were mostly laid down as subaqueous ash fall deposits, and many became reworked, in part even forming turbidite beds (Königer 1999, 2000).

In the Donnersberg-Formation, in contrast, there exist six relatively thick widespread tuff horizons, which represent excellent marker beds (Fig. 3): T1–T6 (Heim 1960, 1961; Lorenz 1971*c*, 1974; Haneke *et al.* 1979; Stollhofen 1991, 1994*a*; Haneke & Lorenz 2000). They have been first realized and traced along the major part of the southeastern flank of the Palatinate Anticline. T1, T2 and T3 end abruptly at faults, which had been active synsedimentarily (Haneke *et al.* 1979; Stollhofen 1991, 1994*a–c*, 1997, 1998). With interruptions, T1, T3 and T4 can be traced around the NE-plunging Palatinate Anticline into the Nahe Syncline (Lorenz 1973; Dörhöfer 1989). In particular, T4 can be traced southwestwards along both flanks of the Nahe Syncline and, via the coeval ignimbrite at Kirn, can be correlated with the ignimbrite level in the Prims Syncline (see below).

The tuff beds vary in their thickness, but also individually along strike. In specific areas, close to their assumed source, the tuff horizons T1 to T4 reach several tens of metres in thickness. The tuffs are mostly fine-grained ash tuffs, but coarse ash tuffs and lapilli ash tuffs also occur. Some of the largest lapilli are 3–4 cm in size. Some tuffs are finely bedded, in part laminated, whereas others are massive for several centimetres or even decimetres in thickness (Stollhofen 1991, 1992, 1994a, d). The lapilli are in part non-vesicular and angular to subrounded; in part they are finely vesicular, with the vesicles being pipe vesicles. The pumice lapilli experienced early post-sedimentary diagenetic compaction. In several of the tuff horizons (T2–T4) accretionary ash grains and accretionary lapilli occur (Lorenz 1974; Stollhofen 1991). Stollhofen (1991) also described thin ash flow deposits (vesiculated ash tuffs) from T1. From the Hirschberg diatreme, Lorenz (1971b) reported a subsided block of welded ignimbrite related to T1. Careful analysis of the field relationships has demonstrated that the tuffs T1, T2 and T3 were laid down in part in shallow-lacustrine environments, and in part in fluviatile environments with partial reworking (Stollhofen 1991, 1992, 1994a). T1 was derived from at least six petrographically–geochemically different magma sources (Stollhofen 1991). The majority of the six tuffs, i.e. T1–T4, consists of silicic tuffs and are devoid of xenocrysts derived from the underlying sediment material. Phenocrysts of bipyramidal quartz and of biotite occur in specific tuff beds. Biotite phenocrysts have been used for age dating (Lippolt & Raczek 1979, 1981; Lippolt & Hess 1983, 1989; Lippolt et al. 1989).

Advanced erosion veils the relationship between most of the intermediate to basic maar–diatreme volcanoes and their tephra blankets surrounding the original tephra-rings. Only part of the tephra blankets from the Rödern, Hirschberg and Mörsfeld diatremes can still be found between lava flows of the Lava Series 1 (Lorenz 1971a, b), whereas the ash and lapilli tuffs, typical of Saar–Nahe diatremes, found in the area of Falkenstein and Schweisweiler, cannot as yet be related with any certainty to specific diatremes.

Because of the dominantly fine grain size, the accretionary ash grains and accretionary lapilli, the non-vesicular juvenile ash grains and lapilli (and the lacustrine and fluviatile depositional environment) the six tuff marker horizons in the Donnersberg-Formation are considered to have been largely phreatomagmatic in origin (Lorenz 1974; Stollhofen 1991, 1992, 1994a).

The ignimbrite of the Prims Syncline and Nahe Syncline

In 1976 a sequence of non-welded pinkish ignimbrites, up to 110 m thick in total, was identified by Minning & Lorenz (1983) in the Prims Syncline, west of the thick lava pile of Baumholder and Idar–Oberstein (Fig. 14). The juvenile lapilli in the ignimbrites represent finely fibrous (pipe vesicles) pumice clasts, which became flattened not by welding but diagenetically. It was realized that the ignimbrite sheets do not contain any sedimentary or volcanic rock clasts or mineral grains from the Rotliegend sedimentary and volcanic sequence underlying the ignimbrites. In contrast, the ignimbrites contain up to 10–20% xenoliths of reddish or more frequently greenish slates (with two schistosities) and quartzites typical of the deformed and slightly metamorphosed Devonian rocks in the Hunsrück region of the Variscan Rhenohercynian zone to the NW. The rhyolite plug of Veldenz, 30 km north of the Prims ignimbrite, is assumed to have been the feeder of the ignimbrite series (Minning & Lorenz 1983) (Fig. 14). The plug is about 1 km in diameter and consists of a rhyolite intrusion (Negendank 1983) and locally, at its southern margin, of a welded ignimbrite facies – the latter containing the same xenoliths as the Prims ignimbrite. In contrast, Müller (1982), who also described these pyroclastic deposits, assumed the Söterberg volcano inside the Rotliegend Prims Syncline to represent the feeder of the tephra, despite the fact that the tephra of the ignimbrite only contains Devonian rock clasts and lacks clasts from Rotliegend sediments and volcanics. An almost identical ignimbrite in the Lower Permian Wittlich Basin, just 10 km north of the Veldenz plug was assumed to have also been derived from the same vent (Minning & Lorenz 1983). However, Bonn & Stets (2000) have cast doubt on this theory.

In 1996, an ignimbrite with the same facies characteristics as in the Prims Syncline was identified near Kirn in the Nahe Syncline, 30 km to the NE of the Prims Syncline, and thus on the NE side of the thick lava pile of Baumholder and Idar–Oberstein (Fig. 14). Thus, the total volume of magma transported into the Prims and Nahe Synclines could have been of the order of 20 km^3. If the ignimbrite of the Wittlich Basin is part of the eruptive system, then the volume of magma erupted must have been even larger. In the Nahe Syncline it was realized that the ignimbrite occurs at the level of, and, therefore, is coeval with T4.

The Nahe Caldera

Krupp (1984) interpreted the geology of the Kreuznach–Lemberg–Waldböckelheim area (Figs 1 & 12) to have been the result of the formation of a large caldera, 13 km in diameter, and

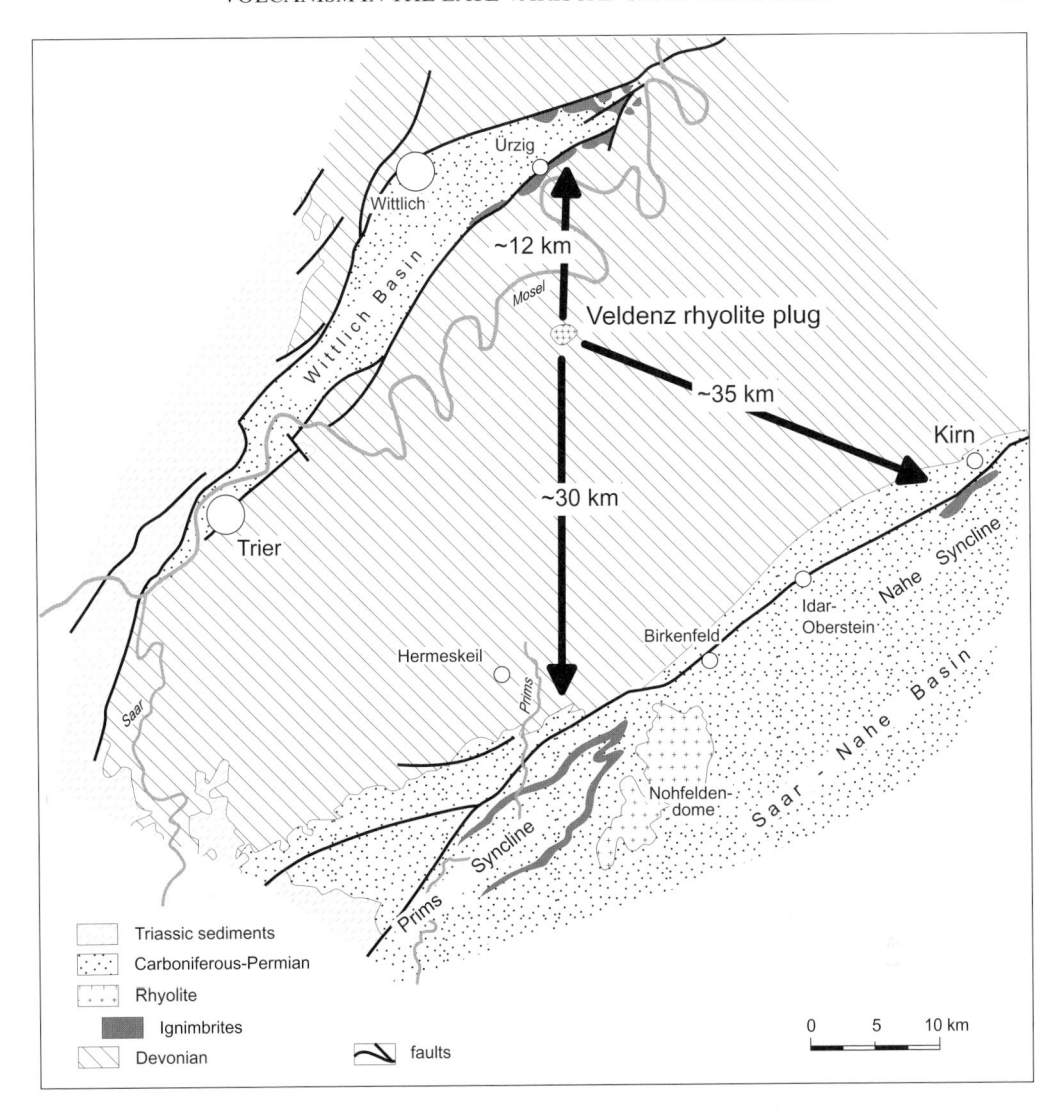

Fig. 14. Schematic map showing the occurrences of ignimbrite in the Prims and Nahe synclines, its feeder vent (Veldenz rhyolite plug) and the occurrence of ignimbrite in the NE part of Rotliegend Wittlich Basin. SHBF represents the Southern Hunsrück Boundary Fault.

called it the Nahe caldera. This caldera is outlined with a distinct caldera fault on the geological map of the Saar–Nahe Basin 1:100 000 by Dreyer et al. (1983) and in the publication of Krupp (1984). Krupp assumed the Kreuznach Dome to have formed on the eastern caldera fault, and the updomed area in the region of Waldböckelheim (the dome structure above the Waldböckelheim laccolith) to be the result of resurgence. According to Krupp, the rhyolite breccia derived from the Kreuznach rhyolite dome occurs only in the depression of the caldera. However, the sequence of the sedimentary, pyroclastic and volcanic units below, at, and above the critical level of the assumed caldera collapse is identical inside and outside the outlined caldera (Bock 1986). Variations in thickness of particular horizons can easily be explained by synsedimentary tectonics at intrabasinal transfer faults, faults that are rather common in the Saar–Nahe Basin (Stollhofen & Stanistreet 1994; Stollhofen 1997, 1998). The crucial rhyolite breccia derived from the Kreuznach rhyolite dome can be shown to occur not only in the area of the assumed caldera floor but even west of the western caldera fault in the

Sobernheim Wald, i.e. outside the caldera too (Bock 1986). Thus, in contrast to Krupp's (1984) assumption, the caldera fault cannot have existed at this time in this particular area. An important aspect not in favour of the assumed caldera model is that, in a graben like the Saar–Nahe Basin which was filled with unconsolidated water-saturated sediments in the uppermost 2000 to 2500 m, caldera collapse would result in a downsag caldera (Walker 1984; Stachel et al. 1994b; Branney 1995; Lorenz et al. 1997, 2000). In such a water-saturated environment, a sag caldera would have induced tremendous sediment flows from the caldera walls. The volcanic rocks interbedded with the sediments would have brecciated and formed blocks floating in the sediment flows. No such rocks exist in the assumed caldera floor area. Thus Krupp's model of a Nahe caldera does not seem to be supported by evidence, and consequently is disputed by us.

Discussion, conclusions and outlook

Near the end of the Variscan orogeny, late-orogenic crustal extension led to the formation of many intermontane basins. One of these late Variscan intermontane basins is the Saar–Nahe Basin. In its c. 20-Ma long non-volcanic syn-rift phase, it was accumulating approximately 5000–5500 m of lacustrine and fluviatile sediments of the Westphalian and the Stephanian Groups and overlying Glan-Subgroup. Thus, the basin was an actively subsiding depocentre. Synsedimentary tectonic activity occurred along the Southern Hunsrück Boundary Fault, which represents the major detachment fault of the basin, and along many transfer and associated faults, and thus influenced sedimentation patterns. During the following volcanic syn-rift phase, i.e. during the deposition of the Donnersberg-Formation, intensive volcanicity affected the basin. With a compositional range stretching from tholeiitic basalts to rhyolites and alkali-feldspar trachytes, magmas rose from their source levels (von Seckendorff et al. 2004) into the Saar–Nahe Basin and became involved in intensive volcanic and subvolcanic activities. The volcanic activity led to emplacement of lava flows, extrusive domes, widespread tephra beds, and the formation of tuff-rings and maar–diatreme volcanoes. The subvolcanic activity, in contrast, led to emplacement of a few dykes, many sills, laccoliths and diatremes. Thus, in both environments there was non-explosive and explosive activity. During these volcanic and subvolcanic activities, fluviatile, alluvial and lacustrine sedimentation continued on the basin floor. Therefore, it can be safely concluded that, in syn-volcanic times (Donnersberg-Formation), the sediments underlying the respective sedimentation surfaces were unconsolidated and water-saturated, i.e. the sediments of the Donnersberg-Formation must have been unconsolidated and water-saturated. It is also obvious that in a basin with such a high sedimentation rate, a certain amount – but of unknown thickness – of the underlying sediments (the Glan-Subgroup and Stephanian Group) of the sedimentary fill of the Saar–Nahe Basin (Fig. 2) must also have been in an unconsolidated state and thus water-saturated.

Maar–diatreme volcanoes

With respect to the above-discussed unconsolidated, water-saturated sediments of the sedimentary fill of the Saar–Nahe Basin, the pyroclastic rocks of the many maar–diatreme volcanoes give supporting evidence: at the time of formation of the maar–diatremes in the Saar–Nahe Basin, the diatremes not only ejected juvenile ash grains and lapilli, xenoliths of fragmented volcanics, and some clasts of pelitic sediments, but they also ejected large quantities of individual mineral grains (and pebbles) constituting the original medium- to coarse-grained sediment volume now occupied by the deep, cone-shaped diatremes. No sedimentary rock clasts formed by fragmentation of indurated sediments were ejected. Thus, the medium- to coarse-grained sediments of the basin fill involved in the eruptions cannot have been in a diagenetically indurated state but, in contrast, they must have been in an unconsolidated state and consequently water-saturated. Diatreme root zones penetrate downward, and are the sites from which the country-rock material constituting the allothigenic fraction of the tephra is ejected. The sedimentary country rocks of the basin-fill in the Saar–Nahe Basin down to diatreme root-zone level must, therefore, have been in an unconsolidated state and consequently water-saturated. Thus, in the Saar–Nahe Basin, the uppermost 1500- to possibly 2500-m thick sediments of the sedimentary basin fill with respect to the formation of the maar–diatreme volcanoes must have represented a typical soft-rock environment (Lorenz 1985, 1986, 1998, 2003b). Into the same 1500 to 2500-m thick soft-sediment package the other subvolcanic bodies of the Saar–Nahe Basin (sills and laccoliths, see below) became emplaced, possibly because further rise of the respective magma batches was hampered by the water-saturated unconsolidated state of the sediments, i.e. by their low shear strength, by

Dykes

There is a paucity of dykes in the Saar–Nahe Basin, and, in the exposed rock sequences, they mostly occur only in the SW of the basin. At greater depths a larger number of dykes must exist in order to have fed the many subvolcanic bodies (diatremes, sills, laccoliths) and volcanic bodies (intrusive–extrusive domes, lava flows). A number of dykes in the SW are up to 10 m thick, which is assumed to have been the result of inflation during ongoing syn-intrusion magma supply (Wright & Fiske 1971; Delaney & Pollard 1981; Wadge 1981).

The dykes of the Saar–Nahe Basin have not been studied with regard to their emplacement processes, and neither have the neighbouring sediments been studied with respect to their physical state at the time of dyke emplacement. Thus the question of why the respective magmas were able to rise in dykes through the assumed soft-sediment environment and to reach the surface, remains unanswered for the time being.

Sills

Basalt to basaltic andesite sills are very common in the Saar–Nahe Basin, and they may extend laterally for up to 12 km. In a number of cases they occur in groups stacked vertically above each other. Their feeder dykes or plugs are usually not exposed. Some sill–dyke complexes occur in the NE of the basin. A number of sills are thicker in the central part of their lateral extent. Thus, they represent flat laccoliths (Johnson & Pollard 1973; Pollard & Johnson 1973; Corry 1988). Some sills are in excess of 100 to 200 m thick. The remarkable thickness of these sills is assumed to be the result of inflation of the sills during prolonged magma supply.

The depth range of the intrusion level of the sills in the Saar–Nahe Basin is from several thousand metres to close to the original surface, and thus, it is in about the same depth range as the vertical downward penetration depth of the maar–diatremes. Without any doubt, the model of the soft-sediment environment advocated for the formation of the diatremes can also be applied with respect to the emplacement of the sills, and in all probability explains the occurrence of the many sills in the Saar–Nahe Basin. The basalt to basaltic andesite magmas which became emplaced in the sills must certainly have been rising in dykes through the basement and the deeper sediment levels of the basin fill. At various levels within the depth range of the uppermost 1500 to 2500-m thick unconsolidated, water-saturated sediments the magma rising within the dykes must have been hampered by the physical state of the sediments and prevented from continuing its rise towards the surface. Despite continuing supply, the magma intruded sideways into the neighbouring sediment pile. It is conceivable that, in contrast to a hard-rock environment with many joints in the hard rocks, no open fissures could form in this water-bearing soft-sediment environment. The sill magmas had to underflow the sediments because the latter had a lower density and a relatively low shear strength, thus the sediments had to deform in a mostly ductile fashion.

The depth range of the sills in the Saar–Nahe Basin is in contrast to the depth range for emplacement of sills only at depths of $c.$ 900 m to $c.$ 2300 m below the original surface, as suggested by Mudge (1968). However, Pollard & Johnson (1973) have already cast doubt on this model. The shallow depth boundary ($c.$ 900 m) of Mudge's model certainly does not exist in the Saar–Nahe Basin, as the shallowest sills occupy a depth below the original surface of less than 100 m to even just a few metres only. According to Mudge (1968) many, but not all, sills intrude beneath shales, which in his opinion represents a fluid barrier. In the Saar–Nahe Basin, however, many sills intruded within bituminous laminated pelites, i.e. they are both overlain and underlain by these sediments.

In most models of emplacement of basalt to basaltic andesite sills intruded into sediments, the diagenetic state of these sediments at the time of sill intrusion has not been taken into consideration. Pollard (1973) distinguished three propagation mechanisms of sills:

1 in brittle host rocks at low confining pressures and temperatures;
2 in brittle rocks at somewhat higher confining pressures, and
3 for the third propagation mechanism he stated: 'at relatively high confining pressures and temperatures ductile faulting occurs near the termination, by rupturing the host rock.'

In his study, the case of the largely ductile deformation of unconsolidated water-saturated sediments had not been considered. Future field and physical modelling investigations of sills emplaced into sediments should study the important aspects of the diagenetic state of the sediments that they were in at the time of sill intrusion and the effects that an appreciable sediment water content had on sill emplacement.

Laccoliths and intrusive–extrusive domes

In summary, the silicic laccoliths and intrusive–extrusive domes of the Saar–Nahe Basin show the following characteristic features relevant for understanding their emplacement (Figs 12 & 13):

1 In the Saar–Nahe Basin neither rhyolite nor alkali-feldspar trachyte magma managed to rise via dykes through the several 1000 m thick Carboniferous–Permian sedimentary basin fill straight to the surface. In contrast, several 100 to several 1000 m below the basin floor, all these magmas were intruded within the sediments of the basin fill. Initially, following the standard model (Johnson & Pollard 1973; Pollard & Johnson 1973; Hyndman & Alt 1987; Jackson & Pollard 1988; Corry 1988; Henry et al. 1997), they were emplaced as sills (Corry 1988: protolaccoliths) and then inflated into laccoliths. More than 12 laccoliths formed initially (but at different times). Some deeply located laccoliths with limited magma supply did not inflate any further and solidified at depth. Deep laccoliths which have updomed their roof sediments but which are not exposed (Selberg, Herrmannsberg, Potzberg) do not have a flat roof, as described and discussed by Pollard (1972, 1973) and Corry (1988), because the overlying sediments clearly form a cupola structure.

The initial sill-like intrusion level varied between the diverse laccoliths. The deepest laccoliths intruded Stephanian sediments (Lemberg, Königsberg, Herrmannsberg, Potzberg, Selberg; Dreyer et al. 1983). Because of a lack of nearby drill holes and data on the exact thickness of the eroded updomed roof rocks, it is not known exactly at what depth below the basin floor these laccoliths became emplaced. The depth of some initial sill intrusion, however, could have been 2000 to 2500 m. For the Donnersberg and Kreuznach rhyolite domes, the initial intrusion depth below the original surface is better known, and amounts to about 1500–2000 m. The western Kuhkopf laccolith intruded into basal sediments of the Donnersberg-Formation (basal Nahe-Subgroup) and thus it intruded at a depth of only c. 450 m below the original surface. Thus the emplacement depth varies by about 2000 m (Fig. 2). The western Kuhkopf laccolith also shows that this particular laccolith was horizontally floored (Henry et al. 1997), was rather concordant with the underlying sediments, and was tilted only later. Probably all laccoliths and also intrusive–extrusive domes, at their initial intrusion level, are less than 10 km in diameter, as seems typical for most laccoliths (Corry 1988; Henry & Price 1989; Henry et al. 1997). It should be noted that the assumed maximum diameter of the floor of the silicic laccoliths is in the same range as the maximum diameter (12 km along strike) of the basalt to basaltic andesite sills. During or after lateral intrusion the largest sills started to inflate and formed flat laccoliths, as did the initial silicic sills (the protolaccoliths of Corry 1988).

2 Within some laccolithic complexes the sub-units intruded almost at the same depth (e.g. Donnersberg), whereas at others the sub-units intruded at different depths, as is very distinct at the Kuhkopf laccolithic and intrusive–extrusive complex (Haneke & Lorenz 2000). Thus, the initial sills were emplaced next to each other at slightly different levels, and not one above each other. Except for the stacked basalt and basaltic andesite sills there is no evidence for the existence of Christmas-tree type laccoliths in the Saar–Nahe Basin (one end-member of the laccolith classification of Corry (1988), who studied many laccoliths in the Western USA, including the laccoliths in the Henry Mountains in Utah). At the Donnersberg, the intrusive–extrusive dome may be underlain by a deeper plutonic body (Haneke 1987) with its surface, according to seismic investigations, at a depth of about 2700–3000 m below the surface. Alternatively, this deep intrusion might represent a slightly earlier laccolithic intrusion. Consequently, the deep intrusion and the Donnersberg dome might jointly represent a Christmas-tree type laccolith/dome complex.

3 Intrusion and inflation of the silicic magmas led to differential updoming of roof sediments, as well as of pre-existing basic to intermediate sills, lava flows, and tephra beds. There is no clear evidence of pronounced punching of the overlying sediments by the laccoliths as seems to be typical for many punched laccoliths: the other endmember of the laccolith classification of Corry (1988). In some cases, however, probably pre-existing synsedimentary faults have been used during inflation of magma and updoming of roof sediments (Potzberg, Herrmannsberg, Königsberg, Kuhkopf, Bauwald, Lemberg).

4 In several cases, updoming and thus 'stretching' of overlying sediments allowed magma from an inflating laccolith to intrude even farther upwards in thick dykes (Herrmannsberg, Selberg, Bauwald). Such dykes could

have been responsible for tephra eruptions prior to the main part of the intrusive dome becoming extrusive. On such dykes, or possibly rooting locally directly on or even in a laccolith, maar–diatreme volcanoes formed by phreatomagmatic explosions (Waldböckelheim cupola: Fig. 12).

5 During further inflation of large laccoliths, the intrusions not only increased in volume upwards but, with decreasing distance to the surface, also radially sideways. This radial growth is due to the decreasing yield strength of the water-saturated sediments the closer the unconsolidated water-saturated sediments are to the original surface (Donnersberg, Kreuznach). This allowed the laccoliths to thrust uplifted sediments from deeper stratigraphic levels on to sediments and volcanics from higher stratigraphic levels, with intermediate ones left behind (Donnersberg, Kreuznach domes: Fig. 13). At the southern margin of the Kreuznach Rhyolite the shales of the Quirnbach-Formation, originally overlying the laccolith, in part became brecciated and many slickensides formed. Consequently, the deeper levels which participated in the updoming and final overthrusting (Lorenz 1973) had already been indurated to some degree.

6 Due to the ongoing inflation and upward and radial growth, several laccoliths breached the surface and extruded. They developed into huge intrusive–extrusive domes of a volume of up to several 10 km^3 (Donnersberg, Kreuznach and Nohfelden domes). As the near-surface sediments were unconsolidated and water-saturated, prolonged updoming led to extensive erosion (slumping, flowing and washing away) of these sediments, with the result that the roof of the inflating dome was increasingly eliminated. The siliciclastic minerals and diverse pebbles of the roof sediments in part were mixed with the ordinary regional sedimentary influx into the Saar–Nahe Basin, in part they were mixed with the rhyolite clasts deposited by rock falls and block-and-ash flows (see below). This explains why breccias consisting of clasts of indurated siltstones, sandstones and conglomeratic arkoses are absent in the vicinity of the extrusive domes. Thus, the effective initial overburden decreased with time by this particular process of unroofing, and, upon further inflation, the intrusive dome became extrusive and in part flowed across the surrounding upthrusted sediments on to the basin floor (Kreuznach, Nohfelden, ?Lemberg). In this respect it is worth noting that Corry (1988, p. 68) stated: 'Laccoliths intruded into weak layers of water-saturated clays or shales may shed their roofs by gravitationally driven slumping as they grow upward. Such a process would allow them to become indefinitely thick or even extrude onto the surface.' Thus, without describing a particularly studied laccolith, Corry (1988) had already deduced the relationship between an inflating laccolith and unconsolidated sediments, as was typical for the intrusive–extrusive domes in the Saar–Nahe Basin.

According to Schmincke (2000, pp. 120–121) the Showa Shinzan intrusive–extrusive Dome of Hokkaido, Japan uplifted lakebeds in 1944 by about 200 m, breached them and grew until 10 September 1945 by 100 m in height (see also Corry 1988). The rhyolitic dome-top tuff and pumice cone successions in the Devonian Bunga beds in SE Australia (Cas *et al.* 1990) and possibly part of the Tuluman volcano in St Andrews Strait, Bismarck Sea, Papua New Guinea (Reynolds & Best 1976) represent to some extent the subaqueous analogues of the Saar–Nahe intrusive–extrusive domes. According to Cas *et al.* (1990) the domes in the Bunga beds also intruded unconsolidated sediments and breached the overlying sediments.

7 When in the Saar–Nahe Basin such an extrusive dome inflated further (as an endogenous dome) and reached a certain height above ground, slope instability should have led to flank collapse as has been frequently the case with recent extrusive domes (Schmincke 2000). Consequently, rock falls and probably even block-and-ash flows came down to low altitudes, and associated fine ash-fall deposits formed (Donnersberg, Kreuznach, Nohfelden). The rhyolite clasts of the breccias completely lack vesicles (Donnersberg). Only a few rhyolite clasts from the Kreuznach rhyolite breccia display some vesicles. This indicates that the carapace of the domes was either not pumiceous or vesicular at all, or it was only slightly vesicular. Since 1956 extrusive domes grew, for example, at Bezymianny, Kamchatka; Mount St Helens, Washington, USA; the Soufrière Hills of Montserrat; Merapi, Java, Indonesia and Unzen on Honshu, Japan, and rock falls and block-and-ash falls were frequent during growth of the respective domes (Swanson & Holcomb 1990; Sato *et al.* 1992; Schmincke 2000). Over the two years from 1902 onwards, the Santiaguito Dome in Nicaragua

grew to a size of 1.2 km in diameter and 500 m in height. And from 1904 to 1967 it doubled its height (Schmincke 2000).

8 When silicic magma had reached the surface in the Saar–Nahe Basin, phreatomagmatic eruptions in all probability gave rise repeatedly to thick but finely bedded tephra deposits. The tephras that are thickest close to the rhyolite domes (Haneke et al. 1979) almost exclusively consist of juvenile clasts. Siliciclastic sedimentary material was rarely part of the ejected material (see below). Thus, the vents ejecting the silicic tephra could not have cut through the unconsolidated sediments, which had been updomed above the intrusive domes, and, therefore, the tephra cannot have been ejected prior to the domes reaching the surface. Consequently, the vents ejecting the silicic tephra in all probability were located on the extrusive domes and, intermittently at least, might have been represented by tuff-rings or tuff-cones (see below and also Wohletz & Heiken 1990).

9 In some cases, marginal extrusion was possible from the respective dome of thick volatile-poor rhyolite lava (Kreuznach, Nohfelden, possibly Lemberg). This lateral extrusion was possibly facilitated by the height of the respective extrusive dome and the heat and large volume of lava involved.

10 At least at two extrusive domes (Donnersberg and Kuhkopf) final inflation led to the formation of fractures in the thick cooling and thus brittle carapace, and magma from the interior of the dome was squeezed upward into these fractures and thus formed rather thin (cm–dm thick) late dykes. At the Bauwald, Selberg and Herrmannsberg laccoliths, magma was even squeezed upward into the updomed roof sediments and formed dykes tens of metres thick.

11 A number of laccoliths are composite laccoliths (Kuhkopf W, Lemberg, Selberg, Herrmannsberg, Himmelberg). At the Lemberg, Selberg and Herrmannsberg domes it may even have been possible that, due to further inflation, the interior magma unit of the composite dome breached the exterior magma unit and became effusive when reaching the surface.

12 In several intrusive and intrusive–extrusive domes, basic magma mingled with the silicic host magma at depth and then this mingled magma batch rose upwards towards the surface (Kreuznach, Lemberg, Bauwald). The composite and mingled dome magmas in all probability indicate that the rhyolite and alkali-feldspar trachyte host magmas which must have been crystallizing phenocrysts in a magma reservoir at depth, were pushed upward by underlying basaltic andesite magma and ultimately possibly by magma rising from the upper mantle into the lower crust of the Variscan late-orogenic belt (Lorenz 1987, 1992a, b; von Seckendorff et al. 2004).

13 At the Donnersberg intrusive–extrusive dome, septa of roof sediments are present between several dome units, i.e. they are hanging down from the former roof cover of the laccolith. They may have evolved by ongoing inflation of laccolith units from former synclines between individual intrusive laccolith units, as exist between the Königsberg, Selberg, Herrmannsberg and Potzberg cupolas and laccoliths.

14 The growing huge extrusive domes of the Saar–Nahe Basin gave rise to clastic deposits on their lower slopes and surrounding low-lying ground. Because of the probably rather cool, thick and very blocky carapace of the growing domes, flank instability and consequent slope failure gave rise to rock falls and possibly even to block-and-ash flows. No evidence for hot block-and-ash flows has yet been recognized in the deposits yet. In contrast, they might be considered to have been relatively cool block-and-ash flows (Sparks pers. comm. 2000). Farther down, on lower ground, debris flows and then fluvial processes took over transporting the clastic material into the surrounding basin-floor environment.

15 It is very intriguing that the laccoliths and intrusive–extrusive domes are characterized by a lack or at least by an obvious paucity of vesicles. The rhyolite of the Donnersberg intrusive–extrusive dome and its rhyolite breccia are completely devoid of vesicles. The Kuhkopf and Bauwald domes only locally display some vesicles, whereas the intrusive–extrusive Kreuznach dome in a few areas is vesicular to some extent (but not very vesicular). The largest vesicles in the Traisen Quarry of the Kreuznach rhyolite are 5 cm in diameter (see above). The Kreuznach rhyolite lava flow also shows some vesicles, but only locally. In the Kreuznach rhyolite, many quartz phenocrysts display embayments and associated fractures (Arikas 1986). This feature, according to Best & Christiansen (1997) and Mock et al. (2002b), suggests that vesiculation occurred at depth and the volatiles were released into the surrounding country rocks, at a depth

greater than the emplacement depth of the laccoliths. This aspect has not yet been studied in the laccoliths and intrusive–extrusive domes of the Saar–Nahe Basin. However, as Eichelberger *et al.* (1986) stated already, the hydrous phenocryst phases of biotite and hornblende suggest elevated volatile contents at depth, which may have been released during rise of the respective silicic magmas into the Saar–Nahe Basin fill.

16 Future field and physical modelling investigations of laccoliths and intrusive–extrusive domes emplaced into sediments should study the important aspects of the diagenetic state of the sediment package and its water content at the time that the laccoliths intruded and, if it happened, inflated into intrusive–extrusive domes.

Lava flows

In this investigation of the volcanics, subvolcanics and pyroclastic deposits of the Saar–Nahe Basin, it is assumed that the emplacement of maar–diatreme volcanoes, sills, laccoliths and intrusive–extrusive domes, as well as formation of most tephra layers in the Donnersberg Formation, is related to the unconsolidated water-saturated uppermost 1500–2500 m of the sedimentary basin fill. Nevertheless, the large volume of basic to intermediate magmas, which managed to reach the surface and formed lava flows, especially in the SW of the Saar–Nahe Basin, seems to contradict the above model. At least it has to be concluded that the unconsolidated sediments cannot have hampered the rise of the respective magmas towards the surface everywhere. Consequently, groundwater cannot have been distributed evenly through the sedimentary basin fill of the Saar–Nahe Basin – a fact that has to be expected anyway – or other unknown aspects were involved that allowed the rise of these magmas to the surface. It is probably not a coincidence that the most prominent dykes in the Saar–Nahe Basin also occur in the SW of the Saar–Nahe Basin, close to the thick lava-flow pile of Baumholder–Idar-Oberstein.

The basalt and basaltic andesite lava flows in the Saar–Nahe Basin are, in part, very thick, i.e. up to 20 and even up to 40 m. In comparison with lava flows from Hawaii and basalts and basaltic andesites from flood-basalt provinces (Large Igneous Provinces), the thick lava flows in the Saar–Nahe Basin, with their pronounced thickness, medium grain size in the core facies and other internal facies textures, as described above, require a rather extended lava production at the source. Analyses of these features suggest that originally thin pahoehoe flows were inflating below their slowly thickening and solidifying upper crust, while advancing forward and laterally. Consequently, in terms of physical emplacement, the large volumes and internal and surface textures of these tholeiitic basalt and basaltic andesite lava flows of the Saar–Nahe Basin allow them to qualify for the term 'flood basalts' (Hon *et al.* 1994; Self *et al.* 1996, 1997).

Tephra deposits

With respect to the mode of formation of the voluminous tephra beds in the Donnersberg-Formation of the Saar–Nahe Basin, it has been suggested already that the eruptions had been largely phreatomagmatic (Lorenz 1974; Lorenz & Haneke 1986; Stollhofen 1991, 1994*a*). Arguments cited are the dominantly very fine grain-sizes of the ash involved (Zimanowski *et al.* 2003); the presence of accretionary ash grains and accretionary lapilli, many angular non-vesicular juvenile ash grains and lapilli, vesiculated tuffs, finely vesicular fibrous pumice clasts, and, finally, the frequent deposition in lacustrine environments, suggesting widespread groundwater and some surface water in the volcanic source areas. The angular juvenile clasts are assumed to be phreatomagmatic in origin. On the other hand, the solidified dome carapace may have been marginally involved in phreatic or phreatomagmatic fragmentation (Heiken & Wohletz 1987). The vesicular pumice clasts seem to contradict the lack or paucity of vesicles in the laccoliths, the intrusive–extrusive domes and their lava flows. These pumice clasts, nevertheless, might indicate magmatic eruptions. As an alternative model, the finely vesiculated state of the pumice clasts might indicate a phreatomagmatic component in a magmatic pumice eruption: a phreatomagmatic explosion emits shock-waves (Zimanowski 1998). Shock-waves emitted by a phreatomagmatic explosion generated by melt–water interaction in a relatively small amount of a larger silicic magma body might trigger sudden vesiculation in the surrounding melt body in the case where this melt envelope had been volatile-oversaturated. Then, in turn, the vesiculating magma will have to rise towards the surface and eject pumice in this phreatomagmatically triggered magmatic eruption. The steam generated by the phreatomagmatic eruption, however, might chill the pumiceous clasts and freeze their finely vesiculated state (Lorenz *et al.* 1993).

The vents of these silicic ash and lapilli ash tuffs are not known as yet. However, since the tephra hardly contains any of the underlying sediments (with the exception of the tephra horizons T5 and T6) it is quite conceivable that the major volumes of the silicic tephra were erupted from the extrusive domes themselves, as already discussed above. Studies of zircons might establish cogenetic emplacement of domes and tephra beds in the Saar–Nahe Basin. Groundwater and rainwater may have penetrated along joints and fissures of the carapace of the growing extrusive domes and resulted in thermohydraulic explosions (see also Wohletz & Heiken 1990). The vents may have evolved intermittently into tuff-rings and/or tuff-cones on top or at the margins of the intrusive–extrusive domes. The feeders of some tephra could have also been thick dykes from intrusive domes like, for example, the thick dykes on the Herrmannsberg, Selberg or Bauwald laccoliths – in case they reached the surface. In contrast, the rhyolitic diatremes at the northern margin of the Waldböckelheim cupola, i.e. the updomed sediments on top of the Waldböckelheim laccolith (Geib 1956, 1972, 1975; Habicht 1966), are definitely phreatomagmatic in origin, as they penetrated the updomed sediments of the cupola.

Ignimbrite

With respect to the ignimbrite sequence in the Prims and Nahe synclines (Fig. 14) it is important to realize that the vent of the ignimbrite eruptions is located outside the thick water-saturated sediment fill of the Saar–Nahe Basin, i.e. in the neighbouring horst of the Hunsrück. Inside the Saar–Nahe Basin the rhyolite and alkalifeldspar trachyte magmas only reached the surface of the basin after formation of intrusive–extrusive domes (or dykes and diatremes rising from the intrusive domes). In all probability the unconsolidated water-saturated, low-density sediments (of increasingly reduced yield strength towards the surface) prevented the silicic magmas in the Saar–Nahe Basin from reaching the surface of the basin floor directly, erupting and forming ignimbrite sheets – in contrast to many ignimbrite eruptions in other Carboniferous–Permian intermontane troughs of the Variscan orogenic belt (Lorenz & Nicholls 1976).

Heat flow, groundwater and diagenesis

An important aspect with respect to the late Variscan intermontane Saar–Nahe Basin is the elevated heat flow, which affected the sedimentary, volcanic and pyroclastic deposits as well as the subvolcanic bodies (Lorenz & Nicholls 1976; Buntebarth 1983; Teichmüller et al. 1983; Henk 1990, 1993a–c; Lorenz 1992a). The elevated heat flow is indicated by the alteration of the primary deposits and rocks, e.g. by the high degree of coalification of the organic matter, the secondary minerals in the volcanics and subvolcanics (including albite, chlorite, calcite, epidote, prehnite, epidote, pumpellyite, hematite, kaolinite and zeolites), and a secondary stable component of magnetization (Berthold et al. 1975; Lorenz & Nicholls 1976; Teichmüller et al. 1983). The rather elevated heat flow was established in the late-orogenic intermontane basin because of (1) crustal extension, thus thinning of the crust, (2) emplacement of large amounts of upper-mantle magmas at or above the base of the crust, thus formation of large magma reservoirs in the lower and possibly also in the middle crust, (3) rise of large amounts of magma from these reservoirs into the basin fill (von Seckendorff et al. 2004), and (4) finally because of the groundwater-rich sediment fill of the basin which could easily 'conduct' heat. The elevated heat flow and the groundwater-saturated sediment fill would have resulted in alteration processes in the range of 'high-temperature' diagenesis, i.e. very low-grade metamorphism, or anchimetamorphism (Teichmüller et al. 1983).

Despite the potential for the diagenesis of the water-saturated sediments to be induced by the elevated heat flow, a high sedimentation rate, as has been assumed for the Saar–Nahe Basin by Henk (1990, 1993a–c) and Stollhofen (1991, 1998, 2000), however, would prevent the high heat flow from rapidly achieving high diagenetic grades very close to the active sedimentation surface. The pyroclastic rocks in the maar–diatreme volcanoes clearly demonstrate that the uppermost 1500 to possibly 2500 m of sediments (at least the medium- to coarse-grained siliciclastic sediments) underlying the sedimentation surface of the basin during the time period equivalent to the Donnersberg-Formation were not yet indurated. Of course, it has to be realized that induration of the sediments increases gradually with depth and depends on overburden, sediment composition, permeability, water content, water chemistry and heat flow. At the time of the volcanicity in the Saar–Nahe Basin, at greater depth in the c. 5000–5500-m thick basin fill, diagenesis had certainly led already to reduced porosity, reduced permeability, mineral alteration processes and formation of cements, and thus to reduced groundwater availability. Thus the sediments at depth would have changed gradually from unconsolidated sediments to indurated rocks, which upon stress variations in the regime of late-orogenic crustal extension could react, more or less, by brittle deformation, thus forming joints and even fissures. Crustal extension exceeding a critical value would allow magmas to rise and make use of joints and fissures in both the hard rocks of the basement and the deeper stratigraphic levels of the basin fill.

Volcanicity in the Saar–Nahe Basin occurred during the Donnersberg-Formation and was intensive, synsedimentary and syntectonic with respect to crustal extension. After having traversed the basement and deeper stratigraphic levels of the basin fill, the magmas involved became emplaced not only at the surface but also in the uppermost 2500-m thick level of the thicker basin fill. The relationships between the subvolcanic features, i.e. diatremes, dykes, sills, laccoliths and the intrusive part of intrusive–extrusive domes, and the surrounding sediments allow the study of the state of diagenesis that the sediments were in at the specific time of volcanicity and subvolcanicity, and thus at a specific time in the evolution of such a basin. The diatreme fills and the contact-metamorphosed and neighbouring sediments in the contact and vicinity of the sills and laccoliths, as well as of the intrusive part of the intrusive–extrusive domes, should be investigated with respect to their status of diagenetic maturity, i.e. their lack of maturity at the time of intrusion of the subvolcanic bodies and diatremes. The relationships between the subvolcanic features and the surrounding sediments also allow the study of the different interaction processes that controlled the emplacement of the magmas within the upper levels of the basin fill, i.e. especially within the unconsolidated, water-saturated sediments.

A similar situation with respect to elevated heat flow and diagenesis is known from the Plio-Pleistocene sediments of the Rhine Rift Graben, where the Pliocene and Quaternary sediments are to a large extent not yet indurated.

Similar conditions of synsedimentary volcanicity occurred, for example, in the:

1. Late Variscan intermontane basins all over Variscan Europe and, in part, north of the orogenic front in northern Germany;
2. Tertiary Rhine Rift, Germany;
3. Tertiary Limagne Rift, France;
4. Mio-Pliocene Ohre Rift, Bohemia, Czech Republic (Ulrych et al. 2002);
5. Mio-Pliocene central and southern Hegau in the Alpine Molasse basin in southern Germany (Keller et al. 1990);
6. Mio-Pliocene/Pleistocene of southern Slovakia (Konecný et al. 1995) and northern Hungary;
7. Pannonian Basin, Hungary (Martin et al. 2002);
8. Carboniferous Midland Valley in Scotland;
9. Tertiary to Quaternary Basin-and-Range Province in the western USA and northern Mexico;
10. Miocene Ellendale Volcanic Field, Western Australia (Smith & Lorenz 1989; Stachel et al. 1994a); and
11. the Proterozoic Argyle pipe, Western Australia (the most diamondiferous pipe in the world: Boxer et al. 1989).

Groundwater and emplacement of volcanic bodies

The various subvolcanic bodes emplaced in the uppermost 2500 m of more-or-less unconsolidated sediments in the Saar–Nahe Basin differ in their relationship with these sediments. When magma, no matter what its chemistry, rose in a dyke-like fashion along hydraulically active faults or fault intersections, it frequently resulted in phreatomagmatic explosive activity and formation of maar–diatreme volcanoes. Since diatremes penetrate downward during ongoing activity (Lorenz 1985, 1986, 1998) this implies that magma had to rise along these faults to levels close to the surface, interact at shallow depth explosively with groundwater, and then the explosive activity penetrated downward. At other feeder dykes, basalt to basaltic andesite magmas intruded as sills into the neighbouring water-bearing sediments at depths mostly already greater than those of the initial explosions of maar–diatreme volcanoes. Nevertheless, some sills intruded at very shallow depths. The silicic laccoliths in the Saar–Nahe Basin evolved from sills and intruded in the same depth range as the basalt and basaltic andesite sills. At even greater depths in the basin fill, diagenesis would have already been in a more advanced state. This may be the reason why sills and laccoliths – judging from the present erosion surface in the Saar–Nahe Basin – appear to have intruded only sediments of the Stephanian Group and Glan-Subgroup, and apparently did not intrude the sediments of the Westphalian Group.

The major dykes in the SW Saar–Nahe Basin and the thick lava flows of the Donnersberg Formation prove, however, that in some areas groundwater did not hamper the rise of the magmas towards the surface.

Because of the thick, and in its upper level – water-saturated, basin-fill, silicic magma could not rise in a dyke-like fashion to the surface unaffected and erupt directly to form ignimbrites or highly vesicular pumice fall deposits. Conforming to the model given above, the only ignimbrite deposit in the Saar–Nahe Basin: the Prims and Nahe Syncline ignimbrite, had its source not inside the Saar–Nahe Basin but outside, i.e. in the neighbouring Hunsrück Mountains. If there had been only thin sediments filling the Saar–Nahe Basin, but the volcanism had been as voluminous as in the Saar–Nahe Basin then there would certainly have been many more lava flows and extensive ignimbrite sheets, as well as more pronounced pumice tephra beds.

Growth of subvolcanic and volcanic bodies

The size, i.e. the thickness of all subvolcanic and volcanic bodies varies remarkably in the Saar–Nahe Basin. Large and thus deep diatremes evolved by growth from small diatremes by continuing magma supply, ongoing phreatomagmatic explosions, and thus ongoing involvement of the sediments occupying the space of the diatreme (Lorenz 1985, 1986, 1998). Thick dykes are assumed to have evolved from thin dykes via inflation, because of a tensional stress field and ongoing magma supply (Wright & Fiske 1971; Delaney & Pollard 1981; Wadge 1981). Likewise, thick sills and voluminous laccoliths inflated because of ongoing magma supply and uplift of the overlying sediments and volcanics. Via such ongoing inflation, several laccoliths in the Saar–Nahe Basin even evolved into huge intrusive–extrusive domes. At the surface, the thick basalt and basaltic andesite lava flows in the Saar–Nahe Basin conform to the model of Hon et al. (1994) and Self et al. (1996, 1997) on inflation of flood basalts due to ongoing magma supply. And, likewise, a thick-bedded tephra sequence also implies eruptions continuing, because of ongoing magma supply, for a prolonged period, resulting in 'inflation' of the cumulate thickness of the tephra beds. The latter aspect is also relevant for the emplacement of thick ignimbrites. Thus, in conclusion, the various large sizes, i.e. thicknesses, of the subvolcanic and volcanic bodies in the Saar–Nahe Basin are the result of inflation in size and thickness with ongoing intrusion, effusion, explosions and eruptions due to ongoing magma supply through the individual feeder dykes at depth.

The above study and synthesis are based on our own research, the supervision of many field mapping courses, numerous student and Dr. rer. nat. thesis projects, performed during the past several decades. They are also based on evaluation of the mostly German literature dealing with the Saar–Nahe Basin, as well the international literature dealing with the various volcanic, subvolcanic and pyroclastic features and their physical volcanology. The interest of many students in the geology and volcanology of the Saar–Nahe Basin is gratefully acknowledged. During the past 3 decades, research on the Saar–Nahe Basin was supported by the Geologische Institut and its successor, the Institut für Geowissenschaften of the Universität Mainz; the Institut für Geologie of the Universität Würzburg; the Geologische Landesamt Rheinland–Pfalz and its successor, the Landesamt für Geologie und Bergbau Rheinland–Pfalz at Mainz; the DFG, Bonn; and, with respect to the research drill holes in the Saar–Nahe Basin, the Institut für Geowissenschaftliche Gemeinschaftsaufgaben, Hannover. The financial support of all these institutions is gratefully acknowledged. Over the years, discussions with colleagues and friends in and outside Germany have helped to clarify and formulate the ideas expressed in this publication. Last but not least, we are very grateful for the thorough, constructive reviews by C. Breitkreuz, Freiberg, Germany, and C. D. Henry, Reno, Nevada, USA, which not only improved both the content and style of the manuscript but also induced new ideas on the emplacement of magmas in the Saar–Nahe Basin. All aspects elucidated in this publication remain, however, the authors' sole responsibility.

References

AMMON, L.v. & REIS, O.M. 1903. *Erläuterungen zu dem Blatte Zweibrücken (Nr XIX) der Geognostischen Karte des Königreichs Bayern 1:100 000.* 1–182 mit einem Blatte (Nr XIX) der Geognostischen Karte des Königreiches Bayern, Piloty & Loehle, Munich.

AMMON, L.v. & REIS, O.M. 1910. *Erläuterungen zu dem Blatte Kusel (Nr XX) der Geognostischen Karte des Königreichs Bayern (1:100 000).* 1–186 mit einem Blatte der Geognostischen Karte des Königreiches Bayern, Piloty & Loehle, Munich.

ANDERSEN, T.B. 1998. Extensional tectonics in the Caledonides of southern Norway, an overview. *Tectonophysics*, **285**, 333–351.

ARIKAS, K. 1986. Geochemie und Petrologie der permischen Rhyolithe in Südwestdeutschland (Saar–Nahe-Gebiet, Odenwald, Schwarzwald) und in den Vogesen. *Pollichia-Buch*, **8**, 1–321.

ARMSTRONG, R.L. & WARD, P. 1991. Evolving geographic patterns of Cenozoic magmatism in the North American Cordillera: the temporal and spatial association of magmatism and metamorphic core complexes. *Journal of Geophysical Research*, **96**, 13 201–13 224.

ARTHAUD, F. & MATTE, P. 1977. Late Palaeozoic strike-slip faulting in southern Europe and northern Africa: results of a right-lateral shear zone between the Apalachians and the Urals. *Geological Society of America Bulletin*, **88**, 1305–1320.

ARZ, C. 1996. *Origin and Petrogenesis of Igneous Rocks from the Saar–Nahe-Basin (SW-Germany):*

Isotope, Trace Element and Mineral Chemistry. Dr.rer.nat. thesis, Universität Würzburg.

ARZ, C. & LORENZ, V. 1994a. An approach to the origin and evolution of magmas in late orogenic collapse basins – an example from the Saar–Nahe-Basin (SW-Germany). *Journal of the Czech Geological Society*, **39/1**, p. 3.

ARZ, C. & LORENZ, V. 1994b. Emplacement dynamics of the Lemberg intrusion, Saar–Nahe-Basin (SW-Germany). *Journal of the Czech Geological Society*, **40/3**, 58–59.

ARZ, C. & LORENZ 1994c. *Magma Mixing and Contamination as a Mechanism to Produce Intermediate Magmas in the Late Variscan Saar–Nahe-Basin*, **10**. Rundgespräch Geodynamik des Europäischen Variszikums, Bayreuth 1994, Abstracts, 28–30.

ARZ, C. & LORENZ, V. 1995a. Magma mixing and contamination as mechanism to produce intermediate magmas in the late Variscan Saar–Nahe-Basin. *Zentralblatt für Geologie und Paläontologie.* **Teil I, 1994, H 5/6**, 537–540.

ARZ, C. & LORENZ, V. 1995b. Aspects of the petrogenesis of late orogenic volcanic rocks: examples from the northeastern part of the Saar–Nahe-Basin. *Terra Nostra*, **95/8**, p. 79.

ARZ, C. & LORENZ, V. 1996. Petrogenese der Vulkanite im Saar–Nahe-Becken: Stabile und radiogene Isotope. *Terra Nostra*, **96/2**, 11–15.

ATZBACH, O. 1976. *Geologische Karte von Rheinland–Pfalz 1:25 000. Erläuterungen Blatt 6311 Lauterecken.* Geologisches Landesamt Rheinland–Pfalz, Mainz.

ATZBACH, O. 1983. *Geologische Karte von Rheinland–Pfalz 1:25 000. Erläuterungen Blatt 6212 Meisenheim.* Geologisches Landesamt Rheinland–Pfalz, Mainz.

ATZBACH, O., DREYER, G. & STAPF, K.R.G. 1974. Exkursion in das Pfälzer Sattelgewölbe und seine Umrandung am 19. April 1974. *Jahresberichte und Mitteilungen des Oberrheinischen Geologischen Verein, N.F.*, **56**, 47–78.

BAHR, C. 1989. *Stratigraphie und Tektonik des Oberrotliegenden im Raum Winnweiler–Schweisweiler/Pfalz unter besonderer Berücksichtigung des 'Hochsteiner Dazit' und seines Gefüges.* Diplomarbeit, Universität Würzburg.

BAKER, B.H. 1987. Outline of the petrology of the Kenya rift alkaline province. *In*: FITTON, J.G. & UPTON, B.G.J. (eds) *Alkaline Igneous Rocks.* Geological Society, London, Special Publications, **30**, 293–311.

BAMBAUER, H.U. 1960. Der permische Vulkanismus in der Nahemulde. 1: Lavaserie der Grenzlagergruppe und Magmatitgänge bei Idar–Oberstein. *Neues Jahrbuch für Mineralogie, Abhandlungen*, **95**, 141–199.

BARGENDA, W. 1988. *Das Kuhkopf-Massiv – zur Geologie eines permokarbonen Rhyolith-Doms im Saar–Nahe-Gebiet (SW-Deutschland) und dessen nördlicher Umgebung.* Diplomarbeit, Universität Mainz.

BEDERKE, E. 1959. Probleme des permischen Vulkanismus. *Geologische Rundschau*, **48**, 10–18.

BENECK, R., KRAMER, W. & MCCANN, T, ET AL. 1996. Permo-Carboniferous magmatism of the Northeast German Basin. *Tectonophysics*, **266**, 379–404.

BENNETT, R.A., WERNICKE, B.P., NIEMI, N.A., FRIEDRICH, A.M. & DAVIS, J.L. 2003. Contemporary strain rates in the northern Basin and Range province from GPS data. *Tectonics*, **22**, doi:10.1029/2001TC001355.

BERTHOLD, G., NAIRN, A.E.M. & NEGENDANK, J.F.W. 1975. A palaeomagnetic investigation of some of the igneous rocks of the Saar–Nahe basin. *Neue Jahrbuch für Geologie und Paläontologie, Monatshefte, Jahrgang 1975*, 134–150.

BEST, M.G. & CHRISTIANSEN, E.H. 1997. Origin of broken phenocrysts in ash-flow tuffs. *Geological Society of America Bulletin*, **109**, 63–73.

BOCK, A.C. 1986. *Petrographische und stratigraphische Untersuchungen an intermediären Vulkaniten und ihren Nebengesteinen im Bereich der Nahe–Mulde (Saar–Nahe-Becken, SW-Deutschland).* Diplomarbeit, Universität Mainz.

BONN, W.J. & STETS, J. 2000. Die Ignimbrite des Wittlicher Rotliegend-Beckens. *Mainzer Geowissenschaftliche Mitteilungen*, **29**, 9–36.

BOXER, G., LORENZ, V. & SMITH, C.B. 1989. The geology and volcanology of the Argyle (AK1) lamproite diatreme, Western Australia. *Proceedings of the Fourth International Kimberlite Conference, Perth, 1986*, **1**, Geological Society of Australia Special Publications, **14**, 140–152.

BRANNEY, M.J. 1995. Downsag and extension at calderas: new perspectives on collapse geometries from ice-melting, mining, and volcanic subsidence. *Bulletin of Volcanology*, **57**, 303–318.

BREITKREUZ, C. & KENNEDY, A. 1999. Magmatic flare-up at the Carboniferous–Permian boundary in the NE German Basin revealed by SHRIMP zircon ages. *Tectonophysics*, **302**, 307–326.

BREITKREUZ, C., AWDANKIEWICZ, M. & EHLING, B.-C. 2002a. Late Palaeozoic andesite sills in the Flechtingen Block, north of Magdeburg (Germany). *In*: BREITKREUZ, C., MOCK, A. & PETFORD, N. (eds) *First International Workshop on the Physical Geology of Subvolcanic Systems – Laccoliths, Sills, and Dykes (LASI) Abstracts and Field Guide.* Geologisches Institut der Universität Freiberg, Germany. Wissenschaftliche Mitteilungen, **20**, 7–8.

BREITKREUZ, C. & MOCK, A. 2002a. Are laccolith complexes characteristic of transtensional basin systems? Examples from Permocarboniferous Central Europe. *In*: BREITKREUZ, C., MOCK, A. & PETFORD, N. (eds) *First International Workshop on the Physical Geology of Subvolcanic Systems – Laccoliths, Sills, and Dykes (LASI) Abstracts and Field Guide.* Geologisches Institut der Universität Freiberg, Germany. Wissenschaftliche Mitteilungen, **20**, 8–9.

BREITKREUZ, C., MOCK, A. & PETFORD, N. (eds) 2002b. *First International Workshop on the Physical Geology of Subvolcanic Systems – Laccoliths, Sills, and Dykes (LASI) Abstracts and Field Guide.* Geologisches Institut der Universität Freiberg, Germany. Wissenschaftliche Mitteilungen, **20**.

BROD, M. 1989. *Stratigraphische und tektonische Verhältnisse des Oberrotliegenden zwischen*

Falkenstein und Heiligenmoschel/Pfalz, unter besonderer Berücksichtigung der Sedimentologie des 'Rhyolithkonglomerates'. Diplomarbeit, Universität Würzburg.

BÜTTNER, R. & ZIMANOWSKI, B. 1998. Physics of thermohydraulic explosions. *Physics Review E*, **57/5**, 1–4.

BUNTEBARTH, G. 1983. Zur Paläogeothermie im Permokarbon der Saar–Nahe-Senke. *Zeitschrift der Deutschen Geologischen Gesellschaft*, **134**, 211–223.

CARLE, J. 1988. *Geologie, Gefüge und Aufbau der Kuhkopf-Intrusion (Süd- und Westteil) bei Kirchheimbulanden/Pfalz (Saar–Nahe-Becken)*. Diplomarbeit, Universität Mainz.

CAS, R.A.F. & WRIGHT, J.V. 1987. *Volcanic Successions: Modern and Ancient: a Geological Approach to Processes, Products and Successions*. Allen & Unwin, London.

CAS, R.A.F., ALLEN, R.L., BULL, S.W., CLIFFORD, B.A. & WRIGHT, J.V. 1990. Subaqueous, rhyolitic dome-top tuff cones: a model based on the Devonian Bunga beds, southeastern Australia and a modern analogue. *Bulletin of Volcanology*, **52**, 159–174.

CHRISTIANSEN, R.L. & LIPMAN, P.W. 1972. Cenozoic volcanism and plate-tectonic evolution of the western United States. II. Late Cenozoic. *Philosophical Transactions of the Royal Society of London, Ser. A*, **271**, 249–284.

CORRY, C.E. 1988. Laccoliths; mechanics of emplacement and growth. *Geological Society of America, Special Papers*, **220**.

COWARD, M.P., DEWEY, J.F. & HANCOCK, P.L. (eds) 1987. *Continental Extensional Tectonics*. Geological Society, London, Special Publications, **28**.

COX, K.G. 1992. Karoo igneous activity and the early stages of the break-up of Gondwanaland. *In*: STOREY, B.C., ALABASTER, T. & PANKHURST, R.J. (eds) *Magmatism and the Causes of Continental Break-up*. Geological Society Special Publications, **68**, 137–148.

CRUDEN, A.R. & MCCAFFREY, K.J.W. 2002. Different scaling laws for sills, laccoliths and plutons. Mechanical threshholds on roof lifting and floor depression. *In*: BREITKREUZ, C., MOCK, A. & PETFORD, N. (eds) *First International Workshop on the Physical Geology of Subvolcanic Systems – Laccoliths, Sills, and Dykes (LASI) Abstracts and Field Guide*. Geologisches Institut der Universität Freiberg, Germany. Wissenschaftliche Mitteilungen, **20**, 15–17.

DALY, R. 1952. The name 'tholeiite'. *Geological Magazine*, **89**, 69–70.

DELANEY, P.T. & POLLARD, D.D. 1981. Deformation of host rocks and flow of magma during growth of minette dikes and breccia bearing intrusions near Ship Rock. *US Geological Survey Professional Papers*, **1202**.

DEWEY, J.F. 1988. Extensional collapse of orogens. *Tectonics*, **7**, 1123–1139.

DEWEY, J.F., RYAN, P.D. & ANDERSEN, T.B. 1993. Orogenic uplift and collapse, crustal thickness, fabrics and metamorphic phase changes: the role of eclogites. *Geological Society Special Publications*, **76**, 325–360.

DIEHL, M. 1981. Basische Gänge im Unterrotliegenden des Saar–Nahe-Gebietes im Raum mittleres Alsenztal und Tiefental–Winterborn. Zweimonatige Kartierung, Universität Mainz.

DIEHL, M. 1982. *Zur Geologie und Genese basischer Sills im SE-Teil des Saar–Nahe-Gebietes*. Diplomarbeit, Universität Mainz.

DITTRICH, D. 1996. Unterer Buntsandstein und die Randfazies des Zechsteins in der nördlichen Pfälzer Mulde. *Jahresberichte und Mitteilungen des Oberrheinischen Geologischen Verein*, N.F., **78**, 71–94.

DÖRHÖFER, S. 1989. *Zur Stratigraphie des Oberrotliegenden in der Umgebung von Neu-Bamberg, Fürfeld und Eckelsheim (Saar–Nahe Gebiet, SW-Deutschland) unter besonderer Berücksichtigung der stratigraphischen Stellung und Genese des effusiven Teils des Kreuznacher Rhyolith Massives*. Diplomarbeit, Universität Würzburg.

DREYER, G. 1970. *Geologische Kartierung im Bereich der Königsberg-Kuppel bei Wolfstein (Pfälzer Sattel)*. Diplomarbeit, Universität Mainz.

DREYER, G., FRANKE, W.R. & STAPF, K.R.G. 1983. *Geologische Karte des Saar–Nahe-Berglandes und seiner Randgebiete 1:100 000*. Geologisches Landesamt Rheinland–Pfalz, Mainz.

EDDINGTON, P.K., SMITH, R.B. & RENGGLI, C. 1987. Kinematics of Basin and Range intraplate extension. *In*: COWARD, M.P., DEWEY, J.F. & HANCOCK, P.L. (eds) *Continental Extensional Tectonics*. Geological Society, London, Special Publications, **28**, 371–392.

EICHELBERGER, J.C., CARRIGAN, C.R., WESTRICH, H.R. & PRICE, R.H. 1986. Non-explosive silicic volcanism. *Nature*, **323**, 598–602.

ELLIOT, D.H. 1992. Jurassic magmatism and tectonism associated with Gondwanaland break-up: an Antarctic perspective. *In*: STOREY, B.C., ALABASTER, T. & PANKHURST, R.J. (eds) *Magmatism and the Causes of Continental Break-up*. Geological Society Special Publications, **68**, 165–184.

FIELD, M. & SCOTT SMITH, B.H. 1999. Contrasting geology and near-surface emplacement of kimberlite pipes in southern Africa and Canada. *In*: GURNEY, J.J., GURNEY, J.L., PASCOE, M.D. & RICHARDSON, S.H. (eds) *Proceedings of the 7th International Kimberlite Conference, Cape Town, April 1998*, **1** (The J.B. Dawson Volume) Red Roof Design, National Book Printers, Cape Town, South Africa, 214–237.

FINK, J.H. 1983. Structure and emplacement of rhyolitic flow: Little Glass Mountain, Medicine Highland, northern California. *Geological Society of America Bulletin*, **94**, 362–380.

FITTON, J.G., JAMES, D. & LEEMAN, W.P. 1991. Basic magmatism associated with late Cenozoic extension in the Western United States: compositional variations in space and time. *Journal of Geophysical Research*, **96**, 13 693–13 711.

FRANCIS, E.H. 1962. Volcanic neck emplacement and subsidence structures at Dunbar, south-east

Scotland. *Transactions of the Royal Society of Edinburgh*, **65**, 41–58.
FRANCIS, E.H. 1970. Bedding in Scottish (Fifeshire) tuffpipes and its relevance to maars and calderas. *Bulletin Volcanologique*, **34**, 697–712.
FRANCIS, E.H. 1982. Magma and sediment – I. Emplacement mechanism of late Carboniferous tholeiite sills in northern Britain. *Journal of the Geological Society of London*, **139**, 1–20.
FRANCIS, E.H. 1991. Carboniferous–Permian igneous rocks. *In*: CRAIG, G.Y. (ed.) *The Geology of Scotland*. The Geological Society of London, 393–415.
FRANKE, W. 1989. Variscan plate tectonics in Central Europe – current ideas and open questions. *Tectonophysics*, **169**, 221–228.
FRANKE, W. & ONCKEN, O. 1990. Geodynamic evolution of the North-Central Variscides – a comic strip. *In*: FREEMAN, R., GIESE, P. & MUELLER, S. (eds) *The European Traverse: Integrative Studies. Results from the 5th Study Centre 26.3.–7.4.1990*. European Science Foundation, Strasbourg, 187–194.
GEHRING, I. 1998. *Geophysik und Petrographie der Diatreme Spitzenberg und Wolfsgalgen (Saar–Nahe-Gebiet, SW-Deutschland)*. Diplomarbeit, Universität Würzburg.
GEHRING, I. & LORENZ, V. 1998. Zuordnung des permokarbonen Wolfsgalgen-Rhyolithes (Saar–Nahe-Gebiet, SW-Deutschland). *Terra Nostra*, **98**, 48–49.
GEIB, K.W. 1956. Vulkanische Schlote mit Schlotbreccien und Eruptivgesteinen im östlichen Nahebergland. *Zeitschrift der Deutschen Geologischen Gesellschaft*, **108**, 265–266.
GEIB, K.W. 1972. Vulkanische Schlote mit Breccien und Eruptivgesteinen in der Umgebung des Lembergs, des Kreuznacher Rhyolithmassivs und im Bereich des Blattes Waldböckelheim (Nahebergland). *Mainzer Geowissenschaftliche Mitteilungen*, **1**, 59–69.
GEIB, K.W. 1975. *Geologische Karte von Rheinland–Pfalz, Erläuterungen Blatt 6112 Waldböckelheim*, Geologisches Landesamt Rheinland–Pfalz, Mainz, 1–146.
GEIB, K.W. & LORENZ, V. 1974. Exkursion zum Kreuznacher Rhyolith und sedimentären Oberrotliegenden nordöstlich von Bad Kreuznach, Hydrogeologie des unteren Nahegebietes, am 20. April 1974. *Jahresberichte und Mitteilungen des Oberrheinischen Geologischen Verein*, N.F., **56**, 95–104.
GERSCHÜTZ, S. 1996. *Geology, volcanology, and petrogenesis of the Kalkrand Basalt Formation and the Keetmanshoop Dolerite Complex, southern Namibia*. Dr.rer.nat. thesis, Universität Würzburg.
GÖPEL, C. 1977. *Mineralogische und geochemische Untersuchungen am Vulkanitkomplex des Bauwaldes/Saar–Nahe-Gebiet*. Diplomarbeit, Universität Mainz.
GRILL, H. 1992. *Geochemische und lithologische Charakterisierung einer vulkanosedimentären Abfolge im Oberrotliegenden des Saar–Nahe-Beckens: das Winnweiler Lager und seine Zwischensedimente*. Diplomarbeit, Universität Würzburg.
GRIMM, K.I. & STAPF, K.R.G. 1991. Die geologische Entwicklung der Rhyolith/Kuselit-Kuppel Herrmannsberg/Pfalz im Rotliegend des Saar–Nahe-Beckens. *Mitteilungen der Pollichia*, **78**, 7–34.
HABICHT, H. 1966. Die permokarbonischen Aufschlußbohrungen der Nahe-Senke, des Mainzer Beckens und der Zweibrücker Mulde. *Zeitschrift der Deutschen Geologischen Gesellschaft*, **115**, 631–649.
HÄFNER, F. 1977. *Die basischen Laven des Oberrotliegenden zwischen Alzey und Odernheim (Saar–Nahe-Gebiet) – Ein Beitrag zu ihrer Geologie, Petrographie und Geochemie*. Dr.rer.nat. thesis, Universität Mainz.
HALLS, H.C. & FAHRIG, W.F. (eds) 1987. Mafic dyke swarms. *Geological Society of Canada Special Papers*, **34**.
HANEKE, J. 1987. Der Donnersberg – zur Genese und stratigraphischen/tektonischen Stellung eines permokarbonen Rhyolith-Domes im Saar–Nahe-Gebiet (SW-Deutschland). *Pollichia-Buch*, **10**, 1–147.
HANEKE, J. 1991. Die Forschungsbohrungen Winnweiler 1050, 1051 und 1052. *In*: GEOWISSENSCHAFTLICHE GEMEINSCHAFTSAUFGABEN, *Tätigkeitsbericht*. 1989/90, Hanover, 70–74.
HANEKE, J. 1997. 625m tiefe Forschungsbohrung bei Gehrweiler/Pfalz im Rotliegend des Saar–Nahe-Beckens. *Geoforum Rheinland–Pfalz*, **1**, 25–29.
HANEKE, J. 1998. In die Erde geschaut: Ergebnisse von sechs geologischen Forschungsbohrungen im Rotliegenden des Donnersbergkreises. *In*: HANEKE, J. & KREMB, K. *280 Millionen Jahre Erdgeschichte: Geowissenschaftliche Forschungen im Donnersbergkreis*. Schriften der KVH Donnersbergkreis Kirchheimbolanden, 10–24.
HANEKE, J. & LORENZ, V. 1987. Emplacement of the Donnersberg Rhyolite, Saar–Nahe-Basin, Palatinate, Germany. *Terra Cognita*, **7**, p. 369.
HANEKE, J. & LORENZ, V. 1990. Das lithostratigraphische Profil der Forschungsbohrungen 'Jakobsweiler 1' und 'Dannenfels 1' am Donnersberg/Pfalz. *Mainzer Geowissenschaftlichen Mitteilungen*, **19**, 297–311.
HANEKE, J. & LORENZ, V. 2000. *Geologische Karte von Rheinland–Pfalz 1:25 000, Blatt 6313 Dannenfels*. Geologisches Landesamt Rheinland–Pfalz, Mainz.
HANEKE, J. & STOLLHOFEN, H. 1994. Das lithostratigraphische Profil der Forschungsbohrung 'Münsterappel I'. *Mainzer Geowissenschaftliche Mitteilungen*, **23**, 221–228.
HANEKE, J., GÄDE, C.W. & LORENZ, V. 1979. Zur stratigraphischen Stellung der rhyolithischen Tuffe im Oberrotliegenden des Saar–Nahe-Gebietes und der Urangehalt des Kohlen–Tuff-Horizontes an der Kornkiste bei Schallodenbach/Pfalz. *Zeitschrift der Deutschen Geologischen Gesellschaft*, **130**, 535–560.
HANEKE, J., LORENZ, V. & STAPF, K.R.G. 1983. Geologie und Grundwasser des Landschaftschutzgebietes Donnersberg. *In*: STAPF, K.R.G. (ed.)

Das Landschaftschutzgebiet Donnersberg in der Nordpfalz. Pollichia-Buch, **4**, 41–66.

HARRY, D.L., SAWYER, D.S. & LEEMAN, W.P. 1993. The mechanics of continental extension in western North America; implications for the magmatic and structural evolution of the Great Basin. *Earth and Planetary Science Letters*, **117**, 59–71.

HEIKEN, G. & WOHLETZ, K. 1987. Tephra deposits associated with silicic domes and lava flows. *In*: FINK, J.H. (ed.) *The Emplacement of Silicic Domes and Lava Flows.* Geological Society of America Special Papers, **212**, 55–76.

HEIM, D. 1960. Über die Petrographie und Genese der Tonsteine aus dem Rotliegenden des Saar–Nahe-Gebietes. *Beiträge zur Mineralogie und Petrologie*, **7**, 281–317.

HEIM, D. 1961. Über die Tonsteintypen aus dem Rotliegenden des Saar–Nahe-Gebietes und ihre stratigraphisch–regionale Verteilung. *Notizblatt des Hessischen Landesamt für Bodenforschung*, **89**, 377–399.

HENK, A. 1990. *Struktur und geodynamische Entwicklung des Saar–Nahe-Beckens.* Dr.rer.nat. thesis, Universität Würzburg.

HENK, A. 1992. Mächtigkeit und Alter der erodierten Sedimente im Saar–Nahe-Becken (SW-Deutschland). *Geologische Rundschau*, **81**, 323–331.

HENK, A. 1993a. Subsidenz und Tektonik des Saar–Nahe-Beckens (SW-Deutschland). *Geologische Rundschau*, **82**, 3–19.

HENK, A. 1993b. Das Saar–Nahe-Becken, eine geodynamische Beckenanalyse. *Die Geowissenschaften*, **11**, Jahrgang für 1993, **H. 8**, 268–273.

HENK, A. 1993c. Late orogenic basin evolution in the Variscan Internides: the Saar-Nahe Basin, southwest Germany. *Tectonophysics*, **223**, 273–290.

HENK, A. 1997. Gravitational orogenic collapse versus plate boundary stresses – a numerical modelling approach to the Permo-Carboniferous evolution of Central Europe. *Geologische Rundschau*, **86**, 39–55.

HENK, A. 1998. Thermomechanische Modellrechnungen zur postkonvergenten Krustenreequilibrierung in den Variscidena. *Geotektonische Forschungen*, **90**.

HENK, A. 1999. Did the Variscides collapse or were they torn apart? A quantitative evaluation of the driving forces for post-convergent extension in Central Europe. *Tectonics*, **18**, 774–792.

HENK, A. & STOLLHOFEN, H. 1994. Struktureller und sedimentärer Ausdruck von Transfer-Störungen erster Ordnung – eine Fallstudie. *Nachrichten der Deutschen Geologischen Gesellschaft*, **52**, p. 95.

HENRY, C.D. & PRICE, J.G. 1989. The Christmas Mountains caldera complex, Trans Pecos Texas: the geology and development of a laccocaldera. *Bulletin of Volcanology*, **52**, 97–112.

HENRY, C.D., KUNK, M.J., MUEHLBERGER, W.R. & MCINTOSH, W.C. 1997. Igneous evolution of a complex laccolith–caldera, the Solitario, Trans Pecos Texas: Implications for calderas and subjacent plutons. *Geological Society of America Bulletin*, **109**, 1063–1054.

HESS, J.C. & LIPPOLT, H.J. 1986. $^{40}Ar/^{39}Ar$ ages of tonstein and tuff sanidines: New calibration points for the improvement of the Upper Carboniferous time scale. *Chemical Geology*, **59**, 143–154.

HESS, J.C. & LIPPOLT, H.J. 1988. Subsidenz und Sedimentation im Saar–Nahe-Becken – die Entwicklung eines Molassetrogs im Lichte isotopischer Altersdaten. *Nachrichten der Deutschen Geologischen Gesellschaft*, **39**, 26–27.

HINZMANN, S. 1987. *Petrographie, Geochemie und Gefüge des Kuhkopfpalatinits.* Diplomarbeit, Universität Mainz.

HON, K., KAUAHIKAUA, J., DENLINGER, R. & MACKAY, K. 1994. Emplacement of and inflation of pahoehoe sheet flows: observations and measurements of active lava flows on Kilauea Volcano, Hawaii. *Geological Society of America Bulletin*, **106**, 351–370.

HOTH, K., HÜBSCHER, H.-D., KORICH, D., GABRIEL, W. & ENDERLEIN, F. 1993. Die Lithostratigraphie der permosilesischen Effusiva im Zentralabschnitt der Mitteleuropäischen Senke. – Der permokarbone Vulkanismus im Zentralabschnitt der Mitteleuropäischen Senke. *Geologisches Jahrbuch*, Reihe A, **131**, 179–196.

HYNDMAN, D.W. & ALT, D. 1987. Radial dikes, laccoliths, and gelatin models. *Journal of Geology*, **95**, 763–774.

JACKSON, M.D. & POLLARD, D.D. 1988. The laccolith–stock controversy: new results from the southern Henry Mountains, Utah. *Geological Society of America Bulletin*, **100**, 117–139.

JOHNSON, A.M. & POLLARD, D.D. 1973. Mechanics of growth of some laccolithic intrusions in the Henry Mountains, Utah, I. Field observations, Gilbert's model, physical properties, and flow of the magma. *Tectonophysics*, **18**, 261–309.

JUNG, D. 1958. Untersuchung am Tholeyit von Tholey (Saar). *Contributions to Mineralogy and Petrology*, **6**, 147–181.

JUNG, D. 1967. Die Mineralassoziationen der Palatinite und ihrer Aplite. *Annales Universitatis Saraviensis, Reihe Mathematisch–Naturwissenschaftliche Fakultät*, **Heft 5** (Geologisch–mineralogisches Sammelheft), 1–130.

JUNG, D. 1970. Permische Vulkanite im SW-Teil des Saar–Nahe-Pfalz-Gebietes. *Aufschluss Sonderband*, **19**, 185–201, Heidelberg.

JUNG, D. & VINX, R. 1973. Einige Bemerkungen zur Geochemie der Magmatite des Saar–Nahe–Pfalz-Gebietes. *Annales Scientifiques de l'Université de Besançon*, 3. Série, **18**, 197–202.

KELLER, J., BREY, G., LORENZ, V. & SACHS, P. 1990. *Volcanism and petrology of the Upper Rhinegraben (Urach–Hegau–Kaiserstuhl). Excursion 2A, Fieldguide.* IAVCEI International Volcanological Congress, Mainz, F.R.G., 1990, 1–63.

KOCH, I. 1938. Die Kuselite des Saar–Nahe-Gebietes. *Neues Jahrbuch für Mineralogie, Geologie und Paläontologie, Abhandlungen, Beilage Band*, **73**, Abt. A, 419–494.

KÖHLER, B. 1987. *Petrographische und geochemische Untersuchungen an intermediären Lagergängen in der Umgebung des Donnersberges.* Diplomarbeit, Universität Mainz.

KONECNÝ, V., LEXA, J., BALOGH, K. & KONECNÝ, P. 1995. Alkali basalt volcanism in Southern Slovakia: volcanic forms and time evolution. *Acta Vulcanologica*, **7**, 167–171.

KÖNIGER, S. 1999. *Distal ash tuffs in the lowermost Permian of the Saar–Nahe Basin*. Dr.rer.nat. thesis, Universität Würzburg.

KÖNIGER, S. 2000. Verbreitung, Fazies und stratigraphische Bedeutung distaler Aschentuffe der Glan-Gruppe im karbonisch–permischen Saar–Nahe-Becken (SW-Deutschland*). Mainzer Geowissenschaftliche Mitteilungen*, **29**, 97–132.

KÖNIGER, S. & LORENZ, V. 2002. Geochemistry, tectonomagmatic origin, and chemical correlation of altered Carboniferous-Permian fallout ash tuffs in SW-Germany. *Geological Magazine*, **139(5)**, 541–558.

KÖNIGER, S. & LORENZ, V. 2003. Petrography and origin of altered Carboniferous–Permian fallout ash tuffs in SW-Germany. *Zeitschrift der Deutschen Geologischen Gesellschaft*, **153**, 209–258.

KÖNIGER, S. & STOLLHOFEN, H. 2001. Environmental and tectonic controls on preservation potential of distal fallout ashes in fluvio-lacustrine settings: the Carboniferous–Permian Saar-Nahe Basin, south-west Germany. *In*: WHITE, J.D.L. & RIGGS, N.R. (eds) *Volcaniclastic Sedimentation in Lacustrine Settings*. International Association of Sedimentology, Special Publications, **30**, 263–284.

KÖNIGER, S., LORENZ, V., STOLLHOFEN, H. & ARMSTRONG, R.A. 2002. Origin, age and stratigraphic significance of distal ash tuffs from the Carboniferous–Permian continental Saar–Nahe Basin (SW-Germany). *International Journal of Earth Sciences (Geologische Rundschau)*, **91**, 341–356.

KÖNIGER, S., STOLLHOFEN, H. & LORENZ, V. 1995. Tuff layers in the 'Lower Rotliegend' (Lebach-Group) of the Saar–Nahe Basin (SW-Germany): occurrences, sedimentation patterns, and significance. *In*: NEGENDANK, J.F.W. & BENEK, R. (eds) *Geodynamics of the European Variscides. Symposium on Permocarboniferous Igneous Rocks*. Terra Nostra, **7/95**, 79–83.

KONRAD, H.-J. & SCHWAB, K. 1970. Ursache und zeitliche Einstufungen lokaler Diskordanzen im Bereich des Pfälzer Sattels. *Abhandlungen des Hessischen Landesamt für Bodenforschung*, **56**, 181–192.

KONZAN, H.P., MÜLLER, E.M. & MIHM, A. 1981. *Geologische Karte des Saarlandes, 1:50 000, GK Saarland*, Geologisches Landesamt des Saarlandes, Saarbrücken.

KONZAN, H.P., MÜLLER, E.M. & MIHM, A. 1989. *Geologische Karte des Saarlandes, GK 50 Saarland, Erläuterungen*, 1–46, Geologisches Landesamt des Saarlandes, Saarbrücken.

KRAWINKEL, J.J. 1988. *Zur Geologie der südlichen Umrandung des Kreuznacher Rhyolithes im Raum Altenbamberg. Zweimonatige Kartierung*, 1–88, Universität Mainz.

KRUPP, R. 1981. Die Geologie des Moschellandsberg-Vulkankomplex (Pfalz) und seiner Erzvorkommen. *Mitteilungen der Pollichia*, **69**, 6–26.

KRUPP, R. 1984. The Nahe Caldera – a resurgent caldera in the Permocarboniferous Saar Nahe Basin. *Geologische Rundschau*, **73**, 981–1005.

LARSEN, H.C. & MARCUSSEN, C. 1992. Sill-intrusion, flood basalt emplacement and deep crustal structure of the Scoresby Sund region, East Greenland. *In:* STOREY, B.C., ALABASTER, T. & PANKHURST, R.J. (eds) 1992. *Magmatism and the Causes of Continental Break-up*. Geological Society Special Publications, **68**, 365–386.

LIPMAN, P.W. & GLAZNER, A.F. 1991. Introduction to middle Tertiary Cordilleran volcanism: magma sources and relations to regional tectonics. *Journal of Geophysical Research*, **96**, 193–199.

LIPPOLT, H.J. & HESS., J.C. 1983. Isotopic evidence for the stratigraphic position of the Saar–Nahe–Rotliegend volcanism. I. $^{40}Ar/^{40}K$ and $^{40}Ar/^{39}Ar$ investigations. *Neues Jahrbuch für Geologie und Paläontologie, Monatshefte*, **1983**, 713–730.

LIPPOLT, H.J. & HESS., J.C. 1989. Isotopic evidence for the stratigraphic position of the Saar–Nahe–Rotliegend volcanism. III. Synthesis of results and geological implications. *Neues Jahrbuch für Geologie und Paläontologie, Monatshefte*, **1989**, 553–559.

LIPPOLT, H.J. & RACZEK, I. 1979. Isotopische Altersbestimmungen an vulkanischen Biotiten des Saar–Nahe-Perm. *Fortschritte der Mineralogie*, **57**, *Beiheft* **1**, p. 87.

LIPPOLT, H.J. & RACZEK, I. 1981. Isotopic age determinations of volcanic biotites from the Saar–Nahe Permian. *International Symposium on the Central European Permian, Jablonna, 1978*, Proceedings, 313, Geological Institute Warsaw.

LIPPOLT, H.J., HESS, J.C., RACZEK, I. & VENZLAFF, V. 1989. Isotopic evidence for the stratigraphic position of the Saar–Nahe–Rotliegende volcanism. II. Rb-Sr investigations. *Neues Jahrbuch für Geologie und Paläontologie, Monatshefte*, **1989**, 539–552.

LISS, D., HUTTON, D.H.W., OWENS, W.H. & THOMSON, K. 2002. Magma flow geometries in Whin Sill derived from field evidence and magnetic analyses – indicators for sill emplacement mechanism. *In*: BREITKREUZ, C., MOCK, A. & PETFORD, N. (eds) *First International Workshop on the Physical Geology of Subvolcanic Systems – Laccoliths, Sills, and Dykes (LASI) Abstracts and Field Guide*. Geologisches Institut der Universität Freiberg, Germany. Wissenschaftliche Mitteilungen, **20**, 28–29.

LIU, M. 2001 Cenozoic extension and magmatism in the North American Cordillera: the role of gravitational collapse. *Tectonophysics*, **342**, 407–433.

LORENZ, V. 1971*a*. Collapse structures in the Permian of the Saar–Nahe-area, South-West-Germany. *Geologische Rundschau*, **60**, 924–948.

LORENZ, V. 1971*b*. Vulkanische Calderen und Schlote am Donnersberg/Pfalz. *Oberrheinische geologische Abhandlungen*, N.F., **20**, 21–41.

LORENZ, V. 1971*c*. Zur Stratigraphie und Tektonik des Oberrotliegenden in der Umgebung von

Schweisweiler und Winnweiler/Pfalz. *Abhandlungen des hessischen Landesamtes für Bodenforschung*, **60**, 263–275.

LORENZ, V. 1972. Sekundäre Rotfärbung im Rotliegenden der Saar–Nahe-Senke, SW-Deutschland. *Neues Jahrbuch für Geologie und Paläontologie, Monatshefte*, Jahrgang für **1972**, 356–370.

LORENZ, V. 1973. Zur Altersfrage des Kreuznacher Rhyolithes unter besonderer Berücksichtigung der Stratigraphie und Überschiebungstektonik an seiner südlichen Umrandung (Saar–Nahe-Gebiet, SW-Deutschland). *Neues Jahrbuch für Geologie und Paläontologie Abhandlungen*, **142**, 139–164.

LORENZ, V. 1974. Die Pyroklastika des Permischen Saar–Nahe-Beckens. 64. Jahrestagung der Geologischen Vereinigung, Bochum, 1974, Kurzfassung der Vorträge, **23**, Bochum.

LORENZ, V. 1976. Formation of Hercynian subplates, possible causes and consequences. *Nature*, **262**, 374–377.

LORENZ, V. 1985. Maars and diatremes of phreatomagmatic origin, a review. *Transactions of the Geological Society of South Africa*, **88**, 459–470.

LORENZ, V. 1986. On the growth of maars and diatremes and its relevance to the formation of tuffrings. *Bulletin of Volcanology*, **48**, 265–274.

LORENZ, V. 1987. Bedeutung des basisch-sauren Vulkanismus im permokarbonen intermontanen Saar–Nahe-Trog/SW-Deutschland. 3. Rundgespräch 'Geodynamik des Europäischen Variszikums'. Kaledonisch–Variszische Strukturen in den Alpen. Fribourg, Schweiz, Oktober 1987, Abstracts, 12.

LORENZ, V. 1992a. *Spätorogene Zustände und Prozesse in den Varisciden*. Vortragskurzfassungen, Einführungskolloquium zum Schwerpunktprogramm der DFG in Gießen, 1992, 'Orogene Prozesse: ihre Quantifizierung und Simulation am Beispiel der Varisciden', Institut für Geowissenschaften, Universität Gießen, 50–51.

LORENZ, V. 1992b. *Petrogenese Bimodaler Vulkanite in Spätorogenen, Intermontanen Becken, am Beispiel Spätvariscischer, Permokarboner Vulkanite*. Vortragskurzfassungen, Einführungskolloquium zum *Schwerpunktprogramm* der DFG in Gießen, 1992, 'Orogene Prozesse: ihre Quantifizierung und Simulation am Beispiel der Varisciden', Institut für Geowissenschaften, Universität Gießen, 51.

LORENZ, V. 1998. Zur Vulkanologie von diamantführenden Kimberlit- und Lamproit-Diatremen. *Zeitschrift der Deutschen Gemmologischen Gesellschaft*, **47(1)**, 19–44.

LORENZ, V. 2003a. Syn- and post-eruptive processes of maar–diatreme volcanoes and their relevance to the accumulation of post-eruptive maar crater sediments. *Földtani Kutatás (Geological Research, Quarterly Journal of the Hungarian Geological Survey)*, (7. Alginite Symposium), **40**, 13–22.

LORENZ, V. 2003b. Maar–diatreme volcanoes, their formation, and their setting in hard rock or soft rock environments. *GeoLines*, **7** (Hibsch 2002 Symposium), 72–83.

LORENZ, V. & HANEKE, J. 1981. Zur Geologie des Donnersberges. *In*: GEMEINDE DANNENFELS (ed.) *Dannenfels, Chronik eines Dorfes*, 136–155, Dannenfels.

LORENZ, V. & HANEKE, J. 1986. Der rhyolithische Vulkanismus im permokarbonen Saar–Nahe-Becken. *Fortschritte der Mineralogie*, **64**, Beiheft **1**, p. 101.

LORENZ, V. & NICHOLLS, I.A. 1976. The Permocarboniferous basin and range province of Europe. An application of plate tectonics. *In*: FALKE, H. (ed.) *The Continental Permian in Central, West, and South Europe*. Nato Advanced Study Institute Series, **C22**, 312–342, Reidel, Dordrecht, The Netherlands.

LORENZ, V. & NICHOLLS, I.A. 1984. Plate and intraplate processes of Hercynian Europe during the late Palaeozoic. *Tectonophysics*, **107**, 25–56.

LORENZ, V., STAPF, K.R.G., HANEKE, J. & ATZBACH, O. 1987. Das Rotliegende des Saar–Nahe-Gebietes in der Umgebung des Donnersberges (Exkursion B am 23. und 24. April 1987). *Jahresberichte und Mitteilungen des Oberrheinischen Geologischen Verein*, N.F., **69**, 53–76.

LORENZ, V., ZIMANOWSKI, B. & FRÖHLICH, G. 1993. Experimental studies on explosive and non-explosive phreatomagmatic eruptions. *In*: ORGANIZING COMMITTEE (eds), *Ancient Volcanism and Modern Analogues*. IAVCEI General Assembly, Canberra, Australia, Abstracts, 64.

LORENZ, V., KURSZLAUKIS, S., STACHEL, T. & STANISTREET, I.G. 1997. Volcanology of the carbonatitic Gross Brukkaros Volcanic Field, Namibia. 6th International Kimberlite Conference, Novosibirsk, Russia, Conference Proceedings, *Russian Geology and Geophysics*, **38**, 40–49.

LORENZ, V., STACHEL, T., KURSZLAUKIS, S. & STANISTREET, I.G. 2000. Volcanology of the Gross Brukkaros Volcanic Field, southern Namibia. *Communications of the Geological Survey of Namibia*, **12** (Henno Martin Volume), 347–352.

MANLEY, C.R. 1995. How voluminous rhyolite lavas mimic rheomorphic ignimbrites: eruptive style, emplacement conditions, and formation of tuff-like textures. *Geology*, **23**, 349–352.

MANLEY, C.R. 1996. In situ formation of welded tuff-like textures in the carapace of a voluminous silicic lava flow, Owyhee County, SW Idaho. *Bulletin of Volcanology*, **57**, 672–686.

MANLEY, C.R. & FINK, J.H. 1987. Internal textures of rhyolite flows as revealed by research drilling. *Geology*, **15**, 549–552.

MARSH, J.S., HOOPER, P.R., REHACEK, J., DUNCAN, R.A. & DUNCAN, A.R. 1997. Stratigraphy and age of Karoo basalts of Lesotho and implications for correlations within the Karoo igneous province. *In*: MAHONEY, J.J. & COFFIN, M.F. (eds) *Large Igneous Provinces. Continental, Oceanic, and Planetary Flood Volcanism*. Geophysical Monograph, **100**, 247–272.

MARTIN, U., AUER, A., NÉMETH, K. & BREITKREUZ, C. 2002. Mio/Pliocene phreatomagmatic volcanism in the western part of the Pannonian Basin, Hungary. *In*: ULRYCH, J., CAJZ, V., ADAMOVIVIC, J. & BOSÁK (eds) *Hibsch 2002 Symposium, Teplá, Ústi, Mariánske Lázne, Excursion guide, abstracts*. Czech Geological Survey, Praha **90**.

MCCAFFREY, K.J.W. & CRUDEN, A.R. 2002. Dimensional data and growth models for intrusions. *In*: BREITKREUZ, C., MOCK, A. & PETFORD, N. (eds) *First International Workshop on the Physical Geology of Subvolcanic Systems – Laccoliths, Sills, and Dykes (LASI) Abstracts and Field Guide*. Geologisches Institut der Universität Freiberg, Germany. Wissenschaftliche Mitteilungen, **20**, 37–39.

METCALF, R.V. & SMITH, E.I. 1995. Introduction to special section: magmatism and extension. *Journal of Geophysical Research*, **100**, 10 249–10 253.

MINNING, M. & LORENZ, V. 1983. Rotliegend-Ignimbrite in der Prims-Mulde (Saar–Nahe-Senke/Südwestdeutschland). *Mainzer Geowissenschaftliche Mitteilungen*, **12**, 261–290.

MOCK, A., BREITKREUZ, C., EHLING, B.-C. & AWDANKIEWICZ, M. 2002a. Field trip to the Flechtingen and Halle Volcanic Complex. *In*: BREITKREUZ, C., MOCK, A. & PETFORD, N. (eds) *First International Workshop on the Physical Geology of Subvolcanic Systems – Laccoliths, Sills, and Dykes (LASI) Abstracts and Field Guide*. Geologisches Institut der Universität Freiberg, Germany. Wissenschaftliche Mitteilungen, **20**, 59–75.

MOCK, A., LANGE, C.-D., BREITKREUZ, C. & EHLING, B.-C. 2002b. Geometry and textures of laccoliths and sills in the Late Palaeozoic Halle volcanic complex (Germany). *In*: BREITKREUZ, C., MOCK, A. & PETFORD, N. (eds) *First International Workshop on the Physical Geology of Subvolcanic Systems – Laccoliths, Sills, and Dykes (LASI) Abstracts and Field Guide*. Geologisches Institut der Universität Freiberg, Germany. Wissenschaftliche Mitteilungen, **20**, 40–41.

MUDGE, M.R. 1968. Depth control of some concordant intrusions. *Geological Society of America Bulletin*, **79**, 315–332.

MÜLLER, G. 1982. Der saure permische Vulkanismus im N-Saarland. *Saarland Vereinigung der Freunde der Mineralogie und Geologie, Sommertagung 1982*, 67–95.

NEGENDANK, J.F.W. 1965. *Das Unterrotliegende im Gebiet Wolfstein–Niederkirchen–Heiligenmoschel (SE-Flanke des Pfälzer Sattels)*. Diplomarbeit, Universität Mainz.

NEGENDANK, J.W.F. 1983. Trier und Umgebung. *Sammlung Geologischer Führer*, **60** (2. Auflage), Gebrüder Bornträger, Berlin, Stuttgart, 1–195.

NEUMANN, E.R., OLSEN, K.H., BALDRIGE, W.S. & SUNDVOLL, B. 1992. The Oslo Rift: a review. *Tectonophysics*, **208**, 1–18.

NICHOLLS, I.A., & LORENZ, V. 1973. Origin and crystallization history of Permian tholeiites from the Saar–Nahe-trough, SW-Germany. *Contributions to Mineralogy and Petrology*, **40**, 327–344.

NICKEL, K.G. 1979. *Geologische und petrologische Untersuchungen im Bereich des Nahetales zwischen Norheim und Staudernheim unter besonderer Berücksichtigung der intermediären Vulkanite*. Diplomarbeit, Universität Mainz.

NICKEL, K.G. 1981. Magma mixing as the probable origin of some Permian volcanic rocks of the Saar–Nahe-Basin (SW-Germany). *Geologische Rundschau*, **70**, 1164–1176.

ONCKEN, O. 1997. Transformation of a magmatic arc and an orogenic root during oblique collision and its consequences for the evolution of the European Variscides (Mid German Crystalline Rise). *Geologische Rundschau*, **86**, 2–20.

OSMUNDSEN, P.T., ANDERSEN, T.B., MARKUSSEN, S. & SVENDBY, A.K. 1998. Tectonics and sedimentation in the hangingwall of a major extensional detachment: the Devonian Kvamshesten Basin, western Norway. *Basin Research*, **10**, 213–234.

PARKER, A.J., RICKWOOD, P.C. & TUCKER, D.H. (eds) 1990. Mafic dykes and emplacement mechanisms. *International Geological Correlation Program Project 257*, Publication Number, **23**, Balkema, Rotterdam, Brookfield.

PLANKE, S., MALTHE-SØRENSSEN, A., SVENSEN, H. & JAMTVEIT, B. 2002. Emplacement of sill intrusions in sedimentary basins. *In*: BREITKREUZ, C., MOCK, A. & PETFORD, N. (eds) *First International Workshop on the Physical Geology of Subvolcanic Systems – Laccoliths, Sills, and Dykes (LASI) Abstracts and Field Guide*. Geologisches Institut der Universität Freiberg, Germany. Wissenschaftliche Mitteilungen, **20**, 48–49.

POLLARD, D.D. 1972. Elastic–plastic bending of strata over a laccolith: why some laccoliths have flat tops. *EOS, Transactions, American Geophysical Union*, **53**, p. 1117.

POLLARD, D.D. 1973. Derivation and evaluation of a mechanical model for sheet intrusions. *Tectonophysics*, **19**, 233–269.

POLLARD, D.D. & JOHNSON, A.M. 1973. Mechanics of growth of some laccolithic intrusions in the Henry Mountains, Utah, II. Bending and failure of overburden layers and sill formation. *Tectonophysics*, **18**, 311–354.

POLLARD, D.D., MULLER, O.H. & DOCKSTADER, D.R. 1975. The form and growth of fingered sheet intrusions. *Geological Society of America Bulletin*, **86**, 351–363.

REIS, O.M. 1921. *Erläuterungen zum Blatte Donnersberg (Nr XXI) der Geognostischen Karte von Bayern (1:100 000)*. 1–320, mit einem Blatte (Nr XXI) der Geognostischen Karte von Bayern (1:100 000), Piloty & Loehle, Munich.

REIS, O.M. 1922. Geologisches Übersichtskärtchen 1:25.000 des Gebietes um und westlich von Kirchheimbolanden (Pfalz). *Geognostische Jahreshefte*, **34**, Jahrgang für 1921, 255–269, München.

REYNOLDS, M.A. & BEST, J.G. 1976. Summary of 1953–1957 eruption of Tuluman volcano, Papua New Guinea. *In*: JOHNSON, R.W. (ed.) *Volcanism in Australasia*. Elsevier, Amsterdam, 287–296.

SATO, H., FUJII, T. & NAKADA, S. 1992. Crumbling of

dacite dome lava and generation of pyroclastic flows at Unzen volcano. *Nature*, **360**, 664–666.

SCHMIDT, C. & REIS, O.M. 1915. Zur Kenntnis des Donnersberggebiets. *Geognostische Jahreshefte*, **34**, Jahrgang für 1915, 63–90.

SCHMINCKE, H.U. 2000. *Vulkanismus*, Wissenschaftliche Buchgesellschaft, Darmstadt.

SCHOLZ, C.H., BARAZANGI, M. & SBAR, M.L. 1971. Late Cenozoic evolution of the Great Basin, Western United States, as an ensialic interarc basin. *Geological Society of America Bulletin*, **82**, 2979–2990.

SCHOPP, H. 1894. *Das Rotliegende in der Umgebung von Fürfeld in Rheinhessen*. Winter'sche Buchdruckerei, Darmstadt, 14 pp.

SCHUSTER, M. 1914. Neue Beiträge zur Kenntnis der permischen Eruptivgesteine aus der bayerischen Rheinpfalz. III. Die Eruptivgesteine im Gebiet des Blattes Donnersberg (1:100 000). *Geognostische Jahreshefte*, **26**, Jahrgang für 1913, 235–266.

SCHUSTER, M. 1933. Ein Überblick über die Eruptivgesteine der Rheinpfalz. *Jahresberichte und Mitteilungen des Oberrheinischen Geologischen Verein*, N.F., **22**, 27–38.

SCHUSTER, M. & SCHWAGER, A. 1911. Neue Beiträge zur Kenntnis der permischen Eruptivgesteine aus der bayerischen Rheinpfalz. I. Die Kuselite. *Geognostische Jahreshefte*, **23**, Jahrgang für 1910, 43–59.

SCHWAB, K. 1965. Petrographische Untersuchungen an basischen Magmatiten in der Umgebung des Donnersberges/Pfalz. *Neues Jahrbuch für Mineralogie, Abhandlungen*, **102**, 258–290.

SCHWAB, K. 1967. Zur Geologie der Umgebung des Donnersberges. *Mitteilungen der Pollichia*, III, Reihe, **14**, 13–55.

SCHWAB, K. 1968. Die Verbreitung der effusiven Vulkanite auf der SE-Flanke des Pfälzer Sattels und ihre Stellung im Profil des Oberrotliegenden. *Mainzer Naturwissenschaftliches Archiv*, **7**, 105–119.

SCHWAB, K. 1971a. Das Intrusionsalter des Kuselites vom Remigiusberg und seine Beziehung zur Potzberg–Kuppel (Saar–Nahe-Gebiet). *Abhandlungen des Hessischen Landesamtes für Bodenforschung*, **60**, 288–297.

SCHWAB, K. 1971b. Effusivgesteine. *In*: ATZBACH, O. & SCHWAB, K. *Geologische Karte Rheinland–Pfalz 1:25 000. Erläuterungen Blatt 6410 Kusel*, Geologisches Landesamt Rheinland–Pfalz, Mainz.

SCHWAB, K. 1981. Differentiation trends in Lower Permian effusive igneous rocks from the southeastern part of the Saar–Nahe-Basin/FRG. *International Symposium on the Central European Permian, Jablonna, April 27–29, 1978, Proceedings*, Warsaw, 180–200.

SELF, S., THORDARSON, TH. & KESZTHELYI, L. 1997. Emplacement of continental flood basalt lava flows. *In*: MAHONEY, J.J. & COFFIN, M.F. (eds) *Large Igneous Provinces. Continental, Oceanic, and Planetary Flood Volcanism*. Geophysical Monograph, **100**, 381–410.

SELF, S., THORDARSON, T. & KESZTHELYI, L., ET AL. 1996. A new model for the emplacement of Columbia River Basalts as large inflated pahoehoe lava flow fields. *Geophysical Research Letters*, **23**, 2689–2692.

SKOGSEID, J., PEDERSEN, T., ELDHOLM, O. & LARSEN, B.T. 1992. Tectonism and magmatism during NE Atlantic continental break-up: the Vøring margin. *In*: STOREY, B.C., ALABASTER, T. & PANKHURST, R.J. (eds) *Magmatism and the causes of continental break-up*. Geological Society, London, Special Publications, **68**, 305–320.

SMITH, C.B. & LORENZ, V. 1989. Volcanology of the Ellendale lamproite pipes, Western Australia. *Proceedings of the Fourth International Kimberlite Conference, Perth 1986, 1*, Geological Society of Australia Special Publication, **14**, 505–519.

STACHEL, T., LORENZ, V., SMITH, C.B. & JAQUES, A.L. 1994a. Evolution of four individual lamproite pipes, Ellendale volcanic field (Western Australia). *Proceedings of the Fifth International Kimberlite Conference, Araxá, Brazil 1991. Companhia de Pesquisa de Recursos Minerais-CPRM-Special Publication* **1/A** January 1994, 177–194, Brasilia.

STACHEL, T., LORENZ, V. & STANISTREET, I. 1994b. Gross Brukkaros (Nambia) – an enigmatic craterfill reinterpreted as due to Cretaceous caldera evolution. *Bulletin of Volcanology*, **56**, 386–397.

STEININGER, J. 1819. *Geognostische Studien am Mittelrheine*, Kupferberg, Mainz, 223 pp.

STOLLHOFEN, H. 1986. *Stratigraphisch–tektonische und lagerstättenkundliche Untersuchungen im Bereich der Bauwald- und Lemberg-Kuppeln (Saar–Nahe-Gebiet, SW-Deutschland) unter besonderer Berücksichtigung der Uran- und Quecksilbervererzungen*. Diplomarbeit, Universität Mainz.

STOLLHOFEN, H. 1991. *Die basalen Vulkaniklastika des Oberrotliegend im Saar–Nahe-Becken (SW-Deutschland)*. Dr.rer.nat thesis, Universität Würzburg.

STOLLHOFEN, H. 1992. Lithofaziesanalyse vulkaniklastischer Gesteine des Oberrotliegend: Beispiele für Eruptions-, Transport- und Ablagerungsmechanismen von Pyroklastika im permokarbonen Saar–Nahe-Becken (SW-Deutschland). *Nachrichten der Deutschen Geologischen Gesellschaft*, **48**, 99–100.

STOLLHOFEN, H. 1993. Features and effects of synsedimentary tectonics in fluvial sequences of the Permocarboniferous Saar–Nahe-Basin (SW-Germany). *5th International Conference on Fluvial Sedimentology, Brisbane, Australia 1993, Abstracts*, 127–128.

STOLLHOFEN, H. 1994a. Vulkaniklastika und Siliziklastika des basalen Oberrotliegend im Saar–Nahe-Becken (SW-Deutschland): Terminologie und Ablagerungsprozesse. *Mainzer Geowissenschaftliche Mitteilungen*, **23**, 95–138.

STOLLHOFEN, H. 1994b. Synvulkanische Sedimentation in einem fluviatilen Ablagerungsraum: Das basale 'Oberrotliegend' im Saar–Nahe-Becken. *Zeitschrift der Deutschen Geologischen Gesellschaft*, **145**, 343–378.

STOLLHOFEN, H. 1994c. The complex interplay between bimodal volcanism, fluvial sedimentation and basin development in the Permo-Carboniferous Saar–Nahe Basin (SW Germany).

15th International Association of Sedimentology, Regional Meeting on Sedimentology, Ischia, Italia, 1994, Abstracts, 390-391.

STOLLHOFEN, H. 1994d. Interaktion zwischen Extensionstektonik, Sedimentation und Magmatismus: das vulkanische Rotliegend des Saar–Nahe-Beckens (SW-Deutschland). *Nachrichten der Deutschen Geologischen Gesellschaft*, **52**, 137-138.

STOLLHOFEN, H. 1997. The interaction of strike-slip deformation, sedimentation and magmatism: the Permo-Carboniferous Saar–Nahe-Basin (SW-Germany). *8th International Association of Sedimentology, Regional Meeting 1997, Heidelberg, Germany, Excursion A1 Field Guide – GAEA Heidelbergensis*, **4**, 1-13.

STOLLHOFEN, H. 1998. Facies architecture variations and seismogenic structures in the Carboniferous–Permian Saar–Nahe Basin (SW Germany): evidence for extension-related transfer fault activity. *Sedimentary Geology*, **119**, 47-83.

STOLLHOFEN, H. 2000. Sequence stratigraphy in continental setting – examples from the Permo-Carboniferous Saar–Nahe-Basin, SW Germany. *Zentralblatt für Geologie und Paläontologie, Teil 1*, **1999/3-4**, 233-260.

STOLLHOFEN, H. & HANEKE, J. 1993. Kontrollfaktoren und Fazies vulkaniklastischer Sedimentation: das 'Rhyolithkonglomerat' im Rotliegend des südlichen Saar–Nahe-Beckens. *Sediment 93, Marburg 1993, Geologica et Palaeontologica*, **27**, 301-302.

STOLLHOFEN, H. & HENK, A. 1994. Struktureller und sedimentärer Ausdruck von Transfer-Störungen zweiter Ordnung. – Eine Fallstudie. *Nachrichten der Deutschen Geologischen Gesellschaft*, **52**, 138-139.

STOLLHOFEN, H. & LORENZ, V. 1992. Tektonostratigraphische Analyse und Modellierung der Interaktion von Sedimentation, Tektonik und Vulkanismus in spätorogenen, intermontanen Becken (Kollapsbecken). *Vortragskurzfassungen, Einführungskolloquium zum Schwerpunktprogramm der DFG in Gießen, 1992. 'Orogene Prozesse: ihre Quantifizierung und Simulation am Beispiel der Varisciden'*, Institut für Geowissenschaften, Universität Gießen, 75-76.

STOLLHOFEN, H. & LORENZ, V. 1993a. Vulkanostratigraphie einer bimodalen Post-Kollisionssuite: Beispiele aus dem spätvariszischen Saar–Nahe-Becken. *Terra Nostra*, **1**, 33-34.

STOLLHOFEN, H. & LORENZ, V. 1993b. Entwicklung eines quantitativen tektonostratigraphischen Modelles für Sedimentation, Tektonik und Vulkanismus in spätorogenen Kollapsbecken – vorläufige Ergebnisse des ersten Projektjahres. *Vortragskurzfassungen, Kolloquium im Schwerpunktprogramm der DFG 'Orogene Prozesse, ihre Quantifizierung und Simulation am Beispiel der Varisciden'*, Institut für Geowissenschaften, Universität Gießen, 52-55.

STOLLHOFEN, H. & LORENZ, V. 1994. Interaktion von Sedimentation, Tektonik und Vulkanismus im spätorogenen Saar–Nahe-Becken. *Terra Nostra*, **3**, 98-99.

STOLLHOFEN, H. & STANISTREET, I.A. 1994. Interaction between bimodal volcanism, fluvial sedimentation and basin development in the Permo-Carboniferous Saar–Nahe-Basin (SW-Germany). *Basin Research*, **6**, 245-267.

STOLLHOFEN, H., FROMMHERZ, B. & STANISTREET, I.G. 1999. Volcanics as discriminants in evaluating tectonic versus climatic control on the Permo-Carboniferous continental sequence stratigraphic record, Saar–Nahe-Basin, Germany. *Journal of the Geological Society of London*, **156**, 801-808.

STOREY, B.C., ALABASTER, T. & PANKHURST, R.J. (eds) 1992. Magmatism and the causes of continental break-up. *Geological Society, London, Special Publications*, **68**.

SWANSON, D.A. & HOLCOMB, R.T. 1990. Regularities in growth of the Mount St. Helens dacite dome 1980–1986. *In*: FINK, J.H. (ed.) *Lava Flows and Domes*. Springer-Verlag, Heidelberg, 9/3-9/24.

TEICHMÜLLER, M., TEICHMÜLLER, R. & LORENZ, V. 1983. Inkohlung und Inkohlungsgradienten im Permokarbon der Saar–Nahe-Senke. *Zeitschrift der Deutschen Geologischen Gesellschaft*, **134**, 153-210.

THEUERJAHR, A.-K. 1971. Die Jeckenbacher Schwelle. Ein paläogeographisches Element der Saar–Nahe Senke zur Rotliegendzeit. *Abhandlungen des Hessischen Landesamtes für Bodenforschung*, **60**, 298-307.

THEUERJAHR, A.-K. 1973. *Geochemisch-petrologische Untersuchungen an jungpaläozoischen Rhyolithen des Saar–Nahe-Gebietes*. Dr.rer.nat thesis, Universität Mainz.

THEUERJAHR, A.-K. 1986. Beitrag zur Genese der jungpaläozoischen Rhyolithe des Saar–Nahe-Gebietes (SW-Deutschland). *Geologische Jahrbuch von Hessen*, **114**, 209-226.

THOMSON, K. 2002. Sill complex geometry and growth: a 3D seismic perspective. *In*: BREITKREUZ, C., MOCK, A. & PETFORD, N. (eds) *First International Workshop on the Physical Geology of Subvolcanic Systems – Laccoliths, Sills, and Dykes (LASI) Abstracts and Field Guide*. Geologisches Institut der Universität Freiberg, Germany. Wissenschaftliche Mitteilungen, **20**, 55-57.

ULRYCH, J., CAJZ, V., ADAMOVIVIC, J. & BOSÁK (eds) 2002. *Hibsch 2002 Symposium, Teplá, Ústi, Mariánske Lázne, Excursion guide, abstracts*. Praha, Czech Geological Survey, 1-119.

VETTER, U. & JUNG, D. 1971. Kuselite. *In*: ATZBACH, O. & SCHWAB, K. (eds) *Geologische Karte von Rheinland–Pfalz 1:25 000. Erläuterungen Blatt 6410 Kusel*, 1-96, Geologisches Landesamt Rheinland–Pfalz, Mainz.

VINX, R. 1974. *Die Gesteine der Lembergintrusion/ Naheberland. Ein Beitrag zum Verständnis des permischen Vulkanismus des Saar–Nahe-Beckens (SW-Deutschland)*. Dr.rer.nat thesis, Universität Hamburg.

VON GÜMBEL, C.W. 1846. Geognostische Bemerkungen über den Donnersberg. *Neues Jahrbuch für Mineralogie, Geognosie, Geologie und Petrefaktenkunde*, 534-576.

VON GÜMBEL, C.W. 1848. Nachtrag zu den geognostischen Bemerkungen über den Donnersberg.

Neues Jahrbuch für Mineralogie, Geognosie, Geologie und Petrefaktenkunde, 158–168.

VON SECKENDORFF, V. 1990. Geologische, petrographische und geochemische Untersuchungen an permischen Magmatiten im Saarland (Blatt 6507 Lebach). *Geologisch–Paläontologisches Institut und Museum, Universität Kiel, Berichte-Reports*, **39**, 1–232.

VON SECKENDORFF, V., ARZ, C. & LORENZ, V. 2004. Magmatism of the late Variscan intermontane Saar–Nahe Basin (Germany): a review. *In*: WILSON, M., NEUMANN, E.-R., DAVIES, G.R., TIMMERMAN, M.J., HEEREMANS, M. & LARSEN, B.T. (eds) *Permo-Carboniferous Magmatism and Rifting in Europe*. Geological Society, London, Special Publications, **223**, 361–391.

WADGE, G. 1981. The variation of magma discharge during basaltic eruptions. *Journal of Volcanology and Geothermal Research*, **11**, 139–168.

WAIGHT, T.E., DEAN, A.A., MAAS, R. & NICHOLLS, I.A. 2000. Sr and Nd isotopic investigations towards the origin of feldspar megacrysts in microgranular enclaves in two I-type plutons of the Lachlan Fold Belt, southern Australia. *Australian Journal of Earth Sciences*, **47**, 1105–1112.

WALKER, G.P.L. 1984. Downsag calderas, ring faults, caldera sizes, and incremental caldera growth. *Journal of Geophysical Research*, **89**, 8407–8416.

WHITE, R.S. 1992. Magmatism during and after continental break-up. *In:* STOREY, B.C., ALABASTER, T. & PANKHURST, R.J. (eds) 1992. *Magmatism and the Causes of Continental Break-up*. Geological Society, London, Special Publications, **68**, 1–16.

WILSON, M., NEUMANN, E.-R., DAVIES, G.R., TIMMERMAN, M.J., HEEREMANS, M. & LARSEN, B.T. (eds) 2004. Permo-Carboniferous rifting and magmatism in Europe. *Geological Society Special Publications*, **223**.

WIMMENAUER, W. 1974. The alkaline province of Central Europe and France. *In*: SØRENSEN, H. (ed.) *The Alkaline Rocks*. John Wiley, New York, 238–271.

WOHLETZ, K. & HEIKEN, G. 1992. *Volcanology and Geothermal Energy*. I–XIV, 1–432, University of California Press, Berkeley, Los Angeles, Oxford.

WÖRRLEIN, C. 2002. *Hydrodynamische Vermischungsprozesse – 'Magma Mingling'. Ein Erklärungsansatz für die Entstehung der basischen Einschlüsse im Kreuznacher Rhyolith*. Diplomarbeit, Universität Würzburg.

WRIGHT, T.L. & FISKE, R.S. 1971. Origin of the differentiated and hybrid lavas of Kilauea Volcano, Hawaii. *Journal of Petrology*, **12**, 1–65.

ZIEGLER, P.A. 1990. *Geological Atlas of Western and Central Europe*. Shell Internationale Petroleum Maatschappij, Den Haag.

ZIMANOWSKI, B. 1998. Phreatomagmatic explosions. *In*: FREUNDT, A. & ROSI, M. (eds) From magma to tephra. *Developments in Volcanology*, **4**, 25–54.

ZIMANOWSKI, B. 1998a. Premixing of magma and water in MFCI experiments. *Bulletin of Volcanology*, **58**, 491–495.

ZIMANOWSKI, B., BÜTTNER, R., LORENZ, V., HÄFELE, H.-G. 1997b. Fragmentation of basaltic melt in the course of explosive volcanism. *Journal of Geophysical Research*, **102**, 803–814.

ZIMANOWSKI, B., BÜTTNER, R. & NESTLER, J. 1997c. Brittle reaction of high temperature ion melt. *Europhysics Letters*, **38**, 285–289.

ZIMANOWSKI, B., WOHLETZ, K.H., DELLINO, P. & BÜTTNER, R. 2003. The volcanic ash problem. *Journal of Volcanology and Geothermal Research*, **122**, 1–5.

The Neogene to Recent Rallier-du-Baty nested ring complex, Kerguelen Archipelago (TAAF, Indian Ocean): stratigraphy revisited, implications for cauldron subsidence mechanisms

B. BONIN[1], R. ETHIEN[2,5], M. C. GERBE[2], J. Y. COTTIN[2], G. FÉRAUD[3], D. GAGNEVIN[4], A. GIRET[2], G. MICHON[2] & B. MOINE[2]

[1]*UPS–CNRS–FRE 2566 'Orsayterre', Département des Sciences de la Terre, Université de Paris-Sud, F-91405 Orsay Cedex, France (e-mail: bbonin@geol.u-psud.fr)*
[2]*CNRS–UMR 6524 'Magmas et volcans', Département de Géologie–Pétrologie–Géochimie, Université Jean-Monnet, F-42023 Saint-Etienne Cedex 2, France*
[3]*CNRS–EP 125, Département des Sciences de la Terre, Université de Nice – Sophia Antipolis, F-06108 Nice Cedex, 2 France*
[4]*Department of Geology, University College Dublin, Belfield, Dublin 4, Eire.*
[5]*GEMOC, Macquarie University, Sydney, Australia*

Abstract: The Kerguelen Archipelago is made up of a stack of thick piles of Tertiary flood basalts intruded by transitional to alkaline igneous centres at various times since 30 Ma ago. In the SW, the Rallier-du-Baty Peninsula is mostly occupied by two silicic ring complexes, each with an average diameter of 15 km, comprising dissected calderas cross-cut by sub-volcanic cupolas. Previous radiometric determinations yield ages ranging from 15.4 to 7.4 Ma in the southern centre, and 6.2 to 4.9 Ma in the northern one. The felsic ring dykes were injected by coeval mafic magmas, forming, successively, swarms of early mafic enclaves, disrupted synplutonic cone sheets, and late cone-sheets. After the emplacement and subsequent unroofing of the plutonic ring complexes, abundant and thick trachytic pyroclastic flows and falls were emitted from the younger caldera volcanoes, while hawaiite and mugearite lava flows were erupted from marginal maars and cones. Huge trachyte ignimbritic flows filled the glacial valleys in the central Peninsula, and capped lacustrine deposits and older lava flows, while related pumice falls are widespread throughout the archipelago. This powerful plinian eruption took place after the network of glacial valleys was established, but before the Little Ice Age that occurred during the last centuries. In the south of the peninsula, even younger trachytic formations are exposed, and fumarolic vents are still active.
The growth mechanisms of a caldera-related ring complex can be explained as a repetitive sequence of two eruptive episodes. The first episode of hydrofracturing, induced by volatile exsolution within the evolving magma chamber, creates a vertical circular fracture zone, along which highly vesiculated magmas are emitted during explosive eruptive events occurring at the surface in a caldera volcano. It is followed by a second episode of cauldron-subsidence of a crustal block down to the degassed magma chamber, induced by pressure release. Downward movement of the crustal block favours the emplacement at shallow depths within the older caldera-filling formations, of discrete magmatic sheets characterized by a 16-km mean diameter and a 1-km mean thickness, corresponding to an average unit volume of 200 km^3. Actually, the estimated volumes of the different igneous episodes within the Rallier-du-Baty nested ring complex vary from 60 to 900 km^3, and correspond to the production during 15 Ma of about 2800 ± 850 km^3 of new materials and a net crustal growth of about 100 ± 30 × 10^3 m^3 per year.

Introduction

In the twentieth century, caldera-related ring complexes were the subject of numerous studies. These were initiated by the seminal paper of Clough *et al.* (1909) on the district of Glen coe, Argyllshire, Scotland, where, for the first time, they describe a cauldron, in the sense of an eroded caldera with inverted relief, formed by subsidence having affected an area, roughly oval in shape, delineated by a boundary fault. Observing that uprise of a series of marginal intrusions accompanied the subsidence, they suggest that these intrusions could have been

Fig. 1. The original concept of cauldron subsidence, as redrawn from figure 14 in Clough et al. (1909).

related to subterranean foundering, favouring the emplacement of shallow plutons resembling bell-jars in form (Fig. 1).

The cauldron-subsidence hypothesis was subsequently validated by Richey who interpreted the Mourne Mountains, Northern Ireland (Richey 1928) and afterwards the intrusives (Richey 1932) of all the British Volcanic Tertiary Province (Richey 1935) as ring complexes, most of them emplaced within a coeval volcanic blanket. Anderson (1936), on the same vertical axis, related caldera volcanoes and plutonic ring complexes, and offered the first mathematical theory based on fluctuating magmatic pressure within a deep magma chamber, parabolic in form. The 1930s saw the discovery of numerous ring structures throughout the world (e.g. Kingsley 1931; Bain 1934), so that, since then, cauldron subsidence-related ring complexes have been defined a worldwide class of intrusive plutons (Billings 1943, 1945; Bonin 1986).

From a structural point of view, ring complexes share some characteristics with laccoliths, e.g. their diameters (7–15 km) and thicknesses (0.5–1.0 km) plot in the laccolith field of Mc-Caffrey & Petford (1997), defined by $T = 0.12 L^{0.88}$, where T is the thickness, L the length, 0.12 a constant, and 0.88 (±0.10) the power-law exponent. Taking into account the circular shape of ring complexes, the relationship becomes $T = 0.22 R^{0.88}$, where R is the radius. However, ring complexes differ from laccoliths by having steeply outward-dipping to subvertical cylindrical, not rectilinear, feeder zones, and by providing space by floor subsidence rather than by floor lifting. The dimensional similarity, defined by $T = bL^a$ with $a < 1$, shared by sills, laccoliths and plutons (Cruden & McCaffrey 2001) with ring complexes, suggests that they form by similar processes.

The Rallier-du-Baty ring complex, Kerguelen Archipelago, is located within an oceanic island on top of a plateau characterized by a 22-km thick oceanic crust (Charvis et al. 1995). Abundant field evidence, such as intrusive contacts, chilled margins, xenoliths and screens of country rocks, contact aureoles, etc., was collected during summer expeditions. Like the typical ring structures emplaced in the continental crust, the Rallier-du-Baty igneous complex is made up of several discrete intrusions, associated with coeval volcanic formations. As such, it provides a good area for testing, in an oceanic environment, the current models of pluton emplacement by lateral spreading and self-affine inflation. This paper evaluates the feasibility of the different growth mechanisms by reviewing the field evidence and by assessing how space was created at very shallow depths within the crust.

The Rallier-du-Baty nested ring complex: geodynamic settings

In the Indian Ocean, the Kerguelen Plateau and Broken Ridge were originally contiguous and split off 46–43 Ma ago (Munschy et al. 1994, and references therein). Their estimated combined crustal volume of approximately 57×10^6 km³ is comparable to the Ontong Java Plateau in the Pacific – the largest documented oceanic plateau on Earth (Schubert & Sandwell 1989). Located within the Antarctic plate, the 2000-km long and 200–600-km wide Kerguelen Plateau is located at latitudes between 46 and 64°S and its mean altitude is 2000 to 4000 m higher than the neighbouring oceanic basins (Schlich 1994).

The Kerguelen Archipelago (6500 km²) (Giret 1993) is the third largest oceanic island after Iceland and Hawaii (Giret et al. 1997). It is located at the northern tip of the Kerguelen Plateau at latitudes 48–50°S and longitudes 68.5–70.5°E (Fig. 2). Mont Ross, its highest point, culminates at 1850 m above sea-level. The Rallier-du-Baty Peninsula, with an area of approximately 800 km², corresponds to 12% of the emergent land.

A brief overview of the Kerguelen Archipelago

Coarse-grained intrusive rocks represent a significant proportion of the archipelago, currently covering an area of about 8% of the land, whereas finer-grained intrusive rocks correspond to about 7% and volcanic, dominantly basaltic, units occupy 85% (Nougier 1970). Plutonic and volcanic rocks yield identical isotopic ratios (Dosso et al. 1979; Dosso & Murthy 1980)

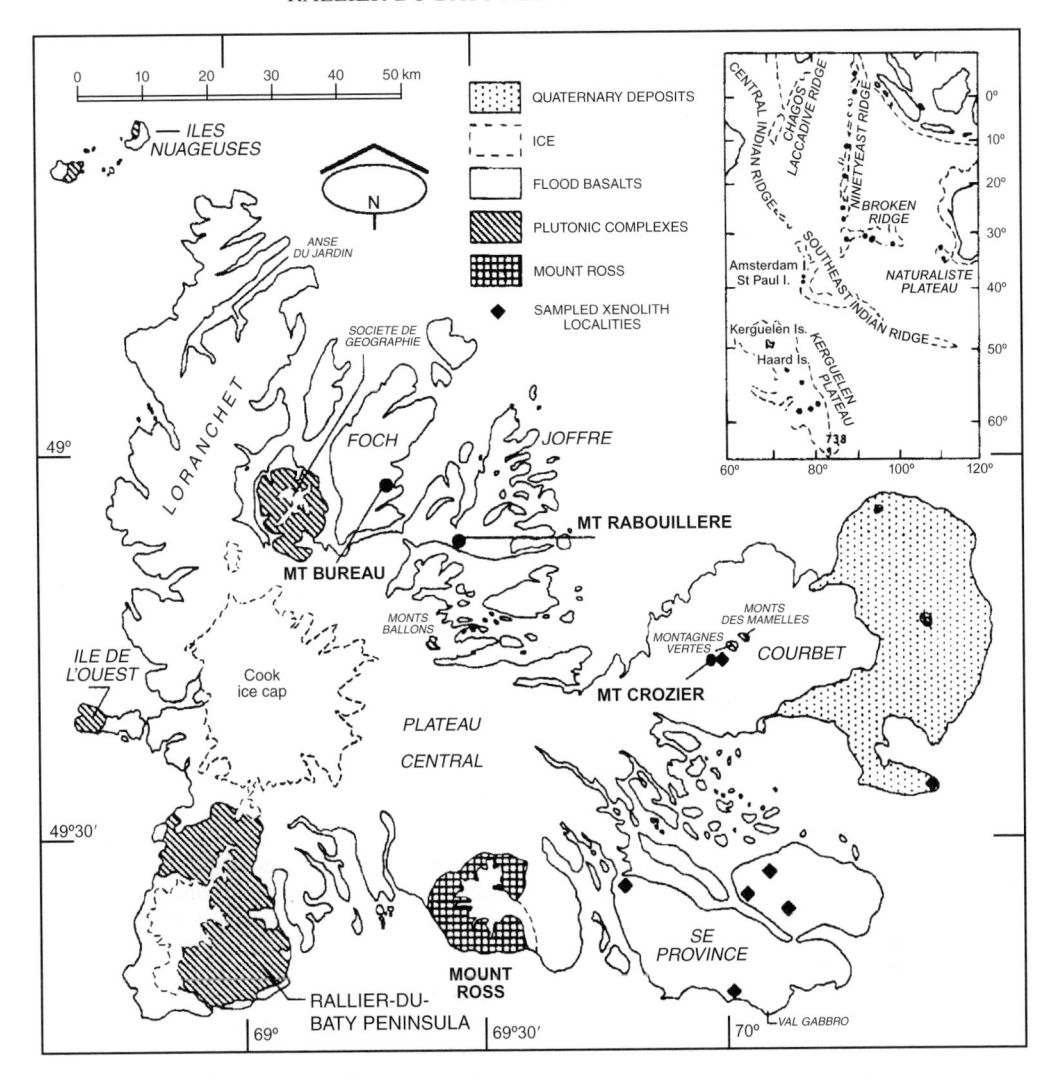

Fig. 2. Map of the Kerguelen Archipelago, showing the location of the Rallier-du-Baty Peninsula and the major features of the main island (after Gagnevin et al. 2003). Diamonds: localities where lower crust xenoliths were sampled (Gregoire et al. 1998); dots: localities where stratigraphic sections of flood basalts were extensively sampled. Inset: position of the Kerguelen Archipelago within the Indian Ocean; dots: drilling and dredging sites.

and no significant evolution of the source composition is recorded for the the last 30 Ma (Yang et al. 1998). The igneous complexes were emplaced within older flood basalts and are cross-cut or capped by younger pyroclastic and lava flows. Their sizes vary from as little as 1 km, up to 18 km. The available radiometric ages range from 39 to 0.13 Ma (Giret 1990; Weis et al. 1998), although younger age determinations are expected (work in progress). The largest igneous complex (about 80% of the intrusive rocks) is located in the Rallier-du-Baty Peninsula, in the SW of the archipelago (Fig. 2).

Previous work on the Rallier-du-Baty igneous complex

After the first study of an aegirine-bearing syenite by Lacroix (1924), the large extension of syenite and granite was observed and mapped by Aubert de la Rue (1931) and Nougier (1970).

The plutonic complex, involving a suite of gabbro, diorite, syenite and granite, cross-cuts Tertiary basalt lava flows and is covered by Quaternary basic to felsic volcanic formations (Marot & Zimine 1976; Lameyre et al. 1981). K–Ar and Rb–Sr isotopic studies in plutonic rocks document ages ranging from 15.4 to 4.9 Ma, short-lived magmatic processes and a pure mantle signature (Lameyre et al. 1976; Dosso et al. 1979). Combined field and laboratory works during the first half of the 1970s resulted in a geological map at the scale of 1:50 000 of the Rallier-du-Baty Peninsula (Giret 1980), where five discrete intrusive centres are distinguished and labelled (Fig. 3), according to their geographical position, as southern (S), intermediate (I), western (W), northern (N) and northeastern (NE).

From 1976 to 1994, no supplementary field studies were performed on the peninsula. As glaciers and ice sheets throughout the archipelago are currently being subjected to protracted reduction, new clean exposures have been revealed, and a reassessment of the geological map of the northern part of the Rallier-du-Baty Peninsula was unavoidable. A 400-km² area south of the Cook ice sheet, corresponding to one-half of the peninsula, was systematically surveyed during the CARTOKER 94–95 summer expedition in order to provide an updated field-controlled map (Fig. 4). CARTOKER is an acronym for 'Cartographie de Kerguelen', which means Kerguelen Mapping.

The Rallier-du-Baty nested ring complex: field notes

Country rocks

The country rocks are made up of an extensive pile of flood basalts that have suffered extensive zeolite alteration. Early Miocene minimum ages are suggested by the 15 Ma maximum age determined for the zeolitization event (Giret et al. 1992). In the nearby Mont Ross area, the unaltered feeder dykes of flood basalts yield K–Ar conventional ages of 24–20 Ma (Weis et al. 1998). The whole of the volcanic pile could not be observed: the lower units being hidden under the sea. About 1000 m thickness of flood basalts can be observed. Two discrete sequences are displayed:

1 The lower sequence forms small hills above the sea and around the large, flat valleys. Lava flows and associated breccias fill up a dense network of channels. The 200-m thick section displays, from bottom to top:

- a 150-m thick series of 5- to 10-m thick basaltic lava flows, with rough columnar jointing. In each lava flow, emplacement within shallow water is indicated by a brecciated lower zone with clasts decreasing in size from 4 cm at the base to 0.5–0.2 mm to the top, overlain by an aphanitic upper zone containing 1-mm to 5-cm wide vesicles, filled or unfilled with late-stage minerals, indicating rapid degassing and air-cooling.
- 0.5- to 1.5-m thick breccias, intercalated in places among lava flows. Scoriaceous bomb fragments provide evidence of renewed volcanic activity after a period of quiescence.
- a 50-m thick chaotic breccia, completing the sequence. It contains a package of metre-scale lava tongues, scoriaceous bomb fragments, some of them with a rounded shape, and lapilli, set within a palagonitic lapilli–ash matrix, suggesting aerial explosive volcanic eruption and emplacement in shallow water.

2 The more monotonous upper sequence forms high plateaux and summits, including the base of the Cook ice sheet. The c. 800-m thick series is composed of a number of extensive 15–20-m thick lava sheets, covering areas of more than 100 km². In this part of the archipelago, outpouring of lava flows occurred at or above sea-level. Intercalated 0.1–0.5-m thick red layers correspond to hydrothermally altered vitreous ash falls (Bellair et al. 1965). Breccias and conglomerates with up to 30-cm diameter boulders are exposed at various levels of the sequence. They were derived from nearby high ground and deposited during periods of volcanic quiescence and erosion.

Vertical dykes and associated horizontal sills were observed locally in the two sequences of the flood basalt pile. Their average widths are less than one metre, and their basaltic composition is identical to that of the lavas. They are, therefore, considered as feeder conduits of the overlying flows and sheets.

Lavas are aphanitic alkali basalt, subsequent alteration is evidenced by late crystals growing in vugs and vesicles. Silica minerals (quartz, chalcedony, and/or opal), and calcite are invariable components of the hydrothermal association, while adularia and celadonite are less widespread. A regional-scale zeolitization event affected the archipelago 15–12 Ma ago (Giret et al. 1992). A heulandite–scolecite–stellerite–stilbite zeolite assemblage was observed in the SE of the

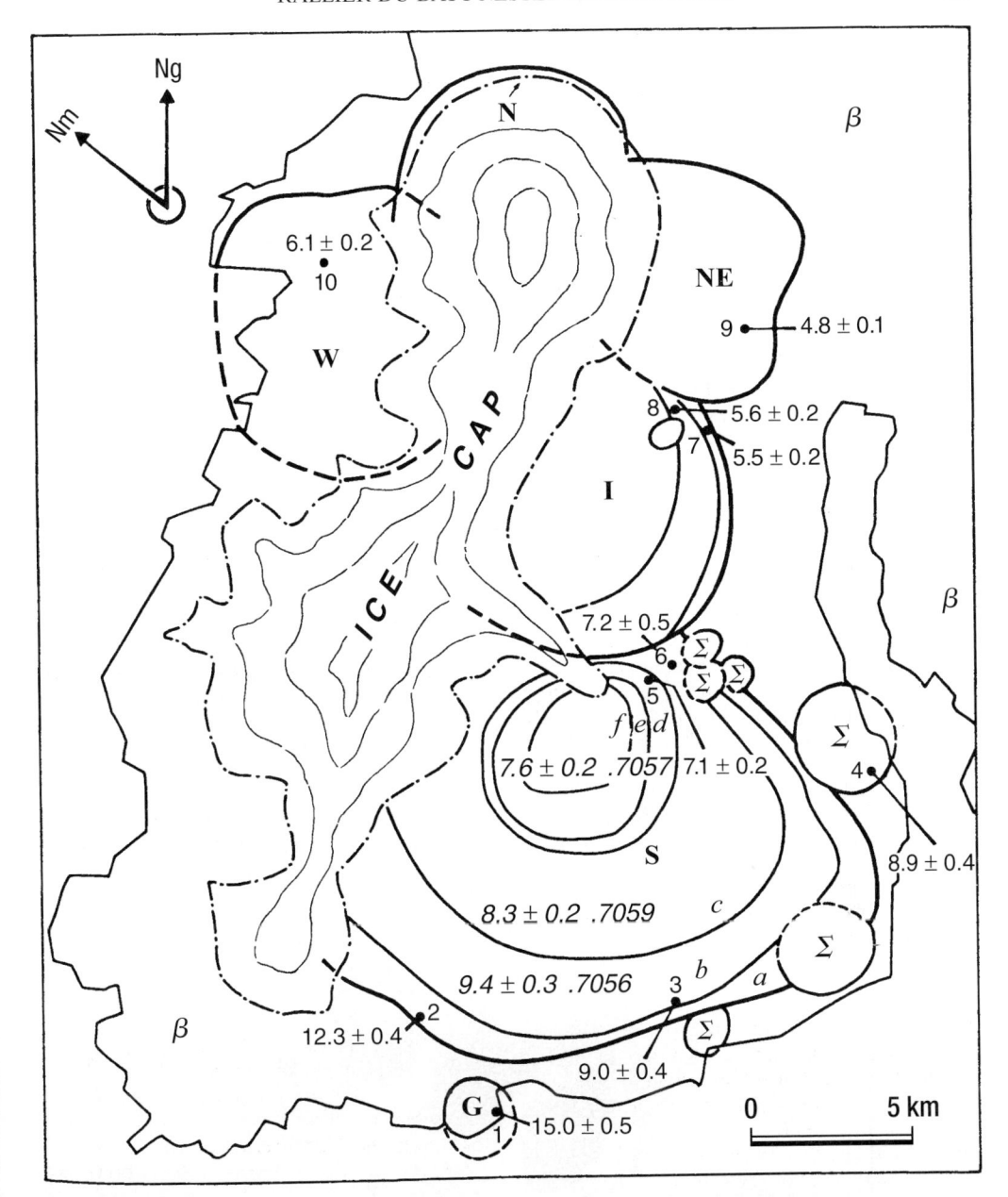

Fig. 3. Preliminary sketch map of the Rallier-du-Baty nested ring complex identifying S, I, NE, N and W massifs, and isotopic data (after Lameyre *et al.* 1976): numbers in italics, Rb–Sr isochron method; other numbers, sampling localities and K–Ar conventional method. Nm, magnetic North; Ng, geographical North. In the southern ring complex: β, basalt; G, gabbro; Σ, syenite satellite cupolas; *a*, *b*, *c* and *d*, syenite–granite units 1, 2, 3 and 4; *e* and *f*, assumed inner units (see text).

Cook ice sheet (Verdier 1989). Estimated temperatures of 110 to 180 °C versus burial depths of 1600 to 2500 m identify high thermal gradients of 70 to 100 °C km^{-1}.

This event was followed locally by the thermal event induced by the emplacement of the Rallier-du-Baty igneous complex itself. No zeolites remain in vugs and vesicles within a 3-km

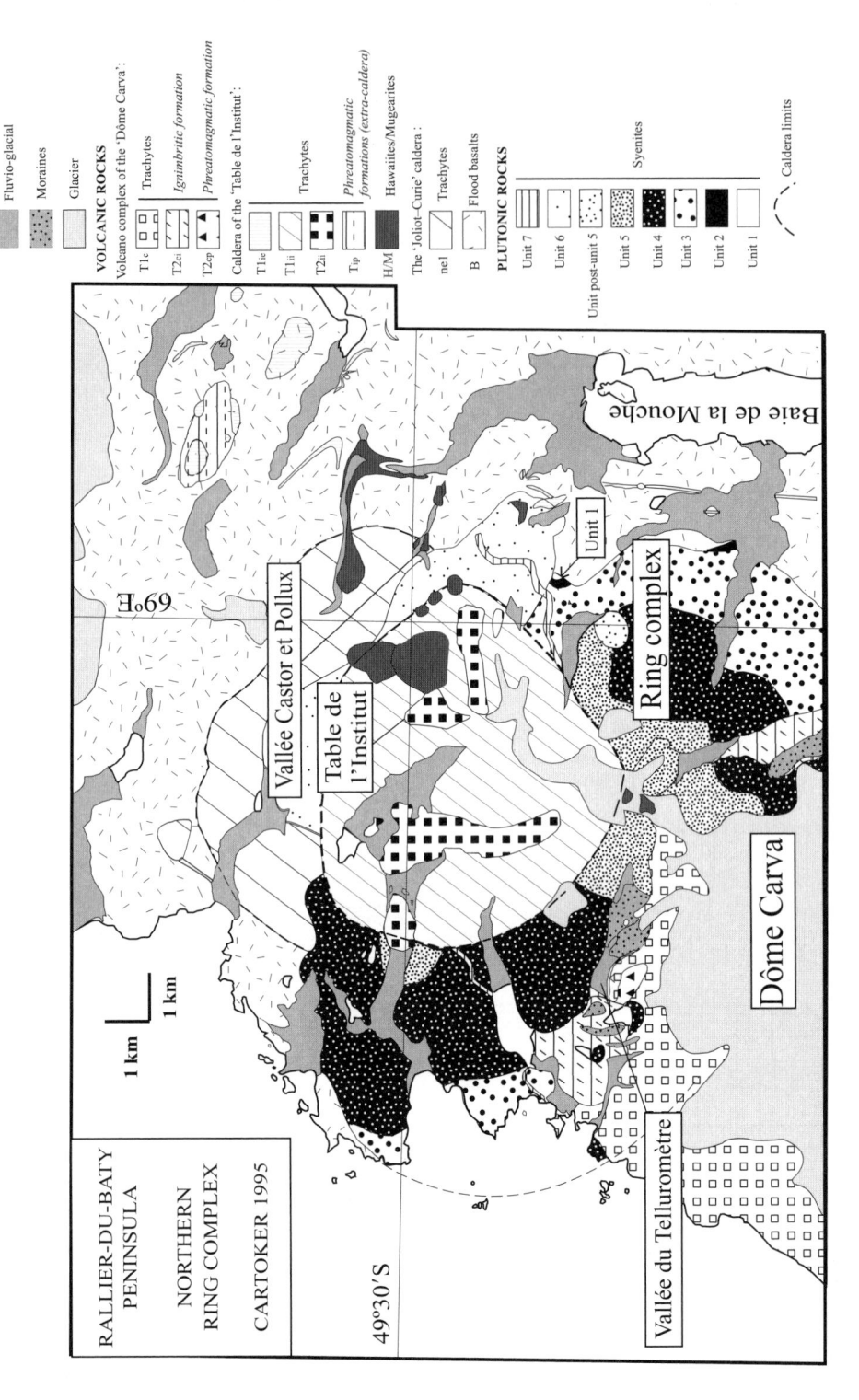

Fig. 4. Geological map of the northern ring complex, as identified during the CARTOKER 94–95 resurvey (slightly modified after Gagnevin *et al.* 2003).

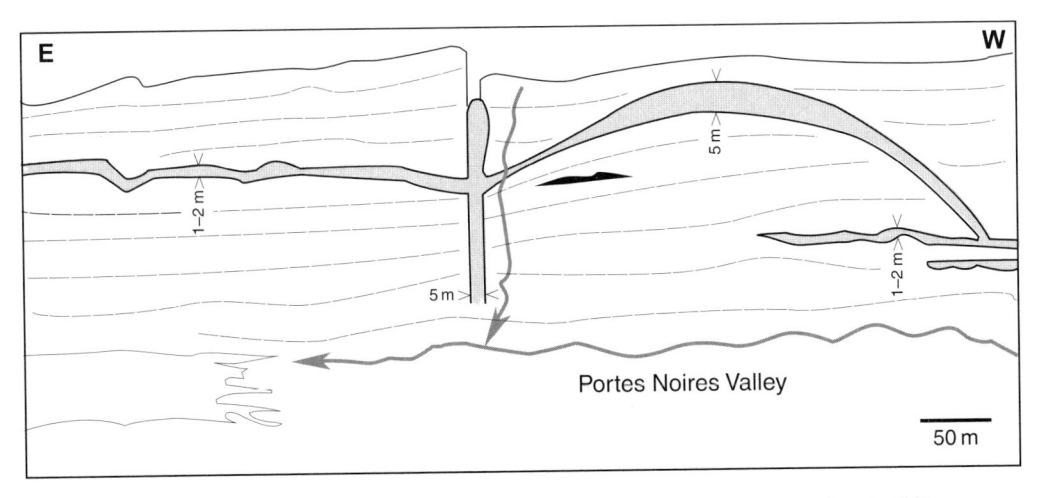

Fig. 5. Sketch of the dyke-and-sill network, on the right side of the Portes Noires Valley (Bonin 1995, unpublished). A dyke is related to curved and jumping sills of trachytic composition (in grey). A smaller sill is made up of obsidian (in black).

wide contact aureole, which is constituted by three zones:

1. the outer zone is characterized by the greenschist-facies association at 3 km to 600 m away from the contact;
2. the middle zone is marked by the albite–epidote hornfels facies association at 600 to 100 m away from the contact;
3. the 100 m-wide inner zone is occupied by the amphibole-hornfels facies. A basalt sampled at 30 m from the contact of the southern centre yields a K–Ar thermal reset age of 11.5 ± 0.2 Ma (Nougier 1970).

External trachytic domes and associated pumice deposits

In the SE of the Cook ice sheet, rounded summits culminating at about 600 m above sea-level are occupied by 0.8- to 2.5-km wide domes of slightly porphyritic trachyte. The volcanic structures were built by a series of contrasting eruptive styles. The early hydro-plinian to plinian episodes resulted in the deposition of thick pumice tuff and breccia rings and ended with the emplacement of multiple extrusions capped by a 250–300-m high main dome. Tuff rings are composed of inward-dipping stratified deposits, composed of white to light-yellow centimetre-scale trachyte pumice, black 1–2-cm wide obsidian balls, and 5-cm wide basalt fragments, set within an ash matrix, and occasionally of breccias composed of volcanic bombs. Well-defined horizontal lamination and vertical columnar jointing indicate that domes were emplaced in highly viscous and even rigid states, and cooled very slowly. A loose network of dykes and sills corresponds to the feeder conduits of the trachyte domes (Fig. 5). North–South-trending 5-m thick dykes and 1- to 5-m thick sills are filled with porphyritic trachyte, while narrower (30–40-cm thick) sills are filled with vitreous rhyolite (Gagnevin *et al.* 2003). Although the eruption ages are unknown, the well-preserved primary volcanic features suggest relatively recent ages.

Centimetre- to decimetre-thick reworked pumice deposits partly fill in valleys and low relief to the east. The primary air-fall deposits driven by westerly winds were removed later on by snow and water. As these deposits are more abundant and thicker at the vicinity of trachytic domes, they are likely to have been derived at least partly from the early plinian episodes. A blanket of pumice deposits of unknown age covers all of the Kerguelen archipelago.

The southern ring complex

The intrusive centre S, the best known, provides an example of a classical ring complex, with ring dykes and satellite intrusions intercalated with screens of older basalt. The central part is occupied by large glacial valleys, lowlands and flat hills, culminating at 549 m above sea level. The border zones constitute an amphitheatre of crests with elevations up to 865 m above sea-level. Access to the area is easy through sandy bays and large breaches opened by Quaternary glaciers. The formations are well exposed on cliffs and hills.

Field evidence was examined in detail during summer expeditions from the 1950s to the 1970s and resurveyed during the CARTOKER 94–95 summer expedition. One single ring complex is identified, constituted by large ring dykes, arcuate dykes and small cupolas aligned along the boundary fault (Fig. 4). Its elliptical outline is defined by a 17-km long east–west major axis and a 15-km long north–south minor axis (Marot & Zimine 1976; Lameyre et al. 1981). The Rb–Sr ratios display such a large range of values that the whole-rock Rb–Sr isochron method can be used successfully for age determination of at least three ring dykes, supplemented by K–Ar ages (Lameyre et al. 1976; Dosso et al. 1979).

Early composite intrusions Two 1.5–2.0-km wide composite intrusions are exposed on the southern side of the complex. At the contact, basaltic wallrocks dip 20–30° southwards and are brecciated by aplitic injections. A bimodal association of gabbro and syenite, each end-member occupying 10–15% of the area, is accompanied by a compositionally expanded monzogabbro–monzodiorite–monzonite hybrid suite, where mafic enclaves of two size ranges: 1–10 m and 50–100 m, are set within a felsic matrix. The oldest rock of the ring complex is the Anse du Gros Ventre gabbro, as shown by K–Ar ages of 15.4 ± 0.5 Ma on whole rock and 13.6 ± 0.4 Ma on biotite (Dosso et al. 1979).

Plutonic ring dykes In the 1:50 000 geological map of the centre S (Giret 1980) eight ring dykes, four late arcuate dykes and two types of satellite cupolas are indicated. From the re-appraisal of field evidence during the 1994–1995 summer expedition, the number of ring dykes was reduced from eight to four. Ring dykes are numbered hereafter from 1 to 4 from older to younger, i.e. from the rim to the core of the igneous intrusion.

1 The hypersolvus alkali-feldspar quartz-syenite unit 1 is directly at the contact with the basalt country rocks. Its preserved thickness ranges from 200 m up to 500 m. In the 50-m thick marginal zone, the texture is porphyritic, with alkali-feldspar megacrysts, and two types of enclaves of wallrocks are exposed: metre-scale rounded blocks occur at 30 m from the contact, whereas smaller, but angular, blocks at only 1–2 m from the contact show no signs of rotation and show limited stoping effects. The rest of the ring dyke is occupied by a coarser-grained rock. The K–Ar age of 12.6 ±

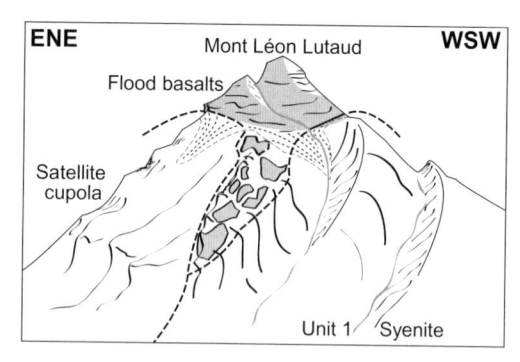

Fig. 6. Relationships between Neogene flood basalts (in grey), unit 1 syenite of the southern ring complex and Anse Syenite satellite cupola (both in white). Note the brecciated aspect of the basaltic screen between the two intrusions (after a sketch by Marot & Zimine 1976).

0.4 Ma is consistent with the Rb–Sr isotope data.

2 The hypersolvus alkali-feldspar quartz-syenite unit 2 is exposed inside and topographically below the unit 1. Its observed thickness ranges from 200 to 500 m. Along the contact, hectametre-scale disrupted screens of basalt are locally preserved. Elsewhere, unit 2 yields a marginal aplitic facies enclosing 0.5–1.0-m elongated enclaves of quartz syenite of unit 1. The Rb–Sr age of 9.7 ± 0.2 Ma substantiates a c. 2.9-Ma period of quiescence between the emplacement of units 1 and 2. A late aplitic dyke yields a consistently lower K–Ar age of 9.2 ± 0.4 Ma.

3 The hypersolvus alkali-feldspar quartz-syenite unit 3 is the most extended intrusion. Its roof has been removed by erosion and its floor by the later ring intrusion, so that its thickness is unknown, but could have exceeded 600 m. At the contact with unit 2, a marginal fine-grained facies has been identified in some places. Mafic (microgabbro) and felsic (trachyte) dykes and cone sheets constitute characteristic features. A Rb–Sr age of 8.6 ± 0.1 Ma substantiates a c. 1.1-Ma period of quiescence between the emplacement of units 2 and 3.

4 The heterogeneous unit 4 encompasses hypersolvus alkali-feldspar quartz syenite to granite rock types. It occupies the central part of the southern ring complex. Its total thickness could exceed the 500 m observed on the cliffs.

The outer contact with unit 3 is almost vertical, with angular outlines, and turns horizontal some 50 m below the Les Deux Frères summit. Within

500 m of the contact, the quartz syenite of unit 3 is hydro-fractured and forms breccias with 10-cm blocks of altered syenite set within a 2-cm wide dark-green aplitic matrix. Blocks of alkali-feldspar quartz syenite of unit 3 have been converted into a secondary fenite assemblage, while the dark aplitic matrix is rockallite, characterized by aegirine needles. The marginal facies of unit 4 is highly oxidized, with joints coated with bluish oxides and hydroxides. Fenitization and hydroxidation processes indicate that fluid saturation occurred during magma cooling of unit 4 and exsolved fluids which escaped through the unit 3 country rocks.

From the contact, inward textural variations include a hypersolvus alkali-feldspar quartz syenite, with less than 20% quartz in volume; a hypersolvus alkali-feldspar granite, with about 25% quartz; and a hypersolvus alkali-feldspar granite, with a radioactive accessory mineralogy. Widespread pegmatitic pockets are filled with quartz and alkali feldspar. Molybdenite platelets occurs within hydraulic breccias, with a matrix composed of quartz + calcite ± aragonite + hematite ± lepidocrocite.

At the inner contact below unit 4, a hypersolvus alkali-feldspar quartz syenite is exposed, which is likely to correspond to a foundered part of the unit 3 ring dyke.

A seven-point whole-rock Rb–Sr alignment has been obtained, including samples from the quartz-syenite–granite association of unit 4 and from the lower quartz syenite. The composite isochron yields a well-defined age of 7.9 ± 0.2 Ma, substantiating the complete thermal reset of the lower quartz syenite and a short 0.7-Ma period of quiescence between the emplacement of units 3 and 4. The unit 4 marginal facies yields a lower K–Ar age of 7.3 ± 0.2 Ma, which is likely to reflect thermal reset. A similar K–Ar age of 7.4 ± 0.2 Ma is given by a dolerite dyke intruding the unit 4 ring dyke.

Plutonic arcuate dykes Four incomplete ring dykes, 4–7 km in length, emplaced at, or near, the contact between units 2 and 3, are elongated along 45° to 80° arcs of circle and referred to hereafter as arcuate dykes. Their 150–300 m thickness has the same order of magnitude as the ring dykes that they intrude. The chief rock type is a porphyritic hypersolvus microsyenite. The first set of dykes, grossly coeval with unit 3, includes a 4-km long, 200-m thick arcuate dyke, with enclaves of unit 2 syenite and forming a breccia at the contact with unit 3 syenite, and another 7-km long 300-m thick arcuate dyke, emplaced within unit 2, crowded with 30% mafic enclaves forming a monzogabbro–monzodiorite–monzonite suite, that could represent dismembered cone sheets. The second set postdates unit 3 with a 6-km long 150-m thick arcuate vertical dyke, with enclaves of unit 3 syenite, and another 5-km long 150-m thick arcuate dyke, with 10-cm wide enclaves of units 2 and 3 syenites, whereas aegirine dykelets cross-cut the wallrocks and the dyke itself.

Satellite cupolas The eastern part of the boundary fault of the southern ring complex is occupied by a bimodal population of cupolas. The first population comprises four small cupolas, with an average diameter of 300–400 m, whereas the second has a diameter of 3.0–3.5 km. Both basalt wallrocks and unit 1 quartz syenite were intruded, with subvertical contacts at the walls changing in a short distance into a horizontal contact at the roof. Brecciated screens of basalt emphasize the contact between the cupolas and the outer unit 1 ring dyke (Fig. 6). The K–Ar age of 9.1 ± 0.4 Ma on the Anse Syénite whole rock is consistent with field evidence that the cupola intrudes the *c*. 12-Ma unit 1. It is identical to the K–Ar age of 9.2 ± 0.4 Ma age obtained on an aplitic dyke cutting the alkali-feldspar quartz syenite of unit 2. The Anse Syénite cupola could therefore be younger than the unit 2 ring dyke and older than the unit 3 ring dyke. No age determinations are available so far for the other cupolas.

The northern ring complex

The northern part of Rallier-du-Baty igneous complex was *terra incognita* before the 1960s. The glacial-carved landscape is dominated by a central ridge of snow-covered or ice-capped summits, with elevations above sea-level varying between 1202 m (Dôme Carva) and 730 m (Table de l'Institut). Deeply dissected valleys, in which torrential rivers are flowing eastwards, westwards, or southwards, show characteristic glacial U- and V-shapes. The west coast is barely accessible by sea, but can be reached from the east by 500-m high passes open through the central ridge. Slopes are covered with scree and valley bottoms are filled with glaciofluvial gravel, whereas cliffs and river banks display reasonably fresh exposures.

Preliminary surveys during the 1974–1976 summer expeditions suggested four nested intrusive centres I, W, NE and N (Fig. 3), comprising no less than 17 different rock types and units (Giret *et al.* 1980; Lameyre *et al.* 1981), although K–Ar ages suggest only a 1.3 ± 0.4 Ma-long period of intrusion (Lameyre *et al.* 1976). Field evidence was carefully re-examined during

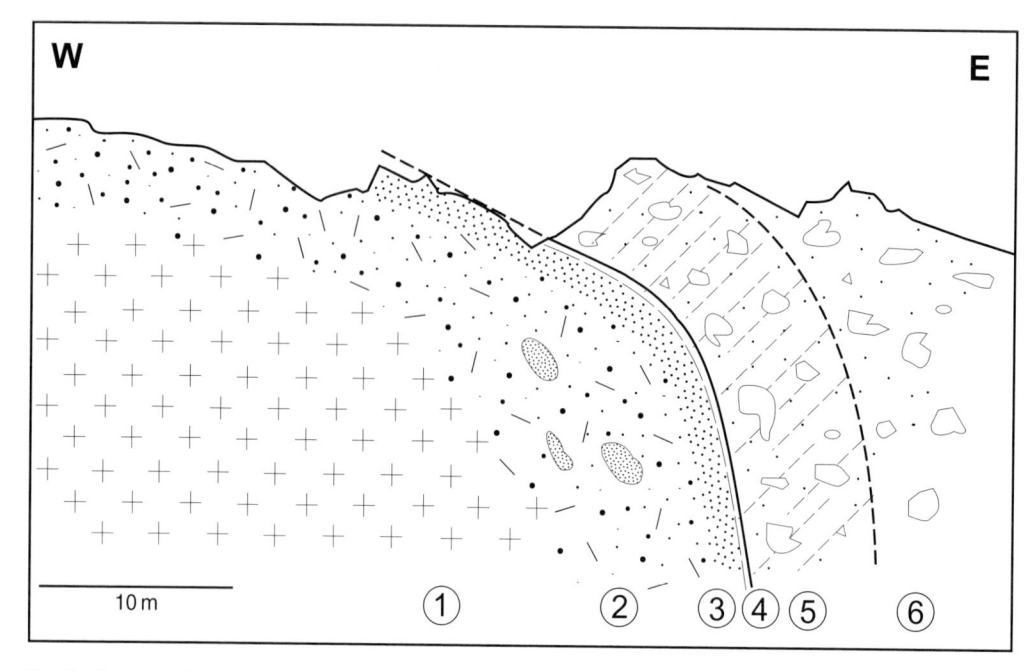

Fig. 7. Cross-section at the contact between Joliot–Curie caldera deposits with unit 6 syenite of the northern ring complex, as observed on the left side of Vallée des Contacts: 1, porphyritic coarse-grained zone; 2, medium-grained zone, with felsic fine-grained enclaves; 3, fine-grained zone; 4, finely fractured zone at the contact; 5, hornfelsed pyroclastic breccia; 6, pyroclastic breccia (sketch by Bonin 1995, unpublished).

the CARTOKER 94–95 summer campaign and critical interpretation of the observations resulted into the identification of only one single ring complex (Fig. 4), slightly larger in size than the southern structure, with a 18-km long N–S major axis and a 16-km long east–west minor axis. Seven discrete ring dykes and cone sheets, which intruded a dissected early caldera, were cut by two later (caldera) volcanoes.

Pre-plutonic caldera-filling volcanic formations: the Joliot-Curie caldera   A 10 km × 1.5 km raft of trachyte pyroclastites was discovered in the northernmost part of the igneous complex, where it occupies the highest drainage basins of the rivers. It is bordered to the north by flood basalts and to the south by quartz syenite. The observed caldera-filling series comprises:

1 a poorly stratified lower sequence of pumice breccias, and
2 an upper sequence of pyroclastic flows with typical ignimbrite features.

The base and the total thickness of the series could not be observed. Intrusive, often autoclastic, domes and related feeder dykes are composed of porphyritic trachyte, which can be highly vesiculated.

The outer contact with the older horizontal flood basalts of the upper sequence consists of a vertical ring fault, culminating at Pic Joliot-Curie (1005 m above sea-level) and often hidden by scree and glaciers. The western segment of the border fault is marked by a double alignment of intrusive domes fed by dykes. The inner contact is outward dipping and displays, from the intrusive centre to the pyroclastites, the unit 6 coarse-grained porphyritic alkali feldspar quartz syenite passing to a 10-m thick medium-grained zone, with felsic fine-grained enclaves, to a 2-m thick fine-grained zone, to a 10-cm thick finely fractured zone at the contact, whereas pyroclastic breccias, converted into hornfels in a 5-m thick zone, are intruded by trachyte intrusions (Fig. 7).

Plutonic ring dykes and cone sheets   Intrusive units were emplaced as either cupolas and ring dykes, or cone sheets. From the reappraisal of field evidence, they are numbered hereafter from 1 to 7, according to their order of decreasing age from the oldest to the youngest (Table 1).

1 The poorly represented hypersolvus alkali-feldspar syenite unit 1 covers a very small area on a 500-m high hill at the eastern side of the

Table 1. *The revised chronological sequence of igneous rocks: Rallier-du-Baty nested ring complex*

Geological units and formations	Emplacement age	Method	Reset age	Remarks	Reference
Pic Saint-Allouarn volcano	Post-Wurm to the seventeenth century	Field evidence			1
Dôme Curva volcano Late ignimbrite and pumiceous falls	Post-Wurm before the seventeenth century	Field evidence		Radiometric dating currently in progress	5
Table-de-l'Institut caldera volcano Trachyte pyroclastic flow in Vallée des Sables	1.15 ± 0.05 Ma	WR K–Ar		Radiometric dating currently in progress	2
Northern ring complex Quartz trachyte unit 7	4.9 ± 0.02 Ma	WR K–Ar			2
Trans-solvus alkali-feldspar quartz syenite unit 6		K-feldspar K–Ar	4.9 ± 0.2 Ma	Sample close to quartz trachyte unit 7	
Hypersolvus alkali-feldspar quartz syenite unit 5	6.2 ± 0.2 Ma	WR K–Ar		Date not consistent with field evidence	
Hypersolvus alkali-feldspar quartz syenite unit 4	5.6 ± 0.2 Ma	K–Ar on gabbro			
Hypersolvus alkali-feldspar syenite–gabbro unit 3	5.7 ± 0.2 Ma	K–Ar on syenite			
Hypersolvus alkali-feldspar granite unit 2					
Hypersolvus alkali-feldspar syenite unit 1					
Joliot–Curie caldera volcano	> 6.2 Ma	Field evidence			5
Southern ring complex Dolerite dyke cutting unit 4	7.4 ± 0.2 Ma	WR K–Ar	7.3 ± 0.2 Ma	Marginal facies close to a dolerite dyke: thermal reset or fenitization effect ?	2
Hypersolvus alkali-feldspar quartz syenite–granite unit 4 floored by hypersolvus alkali-feldspar quartz syenite (foundered units 3 and 2?)	7.9 ± 0.2 Ma	WR Rb–Sr			
Hypersolvus alkali-feldspar quartz-syenite unit 3	8.6 ± 0.1 Ma	WR Rb–Sr			
Satellite cupola: Anse Syénite hypersolvus alkali-feldspar quartz syenite	9.1 ± 0.4 Ma	WR K–Ar			
Aplitic dyke cutting unit 2	9.2 ± 0.4 Ma	WR K–Ar			
Hypersolvus alkali-feldspar quartz-syenite unit 2	9.7 ± 0.2 Ma	WR Rb–Sr			
Hypersolvus alkali-feldspar quartz-syenite unit 1	c. 12 Ma	WR Rb–Sr			
	12.6 ± 0.4 Ma	WR K–Ar			
Early composite intrusion: Anse du Gros Ventre gabbro		Biotite K–Ar	13.6 ± 0.4 Ma	Reset or cooling age?	1
	15.4 ± 0.5 Ma	WR K–Ar	11.5 ± 0.2 Ma	Thermal effects at 30 m from the contact with unit 1 of the southern ring complex	
Flood basalt basement		Field evidence	15–12 Ma	Zeolitization event in the archipelago	3
	24–20 Ma	WR K–Ar		Feeder dykes of flood-basalt basement of the nearby (40 km) Mont Ross volcano	4

(1) Nougier (1970), (2) Dosso *et al*. (1979), (3) Giret *et al*. (1992), (4) Weis *et al*. (1998), (5) Gagnevin *et al*. (2003).

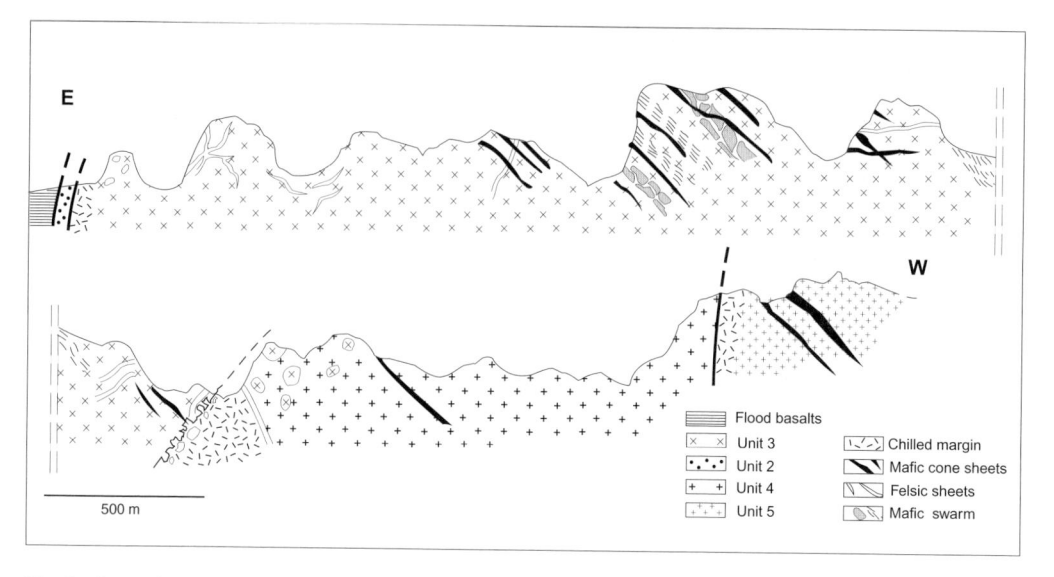

Fig. 8. General cross-section of the Vallée Milady, displaying the relationships between the different intrusive units and with the bimodal network of cone sheets and enclave swarms in the northern ring complex.

ring complex. The outer contact with basaltic wallrocks is marked by a chilled margin in syenite, whereas basalt was converted into amphibole hornfels.

2 The hypersolvus alkali-feldspar granite unit 2, sparsely exposed along the eastern contact, is observed on a hill where it intrudes unit 1, and on a 500-m long, 100-m high whaleback hummock on the left bank of Vallée Milady, where it intrudes basaltic wallrocks (Fig. 10). The outer contact with basaltic wallrocks is marked by quartz-rich pegmatitic and aplitic streaks within quartz-poor (20–25% in volume) granite. From the contact outwards, the recrystallized basalt shows a 1-m wide zone affected by magmatic stoping marked by a network of aplitic dykes, and a 10-m wide zone where columnar jointing was totally destroyed.

3 The composite hypersolvus alkali-feldspar syenite–gabbro unit 3 is one of the most important intrusions and covers a large area in the SE and western parts of the ring complex. The best exposure is located in Vallée Milady, where an almost continuous 2.5-km long horizontal cross-section is displayed (Fig. 10). From east to west, i.e. from outer (=top) to inner (=base) parts of the intrusion, the following facies variations were observed:
- a slightly porphyritic fine-grained marginal facies at the contact with unit 2;
- a porphyritic (0.5-cm long alkali-feldspar phenocrysts) fine- to medium-grained facies;
- a porphyritic (1-cm long alkali-feldspar phenocrysts) medium- to coarse-grained facies: the main facies. It is crowded by places with fine-grained felsic enclaves, interpreted as reworked pieces of chilled margin. Solidus to subsolidus volatile build-up is evidenced by fairly abundant pegmatitic lenses and veins, 5-cm thick amphibole schlieren resembling lindinosite (Lacroix 1922; Orcel 1924) and later millimetre-thick hydraulic breccias. Late 10–20 cm-thick dykelets are composed of alkali-feldspar syenitic aplite and granitic pegmatite.

The porphyritic medium- to coarse-grained main facies is dissected at less than 150 m away from the inner contact with unit 4, and forms 10-m to 1-m diameter angular or rounded stoped blocks set within a dense network of aplitic concentric dykes issued from the unit 4 alkali-feldspar syenite (Fig. 8).

The most striking feature of unit 3 ring intrusion is the supplementary occurrence of bimodal syenite–gabbro suites, emplaced as 0.5- to 10-m thick cone sheets into the solidifying syenite during at least three intrusive episodes (Fig. 8):
- the earliest suite constitutes an inward (westward)-dipping net-veined complex of miarolitic cavity-bearing alkali-feldspar syenite with coeval hybrid enclaves occurring in parallel streaks;
- the subsequent suite is characterized by a second inward-dipping net-veined complex

of fine-grained alkali-feldspar syenite and coeval mafic enclaves with cauliflower-shaped fractal contacts;
- the latest suite comprises a parallel set of inward-dipping 50-cm to 5-m thick felsic and mafic cone sheets, both with columnar jointing.

Net-veined complexes were preferentially emplaced into the porphyritic medium- to coarse-grained main facies, while cone sheets are developed within the marginal facies as well. Similar 5.7 ± 0.2 and 5.6 ± 0.2 Ma K–Ar ages on syenite and gabbro, respectively (Dosso et al. 1979, Table 1), agree with field evidence.

Elsewhere, a nearly vertical outer contact with recrystallized basaltic wallrocks is emphasized by aplitic syenite dyke swarms developed within a 200-m thick zone. Farther to the south, outward-steeply dipping sharp contacts with the intrusive units of the southern ring complex are exhibited, where the unit 3 syenite contains fine-grained enclaves, interpreted as dissected chilled margin, and miarolitic cavity-bearing xenoliths of quartz syenite and granite.

4 The hypersolvus alkali-feldspar quartz syenite unit 4 is one of the largest intrusions. It is exposed in the central and western parts of the ring complex, where it is in contact with basaltic country rocks. The outer contact with basaltic wallrocks, always marked by the finer-grained texture of unit 4 syenite, is nearly vertical at low elevation, outward-dipping and relatively flat, with intercalated basalt screens at higher elevations (about 250 m above sea-level).

Unit 4 comprises two unequally exposed facies. The less-represented facies is composed of coarse (5- to 7-mm long mesoperthitic alkali-feldspar crystals) even-grained alkali-feldspar quartz syenite, but the more extensive facies is a heterogeneous alkali-feldspar quartz syenite, with a medium-grained groundmass enclosing 10-cm large pegmatitic lenses and miarolitic cavities, providing evidence of magma volatile oversaturation. Miarolitic cavities, arranged in a network defined by 30-cm wide intervals, are infilled with alkali feldspar, quartz, fluorite and other minerals. Hydraulic breccias are represented by a 100-m wide cylindrical vent, infilled with a package of cemented syenite blocks, and by several smaller vents filled with silica minerals and sulphides (mainly pyrite), inducing stained and rusty aspect of the heterogeneous syenite.

Cone sheets are less numerous than within unit 3, and almost always composed of microsyenite. They yield sharp contacts and a symmetrical flowage structure derived from the Bagnold effect, with a vesicular core zone crowded with elongate 1-cm long alkali-feldspar phenocrysts, and aphanitic border zones with scarce globular 1- to 1.5-cm diameter alkali-feldspar phenocrysts. Cone sheets were emplaced later than hydraulic breccias.

Relationships with unit 5 are difficult to decipher, as most contacts are covered with scree. In a 10-m wide contact zone, unit 5 medium-grained syenite, containing decimetre-scale round enclaves of unit 4 heterogeneous syenite, develops metre-scale pillow-like rafts within unit 4 syenite. This observation suggests emplacement of unit 5 magma when the still-hot unit 4 was not yet consolidated and perhaps partly liquid.

5 The hypersolvus alkali-feldspar quartz-syenite unit 5 constitutes the most central part of the ring complex. The exposed areas were disconnected, due to subsequent emplacement of the Table de l'Institut caldera volcano. Salient features of unit 5 are abundant quartz (up to 15% in volume) and biotite as the chief mafic mineral component.

Unit 5 shows no contact relationships with basaltic country rocks. It is adjacent to unit 4, and the only exposed contact yields ambiguous features (see above). The unit 5 quartz syenite is intruded by the unit 6 quartz-syenite cupola, and the unit 7 quartz-trachyte dyke.

The K–Ar ages (Table 1), 6.2 ± 0.2 Ma on whole rock and 4.9 ± 0.2 Ma on alkali feldspar, should be reassessed. The 'old' 6.2-Ma age does not agree with field observations that unit 5 quartz syenite is younger than the 5.6–5.7-Ma unit 3 gabbro-syenite. The 'young' 4.9-Ma age was likely to be reset during emplacement of the later unit 7 quartz-trachyte dyke.

Units 4 and 5 syenites are cross-cut by a 1-km wide quartz-poor hypersolvus alkali-feldspar granite stock, which, accordingly, constitutes a post-5 unit. No contact relationships with units 6 and 7 are displayed in the field, so that no more precise chronology of the stock can be ascertained.

6 The porphyritic trans-solvus alkali-feldspar quartz-syenite unit 6 occupies the NE and northernmost parts of the ring complex. The salient feature of unit 6 syenite is the trans-solvus alkali-feldspar assemblage (Bonin 1972; Martin & Bonin 1976), with euhedral mesoperthite crystals set within a groundmass composed of quartz, K-feldspar, and albite graphic intergrowths. At the contact with wallrocks, recrystallized basalt of the lower

sequence hosts a network of aplitic syenite dykes, and unit 6 syenite encloses large basalt screens. Inwards, textural variations within unit 6 syenite include: a chilled margin with acicular alkali feldspar, a fine-grained facies crowded with 0.5–1-cm diameter miarolitic cavities set at 5- to 10-cm wide intervals, a medium-grained facies, and a coarse-grained facies with scarce fine-grained enclaves.

Contact relationships with Joliot–Curie caldera-filling pumice breccias and trachyte domes and with unit 5 quartz syenite confirm that quartz-syenite unit 6 is younger.

7 The quartz trachyte unit 7 is made up of a SE-dipping SW–NE trending, 50- to 200-m thick dyke. The quartz trachyte has rhomb-shaped millimetre-long alkali-feldspar phenocrysts set within a fine-grained to microcrystalline groundmass, and contains devitrification spherulites. It intrudes unit 6 quartz syenite and unit 5 quartz syenite. The K–Ar age of 4.9 ± 0.2 Ma (Table 1) agrees with field evidence, as does the identical alkali-feldspar K–Ar reset age determined on a unit 5 sample collected near the quartz-trachyte dyke.

Later volcanoes

Volcanic formations intrude and/or overlie the Late Miocene plutonic ring dykes and cone sheets in the central and southern parts of the northern ring complex. Two discrete calderas were identified, which document a renewed magmatic activity. The first 10-km diameter caldera is centred on the 730-m high Table de l'Institut. The second 8-km diameter volcano is barely exposed as nunataks under glaciers and centred on the 1202-m high ice-covered Dôme Carva. Pyroclastic and lava flows filled up caldera craters and flowed down outside, along glaciofluvial valleys. Huge amounts of associated air falls were deposited throughout the Kerguelen archipelago.

The Table de l'Institut caldera volcano The Table de l'Institut caldera volcano was discovered by Marot & Zimine (1976). The K–Ar age of 1.15 ± 0.05 Ma (Dosso *et al.* 1979) obtained on an extra-caldera trachyte pyroclastic flow document an apparent 3.75 ± 0.25-Ma long period of magmatic rest. Caldera-related formations occupy a large area within and outside the border ring fault (Gagnevin *et al.* 2003). They are subdivided into:

1 caldera-filling formations,
2 late- to post-caldera trachytic formations, and
3 late- to post-caldera basic to intermediate formations, which were erupted in that order during build-up and subsequent destruction of the volcano.

The border ring fault is frequently hidden by late pyroclastic flows in the valleys and by scree on the summits, but it is marked locally by a 10-m thick porphyritic trachyte ring dyke. Along the near-vertical ring fault, contacts are sharp with older units 4 to 6 alkali-feldspar quartz syenites. In the vicinity of the fault, volcanic formations were tilted, with an inward dip up to 45°, and are cross-cut by a swarm of aplitic dykelets and sills.

1 The caldera-filling formations are composed of pyroclastic flows and falls, lava flows are comparatively rare, and no intercalated epiclastic deposits were identified. Field survey documents a *c.* 600-m thick volcaniclastic pile constituted by three stratigraphic units. As the bottom of the caldera is nowhere exposed, the lowest part of the pyroclastic units remains unknown.

The lower pyroclastic unit was observed on a thickness of 100–120 m. It is composed of conformable coarse layers of trachytic breccia containing angular blocks (5–20 cm) of syenite. Two superimposed series were observed:
(i) a *c.* 20-m thick green lower series and
(ii) a 80–100-m thick black upper series.

The 400–500-m thick upper pyroclastic unit is dominated by barely layered and poorly sorted pumice agglomerates, breccias and tuffs. The sizes of the pumice fragments range from 2 cm to 2 m. Pumice tuffs turn locally into true ignimbrites showing up to 20-cm long and 2-mm thick fiammé, implying high-temperature deposition and welding.

The uppermost Table-de-l'Institut unit is restricted to the top of the 730-m high summit. It is made up of 1-km wide and 50-m thick massive unit of porphyritic trachyte, conformably overlying the upper unit. The horizontal basal contact is completely hidden by scree composed of large boulders of trachyte. The massive lava-like rock is composed of a groundmass, speckled with phenocrysts of anorthoclase–sanidine and rare mafic minerals. This late unit is interpreted tentatively as the result of cooling of an ultimate lava lake at the very end of the syn-caldera volcanic episode.

Mafic rocks occur as hypabyssal units, that could be related for the most part to late- to post-caldera volcanic episodes. The only exception is represented by rare and small (50-m long and 3-m thick) hawaiite–mugearite vesicular lava flows, intercalated within the pumice tuffs of the low part of the upper pyroclastic unit, implying they were erupted during the syn-caldera volcanic episode. Dykes and sills, some of which are related in the form of 'jumping sills', were observed throughout the caldera-filling volcaniclastic pile. Their thicknesses range from 1 to 5 m and they consist of hawaiite–mugearite.

2 The late- to post-caldera trachytic formations are chiefly composed of welded pyroclastic flows. They partly infill the bottom of large U-shaped valleys created by glacial erosion of the volcano. The K–Ar age of 1.15 ± 0.05 Ma was obtained in one of these flows. The flows are arranged as a network radiating from and originating in the central part of the caldera. They are interpreted as the result of late- to post-caldera explosive eruptions occurring during and/or after destruction of the original topography. The eruptive vent could be located in the unexposed icy area south of Table-de-l'Institut.

Trachyte pyroclastic flows are several kilometres long and 30–50-m thick. The most impressive pile inside the caldera displays five discrete superimposed pyroclastic flows, each 50-m thick. They constitute a 250-m high riegel which determined upstream a flat basin, now filled in with volcaniclastic gravel and sand lacustrine deposits, with just a residual lake located in its NW tip. Pyroclastic features, such as up to 30-cm long and 3-cm wide glass shards, and up to 1–2-m diameter blocks of syenite and trachyte set up within a porphyritic groundmass, provide evidence of turbulent to chaotic flow and high-temperature welding. During consolidation, pyroclastic flows developed conspicuous columnar jointing, with the diameters of the prisms about 0.5 m in the upper half, and 1 m in the lower half.

Pyroclastic flows rest generally upon the within-caldera upper pyroclastic unit and units 4 to 6 alkali-feldspar quartz syenites. They overlie seldom alluvial deposits and pumice tuff deposits. Pumice tuff deposits are barely layered at all, and are poorly sorted – also implying turbulent flow, but at lower temperature. Their thicknesses are various and can reach up to 150 m. From stratigraphic evidence, pumice tuff deposits are interpreted as resulting from incipient explosive eruptions, that subsequently emitted the associated pyroclastic flow.

3 The late- to post-caldera mafic units are made up of two discrete rock types, issued from eruptions of contrasting styles: hawaiite, speckled with plagioclase phenocrysts and largely exposed in the NE valleys, and mugearite, bearing ternary feldspar phenocrysts and restricted to the Vallée du Telluromètre.

Within the caldera, and along the border ring fault, hydromagmatic eruptions created maars, and emitted surges and lava flows. Four 0.4- to 1.5-km-diameter maars were identified in the plateau below and to the SE of Table-de-l'Institut. Although eroded, the maar structure is grossly preserved, with a flat-dipping outer tuff-ring and a steeply dipping inner tuff-ring. The nearby basement yields impact figures caused by the ballistic fall of large blocks. From bottom to top, tuff-rings are made up of a 5–10-m thick lower unit of breccia to cinerite, with blocks of basement rocks showing no evidence of fresh magma, and an up to 150-m thick upper unit, characterized by cauliflower-shaped mafic bombs and/or soaked vitreous lapilli set within a palagonitic matrix, providing evidence of the incorporation of fresh magma into a wet environment. The upper unit is intruded by rheomorphic breccia dykes. The individual layers display dune and anti-dune structures, along with highly variable grain sizes and centimetre- to decimetre-scale thicknesses.

Outer tuff-rings are so well developed that they can constitute a nearly uniform blanket on the plateau below the Table-de-l'Institut. A basal unit of breccia containing vitreous lapilli is overlain by an upper unit of breccia with angular blocks of basement rocks (syenite, trachyte) and of bombs of hawaiite, providing evidence of transition from hydromagmatic to strombolian dry magmatic styles of eruption. Downstream, this sequence of eruptive events is illustrated by basal surges capped by 5–15-m thick block (aa) lava flows.

Other hawaiite lava flows issued from cinder cones located outside the caldera. In the Vallée des Chicanes, a highly eroded 1-km large cinder cone constitutes a small riegel on the right bank of the river. It was built by strombolian-type eruptions, that jetted a 100-m thick blanket of bombs and blocks, filling a former valley, and overlain by three superimposed 5–10-m thick hawaiite block-lava flows, with 0.5–1-m-scale columnar jointing. Upstream, a supplementary hawaiite block-lava flow was emitted from a small vent related to a mafic dyke, and overlies a 20–30-m thick spatter of vesiculated bomb fragments.

Mugearite is exposed in the flat area located in the south of the deltaic mouth of the Vallée du Telluromètre. It constitutes a 10–20-m thick lava

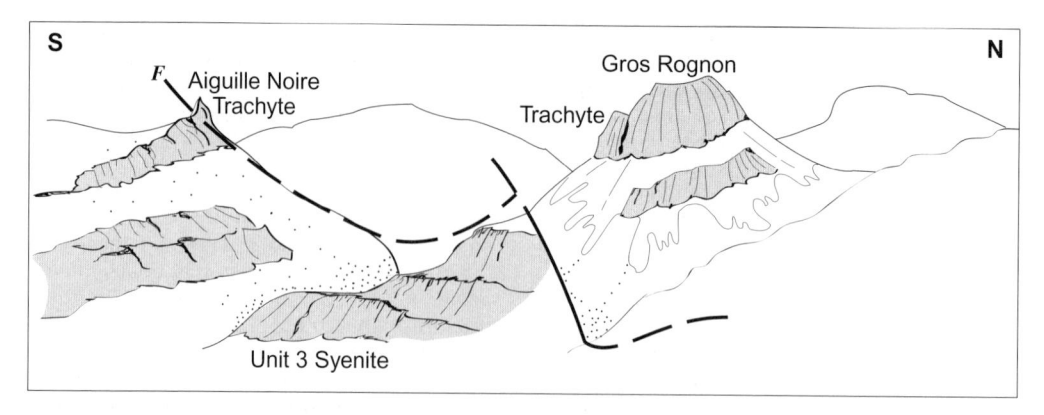

Fig. 9. Sketch of the cliffs above Vallée de Larmor (Bonin 1995, unpublished). The border fault of the Dôme Carva volcano is marked by trachytic pyroclastic flows inside and unit 3 syenite of the northern ring complex outside. The summits in the background are fully covered by ice and snow.

flow filling a former large valley. The related vent was not identified; it is probably hidden by later volcanic formations upstream. The lava flow rests directly upon unit 3 and 4 alkali-feldspar quartz syenites and, locally, alluvial deposits, with no intercalated spatters, and is overlain by pyroclastic formations issued from the Dôme Carva volcano. Columnar jointing is well developed. Steam that issued from the boiling of water contained in the alluvial deposits excavated a hydrothermal chimney through the lower part of lava flow and created a cave used by seal hunters.

Dôme Carva volcano The Dôme Carva volcano was identified during the CARTOKER 94–95 summer expedition. At elevations above sea-level higher than 600 m, the topography is entirely dominated by a 50-km^2 ice-sheet culminating at 1202 m near Dôme Carva, with the exception of scarce nunataks: most of them, but not all, being located close to the border fault (Fig. 9). Glacier tongues radiate from the ice sheet and flow downwards. However, their lengths were strongly reduced during the last 30 years, so that they can no longer reach the seashore and only partly fill the glacial valleys that they created or enlarged.

The border fault could not be observed in the field, because it is hidden by scree and ice. It is marked by sharp cliffs composed of trachyte flows. The trachyte volcanic products comprise volcano-filling and late to post-caldera units, erupted during build-up and subsequent destruction of the volcano. No radiometric dates are available so far.

1 Volcano-filling formations are almost completely capped by the ice sheet, except in the periphery of the volcanic complex. Their total thickness is about 300 m – approximately half of that of Table-de-l'Institut caldera formations. The following stratigraphic sequence, exposed in valleys and banks of glaciers, displays, from bottom to top:
- alkali-feldspar quartz-syenite unit 4, constituting the basement;
- three 50-m thick trachyte pyroclastic flows developing 5–10-cm thick lamination and 1–2-m-scale columnar jointing, and yielding the (biotite + sanidine) phenocryst assemblage;
- 100-m thick layered and poorly sorted breccia, with up to 20-cm diameter blocks of obsidian and pumice set within a palagonitic matrix. Enclaves of coarse-grained leuco- to mesocratic alkaline rocks were found as boulders at the exit of a small defile. With their conspicuous magmatic planar foliation, these rocks are likely to represent ejecta from a deep-seated magma chamber. No mantle nodules were found, which provides further evidence that deep-seated reservoirs stored this kind of enclave.
- 100-m thick layered and better-sorted tuffs and breccia, with blocks of obsidian and pumice set within a vitreous matrix;
- two superimposed 20–30-m thick trachytic units forming a plateau like the Table-de-l'Institut unit and tentatively interpreted as filling up a lava lake at the end of the volcanic episode;

2 Late- to post-caldera units are chiefly composed of trachyte pyroclastic flows. They fill the bottom of glacial valleys that were excavated partly through caldera-filling formations, partly outside the caldera.

- The first major event produced a series of 30–50-m thick pyroclastic flows, exposed close to the border ring fault and outside the caldera in the western coastal area. Pyroclastic flows rest directly upon their basement; intercalated pumice tuffs are scarce or lacking. Lava tubes and 5–10-cm thick lamination developed during flowage, followed by 1–2-m-scale columnar jointing during cooling. Porphyritic trachyte is characterized by sanidine phenocrysts up to 5 mm in size.

  The most complete sequence was observed in the Vallée du Telluromètre. A lower unit, formed by a package of small valley-filling linear pyroclastic flows resting upon alkali-feldspar quartz syenite of unit 4, constituted a riegel in the middle part of the valley, where it produced a naturally dammed lake, now filled with finely layered and well-sorted lacustrine silt deposits. It is overlain by an upper unit of large pyroclastic flows, forming a 100-m high cliff above the valley.

  In the western coastal area, the sea-shore is occupied by the lower unit filling the older glacial valleys and resting upon quartz alkali-feldspar syenite units 3 and 4 as well as upon mugearite lava flows of the Table-de-l'Institut caldera. The upper unit is composed of extensive sheets of pyroclastic flows capping a formerly flat area.

- The second major event was marked by a dramatic explosive eruption, that originated in a vent located to the east of Dôme Carva. Its presumably young age is argued by the following evidence: the related tephras, widespread in the Rallier-du-Baty Peninsula, cover all formations up to the Quaternary alluvial fans and are overlain only by the glacier tongues, that made their maximum extension during the Little Ice Age in the eighteenth century (Nougier 1970). Glacier-forming blue ice contains numerous thin ash layers, suggesting that ice was formed by snow compaction at least partly during ash eruptions. Two discrete types of tephra can be distinguished: ignimbritic flows v. ash and pumice falls.

  Ignimbritic flows are heterogeneously welded and were extensively removed by glaciofluvial erosion. They constitute 3–5-m high residual buttes scattered within the bottom of large valleys, or in saddle-like passes between major drainage basins. They are grossly arranged as a network radiating from the unexposed icy area between the La Pérouse and Arago glacier tongues, where the eruptive vent is inferred to be located. The best preserved 130-m thick ignimbritic flow, observed on the left side of the Vallée du Telluromètre, rests upon quartz alkali-feldspar syenite unit 4, forming a rounded butte, and on the lower trachytic unit of the first late- to post-caldera event, and displays from bottom to top:

- a 50-m thick welded ignimbrite displaying occasional 2–5-cm thick lamination and constant 3–5-m large columnar jointing. Huge fiammé, which can reach up to some metres in length and up to 5–10 cm in width, are associated with large blocks of obsidian and vesiculated pumice set within a vitreous-grained matrix. Overall chaotic and welded texture, as well as poor sorting of matrix grains, suggest high-temperature turbulent flow, compaction, and welding.

- 30-m thick unconsolidated tuffs and breccias, above a ledge used by albatrosses for nesting, corresponding to the transition from welded to unwelded materials. Overall chaotic texture and poor sorting of matrix grains suggest again turbulent flow, but unwelded characteristics imply lower-temperature compaction.

- 50-m thick well-bedded and well-sorted ash and lapilli layers suggest air-fall deposition, with no evidence for turbulent flow. They are covered by blocks of sanidine-bearing ignimbritic obsidian and of floating pumice fragments.

Unwelded ash and pumice falls cover large areas on the whole of the Rallier-du-Baty Peninsula. Where these deposits were not protected by ignimbritic flows, they were removed and subsequently redeposited. Epiclastic alluvial fans comprise a number of 10–20-cm thick cross-bedded layers of 2–3-cm long pumice clast-bearing coarse-grained sandstone, intercalated with 10-cm thick well-bedded layers of finer-grained sandstone, carrying crystal lapilli and 1–3-mm long pumice clasts. Contrasting styles of deposition: glaciofluvial lenticular bedding alternating with aeolian well-sorted sand blanketing, resulted from discrete periodic episodes related to climatic changes. Low-relief areas are covered everywhere throughout the Kerguelen Archipelago by reworked 2–3-cm long matrix-supported pumice clasts.

The Pic Saint-Allouarn volcano The southern ring complex is dominated by high summits covered by a 400-m thick ice sheet extending continuously from Dôme Carva to Pic Saint-Allouarn (1189 m above sea-level). Trachyte pyroclastic flows, emerging from below the ice and probably aided by melting of the ice sheet, diverge radially from a centre located at Pic Saint-Allouarn. The present state of the vegetation on the Grande Coulée and the Wurmian glacial deposits overlain by the Coulée de Vulcain indicates that the eruptions have occurred very recently and could have been as late as during the Little Ice Age of the eighteenth century (Nougier 1970).

Thermal activity

Fumaroles were discovered by seal hunters. Six fumarolic fields are known in the Rallier-du-Baty Peninsula, covering a 50-km^2 area centred on Pic Saint-Allouarn: five on its western side and one on its eastern side. Each field comprises at least 10 vents, each with a diameter of about 10 m. The temperature at the vents ranges from 30 to 105 °C and the pH from 3.9 to 9.5. The geothermal source could correspond to a 2-km deep cooling trachytic body (Verdier 1989). Limited activity is shown by the temperature of the hydrothermal fluids being estimated at less than 350 °C, very low contents of reduced carbon and hydrogen species, and scarce hydrothermal deposits (sulphur, pyrite and hematite). The northern fields lack sulphide deposits and yield five times lower radon abundances than the southern fields, indicating that they are less active (Delorme et al. 1994).

Evidence for current thermal activity elsewhere in the archipelago is rather meagre, with mild (18–20 °C) to hot (63 °C) thermal springs in the Central Plateau, and low-temperature carbonated springs west of the Mont Ross volcano (Verdier 1989).

Time durations and dimensions: implications for growth mechanisms

Erosion has removed most of the caldera volcano-filling formations, which are preserved only along the northern rim of the massif. Below the volcanic cover, a large volume of plutonic rocks was also scavenged by glaciers during the Quaternary ice ages. Thus, the volumes of magmas involved during the last 15 Ma in the build-up of the nested ring complex cannot be evaluated accurately. Only rough estimations will be provided.

Age relationships

The stratigraphic–geochronological chart of the Rallier-du-Baty Peninsula (Table 1) shows that the 15 Ma-long set of igneous episodes, starting with the early composite intrusions and persisting with the latest volcanic manifestations, encompasses the Late Neogene activity in the archipelago:

- The 15.4–12.6-Ma plutonic events, represented by the early composite intrusions and the unit 1 ring dyke of the southern ring complex, yield the same ages as the 16.6–12.4 Ma Ile de l'Ouest complex, the 13.7-Ma Société de Géographie pluton and the late 12.6-Ma trachyte dykes of Monts Ballons. They post-date the faintly zeolitized 17-Ma Monts Ballons and Iles Nuageuses complexes (Giret 1990) and are roughly coeval with the extensive zeolitization event affecting the older flood basalts (Giret et al. 1992). All the coeval complexes are located in the western part of the archipelago and yield silica-oversaturated to slightly undersaturated compositions.
- The 9.7–7.4-Ma plutonic events, represented by the other ring dykes and related cone sheets and dykes of the southern ring complex, are coeval with the 10.2–6.6-Ma tephrite–phonolite, strongly silica-undersaturated suite of Ronarc'h and Jeanne d'Arc Peninsulas, located in the SE part of the archipelago (Leyrit 1992).
- By contrast, the 6.2–4.9-Ma plutonic events of the northern ring complex seem to have no igneous counterparts in the rest of the archipelago.
- The 1.15-Ma to recent volcanic activity, manifested by the eruptions from the Table-de-l'Institut, Dôme Carva and Pic Saint-Allouarn (caldera) volcanoes, is tightly coeval with the 1.02–0.13-Ma slightly silica-undersaturated Mont Ross volcano, located 40 km east of the Rallier-du-Baty Peninsula (Weis et al. 1998). However, the activity on the Rallier-du-Baty Peninsula has persisted well after the extinction of the Mont Ross volcano.

The duration of the igneous activity within each complex decreases drastically with time. The discrete episodes which built the southern ring complex are defined within a 8-Ma period of time. The northern ring complex was emplaced during only 1.3 Ma. The Quaternary (caldera) volcanoes were active during about, or less than, 1 Ma. Now is apparently a period of quiescence, but the persistent fumarolic fields show that the igneous activity should not be considered as definitively extinct.

Dimensions and volumes

The two plutonic ring complexes have similar sizes: 17 km (east–west) × 15 km (north–south) for the southern centre, versus 18 km (north–south) × 16 km (east–west) for the northern one. Each plutonic complex is composed of discrete intrusive units, with a limited number of c. 1-km thick ring dykes and numerous 10-m thick cone sheets and arcuate dykes. According to field observations, each complex is about 2–4-km thick and the estimated volumes of plutonic rocks range from 400 to 800 km³ in the southern centre, and from 450 to 900 km³ in the northern one.

Although the coeval volcanic rocks emitted from the now-eroded overlying calderas have almost completely disappeared, their volumes could be of the same order of magnitude, ignoring the huge volumes of volatiles released. The southern ring complex would then correspond to 800–1600 km³ of magmas and the northern one to 900–1800 km³. As it took 8 Ma for the build-up of the southern centre, and only less than 1.5 Ma for the northern one, the average rates of magma supply differed by an order of magnitude: 0.003–0.004 m³ s^{-1} and 0.02–0.04 m³ s^{-1}, respectively.

With the mean diameter of the Table-de-l'Institut caldera volcano being 8 km, the total volume of the 600-m thick caldera-filling trachytic formations is at least 30 km³. This figure represents a minimum, because the base of the volcanic units was not observed. Hawaiite and mugearite products are comparatively more restricted in volume.

The Dôme Carva volcano has a mean diameter of 5 km; the total volume of the 300-m thick volcano-filling formations is about 6 km³. The volume of the 80-m thick ignimbritic flows, which are accompanied by ash air-fall deposits blanketing the whole Kerguelen Archipelago, is estimated at c. 1 km³. Several cubic kilometres of products jetted during this single eruptive episode can be envisaged, suggesting that the whole archipelago could have been covered by an up to 1-m thick blanket of pumice clasts driven from the vent by westerly winds. This explosive event was certainly one of the most powerful eruptions in the recent times in the Kerguelen Archipelago.

In the Yellowstone volcanic system, which yielded sizes of the same order of magnitude as the Rallier-du-Baty nested ring complex, it is assumed that 67 to 75% of the volumes of volcanic products were ejected in the atmosphere and lost during the explosive eruptions (Christiansen & Blanck 1972). If this was the case in the Rallier-du-Baty Peninsula, the Table-de-l'Institut caldera volcano could have erupted a volume of trachyte of about 90–120 km³, and a total volume of 100–150 km³. Although no plutonic equivalents have been identified so far, their volumes could be of the same order of magnitude. For the same reasons, the erupted volume of about 6–7 km³ in the Dôme Carva volcano and the southernmost centres could reflect a more likely figure of 25–30 km³.

Considering the volcanic formations coeval with the plutonic centres, but now vanished; the volcanic projections lost in the atmosphere; and the plutonic rocks coeval with the caldera volcanoes, but not yet exposed, the total volume of magmatic products emitted over 15 Ma could be estimated roughly at 2800 ± 850 km³. Large volumes of magmatic products were removed early during the eruptions and later on by erosion, so that the Rallier-du-Baty Peninsula is occupied now by 1450 ± 450 km³ of new materials, either exposed, or hidden – added to the Kerguelen crust. Their average thickness of 3.5 ± 1.0 km implies a crustal growth of 100 000 ± 30 000 m³ per year (c. 0.003 ± 0.001 m³ s^{-1}).

These values should be considered as minimum numbers. Cumulative xenoliths occur in volcanic breccias, indicating that mafic and intermediate magmas were emplaced within the lower crust, where by various differentiation processes they generated the highly evolved silicic magmas of the Rallier-du-Baty nested ring complex (Gagnevin et al. 2003). So far, their volumes have not been assessed precisely. Seismological studies indicate a 15–20-km thick crust beneath the nearby Mont Ross volcano (Recq et al. 1994). It is speculated that the Kerguelen Archipelago and even the surrounding plateau (Charvis et al. 1995) represent a continental nucleation process (Grégoire et al. 1998).

Growth mechanisms

The Rallier-du-Baty nested ring complex is dominated essentially by large ring intrusions, composed on top by a shallow 500-m to 1-km thick subhorizontal sheet and, at depths, by a vertical to steeply outward-dipping 50-m wide ring dyke. The upper sheet is occupied by a volume of magma of 100–200 km³, whereas the ring dyke, with a 16-km diameter and an assumed 10–15 km vertical height, has a volume of only 12–18 km³. As for all plutons, there is a 'room problem' to solve. Because neither flowage, nor melting of wallrocks, have been observed, the silicic magmas were likely to have been emplaced in a brittle fashion. The classical mechanisms of emplacement that have been identified so far in a brittle environment include: roof lifting, stoping and floor subsidence.

Tilting and doming of the basaltic wallrocks
Radial tilting of the flood basalts, centred on the Rallier-du-Baty Peninsula, is observed up to 20 km away (Fig. 10). Near the contacts with the

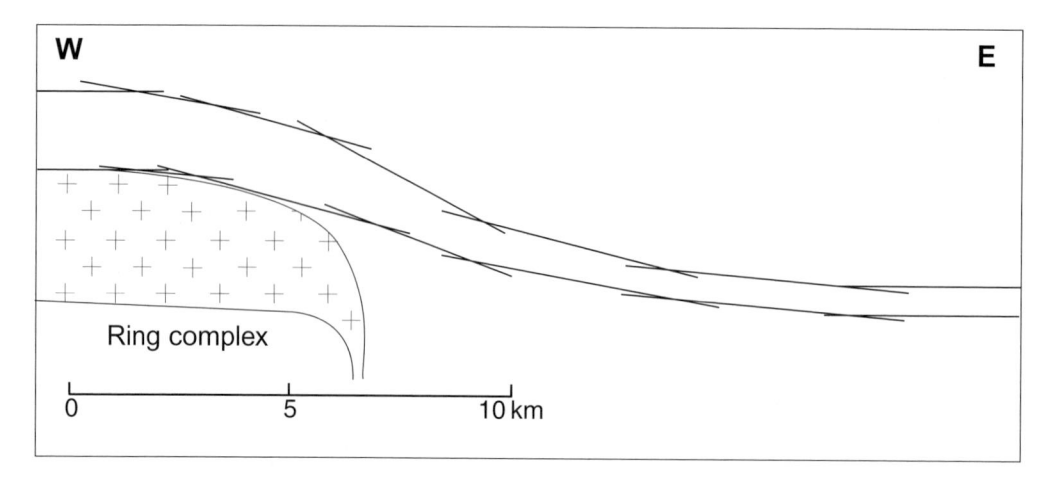

Fig. 10. Uplift and tilting of Neogene flood basalts above the Rallier-du-Baty ring complex, as deduced from the variation of measured dips of individual flows (after Marot & Zimine 1976).

ring complex, outward dips can reach 30°, and vertical uplift of the country rocks is estimated to range from 2.5 to 3.5 km (Nougier 1970; Marot & Zimine 1976), suggesting that roof lifting could be one of the major mechanisms of emplacement. However, two lines of evidence argue against a significant role played by roof lifting. The sidewall contacts with the country rocks are represented by a sharp unconformity, yielding an outward dip higher than that of the flood basalts. The c. 40-km diameter of this structural feature, approximately twice the 16-km mean diameter of the intrusions, cannot indicate a simple roof lifting mechanism, such as that advocated for laccoliths (Corry 1988). It is suggested that the structure could have originated by emplacement within the lower crust and progressive inflation of a dense network of magma chamber(s) (Grégoire et al. 1998). Such a network would be responsible for the protracted magmatic activity in the peninsula during the last 15 Ma.

Extent of stoping process The process of blocks of country rocks sinking from the roof of a pluton into a magma is described as 'stoping'. The principal line of evidence for the feasibility of such a process in the Rallier-du-Baty nested ring complex is the common occurrence of blocks of nearby wallrocks in almost all marginal facies of the plutonic intrusive units. The problem with the stoping mechanism is not whether it could act. It did, but at what scale? As noted by Clarke et al. (1998), a major question is: what proportion of the total ascent of a magma can stoping account for? Although producing spectacular exposures, stoping in the Rallier-du-Baty nested ring complex is always a process limited to a thickness of some tens of metres constituting the external marginal zones of the 500–1000-m thick ring intrusions. These field observations substantiate that stoping alone could not provide more than 5–10% of the space accommodated by the ascending magmas. However, the mechanism could play a role in the final shape of the intrusion and explain, e.g. the polygonal outline of the external contact of the alkali-feldspar syenite–granite unit 4 of the southern complex.

Cauldron subsidence Emplacement of a ring intrusion is basically governed by subterranean subsidence of a piece of brittle upper crust (Fig. 11). Space for the shallow intrusive sheet is provided along a flat tensile fracture opened through vertical translation of the sinking block moving downwards as a piston along a cylindrical boundary fault into the underlying magma chamber (Clough et al. 1909; Billings 1943). Although differing by an order of magnitude, a similar sequence of space-making processes, involving successively minor roof uplift and regional doming; floor subsidence by a major piston mechanism; and local stoping, favoured emplacement of the Coastal Batholith of Peru (Haederle & Atherton 2002).

An upper-crustal-scale balance in the rates of magma extraction from the chamber, magma ascent and intrusion-filling is required, while mass transfer from the chamber to the intrusion is accommodated by upper-crustal embrittlement. Magmas cooled and crystallized in a fairly static environment, as evidenced by weak

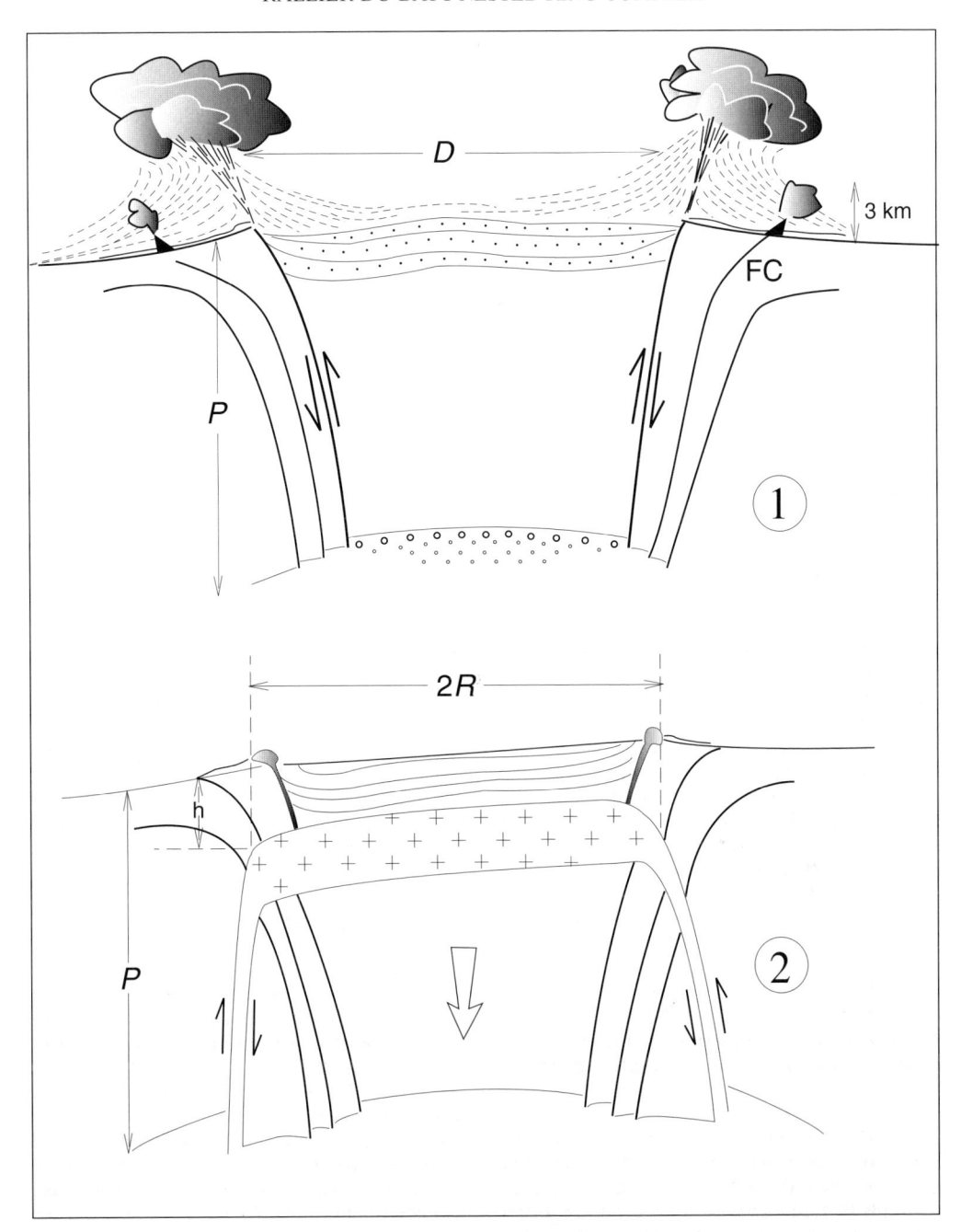

Fig. 11. Schematic cross-sections of an active caldera-related ring complex structure (adapted from Bonin 1986). The sequence of events 1 and 2 may be repeated several times before completion of the magmatic episode. $D = 2R$, diameter of the caldera; P, depth of the roof of the reservoir; h, depth of emplacement of the ring complex. (**1**) Magma overpressure, mostly due to exsolution of a vapour phase below the roof of a deep-seated reservoir, induces upward movement of the overlying crustal block (uplift can reach 3 km), initiation of inward-dipping reverse faults, explosive eruptions, deposition of pyroclastic formations (dots) within and outside the caldera, and emplacement of cone sheets and lateral sills (FC). (**2**) After the explosive eruptive event, a crustal block sinks into the reservoir, inducing outward-dipping normal faults, void space between caldera-filling volcanic formations and the subsiding block, and emplacement of a ring complex (crosses).

magmatic and magnetic fabrics (Geoffroy et al. 1997). Cauldron, or floor, subsidence differs from lopolith floor-depression mechanism by the lack of broadly distributed deformation of low strain magnitude (Cruden 1998). Simple shear deformation was, instead, localized on the boundary fault.

The dimensional parameters: T = thickness and L = length (here the diameter), of a pluton are governed by a power-law relation in the form of $T = bL^a$ (McCaffrey & Petford 1997). From the observation of c. 10-m thick cone sheets and c. 1-km thick ring dykes, the growth of the upper sheet of a ring intrusion can be envisaged as a two-stage mechanism (Figs 11 & 12) including: (i) self-affine propagation along a 16-km wide tensile fracture, with the parameters $a \approx 0$ and $b \approx 10$, until the definitive diameter is reached, and (ii) self-affine vertical inflation, with the parameter a tending to infinity and T reaching about 1 km. It differs from the current models of pluton growth, in which the vertical inflation episode is marked by lower values of the parameter a varying from 6 (Cruden & McCaffrey 2001) to 1.36 (Rocchi et al. 2004).

The time duration for the inflation episode by floor subsidence was extremely short. The Rb–Sr isotopic determinations of the successive intrusions of the southern centre reveal that these high Rb–Sr magmas could not remain isotopically homogeneous after 0.1 Ma (Lameyre et al. 1976; Dosso et al. 1979). The 0.1-Ma period of time constitutes the maximum duration for emplacement, crystallization and cooling of a ring intrusion. Accordingly, the calculated rates of magma supply should be at least 0.03–0.07 $m^3 s^{-1}$, in the lowest range of values suggested by Petford et al. (2000). Taking these values gives a minimum sinking velocity of the subsiding block of 5 to 10 mm yr^{-1}.

Possible causes for cauldron subsidence In the Rallier-du-Baty nested ring complex, each unit records subsidence of the sinking block along a vertical distance ranging from 500 m to 1 km. If the subsided block is cylindrical, a space of the same order of magnitude should be open at depths. According to the geophysical evidence, currently active magma chambers contain a very limited quantity of liquid – less than 100–150 m thick. Thus, emptying of the top of the magma chamber by ascent of the residual liquid up to the ring complex level can accommodate only a small proportion of the observed vertical movement – implying a supplementary process.

We speculate that cauldron subsidence was accompanied by floor-depression of the underlying magma chamber. Basically, the density of a magma increases by c. 10% upon crystallization. A mafic magma, stalled in a magma chamber, may cool and evolve by fractional crystallization, as in the case for the Rallier-du-Baty area. Residual liquids yield lower densities, while denser cumulative rocks become negatively buoyant. If trapping occurs within the lower crust below the strength maximum, the crystallizing chamber can sink at velocities governed by the effective viscosity of the crust. Mafic plutons can sink at 0.5 to 50 mm yr^{-1} (Glazner 1994), consistent with the minimum sinking velocity of 5–10 mm yr^{-1} estimated for the subsiding block of the Rallier-du-Baty nested ring complex.

A relevant example is the Bjerkreim–Sokndal layered intrusion, Rogaland, Norway, where a thick sequence of cumulative mafic rocks is topped by an A-type silicic suite (Duchesne & Wilmart 1997). Magmatic fabrics are marked by mineral foliations and lineations converging and plunging into the centre of the massif. They are interpreted as resulting from centripetal gravitational collapse (Bolle 1998). Similarly, in the Coastal Batholith of Peru, floor subsidence could have occurred by foundering of crustal blocks into a deflating layer of partially melted rocks (Haederle & Atherton 2002).

Chemical modelling (Gagnevin et al. 2003) provides evidence that one kilogram of trachyte–syenite can be produced by differentiation of 5–6 kg of trachyandesite–trachybasalt, leaving 4–5 kg of cumulates, in agreement with the respective thicknesses of the ring complex and the underlying crust. We suggest, therefore, that the Rallier-du-Baty nested ring complex is underlain by a thick package of collapsed mafic chambers. A similar picture is revealed by geophysical imagery (reflection seismic, gravity and aeromagnetic data) in the African continent, where the 15-km wide Messum ring complex, Namibia, rests on top of a network of multiple intrusions of basaltic magmas (Bauer et al. 2003).

Summary and conclusions

The Rallier-du-Baty Peninsula is composed of Neogene to Recent volcanic and plutonic rocks. Four major episodes of magmatic activity have been identified:

1 Tertiary flood basalts were erupted during and/or before the Early Miocene and constitute plateaux that characterize the overall physiography of Kerguelen Archipelago. The volcanic pile, the base of which is now under sea-level, has an observed thickness of at least 1000 m. An extensive zeolitization event took place 15–12 Ma ago.

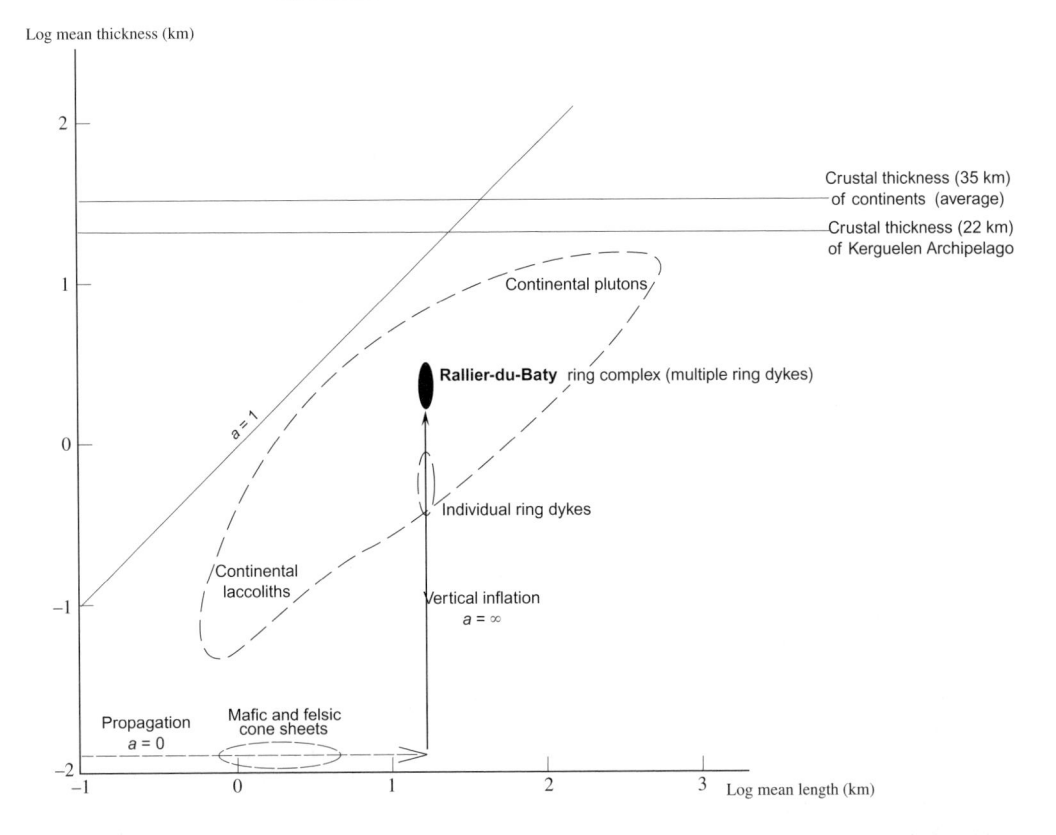

Fig. 12. Thickness v. length plot. Mafic and felsic cone sheets plot along a propagation trend ($a = 0$), individual ring dykes and the entire Rallier-du-Baty ring complexes plot along a vertical inflation trend ($a \to \infty$). Note that the oceanic Rallier-du-Baty ring complex yields dimensions of the same order of magnitude as continental intrusions, and plots in between laccoliths and plutons. Crustal thicknesses of continents and the Kerguelen Archipelago are shown for comparison.

2 Two ring complexes comprise early volcanic rocks intruded by younger plutonic rocks, both sharing quartz alkali-feldspar trachyte–syenite compositions. Remnants of a large caldera are preserved on the northern tip of the northern ring complex. The total volume of plutonic rocks is estimated at 400–800 km³. If the volume of volcanic rocks emitted from the caldera is assumed to be of the same order of magnitude, each ring complex would correspond to the production of 800–1600 km³ of magmatic rocks.

3 The Table-de-l'Institut caldera volcano is composed of trachyte, hawaiite and mugearite. The total volume of trachytic formations is estimated at more than 90 km³. Hawaiite and mugearite products are comparatively more restricted in volume. The Table-de-l'Institut caldera volcano is likely to have erupted a total volume of 100 to 150 km³. No plutonic equivalents have been found so far.

4 The Dôme Carva volcano is entirely composed of trachytic volcanic products, sharing hydro-plinian to plinian characteristics, culminating in voluminous ignimbrites and related pumice and ash falls. The Dôme Carva volcano is likely to have erupted a total volume of 25 to 30 km³.

The plutonic ring complexes were emplaced through a two-stage mechanism involving: (1) horizontal elongation along a tensional fault plane, (2) vertical inflation aided by cauldron subsidence of a crustal block sinking into a deeper magma chamber along a cylindrical fault. They cross-cut a pre-existing caldera volcano, suggesting that the younger volcanoes themselves overlie unexposed ring complexes. Cooling of recent plutonic rocks induces the persistent fumarolic activity observed around Pic Saint-Allouarn.

Over the last 15 Ma, the volcanic and plutonic

rocks of the Rallier-du-Baty Peninsula have recorded estimated volume 2800 ± 850 km^3 of magmas that passed up though the crust of the Rallier-du-Baty Peninsula. A residual volume of 1450 ± 450 km^3 of rocks is exposed, or inferred in the substratum, corresponding to a magmatically induced crustal thickening of 3.5 ± 1.0 km. It is only partly accommodated by the average elevation of 1000 m above the sea-level of the peninsula.

We are grateful to the French Institute for Polar Research and Technology (IFRTP), and especially to the Kerguelen technical staff for continuing support and assistance during the 1994–1995 Austral summer field expedition of the CARTOKER mapping project. The organizers of the LASI workshop at Freiberg: A. Mock, C. Breitkreuz and N. Petford, afforded a welcome opportunity for us to reconsider the role played by different mechanisms in the emplacement of one of the largest silicic ring complexes exposed in an oceanic island. Finally, thanks are due to R. Werner and L. Truelove for the astute comments in their reviews, and to the editors of the issue for their patience and efficient editorial handling of the manuscript.

References

ANDERSON, E.M. 1936. The dynamics of the formation of cone sheets, ring-dykes, and cauldron-subsidence. *Proceedings, Royal Society of Edinburgh*, **56**, 128–163.

AUBERT DE LA RUE, E. 1932. Etude géologique et géographique de l'archipel de Kerguelen. D.Sc. Thesis, Rev. Géogr. Phys. Géol. Dyn., **5**, 231 pp.

BAIN, A.D.N. 1934. The younger intrusive rocks of the Kudaru Hills, Nigeria. *Quarterly Journal of the Geological Society of London*, **90**, 201–239.

BAUER, K., TRUMBULL, R.B. & VIETOR, T. 2003. Geophysical images and a crustal model of intrusive structures beneath the Messum ring complex, Namibia. *Earth and Planetary Science Letters*, **216**, 65–80.

BELLAIR, P., CARRON, J.P., NOUGIER, J. & TRICHET, P. 1965. Niveaux intercalaires dans les strates de basalte des plateaux de l'Archipel de Kerguelen. *Geologische Rundschau*, **55**, 342–354.

BILLINGS, M.P. 1943. Ring-dikes and their origin. *New York Academy of Sciences, Transactions, Series. II*, **5**, 131–144.

BILLINGS, M.P. 1945. Mechanics of igneous intrusion in New Hampshire. *American Journal of Science, Daly Volume*, **243-A**, 40–68.

BOLLE, O. (1998). *Mélanges magmatiques et tectonique gravitaire dans l'apophyse de l'intrusion de Bjerkreim–Sokndal (Rogaland, Norvège): pétrologie, géochimie et fabrique magnétique*. Unpublished Ph.D. thesis, Université de Liège (Belgium), 204 pp.

BONIN, B. 1972. *Le complexe granitique subvolcanique de la région de Tolla-Cauro (Corse)*. Ph.D. thesis, Université Pierre et Marie Curie, Paris, published by Lab. Géol. ENS, no. 7, 127 pp.

BONIN, B. 1986. *Ring Complex Granites and Anorogenic Magmatism*. Studies in Geology, North Oxford Academic Publications, Oxford, 188 pp.

CHARVIS, P., RECQ, M., OPERTO, S. & BREFORT, D. 1995. Deep structure of the northern Kerguelen plateau and hotspot-related activity. *Geophysical Journal International*, **122**, 899–924.

CHRISTIANSEN, R.L. & BLANCK, H.R. JR 1972. Volcanic stratigraphy of the Quaternary rhyolite plateau in Yellowstone National Park. *US Geological Survey Profession Paper*, **729-B**, 18 pp.

CLARKE, D.B., HENRY, A.S. & WHITE, M.A. 1998. Exploding xenoliths and the absence of 'elephant's graveyards' in granite batholiths. *Journal of Structural Geology*, **20**, 1325–1343.

CLOUGH, C.T., MAUFE, H.B. & BAILEY, E.B. 1909. The cauldron-subsidence of Glen Coe, and the associated igneous phenomena. *Quarterly Journal of the Society of London*, **65**, 611–678.

CORRY, C.E. 1988. Laccoliths: mechanics of emplacement and growth. *Geological Society of America Special Publications*, **220**, 110 pp.

CRUDEN, A.R. 1998. On the emplacement of tabular granites. *Journal of the Geological Society of London*, **155**, 853–862.

CRUDEN, A.R. & MCCAFFREY, K.J.W. 2001. Growth of plutons by floor subsidence: implications for rates of emplacement, intrusion spacing and melt-extraction mechanisms. *Physics and Chemistry of the Earth A*, **26**, 303–315.

DELORME, H., VERDIER, O., CHEMINÉE, J.L., GIRET, A., PINEAU, F. & JAVOY, M. 1994. Etude chimique et rapport isotopique du carbone des fumerolles de la péninsule Rallier du Baty (îles Kerguelen). *In:* SCHLICH, R. & GIRET, A. (eds), *Géologie et géophysique des Kerguelen*, Mémoires de la Société Géologique de France, Nouvelle Série, **166**, 25–30.

DOSSO, L., VIDAL, P., CANTAGREL, J.M., LAMEYRE, J., MAROT, A. & ZIMINE, S. 1979. 'Kerguelen: continental fragment or oceanic island?': petrology and geochemistry evidence. *Earth and Planetary Science Letters*, **43**, 46–60.

DOSSO, L. & MURTHY, V.R. 1980. A Nd isotopic study of the Kerguelen Islands: inferences on enriched oceanic mantle sources. *Earth and Planetary Science Letters*, **48**, 268–276.

DUCHESNE, J.C. & WILMART, E. 1997 Igneous charnockites and related rocks from the Bjerkreim–Sokndal layered intrusion (Southwest Norway): a jotunite (hypersthene monzodiorite)-derived A-type granitoid suite. *Journal of Petrology*, **38**, 337–369.

GAGNEVIN, D. *ET AL.* 2003. Open-system processes in the genesis of silica-oversaturated alkaline rocks of the Rallier-du-Baty peninsula, Kerguelen Archipelago (Indian Ocean). *Journal of Volcanology and Geothermal Research*, **123**, 267–300.

GEOFFROY, L., OLIVIER, P. & ROCHETTE, P. 1997. Structure of a hypovolcanic acid complex inferred from magnetic susceptibility anisotropy measurements: the Western Red Hills granites (Skye,

Scotland, Thulean Igneous Province). *Bulletin of Volcanology*, **59**, 147–159.

GIRET, A. 1980. *Notice et carte géologique au 1:50 000 de la Péninsule Rallier du Baty, Iles Kerguelen.* CNFRA, **45**, 14 pp.

GIRET, A. 1990. Typology, evolution, and origin of the Kerguelen plutonic series, Indian Ocean: a review. *Geological Journal*, **25**, 239–247.

GIRET, A. 1993. Les étapes magmatiques de l'édification des îles Kerguelen, océan Indien. *Mémoire de la Société Géologique de France*, **163**, 273–282.

GIRET, A., VERDIER, O. & NATIVEL, P. 1992. The zeolitisation model of Kerguelen Islands, Southern Indian Ocean. *In*: YOSHIDA, Y. *et al.* (eds), *Recent Progress in Antarctic Earth Science*, TERRAPUB, Tokyo, 457–463.

GIRET, A., GRÉGOIRE, M., COTTIN, J.Y. & MICHON, G. 1997. Kerguelen, a third type of oceanic island? *In*: RICCI, C.A. (ed.) *The Antarctic Region: Geological Evolution and Processes*. Terra Antarctica Publication, 735–741.

GLAZNER, A.F. 1994. Foundering of mafic plutons and density stratification of continental crust. *Geology*, **22**, 435–438.

GRÉGOIRE, M., COTTIN, J.Y., GIRET, A., MATTIELLI, N. & WEIS, D. 1998. The meta-igneous granulite xenoliths from Kerguelen Archipelago: evidence of a continent nucleation in an oceanic setting. *Contributions to Mineralogy and Petrology*, **133**, 259–283.

HAEDERLE, M. & ATHERTON, M.P. 2002. Shape and intrusion style of the Coastal Batholith, Peru. *Tectonophysics*, **345**, 17–28.

KINGSLEY, L. 1931. Cauldron-subsidence of the Ossipee Mountains. *American Journal of Science, 5th Series*, **22**, 139–168.

LACROIX, A. 1922. *Minéralogie de Madagascar*. Challamel, Paris, Vol. 2, 694 pp.

LACROIX, A. 1924. Les roches éruptives grenues de l'archipel de Kerguelen. *Comptes Rendus de l'Académie des Sciences, Paris*, **179**, 113–119.

LAMEYRE, J., MAROT, A., ZIMINE, S., CANTAGREL, J.M., DOSSO, L. & VIDAL, P. 1976. Chronological evolution of the Kerguelen islands syenite–granite ring complex. *Nature*, **263**, 306–307.

LAMEYRE, J. *ET AL.* 1981. *Etude Géologique du Complexe Plutonique de la Péninsule Rallier du Baty, Iles Kerguelen.* CNFRA, **49**, 176 pp.

LEYRIT, H. 1992. *Kerguelen: cartographie et magmatologie des presqu'îles Jeanne d'Arc et Ronarc'h.* Ph.D. thesis, Université de Paris-Sud, Orsay, 236 pp.

MCCAFFREY, K.J.W. & PETFORD, N. 1997. Are granitic intrusions scale invariant? *Journal of the Geological Society of London*, **154**, 1–4.

MAROT, A. & ZIMINE, S. 1976. *Les complexes annulaires de syénites et granites alcalins dans la péninsule Rallier du Baty, Iles Kerguelen (T.A.A.F).* Ph.D. thesis, Université Pierre et Marie Curie, Paris, 172 pp.

MARTIN, R.F. & BONIN, B. 1976. Water and magma genesis: the association hypersolvus granite–subsolvus granite. *Canadian Mineralogist*, **14**, 228–237.

MUNSCHY, M., FRITSCH, B., SCHLICH, R. & ROTSTEIN, Y. 1994. Tectonique extensive sur le plateau de Kerguelen. *In*: SCHLICH, R. & GIRET, A. (eds), *Géologie et Géophysique des Kerguelen*, Mém. Soc. Géol. Fr., N.S., **166**, 99–108.

NOUGIER, J. 1970. *Contribution à l'étude Géologique et Géomorphologique des Iles Kerguelen*. CNFRA, **27(1)**, 440 pp., **27(2)**, 246 pp.

ORCEL, J. 1924. Notes minéralogiques et pétrographiques sur la Corse. *Bulletin de la Société des Sciences Historiques et Naturelles de la Corse, Bastia*, **461–464**, 65–127.

PETFORD, N., CRUDEN, A.R., MCCAFFREY, K.J.W. & VIGNERESSE, J.L. 2000. Granite magma formation, transport and emplacement in the Earth's crust. *Nature*, **408**, 669–673.

RECQ, M., LE ROY, I., CHARVIS, P., GOSLIN, J. & BREFORT, D. 1994. Structure profonde du Mont Ross d'après la sismique (îles Kerguelen, océan Indien austral). *Canadian Journal of Earth Sciences*, **31**, 1806–1821.

RICHEY, J.E. 1928. The structural relations of the Mourne Granites (Northern Ireland). *Quarterly Journal of the Geological Society of London*, **83**, 653–687.

RICHEY, J.E. 1932. Tertiary ring structures in Britain. *Transactions of the Royal Society of Glasgow*, **19**, 42–140.

RICHEY, J.E. 1935. *Scotland: the Tertiary Volcanic Districts*. British Regional Geology, HMSO, Edinburgh, 120 pp.

ROCCHI, S., WESTERMAN, D.S., DINI, A., INNOCENTI, F. & TONARINI, S. 2004. Two-stage laccolith growth at Elba Island (Italy). *Geology*, in press.

SCHLICH, R. 1994. Introduction. *In*: SCHLICH, R. & GIRET, A. (eds), *Géologie et Géophysique des Kerguelen*, Mémoires de la Société Géologique de France, Nouvelle Série, **166**, 5–6.

SCHUBERT, G. & SANDWELL, D. 1989. Crustal volumes of the continents and of oceanic and continental submarine plateaus. *Earth and Planetary Science Letters*, **92**, 234–246.

VERDIER, O. 1989. *Champs géothermiques et zéolitisation des îles Kerguelen: implications géologiques (Terres Australes et Antarctiques Françaises, Océan Indien Austral)*. Ph.D. thesis, Université Pierre et Marie Curie, Paris, 271 pp.

WEIS, D., FREY, F.A., GIRET, A. & CANTAGREL, J.M. 1998. Geochemical characteristics of the youngest volcano (Mount Ross) in the Kerguelen Archipelago: inferences for magma flux, lithosphere assimilation and composition of the Kerguelen Plume. *Journal of Petrology*, **39**, 973–994.

YANG, H.-J., FREY, F.A., WEIS, D., GIRET, A., PYLE, D. & MICHON, G. 1998. Petrogenesis of the flood basalts forming the northern Kerguelen Archipelago: implications for the Kerguelen Plume. *Journal of Petrology*, **39**, 711–748.

Tectonic control on laccolith emplacement in the northern Apennines fold–thrust belt: the Gavorrano intrusion (southern Tuscany, Italy)

FRANCESCO MAZZARINI[1], GIACOMO CORTI[2], GIOVANNI MUSUMECI[2] & FABRIZIO INNOCENTI[2]

[1]*Istituto di Geoscienze e Georisorse CNR, Via G.Moruzzi 1, 56100, Pisa, Italy*
(e-mail: mazzarini@igg.cnr.it)
[2]*Dipartimento di Scienze della Terra, Via S. Maria 53, 56126 Pisa, Italy*

Abstract: Mechanical discontinuities within the crust, represented by tectonic structures (faults) or lithological heterogeneities, strongly control the emplacement of magmas as tabular intrusions within the middle–upper crust. The occurrence of mechanical layering is a common feature in fold and thrust belts. In the northern Apenniness, a Cenozoic fold–thrust belt affected in its inner part by Neogene magmatism, the Gavorrano laccolith (southern Tuscany) is a particularly suitable example for studying the relationships between magmatism and tectonic structures. New geological mapping, together with a large amount of subsurface data available from historical mining activity in the area, have allowed the reconstruction of:

1 the original relationships of the intrusion within the nappe pile, and
2 the laccolithic shape of the intrusion.

Using the Gavorrano laccolith as an example, we propose that the emplacement of Neogene intrusions in southern Tuscany was strongly controlled by the occurrence of mechanical discontinuities represented by thrust zones in the nappe pile.

Several authors have proposed on the basis of geological, structural and geophysical data, that intrusions in the middle and upper crust are tabular in shape (e.g. Ameglio *et al.* 1997). Most of these intrusions are represented by bodies in which the space needed for lateral spreading and vertical growth of magma was created by roof lifting (laccoliths) or floor depression (lopoliths) (Corry 1988; Cruden 1998). Laccolithic intrusions mainly occur in the upper levels of the crust, and their development is generally favoured by the occurrence of crustal magma traps in the form of mechanical discontinuities such as ductile or brittle shear zones or original lithological heterogeneities (Cruden 1998 and references therein).

The inner zone of the northern Apennines represents a Cenozoic fold–thrust belt pervasively affected by magmatism, and is presently characterized by anomalously thin crust and high heat flow (Cassano *et al.* 2001; Serri *et al.* 2001; Nicolich 2001; Della Vedova *et al.* 2001). These characteristics have developed since the Late Miocene, due to the onset of ensialic back-arc extension related to the retreat of the northern Apennine subduction zone (Royden *et al.* 1987).

Evidence of Neogene magmatism includes outcrops of intrusive and effusive rocks and large buried plutons within the Larderello and Monte Amiata geothermal fields (Gianelli *et al.* 1997; Franceschini 1998). Intrusive rocks crop out mainly in the Tuscan Archipelago (Elba, Montecristo and Giglio islands) and along the coastal area of southern Tuscany in small exposures in the Campiglia and Gavorrano areas (Fig. 1).

The Gavorrano intrusion is a particularly suitable example for studying the relationships between magmatism and tectonic structures, because of its position within the nappe pile and the large amount of subsurface data available from historical mining activity in the area.

The new geological data on the Gavorrano intrusion presented in this paper have allowed us to construct an emplacement model for the Pliocene intrusions of southern Tuscany – as laccolith bodies exploiting tectonic structures.

Geological framework

During the Alpine orogeny of the central Mediterranean, the northern Apennine fold–thrust belt was formed as a result of the Cenozoic continental collision between the European margin and the Adria microplate. This belt consists of NE-directed thrusts of allochthonous oceanic and continental nappes

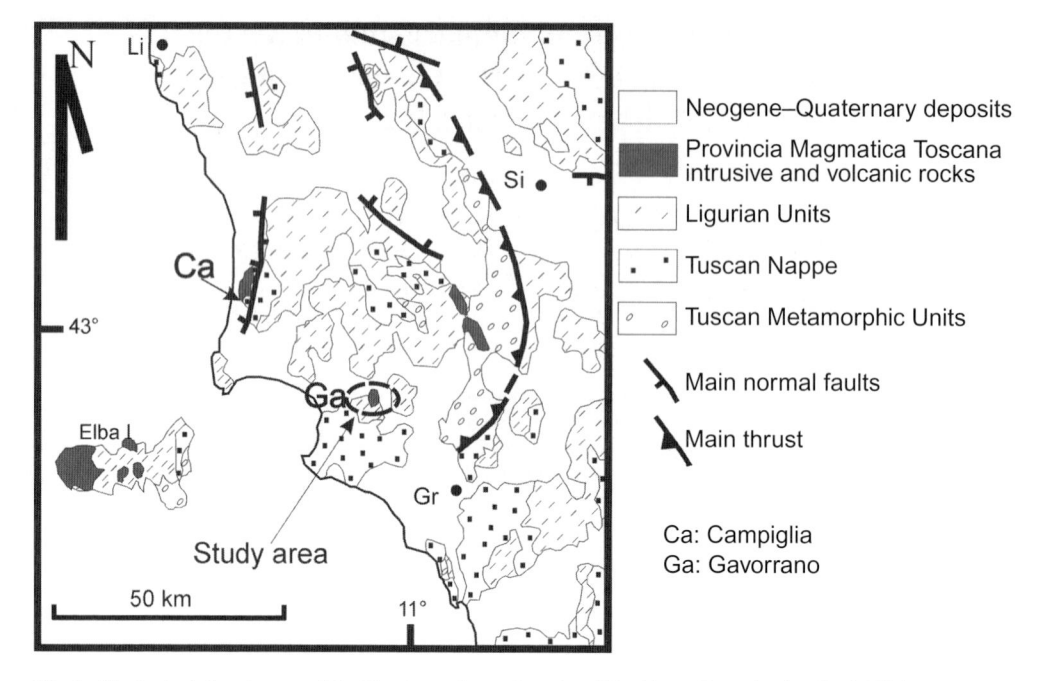

Fig. 1. Geological sketch map of the Northern Apennines (modified from Boccaletti & Sani 1998).

derived from the Alpine Tethyan Ocean and the Adriatic plate, respectively. Crustal deformation started early in the Late Oligocene in the internal zone and migrated eastwards, affecting the foredeep deposits during the Middle–Late Miocene (Carmignani *et al.* 1995). Since the Late Tortonian, the internal zone of northern Apennines has undergone extension, leading to the development of the northern Tyrrhenian Basin (Bartole 1995). This extensional phase is either related to mantle delamination (Keller *et al.* 1994) or slab roll-back (Serri *et al.* 1993), with consequent upwelling of the asthenosphere beneath the chain.

The formation of the Tyrrhenian Basin was accompanied by spatially diffuse magmatism (Provincia Magmatica Toscana, PMT), extending from the Tyrrhenian Basin in the west to southern Tuscany in the east (Innocenti *et al.* 1992). The age of the PMT magmatism becomes younger from west to east, from 7.3 ± Ma (Montecristo Island) to 0.2 ± Ma at the Monte Amiata volcano (Serri *et al.* 2001). PMT products include granitoids, felsic volcanics and minor high-SiO_2 lamproites, as well as K-rich to ultrapotassic, relatively primitive products. Most of the igneous rocks of the PMT consist of exposed and buried plutons (Franceschini 1998; Franceschini *et al.* 2000), whereas volcanic rocks are subordinate.

PMT magmatism in southern Tuscany is mainly represented by intrusive rocks – the larger volume of which occurs in the subsurface of the Larderello geothermal field. This area is currently characterized by very high heat flow (up to 1000 mW m^{-2}) and a steep geothermal gradient (up to 300 °C km^{-1}); several intrusive bodies have been cored at depths ranging from 1 to 4 km in deep geothermal wells. Geophysical data from tomography, teleseismic and magnetotelluric surveys have shown that these high-level intrusions belong to a large batholith ranging between 6 and 23 km in width, with a volume of at least 20 000 km^3 (Gianelli *et al.* 1997). Two exposures of Pliocene intrusive rocks occur in the Campiglia and Gavorrano areas (Fig. 1).

Gavorrano intrusion

The Gavorrano intrusion (Fig. 2) is a granitic body of Early Pliocene age (4.9–4.4 Ma; Serri *et al.* 2001 and references therein) associated with sulphide ore bodies that have been exploited since 1898 (Arisi Rota & Vighi 1972). The exposed intrusion has a NNW–SSE elongated shape and is about 3 km long (Fig. 2). The intrusive rocks have a composition ranging from monzogranite to alkali-feldspar granite (Table 1; Fig. 3). Two main intrusive facies have been

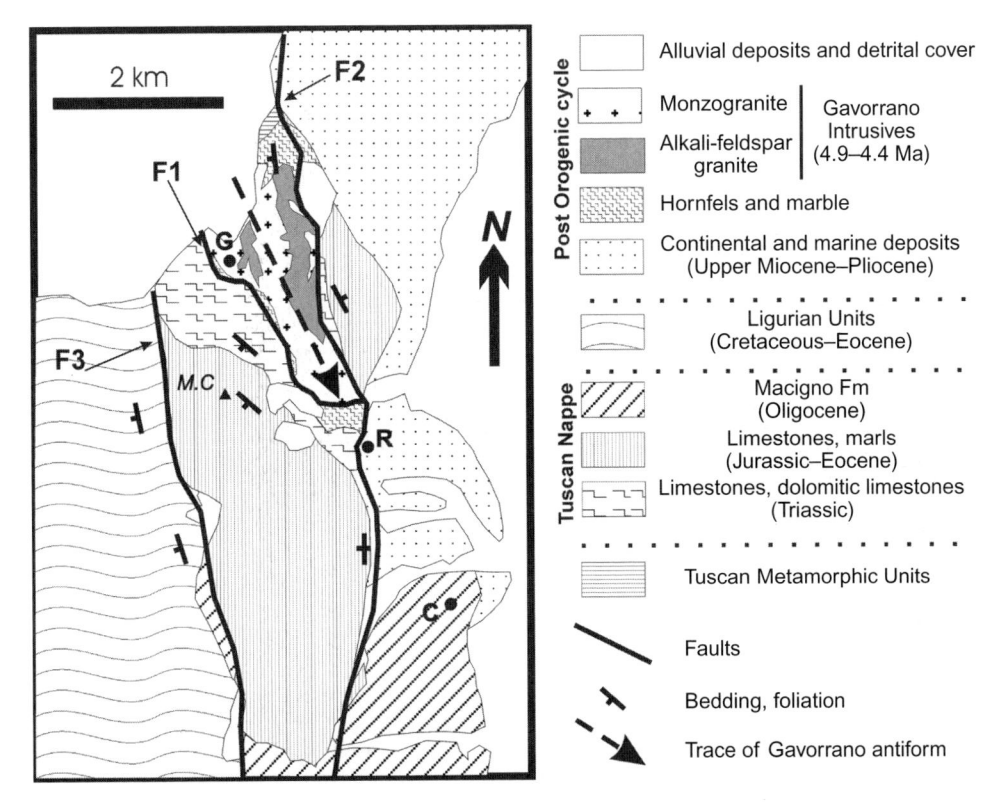

Fig. 2. Geological sketch map of the Gavorrano area. F1, Gavorrano Fault; F2, Monticello Fault; F3, Palaie Fault; G, Gavorrano; R, Ravi; C, Caldana; M.C., Monte Calvo.

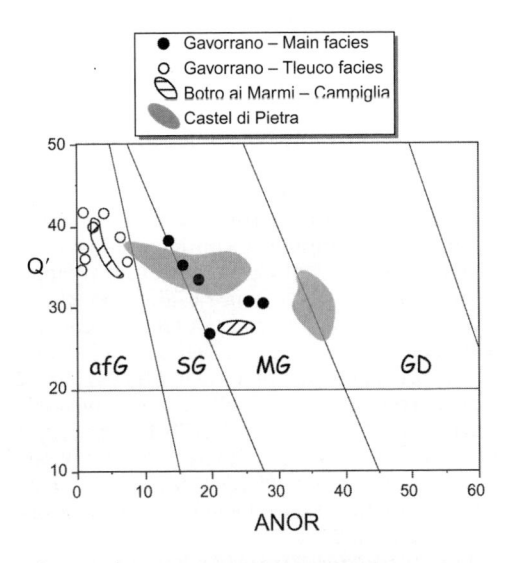

Fig. 3. Normative Q'-ANOR classification diagram (Streckeisen & Le Maitre 1979). Field labels: afG, alkali-feldspar granite; SG, syenogranite; MG, monzogranite; GD, granodiorite. Gavorrano (this work), Castel di Pietra (Franceschini *et al.* 2000) and Botro ai Marmi-Campiglia (Caiozzi *et al.* 1998).

distinguished: a normal facies (NF) which represents about 70% of the exposed pluton, and a leucocratic facies (LF), occurring mainly in the northern part of the intrusive body (Marinelli 1961; Arisi Rota & Vighi 1972).

The normal facies is characterized by a monzogranite with scattered K-feldspar megacrysts up to 8 cm in length. The texture is generally porphyritic with phenocrysts of zoned plagioclase, biotite, K-feldspar and quartz in a medium-grained matrix mainly composed of quartz and alkaline feldspars. Scattered cordierite, generally transformed to pinite, can also be observed. The most common accessory phases are apatite, zircon, monazite, magnetite and pyrite. The leucocratic facies consists of tourmaline-bearing alkali-feldspar granite, often with aplitic texture, occurring in small bodies and dykes. The mineral assemblage is mainly characterized by dominant K-feldspar and minor plagioclase, quartz and Fe^{2+}-rich tourmaline (schorl); muscovite is present locally. The rocks of both facies are generally pervasively altered, with development of sericite, chlorite and calcite in variable amounts.

The relationships observed in the field

Table 1. *Representative major-element analyses of Gavorrano intrusive rocks*

	Normal facies			Leuco facies		
	RV 7	RZ 195	RZ 149	GG 5	RZ 202	RZ 142
SiO_2	66.94	68.30	69.88	74.84	73.32	74.25
TiO_2	0.57	0.55	0.53	0.10	0.22	0.20
Al_2O_3	14.59	15.09	13.89	13.80	13.92	13.46
$Fe_2O_{3\,T}$	4.21	3.16	2.62	1.02	2.00	1.35
MnO	0.06	0.04	0.06	0.01	0.04	0.03
MgO	1.75	1.20	1.18	0.19	0.40	0.32
CaO	1.92	2.19	1.61	0.32	0.34	0.61
Na_2O	2.21	2.81	0.43	2.65	2.22	2.62
K_2O	4.56	4.70	7.44	5.29	5.93	5.55
P_2O_5	0.16	0.16	0.16	0.20	0.12	0.11
LOI	2.13	1.40	3.18	0.84	1.09	0.86
Total	99.10	99.60	100.98	99.26	99.60	99.36
Norm (CIPW)*						
Q	29.05	27.05	33.44	38.43	36.05	35.88
C	2.91	1.78	2.59	3.61	3.52	2.30
or	26.95	27.77	43.96	31.26	35.04	32.80
ab	18.70	23.78	3.64	22.42	18.79	22.17
an	8.48	9.82	6.94	0.28	0.90	2.31
di	0.00	0.00	0.00	0.00	0.00	0.00
wo	0.00	0.00	0.00	0.00	0.00	0.00
hy	7.25	4.95	4.49	1.23	2.47	1.71
mt	1.89	1.42	1.18	0.46	0.90	0.61
il	1.08	1.04	1.01	0.19	0.42	0.38
ap	0.37	0.37	0.37	0.46	0.28	0.25

Major elements determined by XRF (Philips PW1480) on fused discs.
Loss on ignition (LOI) was measured by gravimetry at 1000 °C after pre-heating at 110 °C.
*Norm calculated with Fe_2O_3/FeO ratio, according to Middlemost (1989).

between the normal facies and the leucocratic facies, such as

1. LF dykes intruding the NF,
2. occurrence of angular fragments of NF within the LF,
3. suggest an early emplacement of the monzogranite (NF), followed by alkali-feldspar granite (LF).

At the outcrop and microscopic scales neither facies shows any preferred mineral orientation and/or solid-state foliation.

The intrusion occurs at the core of the Gavorrano Ridge, a north–south-trending antiformal structure whose axis plunges gently SSW. This structure folds a thrust-stack of allochthonous tectonic units represented from the top to the bottom by the Ligurian Units, the Tuscan Nappe and the Tuscan Metamorphic Units. The Gavorrano Ridge, covered on its eastern flank by Pliocene–Quaternary sediments, is bounded by two north–south, NNW–SSE-trending fault zones, the Palaie Fault and the Monticello Fault on the western and eastern side, respectively (Fig. 2; Arisi Rota & Vighi 1972). In the studied area, the Ligurian units consist of Cretaceous to Palaeocene calcareous–marly flysch sequences representing the pelagic sedimentary covers of Jurassic to Cretaceous ophiolitic complexes. The Tuscan Nappe unit consists, starting from the bottom, of Upper Triassic–Cretaceous shallow-marine to pelagic carbonate and siliceous deposits representing a continental passive-margin sequence. The base of the Tuscan Nappe is made up of lower Triassic limestones and evaporites. These latter are in tectonic contact with the underlying Tuscan Metamorphic Units which are made up of Upper Palaeozoic to Upper Triassic metaquarzites, metasandstone and phyllites (Dallegno et al. 1979; Burgassi et al. 1983). The thrust stacks of these units are unconformably covered by Upper Miocene–Upper Pliocene continental to shallow-marine deposits (Bossio et al. 1993).

The intrusion is emplaced within the antiformal structure at the tectonic boundary between the phyllites of the Tuscan Metamorphic Units and the Triassic limestones and evaporites at the base of the allochthonous Tuscan Nappe. The intrusive contact between the granite and

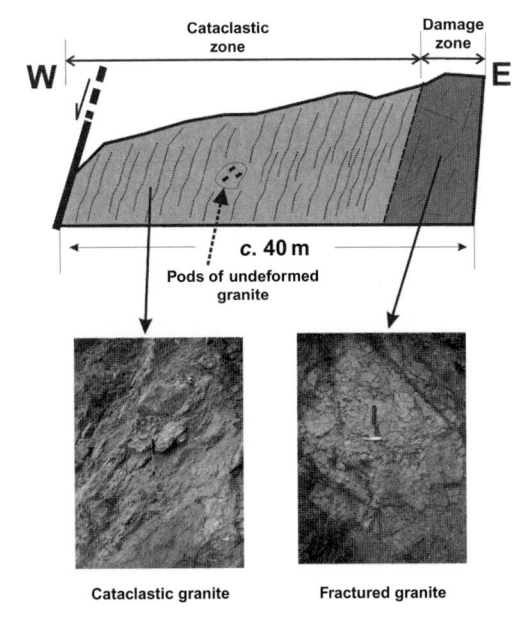

Fig. 4. Line drawings and photographs of the Gavorrano fault (northern termination). Across the section, about 40 m wide, the transition from damaged to strongly cataclastic granite can be observed moving toward the fault plane. Pictures describe different strain regimes into the fault zone.

across granite and host rocks, and are characterized by the development of fault gouge and fault breccia containing clasts of granite and hornfels. The best examples of faulting in granite are exposed in the NW part of the Gavorrano Fault and in its southern termination close to the Ravi mining area. As shown in Figure 4, near the fault the granite is characterized by a metre-wide damage zone, characterized by steeply dipping, regularly spaced, fractures filled by quartz and Fe-rich hydroxides, with a sharp transition into a decametre-wide cataclastic zone. The latter is marked by a westward-dipping cataclastic foliation, small-scale fault planes and metre-scale pods of unfractured granite. Locally, cataclastic rocks are absent and a metre-scale fault gouge characterizes the fault, as recognized in the southern termination of the Gavorrano Fault. On the eastern side of the intrusion, along the Monticello Fault, the granite is intensively deformed with development of fractures and small-scale faults. The kinematic indicators (Riedel fractures, fibre growth, step striations) recognized along the faults (Fig. 5) clearly indicate a down-dip sense of movement for the Monticello Fault and down-dip movements to a strongly oblique strike-slip for the Gavorrano faults – consistent with the overall extensional characteristic of both fault zones.

the metamorphic rocks of the basement is well preserved along the northern and southern sides of the intrusion where a thermal aureole, decametres to hectometres in width, is also present. Conversely, along the eastern and western sides of the granite, the faulted contacts with the carbonate rocks of the Tuscan Nappe do not allow any direct recognition of original relationships. However, the occurrence within the western fault of several slices of marbles and carbonaceous hornfelses indicates that at least the upper portion of the intrusion was emplaced within the base of the Tuscan Nappe. Indeed, close to the fault, the carbonate rocks at the base of the Tuscan Nappe show a weak thermal metamorphism.

Fault systems

The Palaie Fault, bounding the western side of the Gavorrano Ridge (Fig. 2), is a steeply west-dipping normal to oblique fault that brings the Tuscan Nappe sequences into contact with the Ligurian Units (Arisi Rota & Vighi 1972; Rossetti *et al.* 2001). Two main fault systems occur along the contact of the Gavorrano granite: the Gavorrano fault to the west and the Monticello fault to the east (Fig. 2). Both fault systems cut

Thermal metamorphism and hydrothermal alteration

Thermal aureoles characterized by the development of pelitic and calc-silicate hornfelses occur at the northern and southern terminations of the Gavorrano intrusion. Andalusite-bearing spotted schists crop out at the contact with the intrusion and rapidly pass, over a distance of some tens of metres, into low-grade phyllites and metasandstone of the Tuscan Metamorphic Units. Calc-silicate hornfelses, which occur up to 100 m from the intrusion, occur within the carbonaceous formations at the base of the Tuscan Nappe. On the western side, marble and pelitic hornfels occur as tectonized clasts and/or slices along the Gavorrano Fault. The slices of thermal aureole, of a metre to a decametre in size, mainly consist of calcite marbles, often associated with spotted schists, which indicate the development of a thermal aureole within the carbonate rocks of the Tuscan Nappe that crop out west of the Gavorrano Fault (Fig. 2). Hornfelses show fine-grained, banded to massive textures with intense recrystallization. Banded textures, due to alternating layers of different mineralogy, reflect derivation from low-metamorphic-grade foliated rocks such as phyllites and metasandstones,

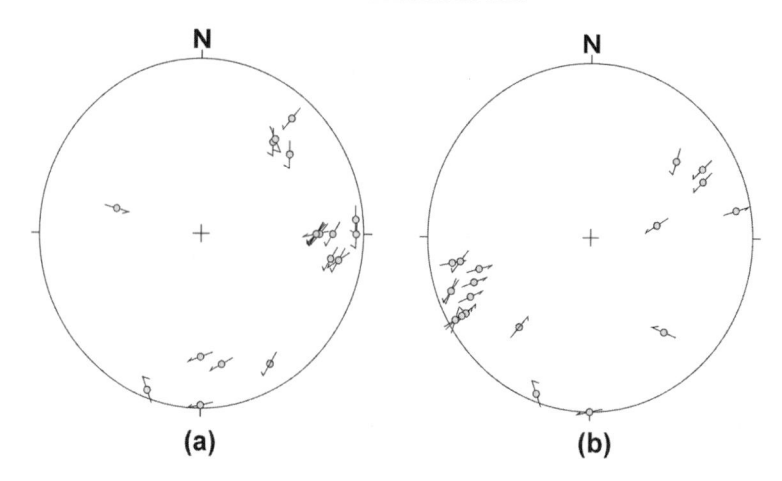

Fig. 5. Stereograms showing fault poles and associated slickenside lineations with slip direction (Wulff net, lower hemisphere projection). (**a**) Gavorrano Fault, (**b**) Monticello Fault.

and the original foliation is preserved by mimetic growth of phyllosilicate grains.

Spotted schists are characterized by biotite and millimetre-size andalusite porphyroblasts – the latter developing close to the contact with the intrusion. The mineral assemblage consists of quartz, plagioclase, K-feldspar, biotite, andalusite. The occurrence of some ovoid blasts completely replaced by fine-grained white mica could also indicate the presence of cordierite, the occurrence of which has been reported by Dallegno et al. (1979) in Mg- and Fe-bearing phyllites close to the intrusion. Calc-silicate hornfelses have mineral assemblages of quartz, muscovite, biotite, diopside, plagioclase, wollastonite, scapolite and calcite.

In the pelitic hornfels, the assemblage And + Bt + K-fs is indicative of a low-pressure thermal metamorphism. The mineral assemblage recognized in the pelitic hornfels can be explained by the following sequence of reactions:

Ms + Chl + Qtz = Crd + Bt + H_2O (Seifert 1970)
Ms + Bt + Qtz = Crd + Kfs + H_2O (Seifert 1976)
Ms + Qtz = And + Kfs + H_2O
(Chatterjee & Johannes 1974).

This prograde sequence of reactions, which led to the progressive disappearance of chlorite and muscovite belonging to the original phyllite, indicates that thermal metamorphism developed at P_{max} < 0.2 GPa and T_{max} ranging between 550° and 600 °C (Pattison & Tracy 1991).

These indications are in agreement with structural data that suggest a depth of magma emplacement of about 4–5 km as derived from (1) the maximum estimated thickness of nearly 3.5 km for the Ligurian Units and Tuscan Nappe overlying the intrusion, and (2) the Pleistocene uplift of about 650 m in southern Tuscany (Marinelli et al. 1993). Assuming an average density of 2.7 g cm^{-3}, a pressure of at least 0.1 GPa can be estimated for the thermal metamorphism around the Gavorrano intrusion.

Hydrothermal activity, with retrograde phenomena, largely affected the thermal aureole and the intrusives. Hydrothermal alteration was particularly intense on the northern side of the intrusion (Rigoloccio area in Fig. 6), with argillification of rocks and production of goethite and clay minerals like halloysite (Dallegno et al. 1979). According to Dallegno et al. (1979) late-stage hydrothermal activity, characterized by a temperature range of 300–170 °C, developed at low pressure.

Shape of the Gavorrano intrusion

The revision of the Gavorrano orebodies mining data (Dallegno et al. 1979) and new geological data, allow us to constrain the shape of the granitic body. The distribution of the granite at depth, contour lines on the top of the metamorphic complex and the distribution of the orebodies (Fig. 6) all indicate a westward subsurface development of the intrusion. Conversely, on the eastern side, the mining data indicate the absence of intrusive rocks down to a depth of 1000 m, according to a throw for the Monticello Fault in excess of 500 m (Arisi Rota & Vighi 1972; Dallegno et al. 1979). A large throw is also supported by the absence of any tectonic slices of hornfels along the Monticello fault (Fig. 2).

Profiling demonstrating reconstructed 3D

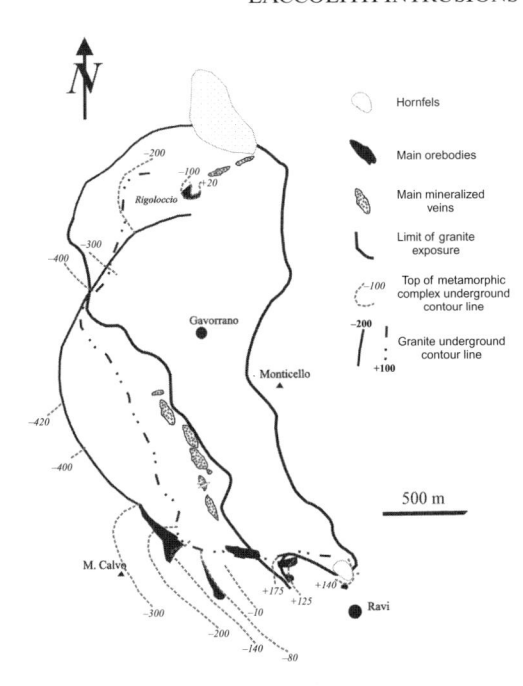

Fig. 6. Map of the Gavorrano intrusion shape deriving from field (this work) and mining data (modified after Dallegno et al. 1979). The limit of granite exposure roughly correspond to the Gavorrano and Monticello fault traces along the western and eastern sides of the intrusion, respectively. See text for explanation.

shape of the granite are shown in Figure 7, providing a three-dimensional view of the underground extent of the Gavorrano intrusion. This analysis confirms the westward subsurface development and the strongly asymmetrical shape of the intrusion. Mining data (western side) and field observation (northern and southern sides), where the original shape of the intrusion is preserved, suggest that the granite is characterized by a rather flat base, corresponding to the contact with the top of the metamorphic complex, and by a moderately to steeply west-dipping contact with the country rocks (Fig. 7). The base of the intrusion rises from depths of about −350 to −400 m a.s.l., in the central part of the pluton (profiles B–B' and E–E' in Fig. 7), up to the surface at its southern and northern margins (profile E–E' in Fig. 7), suggesting that the granite becomes thinner towards its northern and southern terminations. Overall, the granitic intrusion is laccolithic in form, with a maximum thickness of about 700 m, and a maximum (north–south direction) and minimum (east–west direction) dimension in map view of 3000 m and 1700 m, respectively.

The final shape and dimensions (width and thickness) of an intrusion are related to the way in which the host rocks make space for the magmas and their scale-invariant relationships (McCaffrey & Petford 1997; Cruden & McCaffrey 2001). The relationships between intrusion thickness and length can also provide insights on the mechanisms of magma emplacement and intrusion growth (Clemens 1998; Cruden 1998).

The power-law distribution for intrusion shape proposed by McCaffrey & Petford (1997) is defined by the equation:

$$T = kL^n \qquad (1)$$

where T is the intrusion thickness, L the intrusion width, k a normalization constant (0.12 in McCaffrey & Petford 1997) and n is the fractal exponent varying in the range (0.6–1.5; Cruden 1998; Cruden & McCaffrey 2001, 2002).

According to equation (1), several curves with different n values are plotted in Fig. 8. The Gavorrano granite is represented in the diagram by two points falling close to the curve with exponent $n = 1.5$. The first point (filled circle in Fig. 8) is defined as the maximum length, at the map scale (3000 m), and the maximum estimated thickness (700 m); the second point (empty square in Fig. 8) is computed as the equivalent radius for the area mapped in Figure 8, and the average estimated thickness (450 m). The computed exponent (1.5) for the Gavorrano intrusion is remarkably close to the exponent reported by Rocchi et al. (2002) for the shape of the Elba Island laccolith systems and those revised by Cruden & McCaffrey (2002) for several laccolith groups.

In order to determine whether the principal mechanism for space creation during magma emplacement was roof uplift or floor depression, we must consider the relationships between the intrusion geometric parameters and the mechanical properties of country rocks. Indeed, according to Zenzri & Keer (2001), the main processes involved in space creation during magma emplacement can be investigated by using the ratio between the maximum map dimension (L_w) of intrusion and the overburden thickness (h), together with the mechanical properties of overburden and base rocks defined by the parameter (θ) in the following equation:

$$\theta = [G_b(1 - v_o) - G_o(1 - v_b)]/[G_b(1 - v_o) + G_o(1 - v_b)]$$

where G_b, v_b, G_o, v_b are the shear modulus (G) and the Poisson ratio (v) for the overburden

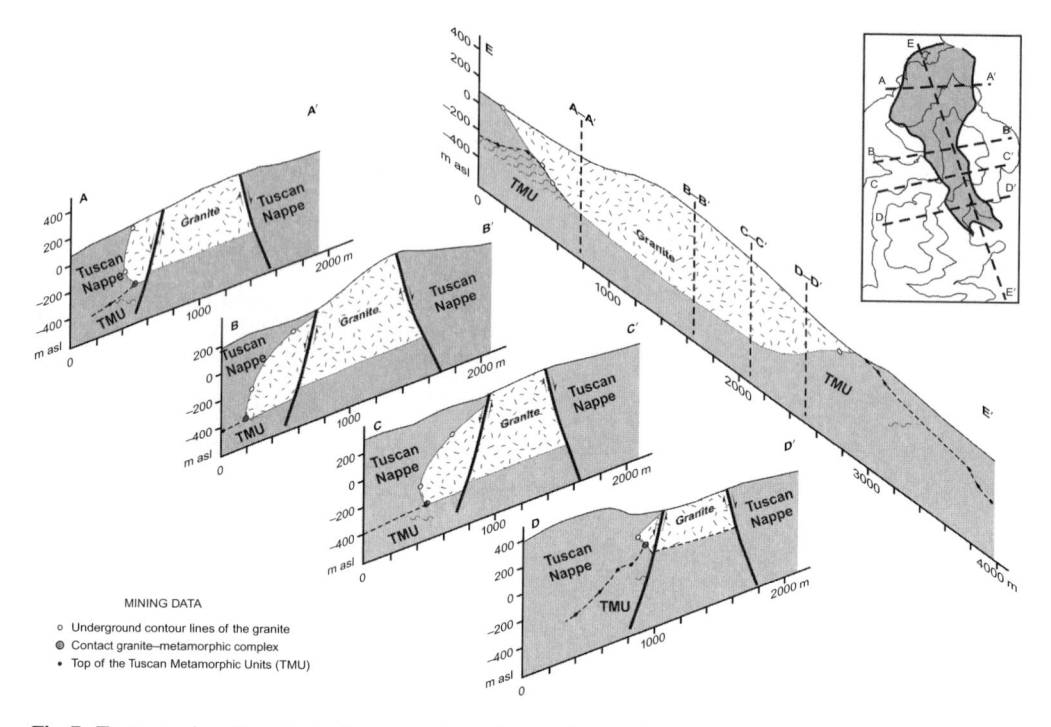

Fig. 7. Transversal and longitudinal cross-sections of the main granitic body reconstructed from the mining data (Dallegno et al. 1979) and from geological and structural data (this work). Where available, the subsurface extent of the Tuscan Metamorphic Units (TMU) is also reported.

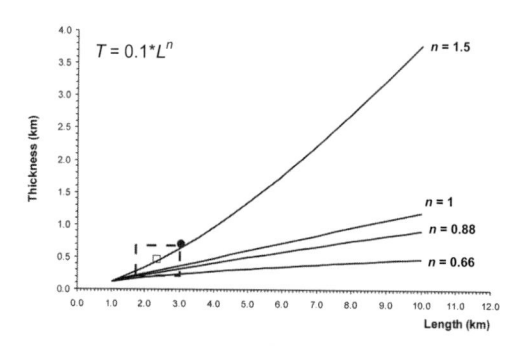

Fig. 8. Intrusion thickness v. intrusion width plot. The curves refer to the power-law distribution (according to McCaffrey & Petford 1997) for different n exponents. Dashed box represents the maximum values for thickness and width relative to Gavorrano. Empty box is for mean thickness (about 450 m) and equivalent radius of the intrusion area; filled circle is for maximum thickness and width of intrusion.

(subscript o) and the base (subscript b), respectively.

The overburden of the Gavorrano intrusion consists of limestone, marls and dolomitic limestones, and the base consists of shale, schist and quartzites. The mechanical properties of such materials (Turcotte & Schubert 2002) and the geometrical parameters of the intrusion reported in Table 2 indicate very low material contrast ($\theta = -0.06$) and intermediate to low geometric ratio ($L_w/h = 0.75$) suggesting a laccolith emplacement mechanism dominated by roof uplift and with minor floor depression (Zenzri & Keer 2001, fig. 5).

Discussion and conclusion

The Gavorrano intrusion represents an asymmetrical laccolith, emplaced at shallow depth (<5 km) within the nappe pile of the Northern Apennines. The reconstructed laccolith shape and its position within the stacked tectonic units indicate that the magma spread along a subhorizontal discontinuity and grew vertically by lifting of the overburden rocks. The exploited discontinuity corresponds to the basal thrust of the Tuscan Nappe on the Tuscan Metamorphic Units, represented by disrupted evaporitic layers. The asymmetrical shape of the intrusion could be the result of:

1 the syn-plutonic roof uplift activity of the Monticello fault, according to the model proposed for the intrusions along the Coastal

Table 2. *Gavorrano laccolith mechanical and geometrical parameters*

Parameters	Overburden*	Base†	θ‡	L_w/h
G (10^2 GPa)	0.30	0.25		
v	0.20	0.25	−0.06	0.75
L_w (km)		3.00		
h (km)		4.00		

L_w: laccolith width; h: overburden thickness; G: shear modulus; v: poisson ratio.
* limestones, evaporites, dolomitic limestones and marls.
† quartzites, shales and schists.
‡ $\theta = [G_b(1 - v_o) - G_o(1 - v_b)]/[G_b(1 - v_o) + G_o(1 - v_b)]$ (Zenzri & Keer 2001).

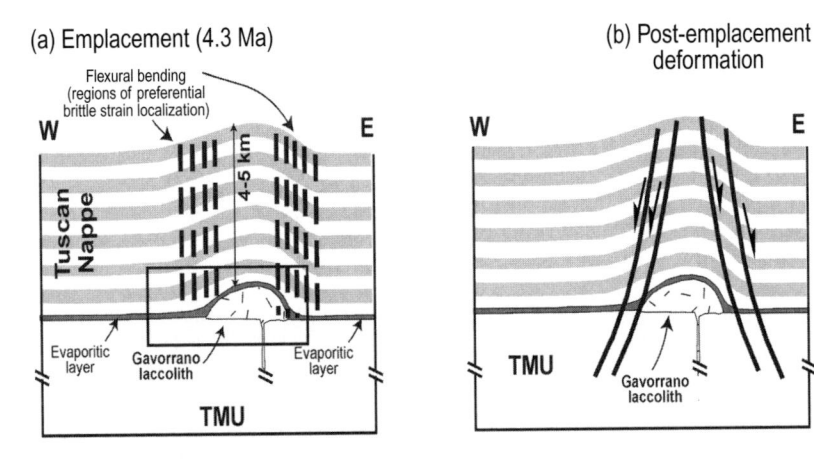

Fig. 9. Cartoon showing the proposed model for the Gavorrano laccolith emplacement (**a**) and post-emplacement deformation (**b**). Vertical dashed lines represent an area with high fracture density for the overburden flexure, TMU, Tuscan Metamorphic Units. Proposed original asymmetry of the laccolith could explain the different position of subsequent border faults.

Cordillera in Chile by Grocott & Taylor (2002), or

2 the post-emplacement activity of the Monticello Fault.

Geological examples and analogue models have stressed the critical role of such a detachment layer between competent units in triggering laccolith formation (e.g. Guillot et al. 1993; Roman-Berdiel et al. 1995). Moreover, analogue models indicated that overburden thickness controls the shape of the laccolith – a thick overburden being associated with a lens-shaped intrusion, and a thin overburden with a bell-shaped laccolith (Roman-Berdiel et al. 1995).

The antiformal shape of the Gavorrano Ridge structure, probably partly inherited from the preceding compressional deformation, was likely enhanced by multi-pulse magma emplacement. The upward flexure of the brittle overburden led, on opposite sides of the antiform, to the formation of fractured zones where faulting was successively partitioned (Fig. 9). The development of fractured zones at the border of the intrusion, in the overburden rocks, has been demonstrated also by analogue models on laccolith growth (Roman-Berdiel et al. 1995).

The post-emplacement deformation in the laccolith and the host rocks (cataclasites and fault zones), the occurrence of slices of thermal aureole along fault planes, and of fault breccias containing clasts of granite, indicate that brittle tectonics (Gavorrano and Monticello faults) acted after the emplacement and cooling of the laccolith. These deformation features observed in the granite and country rocks argue against a synkinematic emplacement of magma in pull-apart structures along major strike-slip zones, as proposed by Rossetti et al. (2000, 2001) and Acocella & Rossetti (2002) for the Neogene intrusions in southern Tuscany.

We therefore suggest that the occurrence of inherited tectonic discontinuities (thrusts) in the inner zone of the northern Apennines, had a strong influence on the emplacement of magmas at upper crustal levels. A similar mechanism has also been envisaged in the western and central part of Elba island, where Late Miocene intrusions were emplaced within a tectonic pile characterized by several mechanical discontinuities represented by thrust surfaces or alternation of sedimentary rocks of different competence (Rocchi et al. 2002).

In conclusion, our proposed model for the Gavorrano intrusion emphasizes that, in fold–thrust belts, the emplacement of magmas at upper-crustal levels is strongly controlled by the occurrence of mechanical discontinuities, mainly represented by tectonic structures separating rocks of different composition and rheological behaviour (schist v. carbonate rocks in our example). In a nappe pile, these structures can act as weaker zones in which rising magmas can be preferentially trapped. The final shape of the intrusion, whether laccolith or loppolith, depends on a number of factors, but seems to be largely controlled by the mechanical properties of the host rocks.

The authors greatly acknowledge S. Cruden and an anonymous referee for their reviews, and N. Petford for helpful editorial assistance.

References

ACOCELLA, V. & ROSSETTI, F. 2002. The role of extensional tectonics at different crustal levels on granite ascent and emplacement: an example from Tuscany (Italy). *Tectonophysics*, **354**, 71–83.

AMEGLIO, L., VIGNERESSE, J.L. & BOUCHEZ, J.L. 1997. Granite pluton geometry and emplacement mode inferred from combined fabric and gravity data. *In*: BOUCHEZ, J.L., HUTTON, D.H.W. & STEPHENS, W.E. (eds) *Granite: from Segregation of Melt to Emplacement Fabrics*. Kluwer Academic Publishers, Dordrecht, 199–214.

ARISI ROTA, F. & VIGHI, L. 1972. Le mineralizzazioni a pirite ed a solfuri misti della Toscana Meridionale. *Rendiconti della Società Italiana di Mineralogia e Petrologia*, **27**, 368–423.

BARTOLE, R. 1995. The North Tyrrhenian–Northern Apennines post-collisional system: constraints for a geodynamic model. *Terra Nova*, **7**, 7–30.

BOCCALETTI, M. & SANI, F. 1998. Cover thrust re-activations related to internal basement involvment during Neogene–Quaternary evolution of the northern Apennines. *Tectonics*, **17**, 112–130.

BOSSIO, A. ET AL. 1993. Rassegna delle conoscenze sulla stratigrafia del Neoautoctono Toscano. *Memorie della Società Geologica Italiana*, **49**, 17–98.

BURGASSI, P.D., DECANDIA, F.A. & LAZZAROTTO, A. 1983. Elementi di stratigrafia e paleogeografia nelle Colline Metallifere (Toscana) dal Trias al Quaternario. *Memorie della Società Geologica Italiana*, **25**, 27–50.

CAIOZZI, F., FULIGNATI, P., GIONCADA, A. & SBRANA, A. 1998. Studio SEM–EDS dei minerali figli nelle inclusioni fluide del granito di Botro ai Marmi (Campiglia Marittima) e possibili implicazioni minerogenetiche. *Atti Società Toscana Scienze Naturali, Memorie – Serie A*, **105**, 65–73.

CARMIGNANI, L., DECANDIA, F.A., DISPERATI, L., FANTOZZI, P.L., LAZZAROTTO, A., LIOTTA, D. & OGGIANO, G. 1995. Relationships between the Tertiary structural evolution of the Sardinia–Corsica Provençal Domain and the Northern Apennines. *Terra Nova*, **7**, 128–137.

CASSANO, E., ANELLI, L., VINCENZO, C. & LA TORRE, P. 2001. Magnetic and gravity analysis of Italy. *In*: VAI, G.B. & MARINI, I.P. (ed.) *Anatomy of an Orogen: the Apennines and Adjacent Mediterranean Basin*, Kluwer Academic Publishers, London, 53–64.

CHATTERJEE, N. & JOHANNES, W. 1974. Thermal stability and standard thermodynamic properties of synthetic 2M1-muscovite, $KAl_2[AlSi_3O_{10}(OH)_2]$. *Contributions to Mineralogy and Petrology*, **48**, 89–114.

CLEMENS, J.D. 1998. Observations on the origins and ascent mechanisms of granitic magmas. *Journal of the Geological Society of London*, **155**, 843–851.

CORRY, C.E. 1988. Laccoliths – mechanics of emplacement and growth. *Geological Society of America Special Papers*, **220**, 110 pp.

CRUDEN, A.R. 1998. On the emplacement of tabular granites. *Journal of the Geological Society of London*, **155**, 853–862.

CRUDEN, A.R. & MCCAFFREY, K.J.W. 2001. Growth of plutons by floor subsidence: implications for rates of emplacement, intrusion spacing and melt-extraction mechanisms. *Physics and Chemistry of the Earth, A*, **26**, 303–315.

CRUDEN, A.R. & MCCAFFREY, K.J.W. 2002. Different scaling laws for sills, laccoliths and plutons: Mechanical thresholds on roof lifting and floor depression. *In*: *First International Workshop on Physical Geology of Subvolcanic Systems – Laccoliths, Sills, and Dykes (LASI)*. BREITKREUZ, C., MOCK, A. & PETFORD, N. (eds) Wissenschaftliche Mitteilungen, Institut für Geologie, Freiberg, **20**, 15–17.

DALLEGNO, A., GIANELLI, G., LATTANZI, P. & TANELLI, G. 1979. Pyrite deposits of the Gavorrano area, Grosseto. *Atti Società Toscana Scienze Naturali, Memorie – Serie A*, **86**, 127–165.

DELLA VEDOVA, B., BELLANI, S., PELLIS, G. & SQUARCI, P. 2001. Deep temperatures and surface heat flow distribution. *In*: VAI, G.B. & MARINI, I.P. (eds) *Anatomy of an Orogen: the Apennines and Adjacent Mediterranean Basin*, Kluwer Academic Publishers, London, 65–76.

FRANCESCHINI, F. 1998. Evidence of an extensive Pliocene–Quaternary contact metamorphism in southern Tuscany. *Memorie della Società Geologica Italiana*, **52**, 479–492.

FRANCESCHINI, F., INNOCENTI, F., MARSI, A., TAMPONI, M. & SERRI, G. 2000. Petrography and chemistry of the buried Pliocene Castel di Pietra pluton (Southern Tuscany, Italy). *Neues Jahrbuch für Geologie und Paläontologie Abh.*, **215**, 17–46.

GIANELLI, G., MANZELLA, A. & PUXEDDU, M. 1997. Crustal models of the geothermal areas of southern Tuscany (Italy). *Tectonophysics*, **281**, 221–239.

GROCOTT, J. & TAYLOR, K.G. 2002. Magmatic arc fault systems, deformation partitioning and emplacement of granitic complexes in the Coastal Cordillera, north Chilean Andes (25° 30′ S to 27° 00′ S). *Journal of the Geological Society, London*, **159**, 425–442.

GUILLOT, S., PECHER, A., ROCHETTE, P. & LE FORT, P. 1993. The emplacement of the Manaslu granite of Central Nepal: field and magnetic susceptibility constraints. *Geological Society, London, Special Publications*, **74**, 413–428.

INNOCENTI, F., SERRI, G., FERRARA, G., MANETTI, P. & TONARINI, S. 1992. Genesis and classification of the rocks of the Tuscan Magmatic Province: thirty years after Marinelli's model. *Acta Vulcanologica*, **2**, 247–265.

KELLER, J.V.A., MINELLI, G. & PIALLI, G. 1994. Anatomy of late orogenic extension: the northern Apennines case. *Tectonophysics*, **238**, 275–294.

MCCAFFREY, K.J.W. & PETFORD, N. 1997. Are granitic intrusions scale invariant? *Journal of the Geological Society of London*, **154**, 1–4.

MARINELLI, G. 1961. L'intrusione Terziaria di Gavorrano. *Atti Società Toscana Scienze Naturali, Memorie – Serie A*, **68**, 117–194.

MARINELLI, G., BARBERI, F. & CIONI, R. 1993. Sollevamenti neogenici e intrusioni acide della Toscana e del Lazio settentrionale. *Memorie della Società Geologica Italiana*, **49**, 279–288.

MIDDLEMOST, E.A.K. 1989. Iron oxidation, norms and the classification of volcanic rocks. *Chemical Geology*, **77**, 19–26.

NICOLICH, R. 2001 Deep seismic transects. *In*: VAI, G.B. & MARINI, I.P. (eds) *Anatomy of an Orogen: the Apennines and Adjacent Mediterranean Basin*, Kluwer Academic Publishers, London, 47–52.

PATTISON, D.R.M. & TRACY, R.J. 1991. Phase equilibria and thermobarometry of metapelites. *Reviews in Mineralogy*, **26**, 105–206.

ROCCHI, S., WESTERMAN, D.S., DINI, A., INNOCENTI, F. & TONARINI, S. 2002. Two-stage growth of laccoliths at Elba Island, Italy. *Geology*, **30**, 983–986.

ROMAN-BERDIEL, T., GAPAIS, D. & BRUN, J.P. 1995. Analogue models of laccolith formation. *Journal of Structural Geology*, **17**, 1337–1346.

ROSSETTI, F., FACCENNA, C., ACOCELLA, V., FUNICELLO, R., JOLIVET, L. & SALVINI, F. 2000. Pluton emplacement in the northern Tyrrhenian Sea area (Italy). *Geological Society, London, Special Publications*, **174**, 55–77.

ROSSETTI, F., FACCENNA, C., FUNICELLO, R., PASCUCCI, V., PIETRINI, M. & SANDRELLI, F. 2001. Neogene strike-slip faulting and pluton emplacement in the Colline Metallifere region (Southern Tuscany, Italy): the Gavorrano–Capanne Vecchie area. *Bolletino Società Geologica Italiana*, **120**, 15–30.

ROYDEN, L., PATACCA, E. & SCANDONE, P. 1987. Segmentation and configuration of subducted lithosphere in Italy: an important control on thrust-belt and foredeep-basin evolution. *Geology*, **15**, 714–717.

SEIFERT, F. 1970. Low temperature compatibility relations of cordierite in the haplopelites of the system K_2O–MgO–Al_2O_3–SiO_2–H_2O. *Journal of Petrology*, **11**, 73–99.

SEIFERT, F. 1976. Stability of the assemblage cordierite + K-feldspar + quartz. *Contributions to Mineralogy and Petrology*, **57**, 179–185.

SERRI, G., INNOCENTI, F. & MANETTI, P. 1993. Geochemical and petrological evidence for the subduction of delaminated Adriatic continental lithosphere in the genesis of the Neogene–Quaternary magmatism of central Italy. *Tectonophysics*, **223**, 117–147.

SERRI, G., INNOCENTI, F. & MANETTI, P. 2001. Magmatism from Mesozoic to Present: petrogenesis, time–space distribution and geodynamic implications. *In*: VAI, G.B. & MARINI, I.P. (eds) *Anatomy of an Orogen: the Apennines and Adjacent Mediterranean Basin*, Kluwer Academic Publishers, London, 77–104.

STRECKEISEN, A. & LE MAITRE, R.W. 1979. A chemical approximation to the modal QAPF classification of igneous rocks. *N. Jb. Miner. Abh.*, **136**, 169–206.

TURCOTTE, D.L. & SCHUBERT, G. 2002. *Geodynamics*. Second Edition, Cambridge University Press, Cambridge, 436 pp.

ZENZRI, H. & KEER, L.M. 2001. Mechanical analyses of the emplacement of laccoliths and lopoliths. *Journal of Geophysical Research*, **106**, 13 781–13 792.

Rate of construction of the Black Mesa bysmalith, Henry Mountains, Utah

GUILLAUME HABERT & MICHEL DE SAINT-BLANQUAT

UMR 5563/ LMTG, Observatoire Midi-Pyrénées, CNRS/Université Paul-Sabatier, 14 Avenue E. Belin, 31400 Toulouse, France
(e-mail: habert@lmtg.obs-mip.fr; michel@lmtg.obs-mip.fr)

Abstract: At shallow-crustal levels, the most efficient process for the accommodation of magma emplacement is roof lifting, which induces an upward vertical displacement of the Earth's surface. Estimates of the rate and duration of this process have rarely been published. One of the most spectacular places where plutons constructed by such mechanisms are exposed is the Henry Mountains in Utah. In this place, Pollard & Johnson (1973) derive a time of 'less than several weeks' for the construction of the Black Mesa bysmalith (BMb), by coupling a mechanical approach with a model for the flow rate of Bingham magma in a tabular conduit with a constant driving pressure at the feeder. The aim of this new study of the BMb is to evaluate the maximum duration of its emplacement and propose a feasible scenario for its construction. Our study of the pluton's internal structures suggest that the BMb is a multi-pulse pluton. We have constrained the duration of BMb emplacement by simulating the thermal evolution of the growing pluton and its wallrocks for different construction scenarios. We have adjusted the number, the thickness and the frequency of the pulses with our textural 'time' constraints, which are the absence of solid-state textures around internal contacts, which implies that a melted zone was maintained in the intrusion during its construction; and the absence of significant contact metamorphism or recrystallization, which means that the increase of temperature in the host rock was relatively small, or short-lived, or both. In accordance with the previous estimates of Pollard & Johnson (1973), we propose that the emplacement of the BMb was a very short geological event, with a *maximum* duration in the order of 60 years, implying *minimum* vertical displacement rates of the wallrocks above the pluton of 4 metres per year. Moreover, our simulations indicate that pulses around 20 metres thick rapidly injected approximately every three months are the most consistent with the constraints from field observations.

The ascent and emplacement of magmas is the principal means by which heat and mass are transferred into and through the Earth's crust. However, the rate and mechanisms of these processes is still under debate (Saint-Blanquat *et al.* 1999; Harris *et al.* 2000; McKenzie 2000; Saint-Blanquat *et al.* 2001). In recent years, attention has focused on dykes as an efficient method to transport large quantities of silicic magma through the crust. Some studies have addressed questions related to the flow rate of magma through dykes (Bruce & Huppert 1990; Petford 1996; Petford *et al.* 2000). Although magma transport and emplacement have been considered somewhat independent processes (e.g. Clemens *et al.* 1997), some degree of *in situ* expansion must accompany dyke-fed pluton growth. This question of growth rate and the subsequent associated deformation rate is still a major problem for the comprehension of magma emplacement mechanisms (Gerbi *et al.* 2004). Yoshinobu *et al.* (1998) discussed the rate of emplacement for fault-controlled emplacement of magmatic bodies, and show that if the pluton is fed by small temporally spaced dyke-fed injections, in accordance with tectonic rates, a sheeted complex, at least on the pluton margins, is likely to occur. As sheeted complexes at pluton margins are only occasionally observed, tectonic rates seem therefore to be too slow to accommodate dyke-fed pluton growth (Gerbi *et al.* 2004), and other mechanisms must be envisioned (Paterson & Tobish 1992).

The Tertiary dioritic intrusions of the Henry Mountains, located on the Colorado Plateau, have experienced no significant regional deformation during and since their emplacement. The country rocks away from the pluton margins are consequently horizontal and undeformed, and constitute ideal markers to quantify translation, rotation and strain associated with forceful emplacement. Most importantly, this allows us to rule out tectonic forces and to unambiguously interpret the emplacement history in terms of magmatic processes only. The Henry Mountains thus constitute an exceptional natural laboratory

From: BREITKREUZ, C. & PETFORD, N. (eds) 2004. *Physical Geology of High-Level Magmatic Systems.*
Geological Society, London, Special Publications, **234**, 163–173. 0305-8719/04/$15.00
© The Geological Society of London 2004.

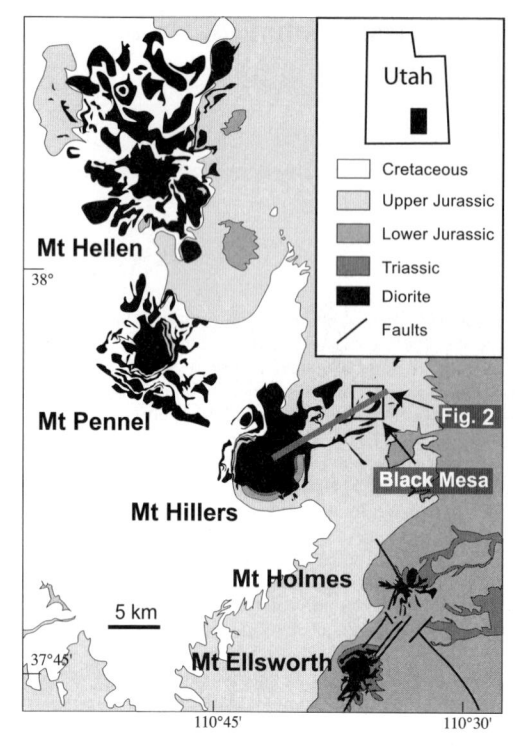

Fig. 1. Simplified geological map of the Henry Mountains and the Black Mesa Bysmalith, modified from Jackson & Pollard (1988).

to study magmatic processes in the absence of any significant regional tectonic strain field. In this study, we choose to constrain the size and frequency of the pulses by combining textural studies of the pluton interior with a numerical simulation of its thermal evolution during non-instantaneous multi-pulse construction.

Geological setting and previous work

The Henry Mountains, in SE Utah are c. 60 km long and trend roughly north–south (Fig. 1). Five main intrusions are surrounded by numerous dykes, sills and minor laccoliths. Intrusions are situated on the gently dipping (1–2° west) eastern limb of a north–south-trending basin, that is bounded on the west by the Waterpocket fold (Hunt 1953). The intrusive rock is a quartz-diorite porphyry (Engel 1959) and it intrudes Permian to Mezozoic formations, essentially sandstones and shales. The ^{40}Ar–^{39}Ar dating on diorites (Nelson 1993) give Middle to Late Oligocene ages (21–31 Ma) and chemical analyses associate the Henry Mountains with subduction (Nelson & Davidson 1993a). Thus, the Henry Mountains and other laccoliths of the Colorado Plateau such as the La Sal and Abajo Mountains appear to be related to a large-scale igneous system with arc-like affinities (Armstrong & Ward 1991; Nelson & Davidson 1993b). The relatively minor volume of igneous rocks emplaced on the Colorado Plateau may be due to the fact that it is underlain by a thick cratonic crust which could have acted as a structural barrier and hampered the ascent of magma to high crustal levels (Nelson & Davidson 1993a).

There are conflicting hypotheses for the emplacement mechanism of the plutonic rocks constituting the Henry Mountains. It is clear that the radially dispersed satellite intrusions are related to the main domes (Figs 1 and 2). Hunt (1953) interpreted the five major intrusions as stocks laterally feeding the satellite intrusions, while Gilbert (1877) and Jackson & Pollard (1988) proposed that the main intrusions are floored intrusions which were emplaced after the small laccoliths. The latter hypothesis is in good agreement with the rotated palaeomagnetic vectors from the sills and the small laccoliths around Mount Hillers, which indicate that these intrusions cooled while still horizontal and were then tilted by the growth of the central main body (Jackson & Pollard 1988).

The Black Mesa bysmalith (BMb) is located on the eastern flank of Mount Hillers and is intrusive between the Jurassic Summerville and Morrison formations. It is surrounded by the Sawtooth ridge laccolith, which is intrusive between the same stratigraphic layers (Johnson & Pollard 1973), and the Maiden Creek sill and Trachyte Mesa laccolith, which both intrude the Entrada Formation.

Based on the stratigraphic section compiled by Jackson & Pollard (1988), who estimated thicknesses obtained through stratigraphic correlations on the whole basin, a maximum of 2.5 km of sedimentary rocks were overlying the Morrison Formation at the time of emplacement, which constitutes the lithostatic load over the roof of the BMb.

BMb has given rise to a great number of studies (e.g. Hunt 1953; Pollard & Johnson 1973; Kerr & Pollard 1998; Zenzri & Keer 2001), because it seems to approximate the laccolithic structural form conceived initially by Gilbert (1877). Pollard & Johnson (1973), used the BMb to validate a three-stage model for forceful emplacement: the sill phase, controlled by lateral propagation of a thin igneous sheet; followed by the laccolith phase characterized by lateral propagation plus bending of the overburden; and finally the bysmalith stage, during which lateral propagation is stopped and the

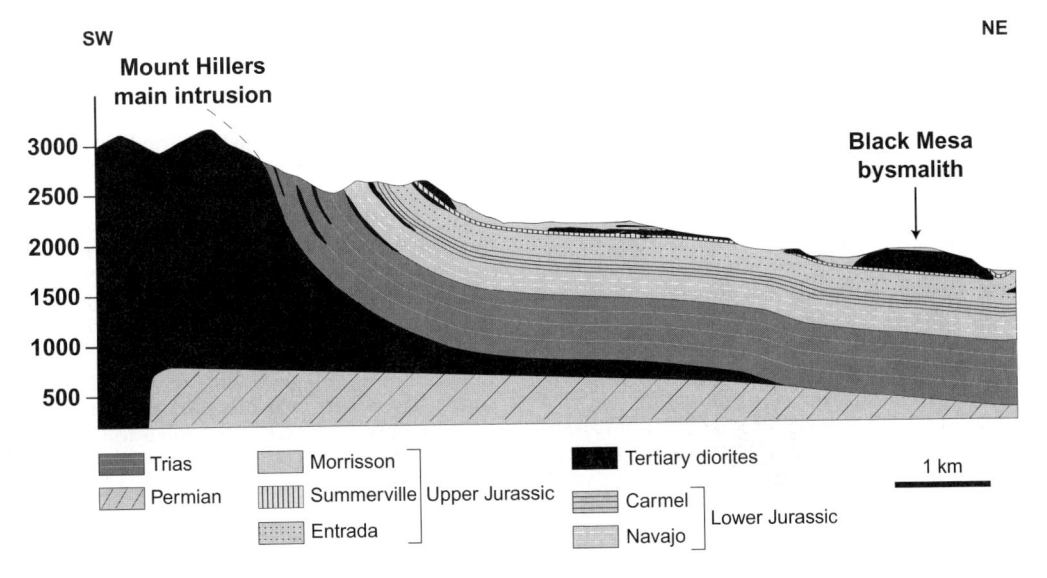

Fig. 2. Interpretative geological cross-section from Mout Hillers to Black Mesa. Note the abundance of intrusions at the contact between the Summerville and Morrison formations.

further injection of magma is accommodated by vertical uplift of the roof permitted by the growth of a peripheral fault. This model is supported by our new data on the internal structure of the pluton (Habert *et al.* in review).

The Black Mesa bysmalith

Geometry of the pluton

The BMb is a cylindrical pluton 1.8 km in diameter and 250 m thick. The intrusion is surrounded by shallowly dipping strata, which abruptly change orientation a few hundred metres from the diorite (Fig. 2), and become subvertical. West of the intrusion, this folding is marked by a syncline; on the eastern part of the intrusion, the contact is characterized by vertical faulting which is marked morphologically by a cliff. The roof of the BMb is flat, slightly north-dipping, and covered by concordant sedimentary strata of the Morrison Formation. The presence of this formation at the base and the top of the exposed diorite in a flat-lying geometry led Hunt (1953), and Jackson & Pollard (1988) to conclude that the current exposed thickness must be close to the real one. The volume of magma is then approximately 0.6 km^3, if an ideal cylinder shape is assumed.

Differential erosion between the sedimentary and the igneous rock has produced a morphologically well-marked intrusion. The presence of valleys all around the BMb, except on the southwest where we found outcrops of diorite, lead Hunt (1953) to envision lateral feeding of the BMb, from the southwest. But a vertical feeder dyke below the pluton is more in agreement with our new data on the fabric pattern of the pluton (Habert *et al.* in review), and with the general model developed by Pollard & Johnson (1973) and Jackson & Pollard (1988).

Field constraints for a multi-pulse pluton construction

In general, two unambiguous field criteria for the recognition of a multi-pulse emplacement are cross-cutting fabrics and/or chilled margins between adjacent layers. At a first glance the main characteristic of Black Mesa is its structural and petrographic homogeneity at the scale of the whole pluton, and we do not observe either of these two criteria. The foliation is parallel to the contacts and the lineation is variable, but the trend seems to be more north–south at the base and east–west on the top of the pluton (Habert *et al.* in review). At the outcrop scale, the lineation is sometimes variable, but without clear discontinuity between adjacent layers with different lineations. In the absence of any regional strain involved in the pluton construction (Pollard & Johnson 1973; Jackson & Pollard 1988), these changes of the orientation of the magmatic lineation at various scales, outcrop to pluton, may be indicative of a multi-pulse history. The flat subhorizontal upper contact of the BMb is cut in some places by late dyke-like bodies of fine-grained diorite, showing clear textural evidence for successive injections. In some places within the main body, we have observed internal contacts, defined by variations in the size of porphyritic minerals (Fig. 3a), and magmatic layering, defined by variations in the relative proportions of phenocrysts. Both structures are subparallel to the magmatic fabric. No solid-state deformation is associated with them, which indicate that the time gap between individual injections was insufficient to allow a complete solidification of the older intrusion before the emplacement of the youngest one. Indeed, the microstructures are magmatic (Fig. 3f) everywhere within the massif, except within rare cataclastic shear bands which are localized:

1 in the first centimetre of the pluton parallel to the adjacent upper contact (Fig. 3c & d),
2 along the lateral margins where they are sub parallel to the margin and associated with vertical faulting in the wallrocks, and
3 within 1–2-metre thick horizontal layers of anastomosing cataclastic bands separated by 10–20-metre thick layers of undeformed diorite (Fig. 3e), which could represent ancient crystallized boundaries between pulses which acted as a preferential planes of cataclastic reactivation during injection of younger pulses at the base of the pluton.

Thin sections from the sediments just above the upper contact revealed no recrystallization of the quartz in the sandstone, and no change in the calcic matrix (Fig. 3b). Close to the lower contact, we observed the growth of metamorphic albite and epidote.

Taken together with the fact that we can directly observe only the external skin of the pluton, all these observations lead us to envision successive injections. We interpret the construction of the BMb with a multi-pulse mechanism, with underaccretion of pulses from below, with a limited and/or short-lived heat transfer to the wallrocks. Due to the petrological homogeneity of the pluton, the structures indicative of this mechanism are very difficult to observe, so we are unable to precisely measure the thickness and, consequently, the number of pulses. We believe that the internal contacts between pulses are poorly developed because the intrusion was constructed very rapidly and stayed hot throughout construction.

Thermal numerical simulations

Principle

We have attempted to constrain the duration of BMb emplacement by modelling the thermal evolution of the growing pluton and its wall-rocks for different construction scenarios, by adjusting the number, the thickness and the frequency of the pulses with our petrostructurally derived 'time' constraints. The two main constraints that we have from our study are:

1 the absence of solid-state textures around internal contacts, which implies that a melted zone was maintained in the intrusion during its construction;
2 the absence of significant contact metamorphism or recrystallization, which means that the increase of temperature in the host rock was relatively small, or short-lived, or both.

The two characteristic parameters that we used to constrain the time and rate of construction are the time between each injection (t_i) and the thickness of each injection (h).

Methodology

Given the relatively simple cylindrical geometry of the pluton, and the ratio of 1/7 between the thickness of the body and its lateral extension, we consider that the approximation of the infinite plane can be used to explore the evolution of the temperature of a vertical line in the middle of the BMb. Therefore, we used a one-dimensional model. The thermal history of the construction of the intrusion and its surroundings has been numerically simulated with the program DF1DEXPL.vba which is a Visual Basic code, based on an explicit finite difference algorithm to solve the one-dimensional conductive heat-transfer equation (1) (Peacock 1990; Gvirtzman & Garfunkel 1996):

$$\frac{\partial T}{\partial t} = \frac{k}{\rho C_p} \frac{\partial^2 T}{\partial z^2} + \frac{A(z)}{\rho C_p} \quad (1)$$

where T is temperature, k is thermal conductivity, t is time, $A(z)$ is internal heat production rate per unit volume, z is depth, c_p is the specific heat per unit mass and ρ is the density. The latent heat (ΔH) is computed using the formulation of Furlong et al. (1991). At each node, if the temperature belongs to the interval of crystallization ΔT ($T_{solidus} - T_{liquidus}$), the latent heat is included in the specific heat with the formula (2) (Furlong et al. 1991).

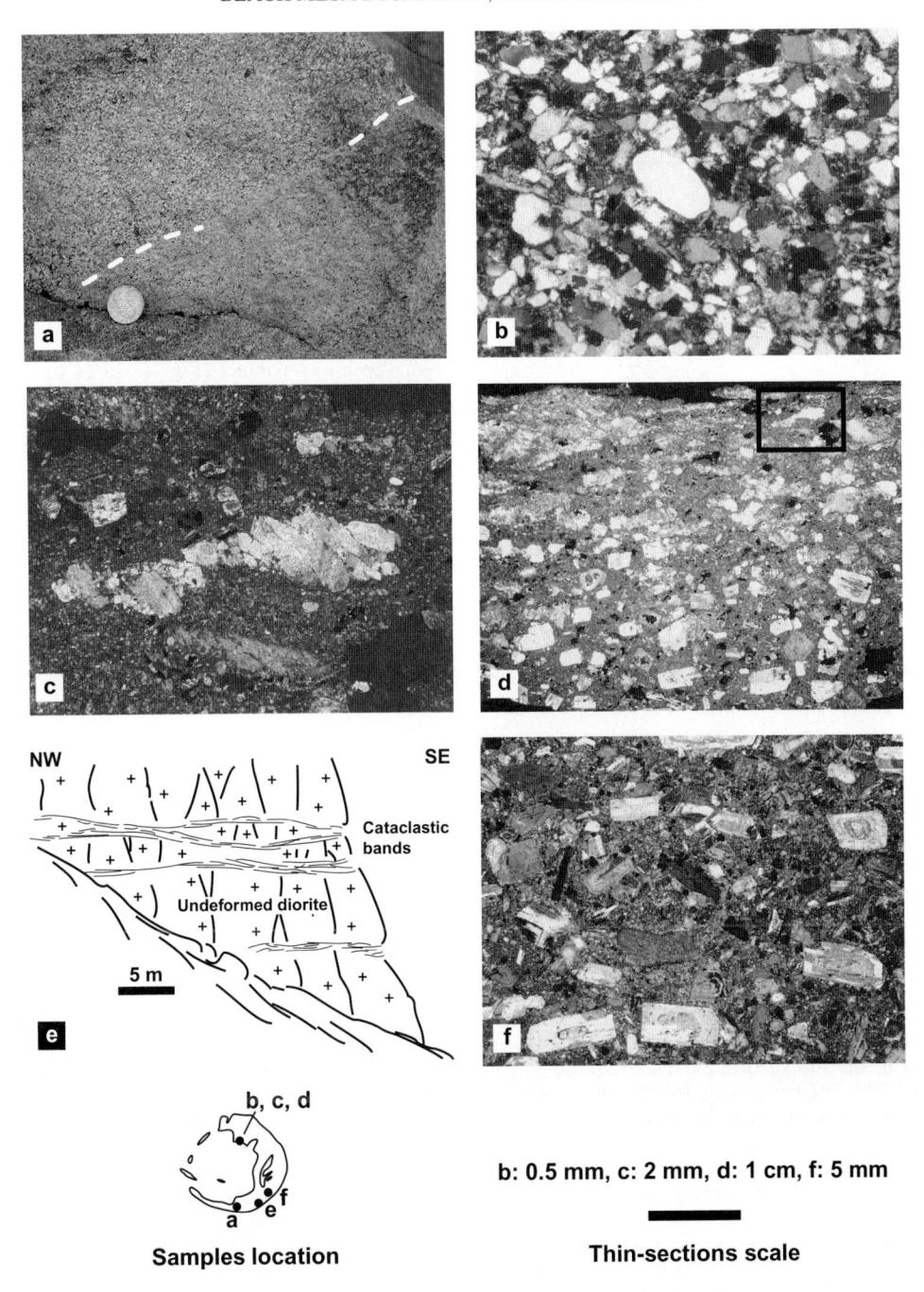

Fig. 3. Field observations. Locations of the different observations are indicated on a sketch of the Black Mesa bysmalith. (**a**) Internal contact (dotted line) marked by a change in the abundance of plagioclase. (**b**) Thin-section of a sandstone of the Morrison Formation, taken 20 cm above the upper contact with the diorite. Note the rounded quartz grains and the absence of recystallization. (**c**) Close-up of a cataclastically deformed diorite. Note the fractures in the plagioclase. (**d**) Thin-section showing the transition from cataclastic to magmatic textures. The contact with the host rock is right at the top of the picture. (**e**) Outcrop where anastomosing cataclastic bands separate layers of undeformed diorite. (**f**) Thin-section showing magmatic texture. Note the foliation marked by the SPO of plagioclase and hornblende.

Table 1. *Thermal parameters used for the numerical modelling.*

Nature of the rock	Sandstone	Diorite
Specific heat (J kg^{-1} K^{-1})	1000	1100
Thermal conductivity (W m^{-1} K^{-1})	2.65	3
Density (kg m^{-3})	2630	2730
Latent heat of crystallization (J kg^{-1})	0	290 000
Crystallization interval (°C)	0	200
Solidus temperature (°C)	0	700

Data used in the model come from Niederkorn & Blumenfeld (1989) and Pavlis (1996).

$$C_{p-\text{modified}} = C_p \frac{\Delta H}{\Delta T} \quad (2)$$

Parameters for the magma and the host rocks are reported in Table 1. The boundary conditions are the same for all simulations:

1. there is no heat flux at the boundaries;
2. boundaries of the model are far enough from the intrusion in order not to be influenced by the heat release from the magma (i.e. the boundaries stay at the same temperature during the entire simulation as there is no internal heat production in the host rock – see Table 1).

However, we keep in mind that the temperatures that we obtain are overestimated, due to the fact that our model heat transfer is only conductive and that there is no 3D dispersion of heat. One way to use this model is then to look at relative differences from one simulation to another. To be able to compare the different simulations, we let the model evolve until external contacts cooled under a temperature of 500 °C.

Previous studies (Johnson & Pollard 1973) suppose that emplacement of new injections are made from below, at the base of the intrusion, with an upward vertical movement of the pluton roof. This is clearly evidenced by the geometry of the wallrocks and by the internal fabric (Habert *et al.* in review). Therefore, we chose to emplace the incoming magmatic pulse below the earlier pulses (underaccretion), but, as there is necessarily a part of the first intrusions which is frozen right at the contact, new injections are emplaced at a certain distance above the lower contact. We fixed this distance by referring to the first stage of the emplacement where a sill propagates between the Morrison and Summerville formations. Injections are emplaced at the middle of that sill. As sills around the BMb, and in the Henry Mountains in general, have a mean thickness of 20 metres (Johnson & Pollard 1973), new injections are placed at 10 metres from the lower contact (Fig. 4a). For models where the thickness h of the individual injections is less than 20 metres, new injections are emplaced in the middle of the older one (Fig. 4b), until the total thickness reaches 20 metres. Each injection is emplaced instantaneously; thickness and time between injections are constant from the beginning to the end of pluton construction. Simulations were made for h equal to 2, 5, 10, 25, 50 and 125 metres, which, for a 250-m thick pluton, correspond respectively to 125, 50, 25, 10, 5 and 2 pulses. For each thickness, we have tested intervals between injection of 1, 2, 5, 10, 25, 50 and 100 years.

Results

Maximum duration of BMb construction The primary goal of our numerical simulation was to estimate the maximum duration of the BMb construction. We chose to approach this problem by taking various thicknesses of individual injections (h), and finding for each thickness the critical time between injection ($t_{i\,cr}$). The $t_{i\,cr}$ is defined as the maximum time between two successive injections for which there is still melt in the feeder zone. In this case, the next injection is emplaced into an unconsolidated material, which does not create solid-state deformation between the two pulses. We call these criteria the *no sheeted body criteria*. As $t_{i\,cr}$ will necessarily increase from the beginning to the end of pluton construction, it is sufficient to check the temperature at the contact between the two first injections, which will be compatible with the absence of chilled margins if higher than the solidus. *The no sheeted body criteria permit us to suggest that the maximum duration of the BMb construction depends primarily on the thickness or number of individual pulses.* We find a maximum time between injections which increases from three days to 17 years for a thickness of injection of 1 metre to 125 metres respectively. This corresponds to a minimum average rate of injection (and roof uplift) of 4 m yr^{-1} for

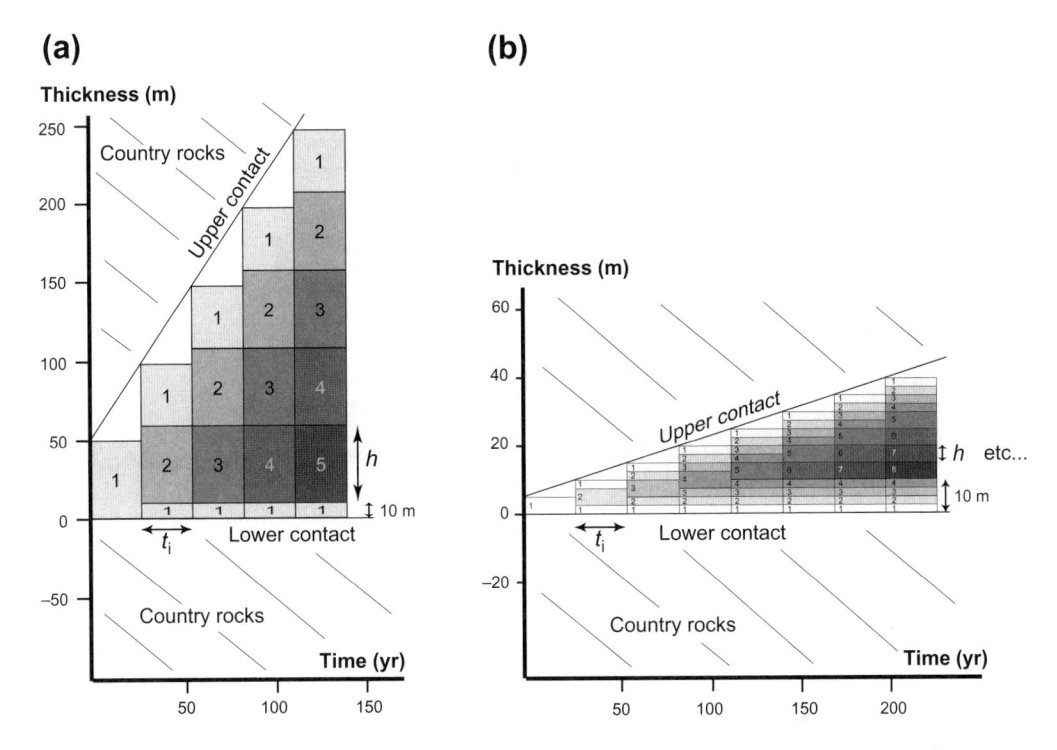

Fig. 4. Outline of the numerical simulation. (**a**) For injections thicker than 20 metres. (**b**) For injections thinner than 20 metres. Thermal parameters used for the simulations are reported in Table 1.

thicknesses of pulses between 25 and 50 m. The rate increases for thinner and thicker pulses, and is very high (>100 m yr^{-1}) for a great number of thin pulses (Fig. 5). These rates imply a maximum duration of the BMb construction of 55 years for pulse thicknesses between 25 and 50 metres.

Respective effects of the time interval and thickness To examine the respective influence of the time between successive injections and injection thickness, we fixed the thickness of the injections h to a certain value and varied the time interval t_i, and calculated the temperature evolution at the lower and upper contacts. We can note that:

1 Due to the constant proximity of the lower contact to the injection zone, the maximum temperature T_{max} at the lower contact, may be above the solidus (between 800 and 500 °C), is always attained just at the end of emplacement, and decreases with increasing injection thickness and/or time interval. The T_{max} and maximum duration of the thermal perturbation (not shown in Fig. 5) were attained for a ratio h/t_i around 1 m yr^{-1} (Fig. 5), which is outside the field of the *no sheeted body criteria*

for a constant growth rate (see above). The T_{max} reaches a maximum for $h = 10$ m and $t_i = 10$ yr (not shown in Fig. 5), which constitutes the parameters for an optimal heating of wallrocks. For a given injection thickness, rocks at the lower contact will be less heated, with either a short time interval between injections (<5 yr), or a very long one (>100 yr), but in the latter case, the permanent magma chamber criteria will not be satisfied. Conversely, for a given time interval, rocks at the lower contact will then be less heated with either thin (<2 m) or thick (>50 m) injections.

2 Due to the screen effect caused by its increasing distance to the injection site, the maximum temperature at the upper contact T_{max} is below 700 °C, is always recorded during the first injection, and is significantly lower than at the lower contact (difference of 200 °C in the $h = 25$ m case). After the first injection, the temperature at the upper contact is significantly lower for long time intervals than for short, and this effect is more pronounced for thin individual injections. Depending on the time interval, a second peak temperature is observed after the end of emplacement due to the heat released by the cooling pluton below

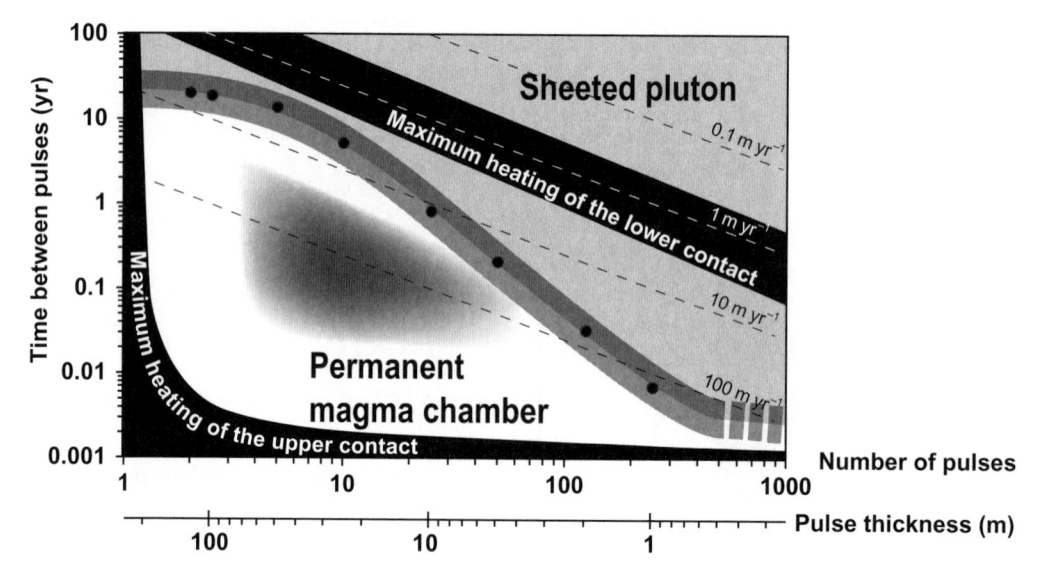

Fig. 5. Summary of our thermal constraints, and area of the most plausible scenario of BMb construction. Black dots represent the critical simulations where the *no sheeted body criteria* are satisified. They separate an upper zone where plutons with chilled margins are built, and a lower zone where a permanent magma chamber is maintained in the feeder zone. Dashed lines outline different rates of construction. The black zones mark the parts of the diagram where the lower or the upper contacts are the warmest. To respect the textural field constraints, the BMb construction can be modelled by a simulation within the permanent magma chamber zone, and as far as possible from the zones where contacts are warmed up. The shaded area represents the most feasible combination of pulse thickness and time interval between pulses compatible with our textural constraints.

the contact. For a given pulse thickness, T_{max} is greater for short time intervals, and conversely, for a given time interval, T_{max} is greater for thick pulses (Fig. 5). The duration of the thermal perturbation follows a similar evolution.

To summarize, the thermal effects at the contacts of the pluton (value of T_{max} and duration of the thermal perturbation) are minimized if the ratio h/t_i is outside the range of 0.5–1, and if the time interval is not too short (>1 yr) and the pulses not too thick (<125 m). If we combine these results with the *no sheeted body criteria*, we can exclude the range of $h/t_i \leq 2$. This leads us to propose that the BMb construction could have involved a combination of pulse thickness and time interval between pulses as outlined in the shaded area in Fig. 5.

If we consider the fact that we have observed no recrystallization of host rocks at the upper contact of the intrusion, and only rare greenschist-facies assemblage minerals (albite, epidote) at the lower contact, the simulations with pulse thickness around 20 metres emplaced approximately every three months seem to best correspond to our constraints. But if we consider the effects of fluid circulation, which are not modelled in our study, the temperature could be drastically reduced, along with the window of the *no sheeted body criteria*. A coupled simulation of conductive and convective heat transfer is necessary to improve upon our results, and could induce a convergence toward the Pollard & Johnson (1973) estimate.

Discussion

Internal structure and thermal simulation

The strongest constraint that we have on the maximum duration of BMb construction is the absence of chilled margins around internal contacts. But our interpretation of the presence of a permanent magma chamber from the beginning to the end of pluton construction could be discussed, as we do not exactly know how an internal contact which was remelted by subsequent magma pulses would appear. We think, however, that these kinds of contacts should have a relatively diffuse texture, which is clearly not the case for the sharp BMb internal contacts that we have observed.

A limitation of our models is the instantaneous injection of individual magma pulses, as the injections are realized in one time-step of the numerical simulation. To test the validity of this assumption, we can calculate an instantaneous magma supply rate by dividing the volume of each injection by the time-step. This instantaneous filling rate has to be equal to the volumetric flow rate in the feeder. Therefore, using the formula of the flow rate in a dyke as a function of its width (Petford 1996; Cruden 1998), we can calculate (3):

$$w = \sqrt[3]{\frac{12\mu Q_E}{gL\Delta\rho}} \quad (3)$$

where w is the dyke width, μ is the viscosity of the magma, Q_E is the volumetric flow rate in the dyke, g is the gravitational acceleration, L is the length of the dyke and $\Delta\rho$ is the density contrast between the crust and the magma. Assuming a length equal to the radius of the intrusion (1 km), we calculated a density contrast of 107 kg m^{-3} and we fixed the viscosity at 10^8 Pa s^{-1} (Pollard & Johnson 1973). The width of the feeding dyke that we obtain is between 10 and 50 metres, when the thickness of the injections are between 1 and 125 metres. Our simulations are then not incompatible with reasonable feeder sizes, except maybe an instantaneous injection of 125 metres.

Another limitation of our simulations is the constant growth rate that we have assumed from the beginning to the end of the pluton construction. Indeed, the results could be slightly different if we consider a non-constant vertical growth rate. For example, a permanent magma chamber is maintained if we simulate the instantaneous emplacement of an initial 30-m thick sill, followed by a regular increase at a rate of 1 m yr^{-1} until the final thickness is achieved.

If we compare our results to those of Petford & Gallagher (2001) who have numerically simulated the partial melting of the lower crust by periodic intrusion of basalts, we note that their line of maximum heating efficiency, i.e. where heat is supplied more or less as quickly as it is lost ($R = t_1/t_d = 1$; t_1 is the time between intrusions and t_d is the characteristic timescale for diffusive heat loss), does not exactly match with ours. In our simulations, optimal heating of wallrocks occurs for an intrusion rate around 1 m per year, which corresponds to an R around 0.2 for the studied time intervals and thickness. One of the reasons for this difference could be the 10-m thick screen of diorite between wallrocks and the new injection at the base of the growing pluton.

Comparison with volcanic systems

One way to evaluate the rate of emplacement, that we have obtained is to compare our results with other natural magmatic systems. Magma production rates can be estimated from volcanic output rates and assumptions about the ratio of intrusive to extrusive products (Crisp 1984; Shaw 1985). Volcanic output rates are typically in the range of 10^{-2} to 10^{-4} km^3 yr^{-1} (Shaw 1985) and a ratio of intrusive to extrusive volume of around two can be assumed for tabular intrusions (Annen & Sparks 2002). The maximum duration of emplacement that we have obtained yields averaged rates between 5 and 12 × 10^{-3} km^3 yr^{-1}, which are clearly comparable with the available data. Recent studies show inflation rates between 5 and 35 × 10^{-3} km^3 yr^{-1} in active volcanic zones (Fialko & Simons 2001; Pritchard & Simons 2002). A better understanding of emplacement mechanisms (time-scale, continuous versus discontinuous feeding) of subvolcanic intrusions will come from detailed observations of changes in vertical elevation in active magmatic zones (Berrino 1998; Nakada & Motomura 1999; Karner *et al.* 2001; Barmin *et al.* 2002). Elevation changes as great as 100 m yr^{-1} due to the growth of an underlying magmatic intrusion have already been observed (Minakami *et al.* 1951). Although we do not know the exact depth of the growing magmatic body in this example, it clearly demonstrates that plutons can be emplaced very rapidly in the shallow crust. In term of size and frequency of pulses, 20 metres every three months, is compatible with the variations of periods of activities in some volcanoes (Nakada & Motomura 1999).

Conclusions

We have constrained the maximum duration of the BMb construction by modelling the thermal evolution of the growing pluton and its wallrocks for different construction scenarios, trying to adjust the number, the thickness and the frequency of the pulses with our petrostructurally derived 'time' constraints, which are:

1 the lack of solid-state deformation around internal contacts and
2 no significant recrystallization of the wallrock at the contact.

We propose that the emplacement of the BMb was a geologically very short event, with a maximum duration of the order of 60 years, implying a minimum vertical displacement rate of the topography above the pluton of 4 metres per year. The simulations with injections of pulse thickness around 20 metres emplaced approximately every three months seem to best correspond with constraints from textural observations. These rates are compatible with observations from active volcanic and subvolcanic systems, but further investigations on possible multiple igneous pulses in these systems are needed.

Fieldwork was funded by CNRS/NSF grant no. 12971, and National Science Foundation grants EAR-0003574. Laboratory work was supported by the LMTG. M. Rabinowicz is greatly acknowledged for valuable and helpful discussion and encouragement during this work. We thank K. Charkoudian and E. Horsman for their help during fieldwork. Comments on drafts of various versions of the manuscript by G. Gleizes, E. Horsman, S. Morgan, P. Olivier, and reviews by N. Petford and D. Pollard, are gratefully acknowledged.

References

ANNEN, C. & SPARKS, R.S.J. 2002. Effects of repetitive emplacement of basaltic intrusions on thermal evolution and melt generation in the crust. *Earth and Planetary Science Letters*, **203**, 937–955.

ARMSTRONG, R.L. & WARD, P., 1991. Evolving geographic pattern of Cenozoic magmatism in the North American Cordillera: the temporal and spatial association of magmatism and metamorphic core complexes. *Journal of Geophysical Research*, **96**, 13 201–13 224.

BARMIN, A., MELNIK, O. & SPARKS, R.S.J. 2002. Periodic behavior in lava dome eruptions. *Earth and Planetary Science Letters*, **199**, 173–184.

BERRINO, G. 1998. Detection of vertical ground movements by sea-level changes in the Neapolitan volcanoes, *Tectonophysics*, **294**, 323–332.

BRUCE, P.M. & HUPPERT, H.E. 1990. Solidification and melting along dykes by the laminar flow of basaltic magma. *In*: RYAN, M.P. (ed.) *Magma Transport and Storage*. John Wiley, New York 87–101.

CLEMENS, J.D., PETFORD, N. & MAWER, C.K. 1997. Ascent mechanisms of granitic magmas: causes and consequences. *In*: HOLNESS, M.B. (ed.) *Deformation-enhanced Fluid Transport in the Earth's Crust and Mantle*. Chapman & Hall, London 144–171.

CRISP, J.A. 1984. Rate of magma emplacements and volcanic output. *Journal of Volcanic and Geothermal Research*, **20**, 177–211.

CRUDEN, A.R. 1998. On the emplacement of tabular granites. *Journal of the Geological Society of London*, **155**, 853–862.

ENGEL, C. 1959. Igneous rocks and constituent hornblends of the Henry mountains, Utah. *Geological Society of America Bulletin*, **70**, 971–980.

FIALKO, Y. & SIMONS, M. 2001. Evidence for on-going inflation of the Socorro magma body, New Mexico, from Interferometric Synthetic Aperture Radar imaging. *Geophysical Research Letters*, **28(18)**, 3549–3552.

FURLONG, K.P., HANSON, R.B. & BOWERS, J.R. 1991. Modeling thermal regimes. *Reviews in Mineralogy*, **26**, 437–498.

GERBI, C., JOHNSON, S.E. & PATERSON, S.R. 2004. Implications of rapid, dike-fed pluton growth for host-rock strain rates and emplacement mechanisms. *Journal of Structural Geology*, **26**, 583–594.

GILBERT, G.K. 1877. Report on the geology of the Henry Mountains, *US Geographical and Geological Survey of the Rocky Mountains Region*, Government printing office, Washington DC.

GVIRTZMAN, Z. & GARFUNKEL, Z. 1996. Numerical solutions for the one-dimensional heat conduction equation using a spreadsheet, *Computers and Geosciences*, **22**, 1147–1158.

HABERT, G. & SAINT-BLANQUAT, M. DE, HORSMAN, E. & MORGAN, S. Rates and mechanisms of non tectonically-assisted emplacement: the Black Mesa bysmalith, Henry Mountains, Utah. In review.

HARRIS, N., VANCE, D. & AYRES, M. 2000. From sediment to granite: timescales of anatexis in the upper crust. *Chemical Geology*, **162**, 155–167.

HUNT, C.B. 1953. Geology and geography of the Henry mountains region, Utah. *US Geological Survey, Professional Papers*, **228**, 234 pp.

JACKSON, M.D. & POLLARD, D.D. 1988. The laccolith-stock controversy: new results from the southern Henry mountains, Utah. *Geological Society of America Bulletin*, **100**, 117–139.

JOHNSON, A.M & POLLARD, D.D. 1973. Mechanics of growth of some laccolithic intrusions in the Henry Mountains, Utah, I. *Tectonophysics*, **18**, 261–309.

KARNER, D.B., MARRA, F., FLORINDO, F. & BOSCHI, E. 2001. Pulsed uplift estimated from terrace elevations in the coast of Rome: evidence for a new phase of volcanic activity? *Earth and Planetary Science Letters*, **188**, 135–148.

KERR, A.D. & POLLARD, D.D. 1998. Toward more realistic formulations for the analysis of laccoliths. *Journal of Structural Geology*, **20**, 1783–1793.

McKENZIE, D. 2000. Constraints on melt generation and transport from U-series activity ratios. *Chemical Geology*, **162**, 81–94.

MINAKAMI, T., ISHIKAWA, T. & YAGI, K. 1951. The 1944 eruption of volcano Usu in Hokkaido, Japan, *Volcanological Bulletin, Series 2*, **11**, 5–157.

NAKADA, S. & MOTOMURA, Y. 1999. Petrology of the 1991–1995 eruption at Unzen: effusion pulsation and groundmass crystallization. *Journal of Volcanology and Geothermal Research*, **89**, 173–196.

NELSON, S.T. 1993. Reevaluation of the Central Colorado plateau laccoliths in the light of new age determination. *US Geological Survey Bulletin*, **2158**, 37–39.

NELSON, S.T. & DAVIDSON, J.P. 1993a. The petrogenesis of the Colorado plateau laccoliths and their relationship to regional magmatism. *US Geological Survey Bulletin*, **2158**, 85–100.

NELSON, S.T. & DAVIDSON, J.P. 1993b. Interaction between Mantle-derived magmas and mafic crust, Henry Mountains, Utah. *Journal of Geophysical Research*, **98(B2)**, 1837–1852.

NIEDERKORN, R. & BLUMENFELD, P. 1989. FUSION:

a computer simulation of melting in quartz–albite–anorthite–orthoclase system. *Computers and Geosciences*, **15**, 715–725.

PATERSON, S.R. & TOBISH, O.T. 1992. Rates of processes in magmatic arcs: implications for the timing and nature of pluton emplacement and wall rock deformation. *Journal of Structural Geology*, **14**, 291–300.

PAVLIS, T.L. 1996. Fabric development in syn-tectonic intrusive sheets as a consequence of melt-dominated flow and thermal softening of the crust. *Tectonophysics*, **253**, 1–31.

PEACOCK, S.M. 1990. Numerical simulation of regional and contact metamorphism using the Macintosh microcomputer. *Journal of Geological Education*, **38**, 132–137.

PETFORD, N. 1996. Dykes or diapirs? *Transactions of the Royal Society of Edinburgh: Earth Sciences*, **87**, 105–114.

PETFORD, N. & GALLAGHER, K. 2001. Partial melting of mafic (amphibolitic) lower crust by periodic influx of basaltic magma. *Earth and Planetary Science Letters*, **193**, 483–499.

PETFORD, N., CRUDEN, A.R., MCCAFFREY, K.J.W. & VIGNERESSE, J.L. 2000. Granitic magma formation, transport and emplacement in the Earth's crust. *Nature*, **408**, 669–673.

POLLARD, D.D. & JOHNSON, A.M. 1973. Mechanics of growth of some laccolithic intrusions in the Henry Mountains, Utah, II. *Tectonophysics*, **18**, 311–354.

PRITCHARD, M.E. & SIMONS, M. 2002. A satellite geodetic survey of large-scale deformation of volcanic centers in the central Andes. *Nature*, **418**, 167–170.

SAINT-BLANQUAT, M. DE, LAW, R.D., BOUCHEZ, J.L. & MORGAN, S. 2001. Internal structure and emplacement of the Papoose Flat pluton: an integrated structural, petrographic & magnetic susceptibility study. *Geological Society of America Bulletin*, **113**, 976–995.

SAINT-BLANQUAT M. DE, LAW, R.D., TIKOFF, B., MORGAN S. & BOUCHEZ, J.L. 1999. The role of rates in emplacement mechanisms of granitic magmas. *Fourth Hutton Symposium*, Document du BRGM, 890, 167.

SHAW, H.R. 1985. Links between magma-tectonic rate balances, plutonism, and volcanism. *Journal of Geophysical Research*, **90(B13)**, 11 275–11 288.

YOSHINOBU, A.S., OKAYA, D.A. & PATERSON, S. R. 1998. Modeling the thermal evolution of fault-controlled magma emplacement models: implications for the solidification of granitoid plutons. *Journal of Structural Geology*, **20**, 1205–1218.

ZENZRI, H. & KEER, L.M. 2001. Mechanical analyses of the emplacement of laccoliths and lopoliths. *Journal of Geophysical Research*, **106**, 13 781–13 792.

Depth, geometry and emplacement of sills to laccoliths and their host-rock relationships: Montecampione group, Southern Alps, Italy

CLAUDIA CORAZZATO[1] & GIANLUCA GROPPELLI[2]

[1]*Dipartimento di Scienze Geologiche e Geotecnologie, Università degli Studi di Milano-Bicocca, Piazza della Scienza 4, 20126, Milano, Italy (e-mail: claudia.corazzato@unimib.it)*

[2]*CNR – Istituto per la Dinamica dei Processi Ambientali, via Mangiagalli 34, 20133, Milano, Italy (e-mail: gianluca.groppelli@unimi.it)*

Abstract: This study focused on a set of shallow subvolcanic bodies, mainly laccoliths and sills, that intruded the Upper Permian–Lower Triassic sedimentary sequence of the central Southern Alps, in the area of Montecampione (Val Camonica, Italy). These intrusions represented a shallow magmatic reservoir probably associated with Triassic volcanism. Based on a detailed stratigraphic reconstruction, this paper presents results dealing with the evaluation of the emplacement depth, the estimated volume of the subvolcanic bodies, a description of their geometries, and their relation to the host rock and response of the sedimentary units to the intrusions. The emplacement depth was estimated using the thickness of the sedimentary overburden at the time of emplacement, and by applying simple equations involving laccolith dimensions. The results are comparable, and support an average emplacement depth of about 1300 m. The minimum volume of the intrusions was obtained using a GIS, and is about 1 km³. Concerning the relationship between the intrusive bodies and the host rock, we observed that sills are mainly emplaced into the Servino Formation, while the laccoliths are emplaced near the contact between the Verrucano Lombardo and the Servino Formation. The two sedimentary units show a different response to the intrusion: the Verrucano Lombardo always appearing fractured and tilted, while the Servino Formation shows a range of deformation patterns, from light ductilization at the contact, to folding, brecciation and foliation. These different responses reflect the mechanics of emplacement and geometry of the intrusions, and local heterogeneities in the host rock. Both units show a local thermal effect close to the contact.

This study deals with a set of shallow subvolcanic intrusions, mainly laccoliths and sills, that intruded the well-documented (e.g. Assereto & Casati 1965; Bianchi *et al.* 1970, 1971; Boni *et al.* 1972; Boni & Cassinis 1973) Upper Permian–Lower Triassic sedimentary sequence of the central Southern Alps, in the area of Montecampione (Val Camonica, Italy, Fig. 1).

The research has been carried out within the Prototype Geological Map Project, supported by the Servizio Geologico Nazionale (SGN) and the Consiglio Nazionale delle Ricerche (CNR), and was aimed at developing new methodologies for the mapping of subvolcanic bodies and the definition of lithostratigraphical units. The research consisted of detailed geological and structural field mapping, by which different lithostratigraphical units were defined, as reported in Corazzato *et al.* (2001), together with a petrographic characterization of the intrusions and their attribution to a geodynamic setting (Armienti *et al.* 2003). This paper represents the progress of the research project and deals with the emplacement characteristics of the subvolcanic bodies.

Many models of laccolith emplacement have already been proposed in the literature (e.g. Johnson & Pollard 1973; Pollard & Johnson 1973; Corry 1988; Roman-Berdiel *et al.* 1995; Friedman & Huffman 1998; Kerr & Pollard 1998), but not all of them could be applied in the present case, due to the objective limitations basically concerning the host rock – often lacking. We thus considered the relations with the host rock and the morphometric characteristics, as in the works of Cruden & McCaffrey (2002) and McCaffrey & Cruden (2002).

The geological dataset and the topographic information have been integrated in a Geographical Information System (GIS) following the SGN's geographical database (Artioli *et al.* 1997). This has allowed us to perform different applications, such as two- and three-dimensional visualizations, aiding in morphological analyses, and elaborations, e.g. thematic maps or the computation of the intrusion volumes. The use of

From: BREITKREUZ, C. & PETFORD, N. (eds) 2004. *Physical Geology of High-Level Magmatic Systems.* Geological Society, London, Special Publications, **234**, 175–194. 0305-8719/04/$15.00
© The Geological Society of London 2004.

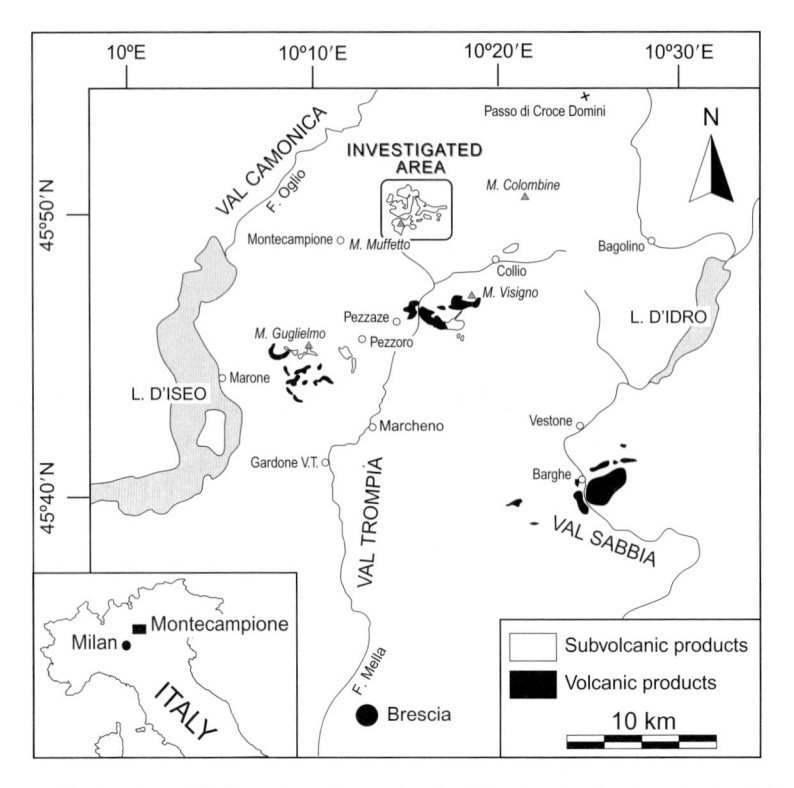

Fig. 1. Geographical setting of the investigated area, showing Triassic subvolcanic and volcanic bodies of the Brescian Prealps (modified after Cassinis & Zezza 1982).

GIS in morphological and structural analyses has already been considered in the study of volcanic areas (e.g. Favalli *et al.* 1998; Aldighieri *et al.* 2002; Norini *et al.* 2004), and the GIS approach to volume estimates of intrusive and volcanic products has been reported in Natoli (2000), Aldighieri *et al.* (2000) and Calvari *et al.* (2004).

After a geological introduction to the Montecampione group, we will present the estimate of the emplacement depth obtained considering the thickness of the sedimentary cover or by theoretical calculations. We will also present the evaluation of the volume of the subvolcanic intrusions and the description of their geometries, and we will analyse their relationships with the host rock, based on detailed field observations over a range of scales.

Geological setting

The Montecampione group, recently defined by Corazzato *et al.* (2001), belongs to the volcanic and subvolcanic rock bodies cropping out in the Brescian Prealps area, described in Cassinis & Zezza (1982) (Fig. 1). Radiometric dating (Rb–Sr, whole rock and biotite) of the studied intrusions (231 ± 5 and 226 ± 4 Ma; Cassinis & Zezza 1982) places them as part of the Middle or Late Triassic volcanism. Coeval pyroclastic fall beds interlayered in a nearby sedimentary succession (Pasquaré & Rossi 1969) were related to this volcanic activity, possibly fed by subvolcanic intrusions similar to those of the study area, and the Arenaria di Val Sabbia volcanic sandstones were related to the erosional reworking of these products. The Montecampione group consists of magmatic bodies (sills to laccoliths) intruding the Upper Permian–Lower Triassic sequence at different stratigraphic levels, namely the Verrucano Lombardo and the Servino Formation.

In the study area, the Verrucano Lombardo (Assereto & Casati 1965; Ori *et al.* 1988; Cassinis 1988; Cassinis *et al.* 2000 with references) is represented by clastic red beds, mainly composed of metre-thick banks formed by fine to very coarse-grained conglomerates, coarse- to fine-grained sandstones and siltstones. In the study area, the Verrucano Lombardo has a maximum thickness of more than 400 m (Perotti & Siletto 1996) (Fig. 2) and mainly rests non-conformably on

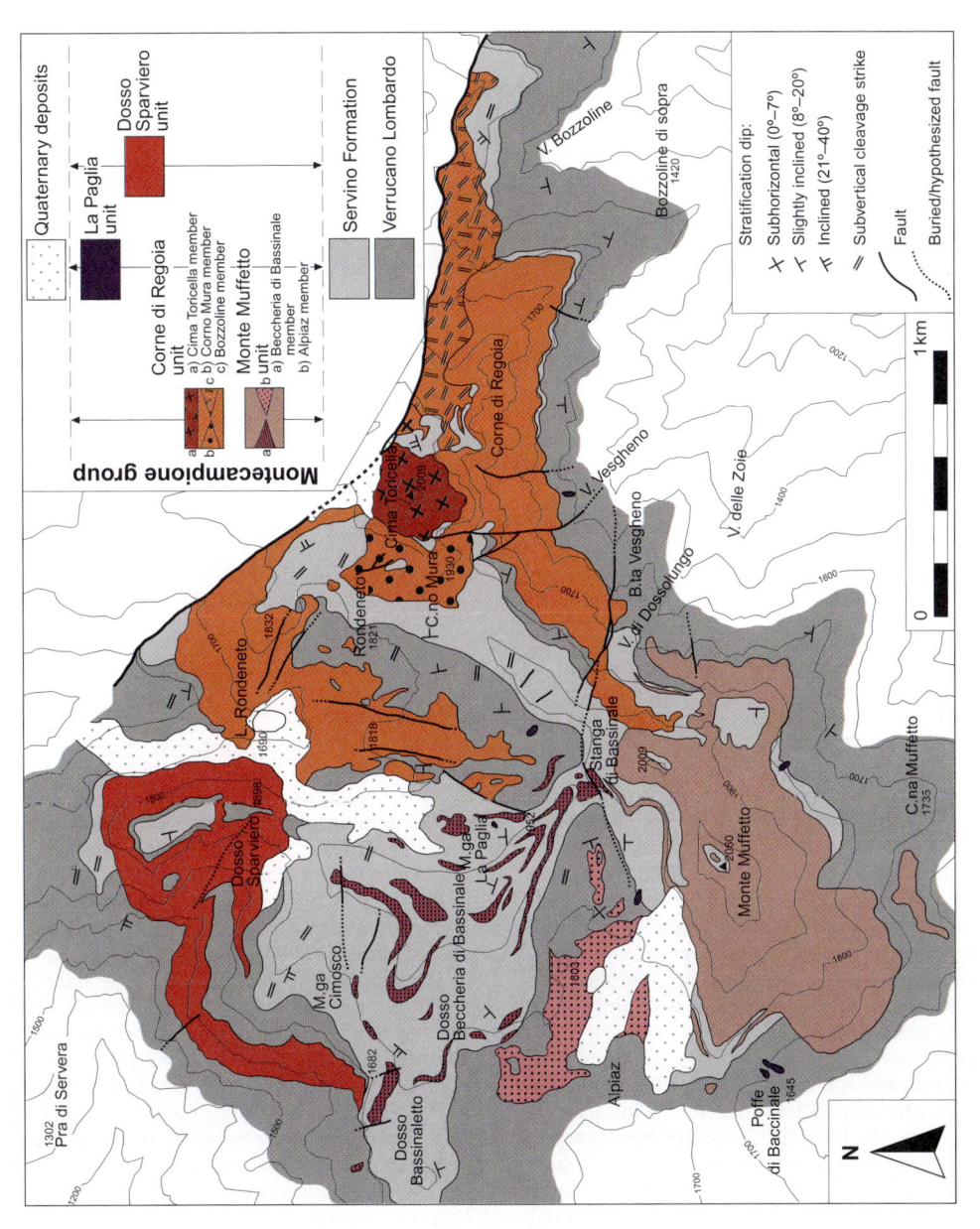

Fig. 2. Simplified geological map of the studied area, including the Montecampione group formations and members, and the host-rock formations.

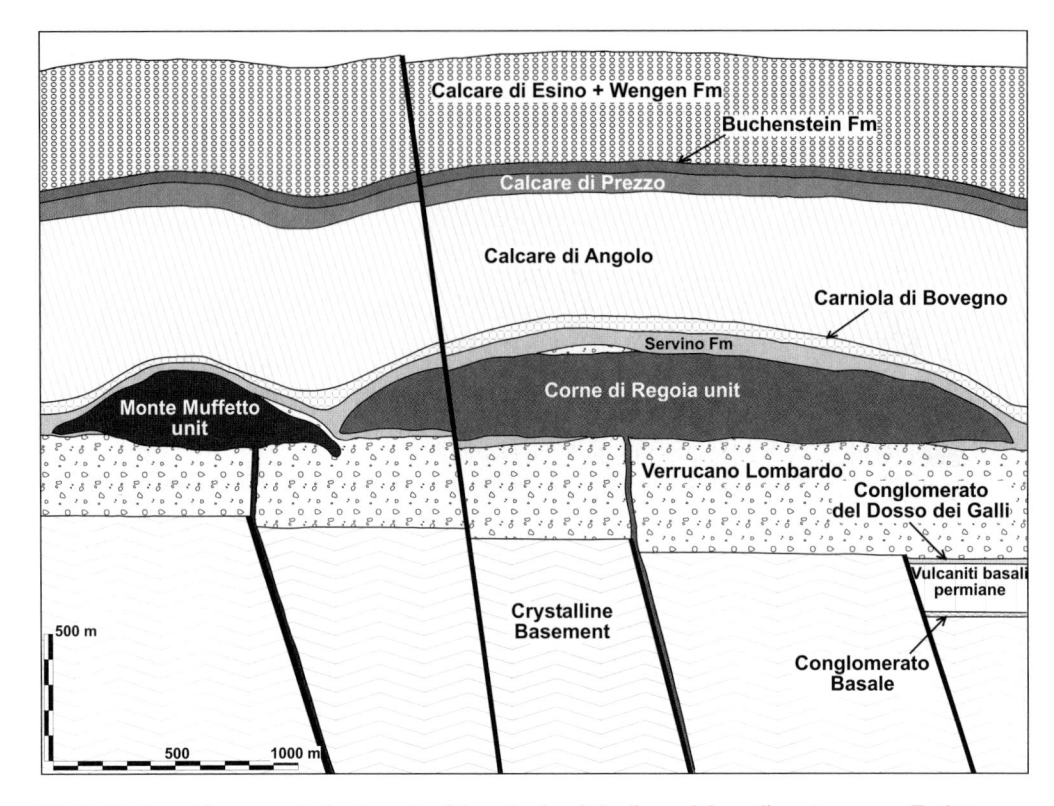

Fig. 3. Stratigraphic representation to scale of the subvolcanic bodies and the sedimentary cover. Fault traces are indicated in black.

the Variscan basement (Benciolini *et al.* 1999) (Fig. 3).

The Servino Formation (Assereto & Casati 1965; Sciunnach *et al.* 1996, 1999; De Donatis & Falletti 1999) is represented by a wide variety of sedimentary lithotypes, comprising mature quartzose sandstones, mudstones, marls, sandstones and dolomitic siltstones, marly and oolitic limestones, and arkosic sandstones. In the study area, the Servino Formation has an average thickness of 100 m and rests in paraconformity on the Verrucano Lombardo.

The Montecampione group bodies show variable thickness from a few centimetres to hundred metres, and extend from metres to kilometres. The magmatic rock has a porphyritic texture, with phenocrysts of quartz, pink feldspar, pyroxene, amphibole and biotite and diffuse mafic microcrystalline enclaves. Armienti *et al.* (2003) classified the Montecampione group subvolcanic bodies following the terminology of volcanic rocks because of the presence of abundant devitrified groundmass and a vacuolar texture: based on the chemical composition, they can be classified in the alkaline series as trachyandesites to trachytes, although it is not possible to determine the sodic or potassic affinity because of the high fluid circulation. These authors, based on trace element studies (pronounced enrichment in LREEs and LILEs), related the genesis of these magmas to a back-arc setting associated with the great depths achieved by a subducting slab (for details on the geodynamic framework, see Armienti *et al.* 2003).

In the simplified geological map of Figure 2, the subvolcanic bodies are organized on the basis of new lithostratigraphical units defined by Corazzato *et al.* (2001) (which we refer to for stratigraphic subdivision). The 'Montecampione group' (Corazzato *et al.* 2001) gathers together four informal lithostratigraphical units, formations in rank: Monte Muffetto, Corne di Regoia, Dosso Sparviero and La Paglia (Fig. 2). A brief description of these units follows (for a detailed description, see Corazzato *et al.* 2001).

Monte Muffetto unit

This unit consists of subvolcanic bodies, located in the western and southwestern parts of the

Fig. 4. Dosso Sparviero unit subvolcanic body and host-rock remnants; view from the SE. The wall height is 265 m, and a talus is present at its base. Location is given in Figure 12.

studied area (Fig. 2). The main body, centred on the Monte Muffetto high, shows a maximum thickness of 310 m in correspondence with the southern slope of Monte Muffetto. Apart from this main body, two members were distinguished (Corazzato et al. 2001): the Alpiaz member and a set of sills, grouped with the name of Beccheria di Bassinale member.

The chronological relationship between the Monte Muffetto and Corne di Regoia units was inferred on the basis of a wide-scale deformation induced by the emplacement of the more recent body of Corne di Regoia on the Monte Muffetto unit body (Fig. 3).

Corne di Regoia unit

This unit is represented by a magmatic body 2.7 km in length in an east–west direction and more than 2 km north–south, and a maximum thickness of 335 m (Fig. 2). It is the unit with the greatest extent in the studied area, although the outcropping part represents just the remnant part of the original body – the southern and eastern ones being removed by erosion (in correspondence with the Corne di Regoia wall) and by the Monte Rosello structural lineament. Apart from the main magmatic body, three lesser bodies defined as members (Corazzato et al. 2001) were recognized within the unit: the Corno Mura member, Bozzoline member and Cima Toricella member.

Dosso Sparviero unit

This subvolcanic body crops out in the northwestern part of the studied area (Fig. 2) with a kilometric extension and a maximum thickness of 265 m. The subvolcanic body shows a main portion where the greatest thickness is recorded (Figs 2 & 4), and from which two apophyses depart westwards. A thick talus at its base hides the contact with the Corne di Regoia unit – interpreted as vertical.

La Paglia unit

This unit links some bodies with limited extent and metre thickness, mostly dykes, found in the outer parts of the area, both in the Verrucano Lombardo and the Servino Formation. Compared to the overall petrography of the group, the unit shows compositional (trachyandesite–trachyte) and mineralogical affinity with the mafic enclaves found in other units.

Emplacement depth

As observed in different regional settings, magma emplacement depth appears to be directly related to the dimensions of the intrusive body (Corry 1988), and should normally correspond to the level of neutral buoyancy between the intruding magma and the host rock. We approached the estimate of the emplacement depth for the Montecampione group subvolcanic bodies by:

(1) evaluating the thickness of the sedimentary cover at the time of the emplacement, and
(2) applying some considerations on the dimensions of the bodies following Johnson's (1970) equations reported and applied in Skármeta & Castelli (1997).

Thickness of sedimentary cover

Taking into account that the radiometric Rb–Sr ages available in the literature are 226 ± 4 Ma and 231 ± 5 Ma for the Dosso Sparviero and Monte Muffetto units respectively (Cassinis & Zezza 1982), we based our estimate on information reported by Assereto & Casati (1965) about the sedimentary succession overlying the Servino Formation in the Ladinian–Carnian from preserved sections in adjacent areas (Darfo–Boario Terme sector) and we obtained a value of about 1300 m as the best estimate of the emplacement depth. The thickness for the sedimentary overburden formational units and their age are the following (Assereto & Casati 1965):

- Carniola di Bovegno (upper Olenekian (?)–Lower Anisian (?)): 50 m
- Calcare di Angolo (Middle–Lower Anisian): 700 m
- Calcare di Prezzo (Upper Anisian): 80 m
- Buchenstein Formation (Lower Ladinian): 60 m
- Calcare di Esino and Wengen Formation (Upper Ladinian–Lower Carnian): 450 m

In more detail, the stratigraphic position and relations with the host rock units and regional structures can be summarized in the stratigraphic sketch of Figure 3. In this sketch, horizontally and vertically scaled, the laccoliths of the Monte Muffetto and Corne di Regoia units and the sedimentary succession mentioned above are represented. Faults in the basement, defining a regional horst structure ('Dorsale della Val Trompia'; Cassinis 1983), were probably the discontinuities used for magma supply to the intrusions. Other faults were active after the emplacement of the bodies, displacing them and the stratigraphic succession. The sedimentary overburden arched in correspondence with the doming due to laccolith growth, and it was assumed that the sedimentary layers were able to slip over one another during bending. The deformation induced on the Monte Muffetto unit body and host rock due to the emplacement of the more recent body of Corne di Regoia unit is represented.

Theoretical emplacement depth

We have analysed the dimensions (base diameter and maximum thickness) of each intrusive body, following Johnson's (1970) equations reported in Skármeta & Castelli (1997). Johnson (1970), Johnson & Pollard (1973), and Pollard & Johnson (1973) applied a mathematical analysis to explain the shape and emplacement depth of the laccoliths of the Henry Mountains, Utah (see also Habert & Saint-Blanquat, this volume). Their equations consider elastic behaviour, and take into account an idealized laccolith, perfectly symmetrical and with a circular or elliptical base. The sedimentary overburden is considered to be isotropic, and constituted by parallel layers that have the capability of slipping over one another during bending due to the growth of the laccolith and associated doming. Skármeta & Castelli (1997) applied the same equations to model the emplacement of the Torres del Paine laccoliths in Chile.

Based on the measured values of maximum thickness of each body (S) and its average diameter (L), we applied Johnson's (1970) equations in order to obtain an estimate of the intrusion depth, that is the thickness of the sedimentary country rock overhanging the intrusion (T). The applied formulae are the following:

$$L^3 = \frac{S}{L} k_2 T^3 \quad (1)$$

with $k_2 = f(E, v, P_m - \gamma T)$, and equal to about 10^2 (Johnson 1970)

$$L^3 = k_2 T^3 \quad (2)$$

where $k_3 = 2.3$ is the constant value for the Henry Mountains (Johnson 1970).

The input data and the results are summarized in Table 1. The obtained depths are consistent with the estimate based on the thickness of the sedimentary cover at the time of intrusion, in spite of the limitations in field measurements of the dimensions (due to erosion of some portions of the bodies), and the fact that originally these equations best apply to ideal laccoliths. We approximated the bodies to circular intrusions in plan view, indicating the average diameter and

Table 1. *Intrusion depth and volume computation of the subvolcanic bodies*

Unit	Measured lengths		k_3	k_2	Calculated emplacement depth		Depth from sediment overburden (km)[‡]	Minimum volume (km³)
	Maximum thickness S (m)	Maximum diameter L (m)			Depth T (km)[†]	Depth T (km)[*]		
Corne di Regoia	335	2750	2.3	100	2.08	1.20	1.34	0.45
Monte Muffetto	310	1350	2.3	100	1.02	0.47	1.34	0.20
Dosso Sparviero	265	650	2.3	100	0.49	0.19	1.34	0.06
							Total volume:	0.71

Intrusion depth computation after Johnson (1970) and Skármeta & Castelli (1997) is based on the measured dimensions of the intrusions and overburden properties, and comparison with the value obtained based on the sedimentary cover. Volumes of the Corne di Regoia, Monte Muffetto and Dosso Sparviero unit intrusions were obtained from ILWIS software elaborations.
*$L^3 = (S/L)k_2 T^3$, with $k_2 = f(E, v, P_m - \gamma T)$, about 10^2 (Johnson 1970).
†$L^3 = k_3 T^3$ where $k_3 = 2.3$ is the value for the Henry Mountains (Johnson 1970).
‡The same depth value is reported for each intrusive unit, since their differences in emplacement depth are smaller than the accuracy in the estimate of the sedimentary overburden.

the maximum thickness observed. This simplification may explain why the depth value for the Dosso Sparviero unit, which least resembles an ideal laccolith (see later for a detailed description), is smaller than those for the Corne di Regoia and Monte Muffetto units. The constants k_2 and k_3 are taken from the example of the Henry Mountains in Utah. No computations were made with regard to the sills, or to the portions where information on the entire shape was lacking (e.g. the Alpiaz member). Because of these assumptions and the introduced simplifications, both equations reflect an order of magnitude value only for the intrusion depth estimate.

Volume estimates

To quantitatively assess the order of magnitude of the intrusions, we computed their minimum volumes using GIS software (ILWIS 2.2, ITC). In order to obtain these results we first prepared a point map representing the altitude of the contact points between the host rock and the intrusion, both for the top and the base of the intrusion. These two maps were then interpolated, applying the 'moving surface' method, in order to obtain the top and the bottom surfaces of the body by fitting them through weighted point values. Each surface is represented by a raster map in which each pixel has a height value calculated by an interpolation on the input map value. These two maps were subtracted to obtain a thickness map, that was used with respect to areas to calculate the volume. Figure 5 shows the thickness map obtained for the Corne di Regoia unit (**a**) and a 3D view of the top of this laccolith (**b**). Table 1 synthesizes the results of the volume computation, while, as a whole, the minimum volume for the preserved portions of the shallow

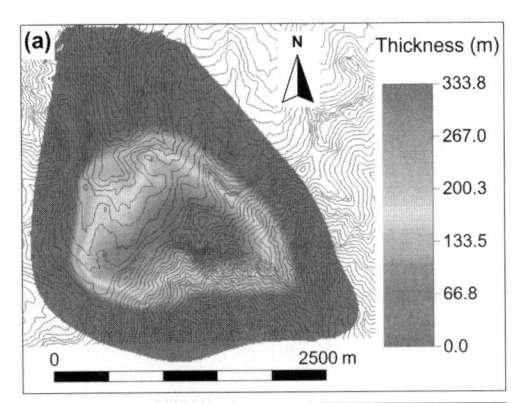

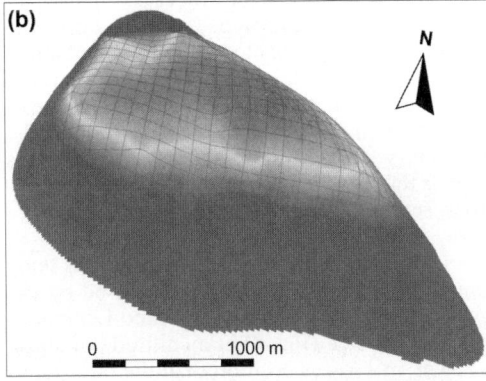

Fig. 5. Thickness map (**a**) and 3D model of the Corne di Regoia laccolith (**b**) obtained by ILWIS GIS software.

magma reservoir represented by the subvolcanic bodies is estimated to be about 1 km^3. This should be regarded as a minimum volume, since neither the eroded portions of the subvolcanic bodies or those truncated by structural lineaments, nor those bodies that are not showing their basal contact with the host rock (e.g. Alpiaz member), are taken into account.

Geometry and relations with the host rock

Geometry of the bodies

A wide variation in the shapes of the intrusive bodies within the Montecampione group was observed. Taking into account the similarity in the petrography of the different units within the group, consistent with Corry's (1988) observations at many different sites in the world, it can be assumed that the magmas forming the intrusions had similar physical and rheological properties at the time of emplacement. If so, the variation in intrusion shape should be related to processes other than variations in magma rheology. This may be ascribed to both different mechanics of emplacement, magma supply rate and interference of nearby intrusions, as well as variations in the mechanical properties of the overburden and the host rock.

A 3D shaded view of the geological map draped on a digital terrain model was prepared in a GIS environment in order to better illustrate the different intrusion shapes and their volumetric entity, together with the overall relations with the host rock succession (Fig. 6).

The majority of the investigated subvolcanic bodies are not classifiable as ideal laccoliths, which are defined as concordant intrusions with a flat floor and convex upper surface (Gilbert 1877), apart from the main parts of the Corne di Regoia and Monte Muffetto units (Fig. 6). They instead range between sets of sills and punched laccoliths, following Corry's (1988) terminology, with more complicated combinations of geometric characteristics. Such a variability is also reflected in a wide dimensional range, with centimetric to hundreds metres thickness and metric to kilometric extent. A detailed description of the observed geometries follows.

Intrusion geometry

At one end, the Montecampione group shows various sets of sills, mainly emplaced into the Servino Formation. These sills, that volumetrically represent a minimum part of the intrusions, crop out widely, mainly in the area of the Dosso Beccheria di Bassinale (Fig. 7a), where a member within the Monte Muffetto unit is defined. Here they show multiple levels of intrusion within an altitude range of about 250 m, extensions up to 700 m and individual thickness up to 10 m (Fig. 7c). Columnar jointing related to cooling and pinch-out structures can be observed.

Other sills are associated with the main laccolithic bodies, especially in the case of Monte Muffetto unit, where they show extensions in the order of 500 m and thickness of about 10 m. It is noteworthy that, as observed elsewhere by Corry (1988), the sills are usually contained within the ductile and anisotropic layers of the Servino Formation and are not formed preferentially at either the top or base of ductile beds, unlike the laccolithic bodies that have been emplaced in the proximity of the contact between Verrucano Lombardo and Servino Formation. A separate case should be considered: i.e. two mainly concordant protrusions to the west of the Dosso Sparviero unit, emplaced in the Verrucano Lombardo and showing a 100 m thickness, that set at the threshold with roof-lifting bodies.

The other geometrical end member is represented by the Dosso Sparviero unit, which can be defined as a 'punched laccolith' according to Corry (1988), who describes this type of intrusion as cylindrical in plan view: bounded by peripheral faults and steep to vertical sides and characterized by a flat top (Figs 3 & 6), mechanically punching its way through the roof rock. This laccolith shows discordant margins (Fig. 6a), while its roof contact with concordant host-rock remnants is exposed. In particular, these remnants clearly indicate both the level of intrusion a few metres below the Servino–Verrucano contact, and the uplifting of this contact due to the intrusion with respect to its altitude at the periphery of the laccolith, whose thickness equals the total deflection of the roof. The host-rock sequence is undisturbed a short distance away from the vertical contact to the north. A possible differential punching among this body could be also taken into account based on the presence of a remnant of Verrucano Lombardo on the NW side (Fig. 8) at an altitude of about 1750 m, in a lower position than the other remnants at the top (about 1890 m). As regards the aforementioned planar intrusions departing from the base of the laccolith to the west, these could be regarded as subsidiary sills–laccoliths representing a lateral extension of the intrusive body, that assumes a more complex shape (Fig. 6a).

The Corne di Regoia unit represents the most

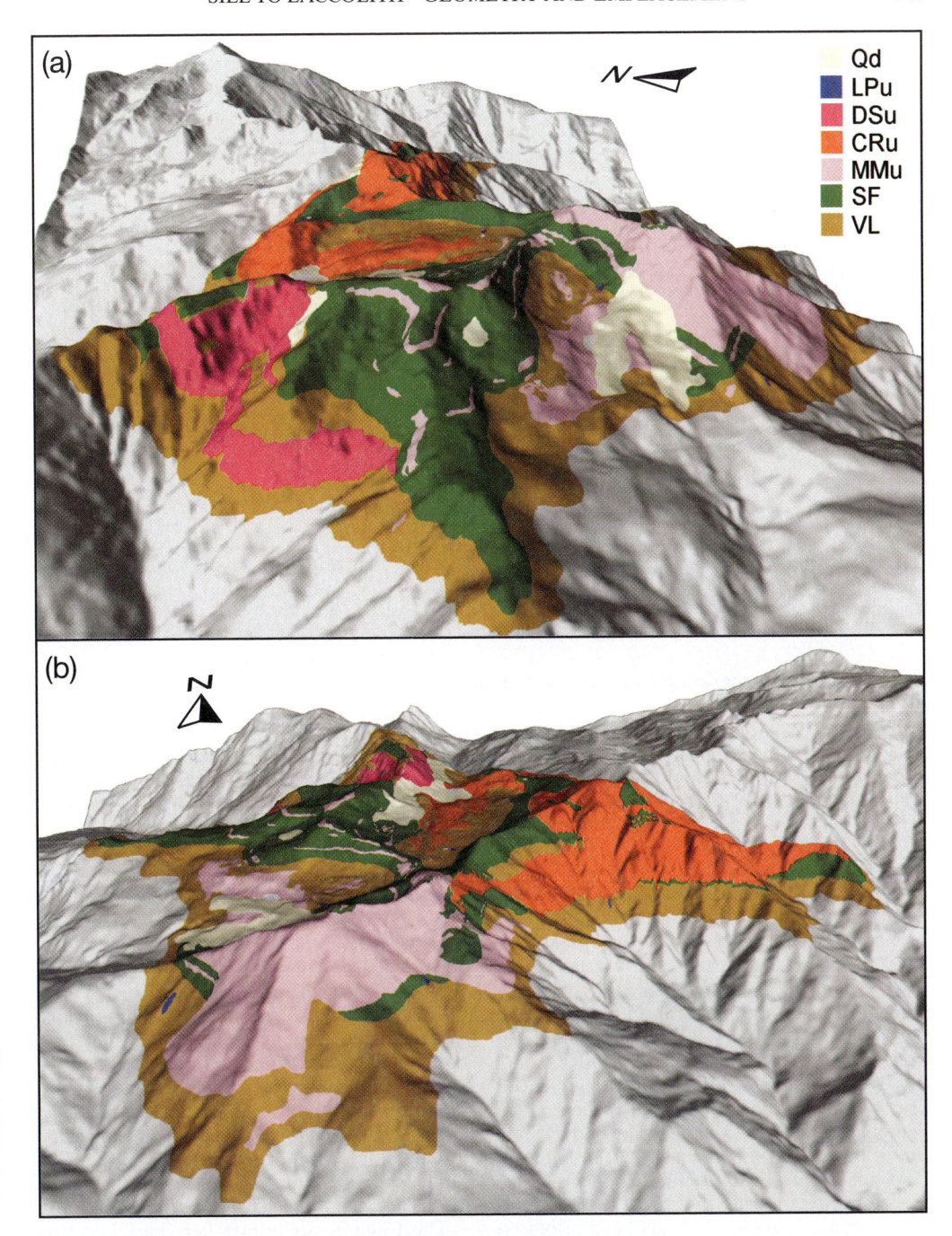

Fig. 6. 3D views of the geological map draped over a digital terrain model. DEM pixel size is 10 m. (**a**) View looking east, with light from the west (270/50); (**b**) view looking north, with light from the east (110/70). VL, Verrucano Lombardo; SF, Servino Formation; MMu, Monte Muffetto unit; CRu, Corne di Regoia unit; DSu, Dosso Sparviero unit; LPu, La Paglia unit; Qd, Quaternary deposits.

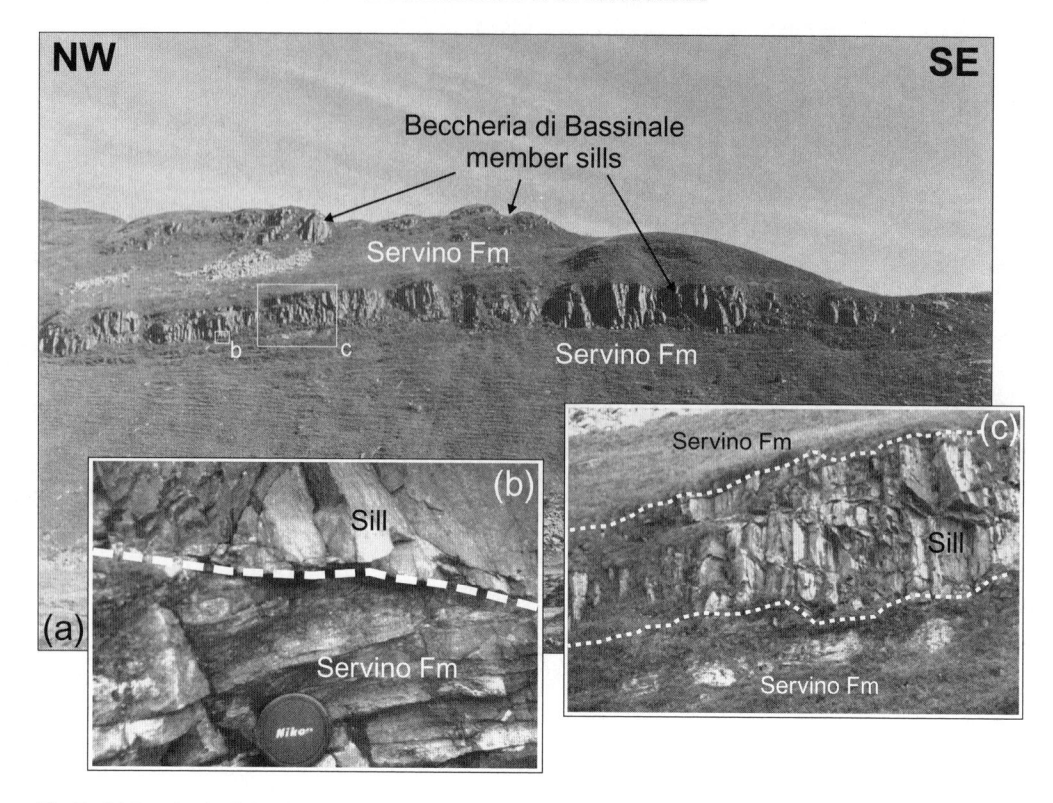

Fig. 7. (**a**) Beccheria di Bassinale member sills, intruded in the Servino Formation. The average thickness of the sills is 10 m. Columnar jointing related to cooling and pinch-out structures are observed. (**b**) Example of how a sill, which is mainly concordant with the sedimentary bedding, locally crosscuts it. A 6-cm objective cap is given for scale. (**c**) Detail of the sill with its basal contact with the Servino Formation host rock. Location is given in Figure 12.

relevant body of the area, both for volume and areal extent. It occupies the central and eastern part of the area, and consists of a laccolithic body with a generally flat base and a convex top surface, emplaced in correspondence with the contact between the Verrucano Lombardo and Servino Formation (Figs 3 & 6b). This emplacement was more complicated than a simple model of intrusion and roof lifting – being influenced also by pre-existing structures in the host rock. A model for the intrusion of the Corne di Regoia unit subvolcanic body is illustrated in Figure 9 (not to scale). From field evidence it was possible to reconstruct that the host-rock sedimentary succession had already been displaced by a fault that had downthrown the eastern block and the contact between the Verrucano Lombardo and the Servino Formation (stage 1). The intrusion first propagated as a wedge in the Servino Formation, uplifting it (2), and then crossed the fault region into the Verrucano Lombardo and partly used this discontinuity as a path to propagate upwards and then set into the Servino Formation layers as Corno Mura member (3): laccolithic in shape but with a discordant base with respect to the host rock (Fig. 10a & 10b). The fault was reactivated after the intrusion emplacement, as evidenced by displacement and fault planes observed in the field both in the host rock and the intrusive body (4).

Much of the Corne di Regoia unit intrusion to the south and NE has been removed by erosion and by a fault that to the east brings the intrusion into contact with the pre-Permian–Permian sequence from the Crystalline Basement to the Verrucano Lombardo (Corazzato *et al.* 2001).

The Cima Toricella member (Figs 2 & 10a) consists of a rounded dome and shows clear chronological relations with Corne di Regoia main unit: it was emplaced later than the main unit, stacking directly over it while probably using the same feeding conduit.

The Monte Muffetto unit is another example of a laccolithic body, complicated by many

Fig. 8. View of the Dosso Sparviero unit subvolcanic bodies, from the NW. There is evidence of a remnant of Verrucano Lombardo located at an altitude of about 1750 m, in a lower position than the other remnants at the top (at about 1890 m).

apophyses represented by thin sills. It was generally emplaced just above the contact between the Verrucano Lombardo and the Servino Formation, nevertheless the southern apophysis unconformably crosses the Verrucano Lombardo. By contrast, the Alpiaz member, whose base is not observed, was entirely emplaced within the Verrucano Lombardo. The chronological relationship between the Monte Muffetto and Corne di Regoia units was inferred on the basis of a wide-scale deformation induced by the emplacement of the more recent body of the Corne di Regoia on the Monte Muffetto unit body and the host rock (Fig. 3).

Field investigations have enabled us to determine the dimensions of the intrusions, but with some limitations due to partial erosion. In order to compare dimensional information with scaling laws for other known intrusions (McCaffrey & Petford 1997; Cruden & McCaffrey 2001; McCaffrey & Cruden 2002; Cruden & McCaffrey 2002), we plotted thickness (T) and width (L) data of the Montecampione group subvolcanic bodies over the logarithmic S-curve defined by Cruden & McCaffrey (2002) (Fig. 11). Our data, representing the different units, fit the S-trend and are located in the lower and middle part of the plot, in accordance with the range in geometries previously described.

The slight shift towards lower values of width (L) with respect to the S-fit can be explained by the fact that, contrary to the measured thickness (T), which is the real one, the extensions are to be considered minimum values measurable in the outcrops, where many parts of the intrusion are eroded or truncated, inducing an underestimate of L.

Relations with the host rock and effects of the intrusions

The Montecampione group subvolcanic bodies were emplaced at different levels in the Verrucano Lombardo and the Servino Formation. We systematically observed that while sills were emplaced within the whole available thickness of the Servino Formation, the main intrusive bodies represented by laccoliths (Monte Muffetto and Corne di Regoia) and a punched laccolith (Dosso Sparviero) were emplaced preferentially a few metres from either the top or base of the contact between the Verrucano Lombardo and the Servino Formation (Fig. 6a & 6b).

The map of contact typologies (Fig. 12) summarizes the different types of contact observed in the field. Apart from the stratigraphic and tectonic boundaries (buried when indicated),

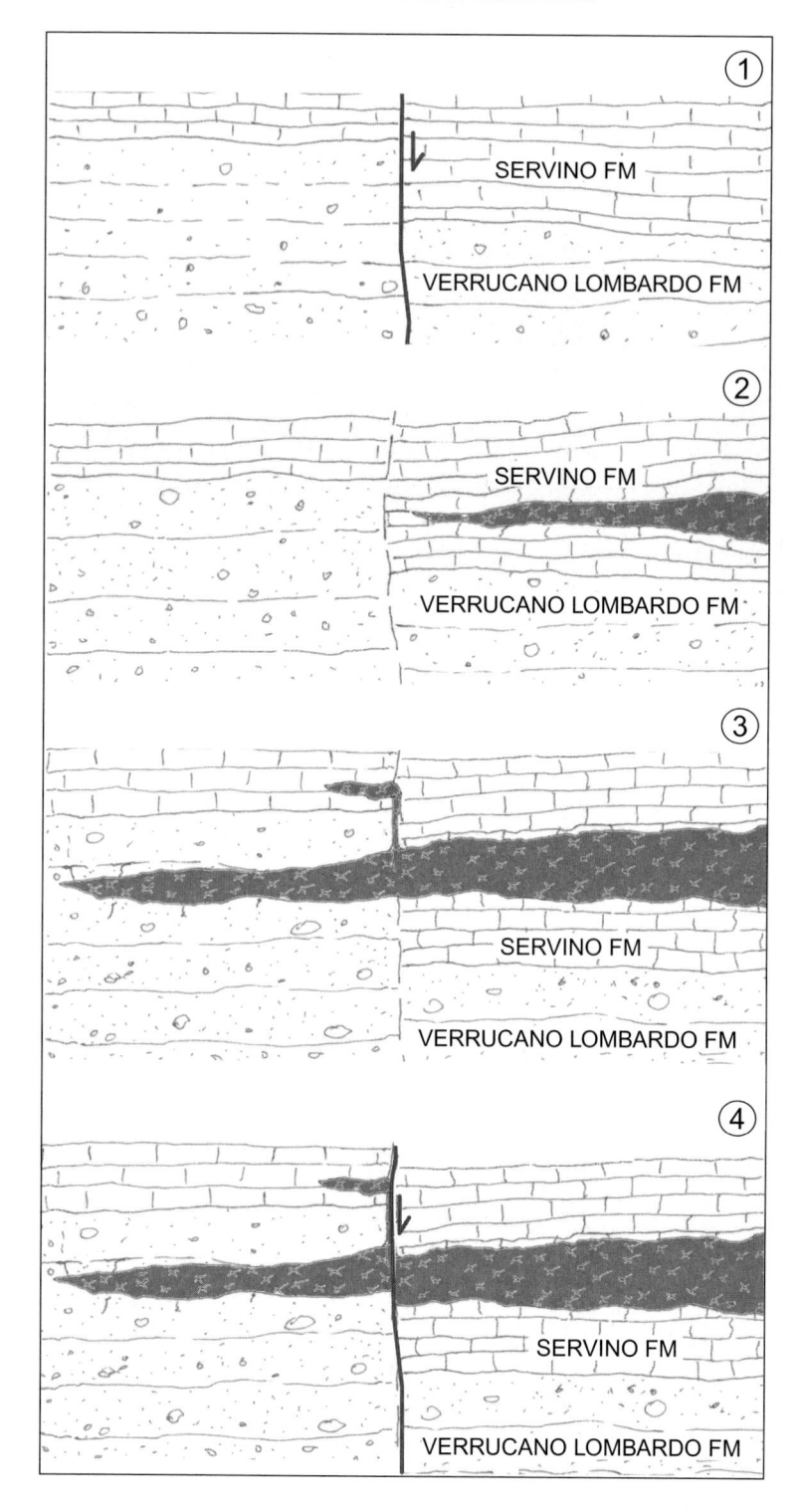

Fig. 9. Model for the intrusion of Corne di Regoia unit subvolcanic body: (**1**) pre-intrusion faulting of the host rock; (**2**) intrusion of the subvolcanic body; (**3**) partially guided by the pre-existing fault; and (**4**) post-intrusion fault reactivation. Location of the described portion is given in Figure 12.

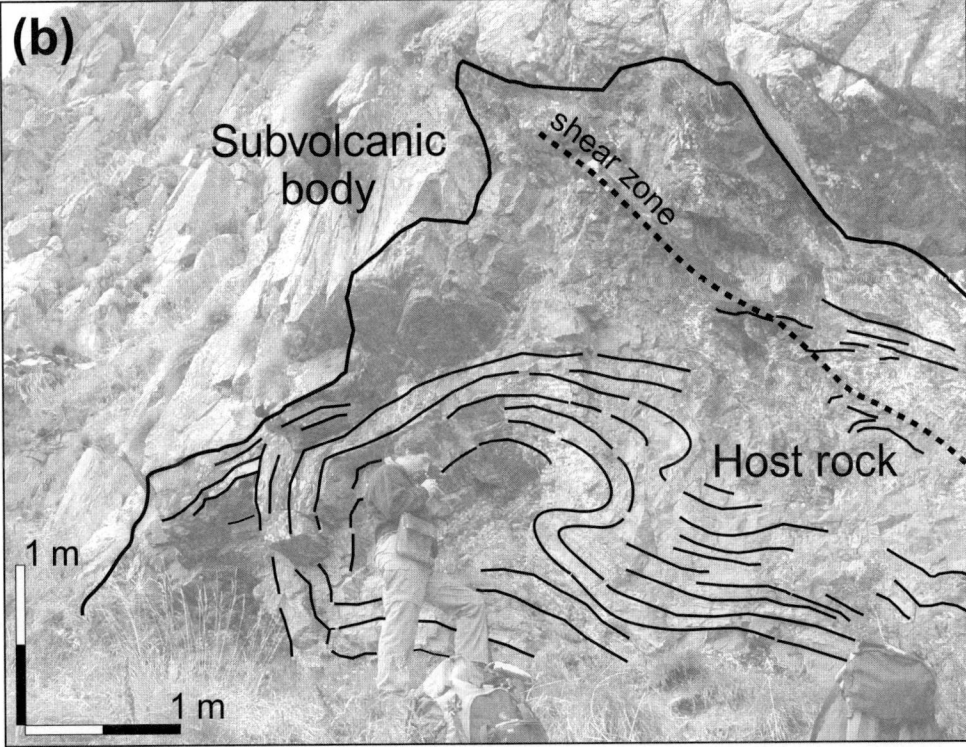

Fig. 10. (a) View of the Corno Mura member body from the SW; it also shows the contact with the Servino Formation at its base, and the Cima Toricella member with respect to Corne di Regoia unit. Horizontal field of view is about 800 m. Rectangle gives the location of the inset; (b) detail of the box-folding developed in the host rock (Servino Formation) at the base of Corno Mura member intrusion. A centimetre-scale shear zone is also present. Location is shown in Figure 12.

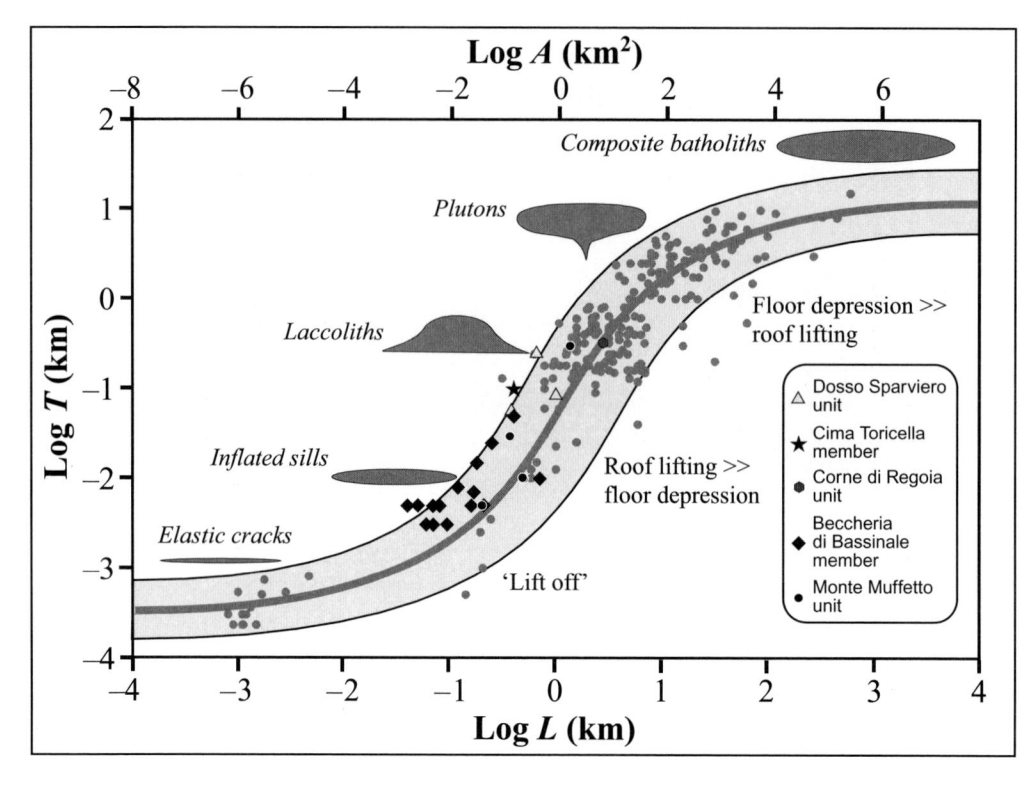

Fig. 11. Plot of data from the Montecampione group subvolcanic bodies: intrusion thickness (T), and width (L), compared with the intrusion-style S-curve by Cruden & McCaffrey (2002). The scale is logarithmic.

the interest is focused on the distinction of the magmatic contacts based on the associated relations with the original stratigraphy of the host rock and the deformation and thermal effects produced.

In more detail, it was possible to distinguish magmatic contacts associated with host-rock deformation and thermal effects, and in both cases specify whether these contacts were unconformable with respect to the host-rock stratification. Figure 12 also gives the locations of all the figures illustrating the different typologies that are described in detail below.

The subvolcanic bodies intruded the sedimentary units with different mechanics and host-rock deformation:

(1) the Verrucano Lombardo is constantly brittly fractured and rigidly tilted, and for this reason only one example of contact will be presented, while
(2) the Servino Formation's response to the intrusions is ductile at variable degrees and intensity of deformation, thus, different examples will be given.

At some sites, both the Verrucano Lombardo and the Servino Formation show evidence of thermal effects and contact metamorphism, but these aspects are presently still being studied in the framework of the research and will not be dealt with here.

The contact between the intrusive bodies and the Verrucano Lombardo host rock is usually sharp, associated with brittle deformation consisting of fracturing, locally with some thermal effects. One example is the top contact between the Corne di Regoia unit laccolith and the sandstone host rock represented in Figure 13, where the contact with this thin, fractured remnant is sharp and subhorizontal. The whitish aspect of the host rock, locally observed at many other sites in the area, can be related to a thermal effect.

With regard to the Servino Formation's response to magma intrusion, structures range from sharp contacts with minor thermal effects to the development of folds, magmatic breccias and foliation. This variability can be related to the geometry of the body and the mechanics of intrusion, but local inhomogeneities in the host rock may also be important.

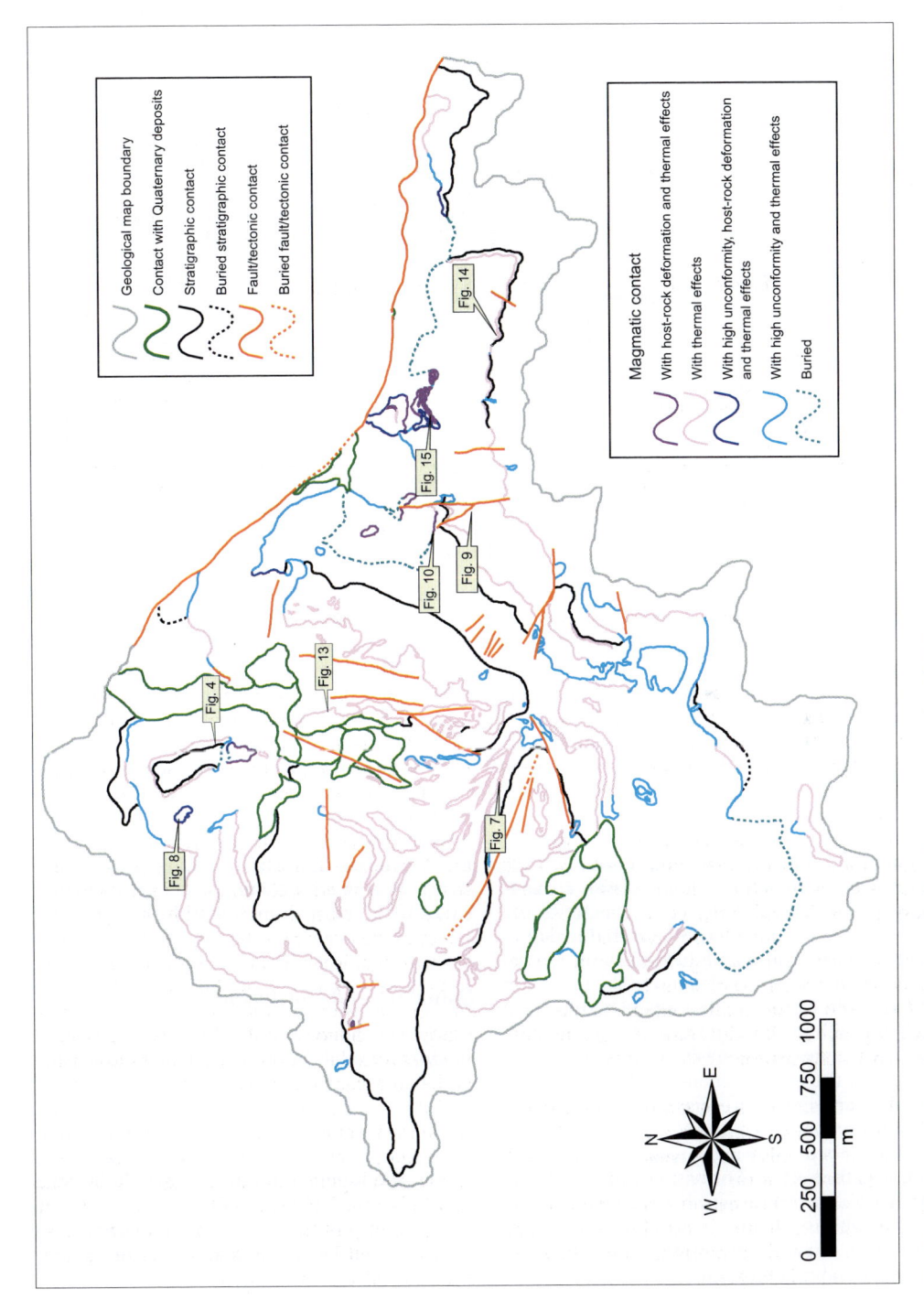

SILL TO LACCOLITH – GEOMETRY AND EMPLACEMENT 189

Fig. 12. Contact typologies map. Locations of the other figures are also shown.

Fig. 13. Contact between the Corne di Regoia unit and the host rock (Verrucano Lombardo). A 32-cm hammer is shown for scale. Location is given in Figure 12.

An example of a sharp contact with the Servino Formation is seen in the sills of the Beccheria di Bassinale member (Fig. 7b). This contact, which is mainly concordant with the sedimentary bedding, can locally crosscut it. A centimetre-scale layer showing thermal effects and obliterated sedimentary structures at the contact is often present.

A magmatic breccia layer (Fig. 14) about 15 cm thick, characterized by flattened elements of intrusive rock floating in a greenish groundmass, was observed at the basal contact of the Corne di Regoia unit laccolith, probably due to extreme ductilization and flowage of the closest portion of the Servino Formation. The transition is sharp both within the underlying bedded host rock and with the base of the massive intrusion. We can relate the genesis of this magmatic breccia to the emplacement mechanism of the laccolith, and similar breccias can be observed at other sites along this contact.

Another site showing strong ductile deformation in the Servino Formation is located at the base of Corno Mura member laccolith, whose wall height is about 100 m (Fig. 10a). The contact is sharp but irregular and strongly unconformable, with metric box folds and several minor folds and drag folds affecting the host rock, along with decimetre-thick shear bands (Fig. 10b). The inner sedimentary structures near the contact are locally obliterated, and new crystals (epidote and white micas) have formed due to contact metamorphism.

Other contact structures between the Corne di Regoia unit and the Servino Formation include highly deformed host rock with an intense foliation, with no primary sedimentary structures recognizable (Fig. 15). Further studies dealing with metamorphism are under way.

Summary and conclusions

Based on a detailed stratigraphic reconstruction (Corazzato *et al.* 2001), the main results of the investigation presented here are:

(1) we have evaluated the magma emplacement depth, using:
   (i) the sedimentary overburden at the time of emplacement, and
   (ii) by applying equations (Johnson 1970) based on measured laccolith dimensions.

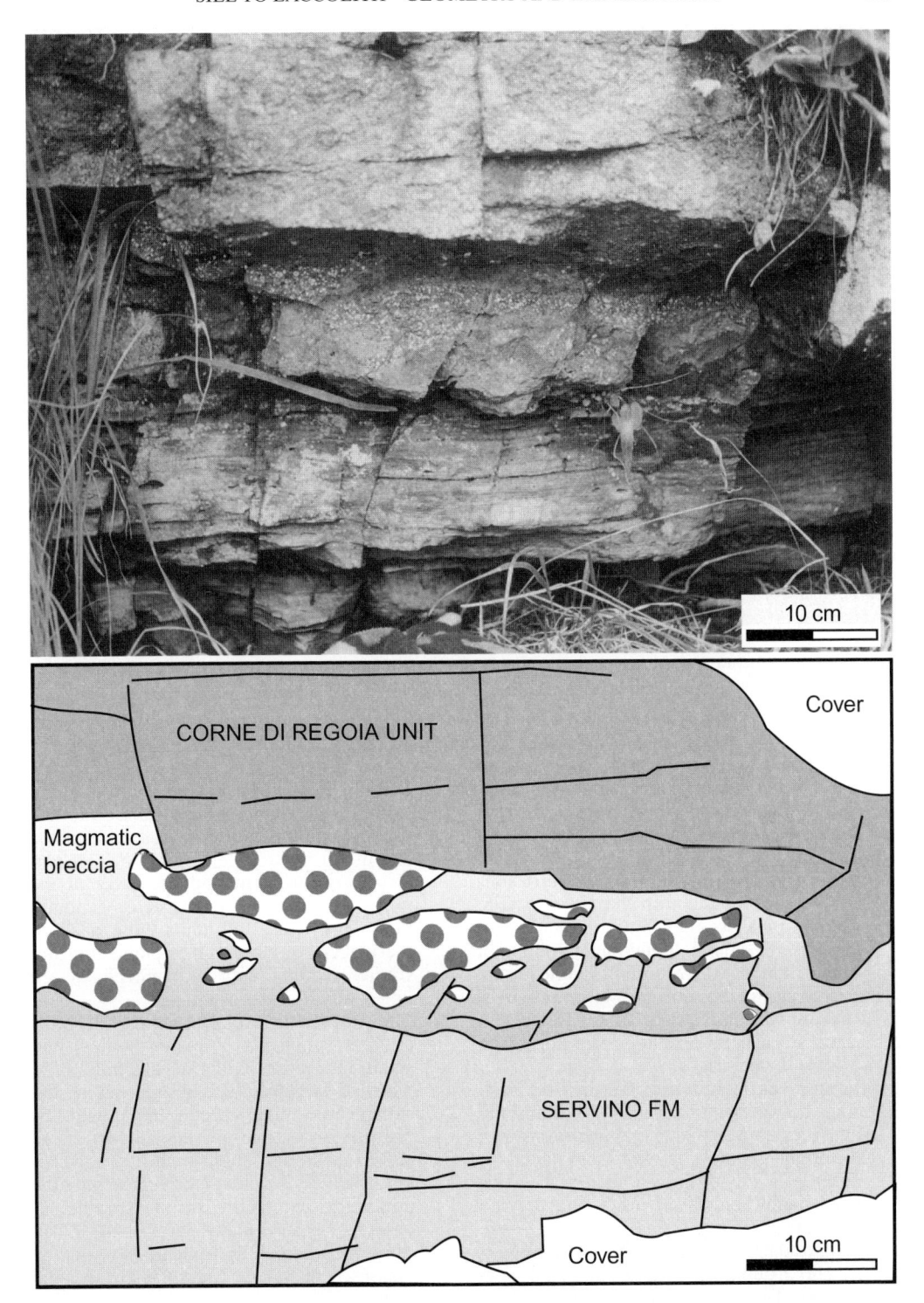

Fig. 14. Magmatic breccia at the basal contact of the Corne di Regoia unit laccolith with the host rock (Servino Formation). Location is given in Figure 12.

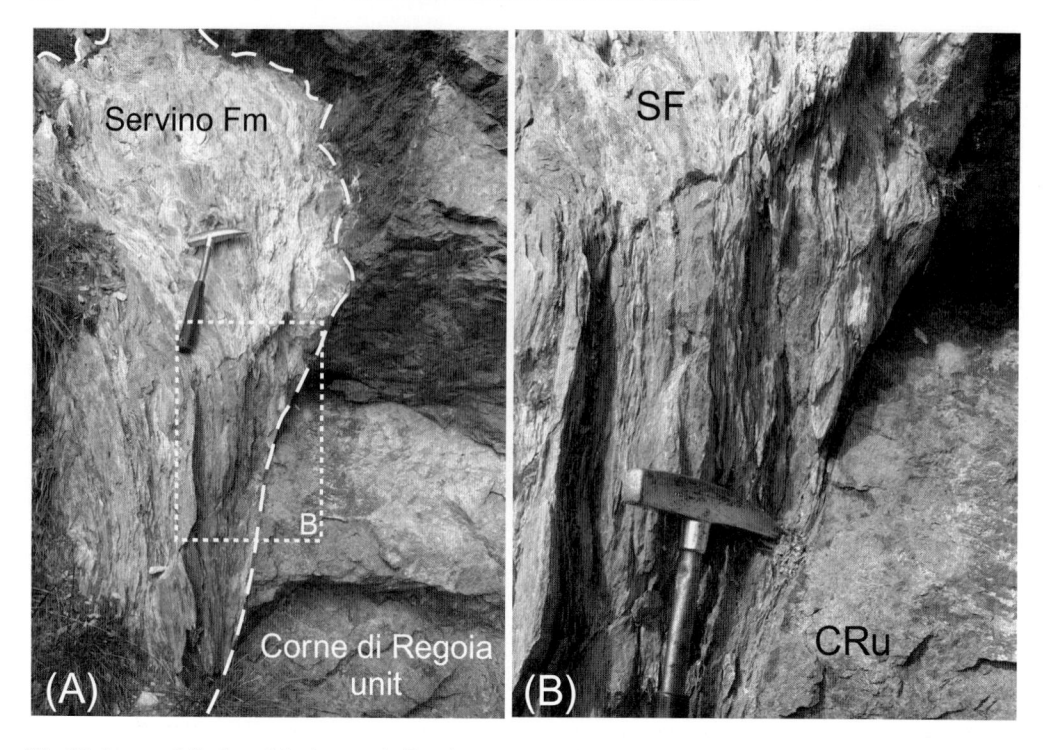

Fig. 15. Intense foliation of the host rock (Servino Formation) at the contact with 'fingers' of Corne di Regoia unit laccolithic body. A 32 cm hammer is shown for scale. Location is given in Figure 12.

The results with both methods are comparable, and taking into account limitations due to field measurements, suggest an estimated average emplacement depth of about 1300 m.

(2) The volume of the subvolcanic intrusions has been estimated with the aid of a GIS. Taking into account limitations due to erosion, we estimate a minimum volume of about 1 km³ for the Montecampione group, representing the preserved section of a shallow magmatic reservoir feeding a volcanic edifice in the Southern Alps during the Middle–Late Triassic.

(3) From a consideration of the field relationships between the intrusive bodies and the host rocks, we conclude that intrusions with different geometries were emplaced at different levels in the succession. The sills were mainly emplaced inside the Servino Formation, while the laccoliths were emplaced near the contact between the Verrucano Lombardo and the Servino Formation. The two sedimentary units show differing responses to the intrusion: the Verrucano Lombardo is constantly fractured and tilted, while the Servino Formation shows a range of deformation responses, from light ductilization of the contact to folding, brecciation and foliation. We propose that these different responses of the Servino Formation depend on the mechanics of emplacement and geometry of the intrusions, along with local heterogeneities in the host rock. Both units show localized thermal effects close to the contact. Further studies dealing with metamorphism are under way.

(4) The intrusions are represented by sills (always apophyses of major bodies) and small laccoliths, with an example of a punched laccolith. The geometries of the subvolcanic intrusions fit the power-law distribution S-curve of Cruden & McCaffrey (2002), although slightly shifted towards lower values of width (L), because only the minimum extension is measurable in outcrop, because many parts of the intrusions are eroded or truncated, while the measured thickness values (T) are the real ones.

This work was financed by the Accordo di Programma SGN–CNR within the Prototype Geological Map

Project, and C.C. benefited from a one-year contract on these funds. The research was also supported by CNR-Agenzia 2000 Giovani grants CNRG00F2A7. We wish to acknowledge P. Armienti, C. Bigoni, F. Forcella, C. Larghi, L. Marinoni, E. Natoli, G. Pasquarè, and D. Sciunnach for fruitful discussion and suggestions during the fieldwork. B. Aldighieri, A. Gigliuto, G. Norini and A. Rovida are acknowledged for their contribution to DEM elaborations. We thank the Azienda Regionale delle Foreste–Ufficio Operativo di Breno (Brescia), Cooperativa Exodus in Sònico (Brescia) and Rifugio Alpini (Monte Cimosco) for assistance and logistical support. Careful reviews and suggestions by N. Petford and A. Ronchi were greatly appreciated.

References

ALDIGHIERI, B., FOGGI, B., GROPPELLI, G., MORELLI, E., TESTA, B. & VICINI, D. 2000. *Cartografia Multitematica: un Esempio di Applicazione all'Isola di Capraia*. ASITA 2000, Atti della 4ª Conferenza Nazionale ASITA, Volume I, 395–400.

ALDIGHIERI, B., BORELLI, E., GROPPELLI, G. & TESTA, B. 2002. I caratteri dell'isola. Assetto morfologico. In MORELLI, E. (ed.) *L'Isola di Capraia. Progetto di un Paesaggio Insulare Mediterraneo da Conservare*, Alinea Editrice, Florence, 26–39.

ARMIENTI, P., CORAZZATO, C., GROPPELLI, G., NATOLI, E. & PASQUARÈ, G. 2003. Geological and Petrographic Study of Montecampione Triassic Subvolcanic Bodies (Southern Alps, Italy). Preliminary Geodynamic Results. *Bollettino della Società Geologica Italiana*, Special Volume, **2**, 67–78.

ARTIOLI, G.P., BONANSEA, E., CARA, P. ET AL. 1997. *Carta Geologica d'Italia – 1:50.000. Banca Dati Geologici. Linee Guida per l'Informatizzazione e per l'Allestimento per la Stampa dalla Banca Dati*. Quaderni Serie III, Vol. 6, Istituto Poligrafico e Zecca dello Stato, Rome, 142 pp.

ASSERETO, R. & CASATI, P. 1965. Revisione della stratigrafia permo-triassica della Valle Camonica meridionale (Lombardia). *Rivista Italiana di Paleontologia e Stratigrafia*, **71**, 999–1097.

BENCIOLINI, L., PASQUARÈ, F.A. & PASQUARÈ, G. 1999. Studio dei contatti fra corpi subvulcanici del Triassico medio e rocce sedimentarie incassanti nel Sudalpino Centrale (Valcamonica–Val Trompia). In: OROMBELLI, G. (ed.) *Studi Geografici e Geologici in Onore di Severino Belloni*. Univ. Studi Milano. Univ. Studi Milano-Bicocca. Glauco Brigati, Genoa, 19–34.

BIANCHI, A., BONI, A., CALLEGARI, E. ET AL. 1970. *Carta Geologica d'Italia Alla Scala 1:100.000. Foglio '34 – Breno'*, Italian Geological Survey, Rome.

BIANCHI, A., BONI, A., CALLEGARI, E. ET AL. 1971. *Note Illustrative della Carta Geologica d'Italia Alla Scala 1:100.000. Foglio '34 – Breno'*. Ministero Industria, Rome, 134 pp.

BONI, A. & CASSINIS, G. 1973. *Carta Geologica delle Prealpi Bresciane a Sud dell'Adamello. Note Illustrative della Legenda Stratigrafica*. Atti Ist. Geol. Univ. Pavia, **23**, 119–159.

BONI, A., CASSINIS, G., ROSSETTI, A. & CERRO, C. 1972. *Carta geologica delle Prealpi Bresciane a Sud dell'Adamello, alla scala 1:50.000*. Atti Ist. Geol. Univ. Pavia, **23**.

CALVARI, S., TANNER, H., GROPPELLI, G. & NORINI, G. 2004. Valle del Bove, eastern flank of Etna volcano: a comprehensive model for the opening of the depression and implications for future hazards. In: CALVARI, S., BONACCORSO, A., COLTELLI, M., DEL NEGRO, C. & FALSAPERLA, S. (eds) *Etna Volcano Laboratory*, Geophysical Monographs AGU, **143**, 65–75.

CASSINIS, G. 1983. Il Permiano nel gruppo dell'Adamello, alla luce delle ricerche sui coevi terreni delle aree contermini. *Memorie della Società Geologica Italiana*, **26**, 119–132.

CASSINIS, G. 1988. *Carta Geologica dei Depositi Continentali Permiani a Sud dell'Adamello*. Atti Ticinensi di Scienze della Terra, **31**, Tav. I.

CASSINIS, G. & ZEZZA, U. 1982. Dati geologici e petrografici sui prodotti del magmatismo triassico nelle Prealpi Bresciane. In: CASTELLARIN, A. & VAI, G.B. (eds) *Guida alla Geologia del Sudalpino Centro-Orientale*. Guide Geologiche Regionali, Servizio Geologico Italiano, 157–171.

CASSINIS, G., CORTESOGNO, L., GAGGERO, L., MASSARI, F., NERI, C., NICOSIA, U. & PITTAU, P. (co-ordinators) 2000. *Stratigraphy and Facies of the Permian Deposits Between Eastern Lombardy and the Western Dolomites, Field Trip Guidebook*. 'The Continental Permian International Congress', 15–25 September 1999, Brescia, Italy. Earth Science Department, Pavia University, 157 pp.

CORAZZATO, C., GROPPELLI, G., NATOLI, E. & PASQUARÈ, G. 2001. Il Gruppo di Montecampione: Stratigrafia dei Corpi Subvulcanici Triassici tra la Val Camonica e la Val Trompia. Atti Ticinensi di Scienze della Terra, Pavia, **42**, 141–152 (in Italian with English extended abstract).

CORRY, C.E. 1988. *Laccoliths: Mechanisms of Emplacement and Growth*. Geological Society of America Special Papers, **220**, 110 pp.

CRUDEN, A.R. & MCCAFFREY, K.J.W. 2001. Growth of plutons by floor subsidence: implications for rates of emplacement, intrusion spacing and melt-extraction mechanisms. In: BROWN, M. (ed.) *Crustal Melting and Granite Magmatism*. Physics and Chemistry of the Earth (A), Solid Earth and Geodesy, **26(4–5)**, 303–315.

CRUDEN, A. & MCCAFFREY, K. 2002. Different scaling laws for sills, laccoliths and plutons: Mechanical thresholds on roof lifting and floor depression. In: BREITKREUZ, C., MOCK, A. & PETFORD, N. (eds) *Physical Geology of Subvolcanic Systems – Laccoliths, Sills, and Dykes (LASI)*. LASI, Freiberg, 12th-14th october 2002. Wissenschaftliche Mitteilungen des Institutes für Geologie der TU Bergakademie Freiberg, **20**, 15–17.

DE DONATIS, S. & FALLETTI, P. 1999. The Early Triassic Servino Formation of the Monte Guglielmo area and relationships with the Servino of Trompia and Camonica Valleys (Brescian Prealps, Lombardy). In: GOSSO, G., JADOUL, F., SELLA, M. & SPALLA, M.I. (eds) *3rd Workshop on Alpine Geological Studies. Biella – Oropa,*

September 29th–October 1st 1997, Memorie Scienze Geologiche Padova, **51**, 91–101.

FAVALLI, M., INNOCENTI, F., PARESCHI, M.T. ET AL. 1998. The DEM of Mt. Etna: geomorphologic and structural implications. *Geodinamica Acta*, **12(5)**, 279–290.

FRIEDMAN, J.D. & HUFFMAN, A.C. (co-ordinators) 1998. *Laccolith Complexes of Southeastern Utah: Time of Emplacement and Tectonic Setting – Workshop Proceedings*. US Geological Survey Bulletin, **2158**.

GILBERT, G.K. 1877. *Geology of the Henry Mountains, Utah*. US Geographical and Geological Survey of the Rocky Mountain Region, 170 pp.

HABERT, G. & DE SAINT BLANQUAT, M. 2004. Multi-pulse emplacement rate of the Black Mesa bysmalith, Henry Mountains, Utah. *In*: Physical Geology of High-Level Magmatic Systems, BREITKREUZ, C. & PETFORD, N. (eds) Geological Society, London, Special Publication, this volume.

JOHNSON, A.M. 1970. *Physical Processes in Geology*. Freeman, Cooper & Co., San Francisco. 577 pp.

JOHNSON, A.M. & POLLARD, D.D. 1973. Mechanics of growth of some laccolithic intrusions in the Henry Mountains, Utah, I. Field observations, Gilbert's model, physical properties and flow of the magma. *Tectonophysics*, **18**, 261–309.

KERR, A.D. & POLLARD, D.D. 1998. Toward more realistic formulations for the analysis of laccoliths. *Journal of Structural Geology*, **20**, 1783–1793.

McCAFFREY, K. & CRUDEN, A. 2002. *Dimensional data and growth models for intrusions*. *In*: BREITKREUZ, C., MOCK, A. & PETFORD, N. (eds) Physical Geology of Subvolcanic Systems – Laccoliths, Sills, and Dykes (LASI). LASI, Freiberg, 12th–14th October 2002. Wissenschaftliche Mitteilungen des Institutes für Geologie der TU Bergakademie Freiberg, **20**, 37–39.

McCAFFREY, K.J.W. & PETFORD, N. 1997. Are granitic intrusions scale invariant? *Journal of the Geological Society, London*, **154(1)**, 1–4.

NATOLI, E. 2000. *Rilevamento geologico e analisi dei corpi subvulcanici di Montecampione (BS): metodologie e proposte finalizzate al progetto CARG*. Unpublished M.Sc. thesis, Università degli Studi di Milano, Milan.

NORINI, G., GROPPELLI, G., CAPRA, L. & DE BENI, E. 2004. Morphological analysis of Nevado de Toluca volcano (Mexico): new insights into the structure and evolution of an andesitic to dacitic stratovolcano. *Geomorphology*, **62(1–2)**, 47–61.

ORI, G.G., DALLA, S. & CASSINIS, G. 1988. Depositional history of the Permian continental sequence in the Valtrompia–Passo Croce Domini area (Brescian Alps, Italy). *In*: CASSINIS, G. (ed.) *Proceedings of the Field conference on: Permian and Permian–Triassic boundary in the Western Tethys, and Additional Regional Reports. Brescia, 4–12 July 1986*. Memorie della Società Geologica Italiana, **34**, 141–154.

PASQUARÉ, G. & ROSSI, P.M. 1969. Stratigrafia degli orizzonti piroclastici medio-triassici del Gruppo delle Grigne (Prealpi Lombarde). *Rivista Italiana di Paleontologia e Stratigrafia*, **75**, 1–87 (in Italian with English abstract).

PEROTTI, C. & SILETTO, G.B. 1996. Le caratteristiche geometriche dei bacini permiani tra la Val Camonica e la Val Giudicarie (Subalpino centrale). *Atti Ticinensi di Scienze della Terra*, 1996 (Serie Speciale), **4**, 77–86.

POLLARD, D.D. & JOHNSON, A.M. 1973. Mechanics of growth of some laccolithic intrusions in the Henry Mountains, Utah, II. Bending and failure of overburden layers and sill formation. *Tectonophysics*, **18**, 311–354.

ROMAN-BERDIEL, T., GAPAIS, D. & BRUN, J.P. 1995. Analogue models of laccolith formation. *Journal of Structural Geology*, **17(9)**, 1337–1346.

SCIUNNACH, D., GARZANTI, E. & CONFALONIERI, M.P. 1996. Stratigraphy and petrography of Upper Permian to Anisian terrigenous wedges (Verrucano Lombardo, Servino and Bellano Formations; Western Southern Alps). *Rivista Italiana di Paleontologia e Stratigrafia*, **102(1)**, 27–48.

SCIUNNACH, D., GARZANTI, E., POSENATO, R. & RODEGHIERO, F. 1999. Stratigraphy of the Servino Formation (Lombardy, Southern Alps) towards a refined correlation with the Werfen Formation of the Dolomites. *In*: GOSSO, G., JADOUL, F., SELLA, M. & SPALLA, M.I. (eds) *3rd Workshop on Alpine Geological Studies. Biella – Oropa, September 29th–October 1st 1997*. Memorie Scienze Geologiche Padova, **51/1**, 103–118.

SKÁRMETA, J.J. & CASTELLI, J.C. 1997. Intrusión sintectónica del Granito de las Torres del Paine, Andes patagónicos de Chile. *Revista Geológica de Chile*, **24(1)**, 55–74.

Rise and fall of a nested Christmas-tree laccolith complex, Elba Island, Italy

D. S. WESTERMAN[1], A. DINI[2], F. INNOCENTI[3] & S. ROCCHI[3]

[1]*Norwich University, Department of Geology, Northfield, Vermont 05663, USA*
[2]*CNR, Istituto di Geoscienze e Georisorse, Pisa I-56127, Italy*
[3]*Università di Pisa, Dipartimento di Scienze della Terra, Pisa I-56126, Italy*
(e-mail: rocchi@dst.unipi.it)

Abstract: In two separate areas of western and central Elba Island (Italy), Late Miocene granite porphyries are found as shallow-level intrusions inside a stack of nappes rich in physical discontinuities. Detailed mapping of intrusive rocks, along with their relations with country rocks, show that outcrops from western and central Elba Island expose the same rock types, with matching intrusive sequence, petrography and geochemical features. Structural and geological data indicate that these layers were originally part of a single sequence that was split by eastward-directed décollement and tilting. The two juxtaposed portions of the original sequence allow the restoration of a 5-km thick sequence, made up of nine main intrusive layers, building three Christmas-tree laccoliths nested into each other to support a structural dome. During their construction, the role of the neutral buoyancy level was of minor significance with respect to the role played by the relatively thin overburden and/or the large availability of magma traps inside the intruded crustal section. Emplacement of the Monte Capanne pluton into the base of the domal structure likely caused oversteepening and initiated decapitation of the complex, with gravity sliding of the upper half off the top.

Mechanisms related to the generation, movement and emplacement of granitic magmas are the subject of intensive study and conflicting hypothesis (Bouchez *et al.* 1997; Castro *et al.* 1999; Petford *et al.* 2000; Brown 2001). Efforts to understand such mechanisms benefit greatly from the collection of well-constrained data on the depth and shape of the granitic intrusions. By their nature, most granite intrusions become directly accessible only after loss of the overburden, and then generally for only two-dimensional observations. In this context, opportunities to examine crustal sections exposing an intrusion's roof and floor, such as from tilted plutons, are invaluable (Wiebe & Collins 1998). These opportunities are scarce for deep intrusions, but a little more common for shallow intrusions such as laccoliths. Additional insights on emplacement mechanisms come from analogue experimental modelling (Roman-Berdiel *et al.* 1995; Benn *et al.* 1998; Roman-Berdiel 1999) and geophysical modelling coupled with geobarometric estimates, all of which are valuable in reconstructing the 3D shape and depth of an intrusion (Améglio & Vigneresse 1999). In addition, understanding of the build-up of igneous complexes requires reconstruction of their tectonic evolution, as well as addressing the problems associated with uplift, oversteepening, tectonic dismemberment (gravity sliding, décollement), and denudation.

On Elba Island, the processes associated with both the construction and destruction of a well-documented complex are clearly illustrated in the field. Serendipitous tectonic–gravitational splitting and tilting of a Miocene igneous complex resulted in roof and floor exposures of several intrusive layers. This allowed direct field observations to determine the emplacement sequence, and to estimate the thickness of the intrusive layers and host rock, and hence emplacement depths. The geometry of the complex as a whole represents a prime example of a nested Christmas-tree laccolith complex. On the other hand, the dimensional parameters of the single layers provide a coherent data-set, suggesting that the filling of intrusive layers was halted during vertical inflation, owing to the widespread availability of crustal magma traps that promoted magma injection in new horizontal planes before the layers were completely filled. Finally, the primary intrusive geometric relationships and the reconstructed tectonic evolution suggest that the instability leading to the dismemberment of the complex was at least in part of gravitational origin, and therefore instigated by construction of the complex itself.

From: BREITKREUZ, C. & PETFORD, N. (eds) 2004. *Physical Geology of High-Level Magmatic Systems.*
Geological Society, London, Special Publications, **234**, 195–213. 0305-8719/04/$15.00
© The Geological Society of London 2004.

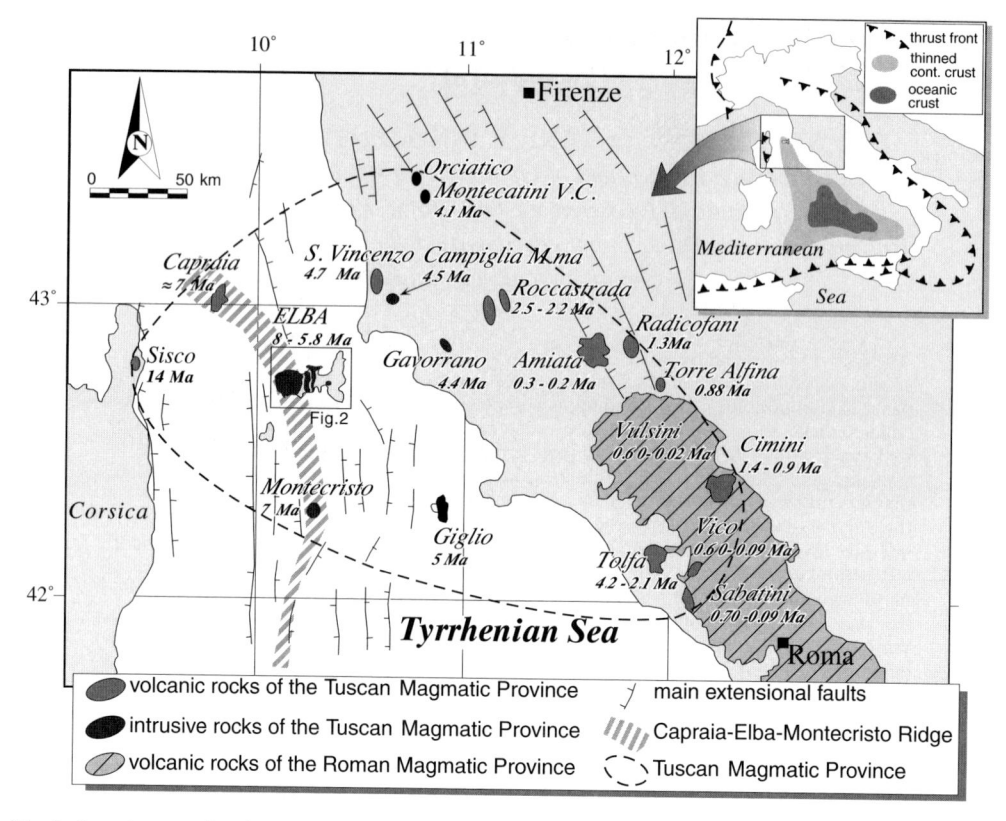

Fig. 1. Location map for the Tuscan Magmatic Province, with outcrops of intrusive–subvolcanic and volcanic rocks. Also reported are the younger potassic–ultrapotassic outcrops of volcanic rocks of the Roman Magmatic Province.

Geological outline

Regional geology

Elba Island is located at the northern end of the Tyrrhenian Sea, a region affected by extensional processes behind the eastward-progressing front of the Apennine mobile belt (Fig. 1). The backbone structure of the Apennines was constructed when the Sardinia–Corsica block collided with the Adria plate (Malinverno & Ryan 1986). This orogenic system evolved diachronously as the extensional regime migrated from west to east, trailing the retreat of the compressive regime (Brunet et al. 2000) and giving way to the opening of the extensional ensialic back-arc Tyrrhenian Basin.

Igneous activity associated with extensional processes also migrated from west (14 Ma) to east (0.2 Ma) as the west-dipping Adriatic plate delaminated and rolled back to the east (Serri et al. 1993). Intrusive and extrusive products of mantle–crustal hybrid composition built the Tuscan Magmatic Province, spreading over about 30 000 km² in southern Tuscany and the northern Tyrrhenian Sea (Poli 1992; Westerman et al. 1993; Innocenti et al. 1997; Dini et al. 2002). Extensional processes and igneous activity affected the area of Elba Island during the Late Miocene (Bouillin et al. 1993; Jolivet et al. 1994).

Local geology

The structure of Elba Island is made up of five complexes (Fig. 2) which were stacked on to each other during the eastward Apenninic compressional event prior to 20 Ma (Deino et al. 1992). The lower three complexes (I–III) have continental features, consisting of metamorphic basement and shallow-water clastic and carbonate rocks, while the upper two (IV–V) are oceanic in character (Trevisan 1950; Keller & Pialli 1990; Pertusati et al. 1993). In more detail, Complex IV consists of Jurassic oceanic

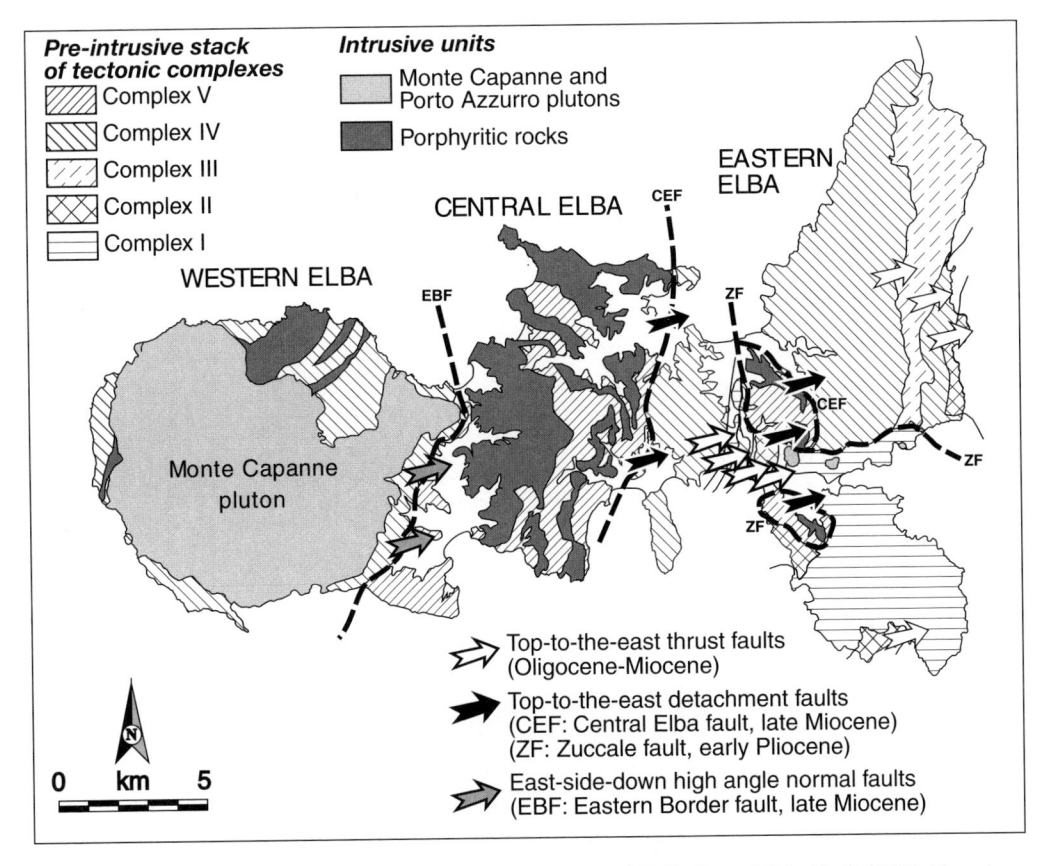

Fig. 2. Tectonic sketch map of Elba Island. Eastern Border Fault (EBF), Central Elba Fault (CEF), Zuccale Fault (ZF).

lithosphere of the western Tethys Ocean (peridotite, gabbro, pillow basalt and ophiolite sedimentary breccia) and its Upper Jurassic–Middle Cretaceous sedimentary cover (chert, limestone, and argillite interbedded with siliceous limestone). These rocks were deformed and metamorphosed during the Apenninic compression, to form east-verging folds. Complex V consists of argillite, calcarenite and sandy marl of Palaeocene to Middle Eocene age, overthrust by an Upper Cretaceous flysch sequence (Keller & Pialli 1990). Several intrusive bodies of various sizes and Miocene ages are exposed within Complex IV in western Elba and within Complex V in central Elba, making up the western intrusive complex. A younger and much smaller eastern intrusive complex is restricted to eastern Elba. Large-scale faults subdivide Elba Island into these three main zones (Fig. 2) and are the key to the reconstruction of the original geometry of the intrusive complex.

Current structural framework

Western Elba and the Eastern Border Fault
Western Elba consists of the Monte Capanne pluton and its thermometamorphic carapace of Complex IV rocks containing hypabyssal porphyry intrusions. It is separated from central Elba by the Eastern Border Fault that parallels the east side of the Monte Capanne pluton and put in contact the pluton's thermally metamorphosed host rock of Complex IV with the unmetamorphosed flysch of Complex V (Figs 2, 3 & 4). The Eastern Border Fault is marked, for the most part, by a distinct plane that dips moderately to steeply to the east. This fault separates a western footwall breccia of hornfelsed Complex IV rocks (ophiolitic material and deep-marine cover rocks) plus fragments of the Monte Capanne pluton (Fig. 4a), locally mineralized by quartz and hematite, from an eastern hanging-wall breccia made of Complex V flysch and megacrystic San Martino porphyry.

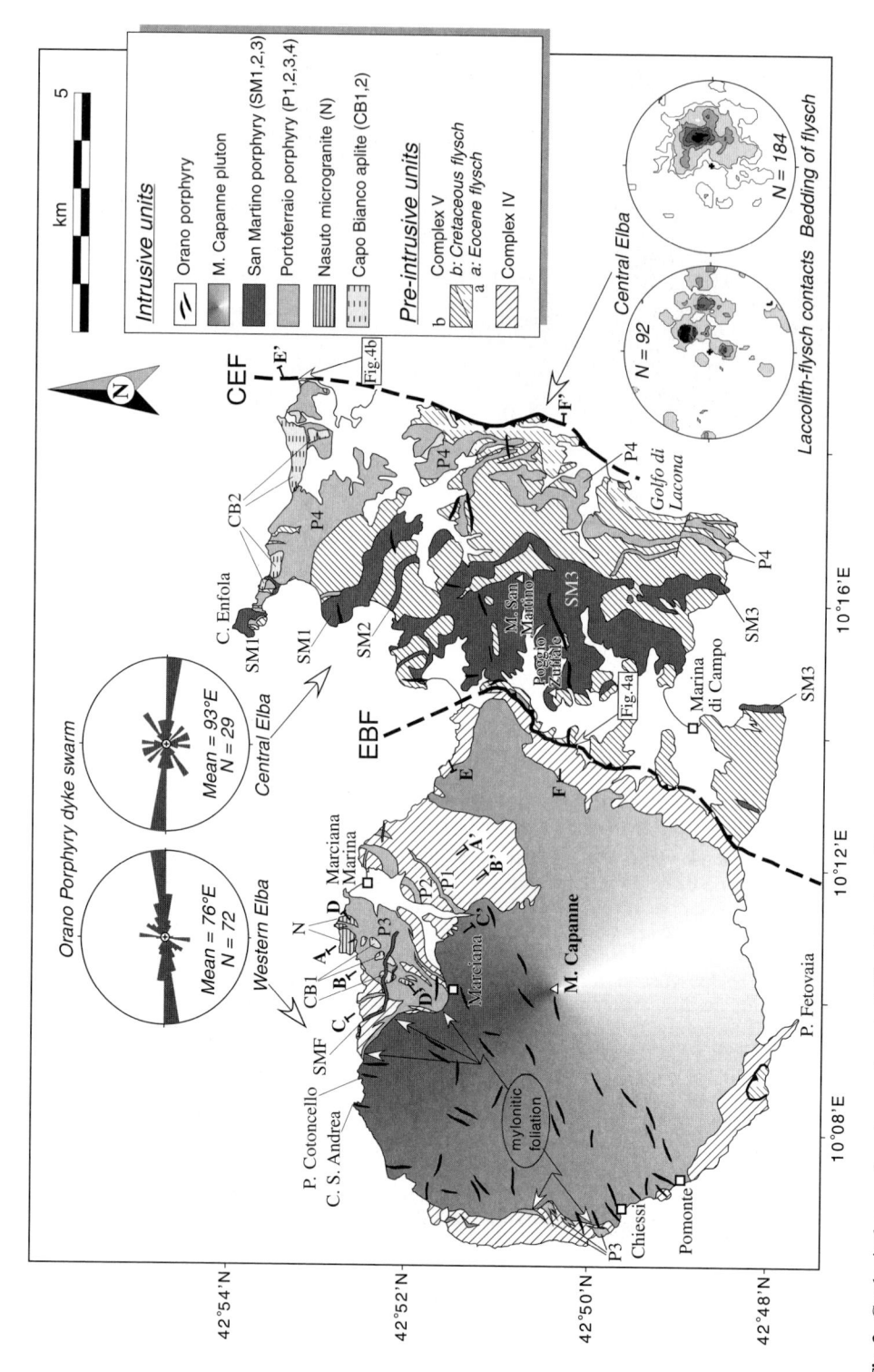

Fig. 3. Geological map of western and central Elba Island, modified after Rocchi *et al.* (2002). Labels on laccolith layers refer to Figure 5. Equal-area stereograms show orientations of bedding, fold axes and faults of the Upper Cretaceous flysch, and laccolith–host rock contacts in central Elba. Rose diagrams show strike of Orano dykes in western and central Elba.

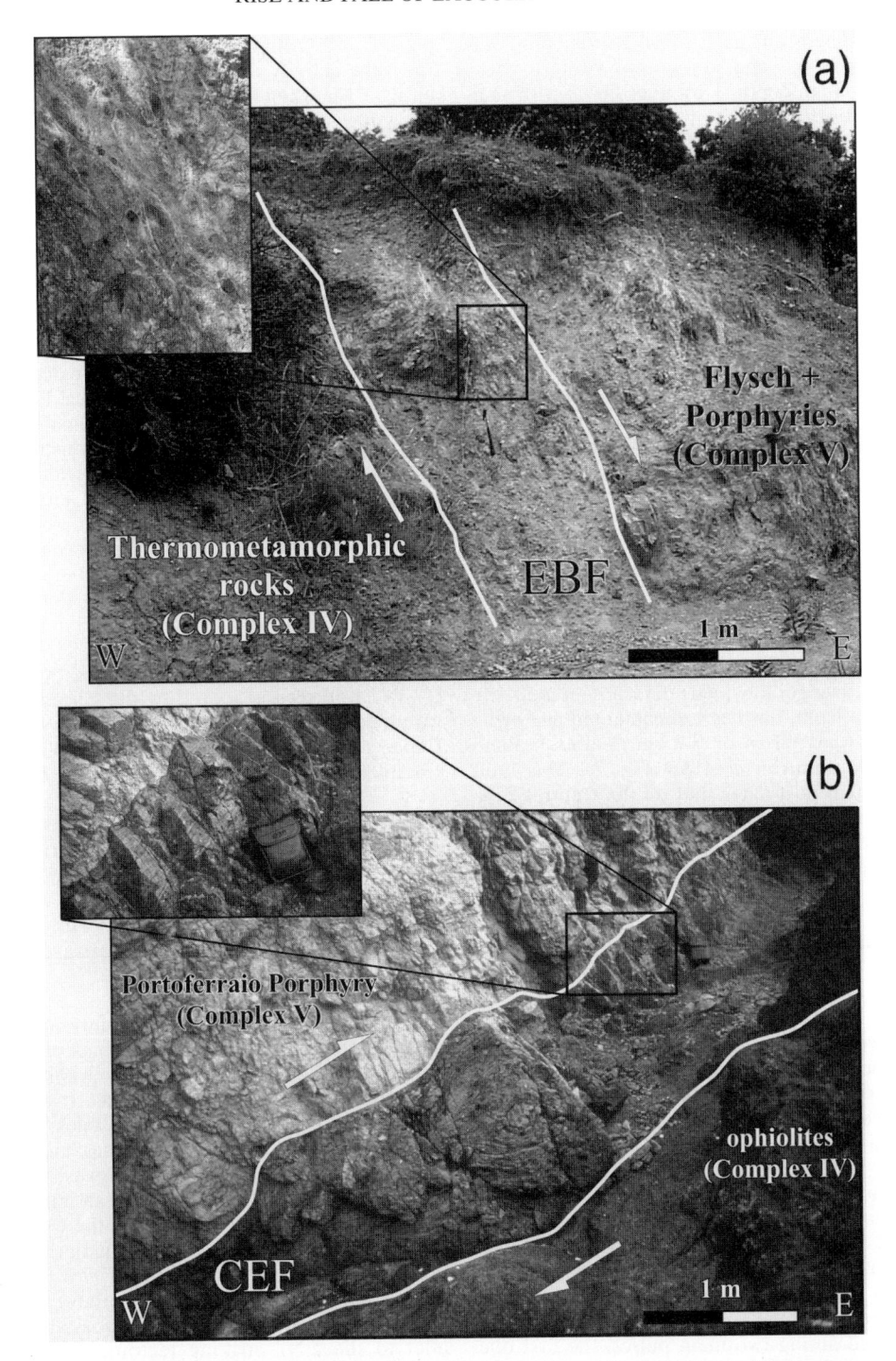

Fig. 4. (a) Chaotic breccia marking the high-angle normal fault separating western and central Elba (Eastern Border fault). Outcrop NW of La Pila, looking north. (b) Tectonic breccia of the Central Elba fault from Le Ghiaie, Portoferraio, illustrating the imbricated low-angle fabric (note the lens of Portoferraio porphyry). Clasts include western Elba lithologies, including Portoferraio porphyry, that constrain time of displacement. Locations of photos are also reported in Figure 3.

Movement on the Eastern Border fault was 'west side up' and juxtaposed western rocks from 4–5 km depth (Dini *et al.* 2002) with shallowly buried sedimentary rocks and their enclosed porphyries on the east side (Fig. 4).

Central Elba and the Central Elba Fault Central Elba is separated from eastern Elba by the low-angle Central Elba Fault, marked by a zone containing a tectonic mélange of rocks from Complexes IV and V (Trevisan 1950; Bellincioni 1958; Perrin 1975), most notably rocks whose equivalents crop out in western Elba (Fig. 4b). These are (1) thermally metamorphosed serpentinite and basalt (Marinelli 1955), and garnet- and wollastonite-bearing marble (Vom Rath 1870), identical to rocks in the Monte Capanne contact aureole, and (2) tourmaline-free aplite porphyry and K-feldspar phenocryst-bearing porphyry. The fault dips gently westwards, as does the dominant fabric of the rocks resting on it (Fig. 3), such that the highest part of the section occurs at the western edge against the steeply dipping Eastern Elba fault.

Eastern Elba and the Zuccale Fault In eastern Elba, a younger low-angle detachment fault, the Zuccale Fault, has been documented as having an eastward throw of 5–6 km (Keller & Pialli 1990; Pertusati *et al.* 1993; Fig. 2). This fault movement post-dates that on the Central Elba Fault, since it sliced off the leading edge of the Central Elba Fault to produce a klippe of Complex V rocks in eastern Elba.

The intrusive sequence

Western–central Elba intrusive complex

The western igneous complex consists of intrusive bodies that crop out within a stacked tectonic complex, almost exclusively over western and central Elba. Field observations, petrographic features (Table 1) and geochemical data, along with intrusive relationships, have been used to define the different intrusive units and correlate them between exposures (Dini *et al.* 2002). The relative chronology in western–central Elba has been firmly established on the basis of crosscutting relations. Isotope chronology points out that magmatism occurred during two main pulses, the first one around 8 Ma that emplaced the three older units (Capo Bianco aplite, Nasuto microgranite and Portoferraio porphyry), and the second one around 7 Ma that emplaced the three younger units (San Martino porphyry, Monte Capanne pluton, and Orano porphyry; Dini *et al.* 2002). Intrusive activity in eastern Elba was delayed until 5.9 Ma when the Porto Azzurro pluton and the associated products were emplaced (Conticelli *et al.* 2001).

Capo bianco aplite The Capo Bianco aplite is a white porphyritic rock with alkali-feldspar granite compositions (Table 1). In western Elba, the outcrops occur on a ridge structurally above ultramafic rocks and below an argillaceous unit of Complex IV, constituting five adjacent but isolated caps (Figs 3 & 5) that were likely emplaced as a single sill subsequently dismembered by younger intrusions. In central Elba, a tourmaline-rich layer of Capo Bianco aplite (Fig. 6a; layer CB2 in Figs 3 & 5) is reconstructed as having originally intruded higher within the pre-intrusive sequence: the pristine intrusion level was along the tectonic contact between Cretaceous flysch and underlying Eocene calcarenite, then the layer was dismembered and encased within the later intrusion of Portoferraio porphyry (P4). This can be seen in the northern part of central Elba (Figs 3 & 5), where layer CB2 is intruded and surrounded by layer P4, which in turn is floored by the Eocene calcarenite and topped by the Cretaceous flysch. Exposures in eastern Elba are within a block of Complex V flysch floored by the Zuccale Fault.

Nasuto microgranite The Nasuto microsyenogranite (Table 1) crops out over an area of 0.5 km^2 along the northern shore of western Elba (Fig. 3). It is entirely surrounded, as well as intruded, by the younger Portoferraio porphyry, such that its primary intrusive contacts are lost.

Portoferraio porphyry The Portoferraio porphyry contains prominent phenocrysts of sanidine (Fig. 6b) and has monzogranite to syenogranite compositions (Table 1). It occurs as four major layers up to 700 m thick, commonly interconnected and accompanied by minor dykes and sills (Fig. 3). East–west trending dykes connected to the floor of an intrusive layer (P4 in Fig. 3) are exposed on the western side of the Golfo di Lacona. Three major layers occur in western Elba. The lowest two layers, with maximum thicknesses of about 75 m (Fig. 5), intruded Complex IV metabasalts parallel to the ENE-striking tectonic fabric. A higher layer of Portoferraio porphyry was emplaced between hornfelsed argillite above and ophiolitic rocks below. It intruded at the same level previously exploited by the Capo Bianco aplite, and now encases and cross-cuts

Table 1. *Summary of petrographic and chronological features of the Late Miocene intrusive units from Elba Island*

Intrusive unit	Rock type	Texture	Paragenesis	Accessories	MME	Xenoliths	Age (Ma)
Monte Castello dyke	Shoshonite	Sub-aphyric	Pheno: Cpx, Ol ghosts xeno: Pl, Kfs, Qtz gm: Cpx, Pl, San	Mag, Chr	No	No	5.83 ± 0.14
Porto Azzurro pluton	Monzogranite	Qtz ± Pl phenocrysts, medium-grained equigranular matrix	Pl, Qtz, Kfs, Bt	Zrc	No	No	5.9 ± 0.2
Orano porphyry dyke swarm	Qtz monzodiorite to monzogranite	Porphyritic (<20–35%) very fine-grained gm	Pheno: Pl, Bt Rare Amph, Cpx relics Xeno: Kfs, Qtz Gm: Pl, Mg–Bt, Qtz, Kfs	Ap, Zrc, Aln, Thr, Mag, Ilm, Per	Common	Rare (1–3 cm)	6.83 ± 0.06 6.87 ± 0.30
Late felsic dykes associated with the Monte Capanne pluton	Leuco-syenogranite	Medium grain size, locally anisotropic	Qtz, Kfs, Pl, Bt, Ms ± Crd or Grt	Ap, Zrc, Mnz, Tur	No	No	Coeval with M. Capanne
Monte Capanne pluton	Monzogranite	Variable % mega Kfs, medium-grained equigranular matrix	Pl, Qtz (up to 15 mm), Kfs (up to 15 cm), Bt (up to 5 mm)	Ap, Zrc, Mon, Aln, Ilm, Tur	Variable vol.% (5 cm–5 m)	Scattered (1–5 cm)	6.9
San Martino porphyry	Monzogranite	Porphyritic (40–50%), fine-grained gm, miarolitic cavities	Mega: San (3–15%; c. 5 cm); pheno: Qtz, Pl, Bt; Gm: Qtz, Kfs ± Pl	Ap, Zrc, Mon ± Aln ± Tur	Common c. 1/10 m² (2 cm–2 m)	Common (1–10 cm)	7.2 ± 0.1 7.44 ± 0.08
Portoferraio porphyry	Monzogranite (minor syenogr.)	Porphyritic (25–50%), fine-grained gm	Pheno: San, Qtz, Pl, Bt; Gm: Qtz, Kfs ±Pl	Ap, Zrc, Aln, Mon, Thr	No	Rare (1–3 cm)	8.4 ± 0.1
Nasuto microgranite	Monzogranite	Porphyritic (25–30%), microgranular gm	Pheno: Qtz, Pl, Kfs, Bt; Gm: Qtz, Kfs, Pl		No	No	–
Capo Bianco aplite	Alkali-feldspar granite	Porphyritic, trachytoid gm	Pheno: Qtz, Pl, Kfs, Ms; gm: Ab	Tur abundant in central Elba; Xen, Mon, Nb–Ta oxides	No	No	7.91 ± 0.1 7.95 ± 0.1 >8.5

Abbreviations: Ab, albite; Aln, allanite; Ap, apatite; Bt, biotite; Chr, chromite; Crd, cordierite; Grt, garnet; Ilm, ilmenite; Mag, magnetite; Mon, monazite; Ms, muscovite; Ol, olivine; Per, perrierite; Pl, plagioclase; Qtz, quartz; San, sanidine; Thr, uraniferous thorite; Tur, tourmaline; Xen, xenotime; Zrc, zircon; mega, megacrysts (>2 cm); pheno, phenocrysts (<2 cm); xeno, xenocrysts; gm, groundmass; MME, mafic microgranular enclaves. For references and discussion on radiometric ages of western and central Elba, see Dini *et al.* (2002). For eastern Elba, see Maineri *et al.* (2003) and Conticelli *et al.* (2001).

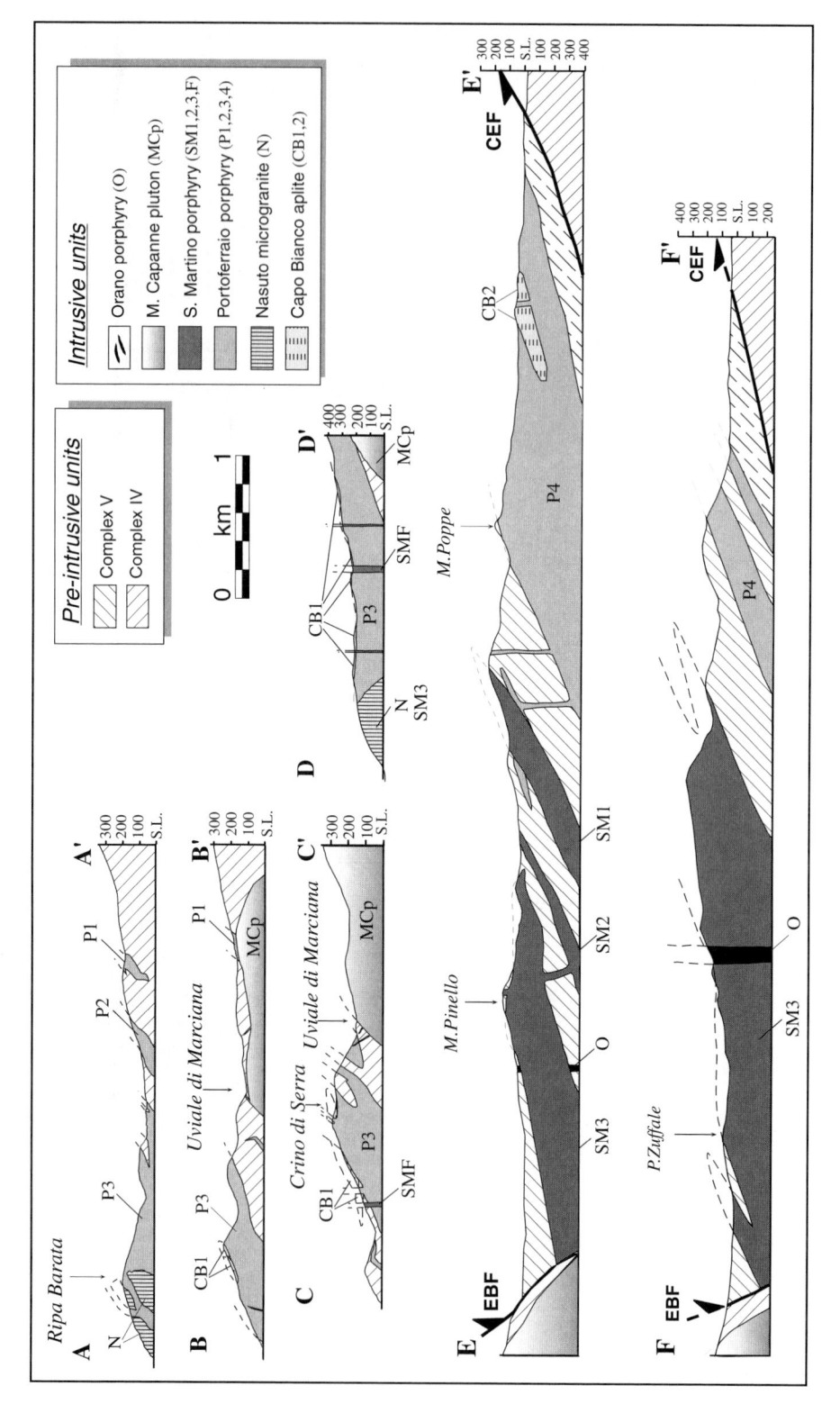

Fig. 5. Cross-sections: A–A′, B–B′ and C–C′ are parallel sections oriented N38°W through the northern margin of western Elba; D–D′ is perpendicular to A, B and C sections and illustrates the continuity of the Capo Bianco aplite laccolith; E–E′ trends N65°E across the northern part of Central Elba; F–F′ trends east–west through central Elba. Location of cross-sections is reported in Figure 3. Labels on intrusive masses correspond to those reported in Figure 3 (see also the legend).

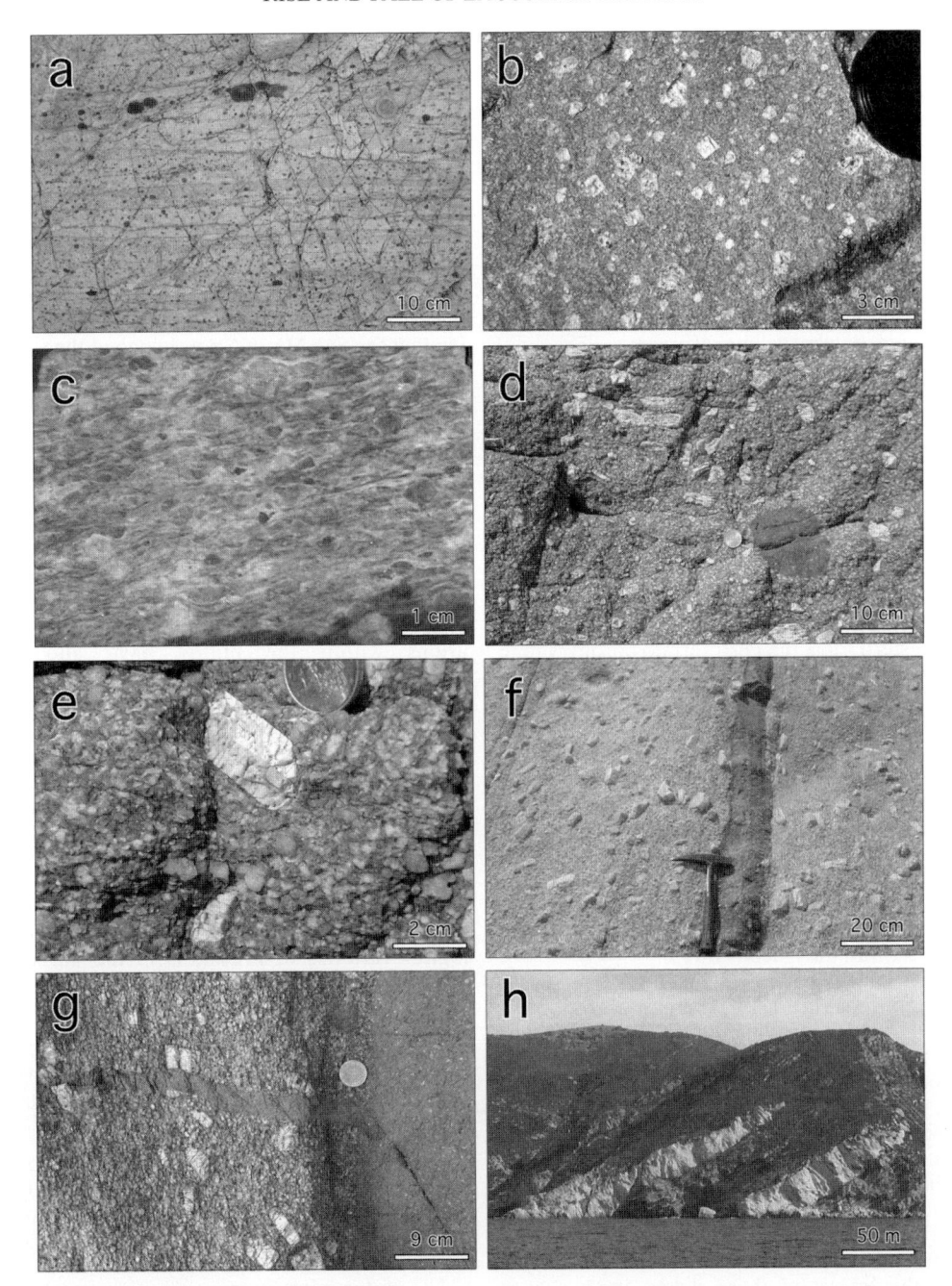

Fig. 6. (a) Capo Bianco aplite – mesoscopic texture from Capo Bianco; (b) Portoferraio porphyry – mesoscopic texture; (c) foliated Portoferraio porphyry, 1.5 km NE of Marciana; the foliation is N52°W, 55°N with stretching lineation N5°E, 50°N; (d) San Martino porphyry – mesoscopic texture from Punta Bardella, also showing a mafic microgranular enclave (right of the coin); (e) San Martino mesoscopic texture showing Carlsbad-twinned K-feldspar megacryst and euhedral quartz phenocrysts, Punta Bardella; (f) mesoscopic texture of the Sant'Andrea facies of the Monte Capanne pluton cut by a dark dyke of the Orano porphyry, Capo Sant'Andrea; (g) Orano dyke and dykelet cutting the Sant'Andrea facies of the Monte Capanne pluton, between Capo Sant'Andrea and Punta Cotoncello; (h) tapering of two terminations of laccolith layer P4 (see Figs 3 & 5), southern shore of central Elba, view from the sea.

the Capo Bianco aplite and the Nasuto microgranite (Fig. 5). This layer terminates to the SW against the Monte Capanne pluton, but was likely connected with the Chiessi outcrops (Fig. 3) before the Monte Capanne pluton intruded and deformed the porphyry. Close to the pluton contact, the groundmass in this layer exhibits a strong mylonitic foliation (Fig. 6c) that strikes parallel to the contact. Quartz phenocrysts here have subgrain boundaries; plagioclase cracks are cemented by micrographic quartz plus K-feldspar; and primary biotite is transformed to oriented polycrystalline aggregates. The total length of the layer was, therefore, in excess of 9 km, yet only 3 km are presently exposed along strike in the NE part of western Elba. The maximum thickness measured in cross-sections is around 700 m.

The fourth layer of Portoferraio porphyry occurs in central Elba, between Cretaceous flysch above and Eocene calcarenites below (Fig. 3). The upper surface of this NNW-striking layer is generally subparallel to bedding in the overlying flysch that dips moderately WSW. As in western Elba, the Portoferraio porphyry here surrounds the older Capo Bianco aplite, with dykes truncating tourmaline laminations and mineralized fracture surfaces in the aplite. The maximum thickness of the layer, as calculated from cross-sections, is about 400 m. Along the southern shore of central Elba, this layer splits into five tapering branches. Only small dykes and sills penetrated the overlying Cretaceous flysch. Visible contact-metamorphic effects in the host rock are restricted to local biotite crystallization over a millimetre-thick layer at the contact with the igneous rocks.

San Martino porphyry The San Martino porphyry is a monzogranite porphyry (Table 1) that occurs as dykes or thick layers, characterized by prominent megacrysts of sanidine (Fig. 6d & 6e) set in a very fine-grained groundmass. Locally, megacrysts are aligned subparallel to intrusive contacts, with orientations steep in dykes and subhorizontal in sills. Close to some contacts the megacryst content is very low. In western Elba, only WNW-striking dykes occur, cutting the older intrusive units and the rocks of Complex IV; the largest of these dykes extends over 2.5 km with a thickness of 25 to 50 m (Fig. 3). In central Elba, dykes of San Martino porphyry cut the fabric of the sedimentary host as well as layers of the older intrusive units, while the main bodies are four parallel, gently westward-dipping layers. They were emplaced above the layers of Portoferraio porphyry, concordant to the fabric of the folded and faulted strata of the host flysch. Bifurcation along the upper surface forms branches, and at Monte San Martino, a large septum of flysch maintains a simple planar geometry over a map distance of 1.5 km (Fig. 3). The maximum thicknesses of these igneous layers are between 100 and 700 m, tapering out towards both the northern and the southern end (Fig. 5). Lengths measured in the north–south direction range between 2.4 and 8.3 km, with the thickest and largest layer exposed over more than 18 km². Miarolitic cavities are found in this unit, but are notably absent in the other units, suggesting that the San Martino porphyry had the shallowest emplacement level. As with the Portoferraio porphyry, visible contact-metamorphic effects are essentially absent in the host rock.

Monte Capanne pluton The Monte Capanne pluton in western Elba is the largest of those exposed in the Tuscan Magmatic Province (Marinelli 1959; Poli 1992). It is roughly circular in plan (Fig. 3) and is mainly composed of a monzogranite (Table 1) with prominent K-feldspar megacrysts (Fig. 6f & 6g). Limited variations within the pluton have allowed description of three internal facies (Dini *et al.* 2002). Contacts with surrounding country rock are mostly intrusive in nature and dip away from the pluton. Host rocks, belonging to the ophiolitic–sedimentary tectonic Complex IV (Fig. 2) and exhibiting shear fabric acquired during the Apenninic compressive phase, were overprinted by thermal metamorphism and deformed by emplacement of the pluton itself (Daniel & Jolivet 1995); this younger deformation cross-cuts the older compressive shear fabric. Lithological varieties of protoliths in the aureole preserve the reactions

1 andalusite = sillimanite,
2 talc + forsterite = anthophyllite + H_2O, and
3 the breakdown of muscovite + quartz (Thompson 1974; Spear & Cheney 1989; Tracy & Frost 1991).

Although these reactions depend on the compositions of fluids and solid-solution phases, taken together they suggest peak contact-metamorphic conditions with temperatures in excess of 600 °C at a pressure of 0.1–0.2 GPa.

The pluton is cut by several leuco-syenogranite dykes, occurring mainly close to the pluton's contact, within both the pluton and its thermometamorphic aureole. These dykes commonly have thicknesses up to tens of metres.

Orano porphyry The Orano porphyry unit is a swarm of nearly 100 dark-coloured granodioritic to quartz monzodioritic dykes (Table 1) that cross-cut all the other intrusive units of the sequence (Fig. 3). Contacts with the host rock are sharp and planar (Fig. 6f), commonly exhibiting abrupt changes in orientation; thicknesses range from less than a metre up to a maximum of 50 m. Orano dykes are restricted in western Elba to the NW portion of the Monte Capanne pluton and its contact aureole (approximately six dykes/km^2). Orano dykes crop out only in the northern half of central Elba, an area bordered on its southern edge by the longest exposed dyke (2.5 km). In both areas, strikes cluster around the east–west direction. Some dykes are internally zoned, testifying to exploitation of the same conduit by succeeding magma pulses (Dini et al. 2002).

Eastern Elba intrusive complex

Porto Azzurro pluton The Porto Azzurro pluton is a megacrystic monzogranite (Table 1), with compositions similar to the most acidic portions of the Monte Capanne monzogranite (Conticelli et al. 2001). These rocks have only limited exposure in the south-central part of eastern Elba (Fig. 2), but a significant size is suggested for the pluton by the abundance of thermometamorphic rocks surrounding the exposed portion. The reconstruction of palaeo-isograds and their offset constrains the extent of movement on the younger Zuccale Fault noted above (Pertusati et al. 1993).

Monte Castello dyke A brownish-grey porphyritic dyke occurs in eastern Elba (Conticelli et al. 2001). The rock is quite altered, and the original phenocryst assemblage consisted of plagioclase, clinopyroxene, olivine and scattered K-feldspar megacrysts of likely exotic origin (Table 1). The dyke had an original shoshonite composition, and an emplacement age of 5.8 Ma (Conticelli et al. 2001). Its petrographic and geochemical features resemble those of the Orano dykes. A pattern of emplacement of mafic dykes after a main pluton occurred in western Elba at 6.9–6.8 Ma, and was apparently repeated in eastern Elba 1 Ma later.

In summary, the igneous sequence of western–central Elba started with emplacement of the Capo Bianco aplite and Nasuto microgranite, followed in succession by the Portoferraio porphyry and, after a time lag of about 1 million years, by the San Martino porphyry. The deepest layers of this complex were then intruded and deformed by the Monte Capanne pluton and its associated late leucocratic dykes and veins. Finally, the Orano dyke swarm was emplaced, cutting through the entire succession.

Approximately 1.5 million years later, the next locus of igneous activity developed further east in eastern Elba, where the Porto Azzurro pluton and the Monte Castello dyke were emplaced.

Discussion

Construction of the intrusive complex

Controls on emplacement Most of the laccolith layers at Elba were emplaced along strong crustal heterogeneities such as thrust surfaces between complexes, secondary thrusts inside complexes, and bedding in the flysch (Fig. 3). Evidence constraining emplacement mechanisms of the laccoliths includes:

1 subvertical dykes of San Martino porphyry observed and mapped at the lowest levels of the complex (Fig. 3), with laccoliths of that unit observed at higher levels;
2 laccolith layers of Portoferraio porphyry exposed along with their east–west trending feeding dykes on the western side of the Golfo di Lacona;
3 tapering branches observed to emanate from the top surfaces of laccolith layers (at Poggio Zuffale and M. San Martino);
4 main sheets tapering out at their visible (eastern) ends.

The general picture is, therefore, that of vertically rising magma that stops and spreads laterally. Such a switch from vertical to horizontal magma movement could be related to reaching the neutral buoyancy level, but in this case magma should not have the energy to lift the overburden during filling of laccoliths (Hogan et al. 1998). An additional problem with the role of neutral buoyancy is represented by the progressively shallower emplacement level over time of magmas with similar densities (at liquidus conditions, calculated densities vary by less than 0.02 g cm^{-3}), with the most dense magma (Orano) going to the shallowest levels. It has to be assumed, therefore, that the rising magma had a residual driving pressure exceeding the vertical stress when it changed its direction of movement from vertical to horizontal. Therefore, the magma spreads laterally when encountering a subhorizontal strength anisotropy that behaved as a crustal magma trap (Hogan & Gilbert 1995; Hogan et al. 1998). Indeed, at the trap, magma driving pressure must be sufficiently greater than the lithostatic load to allow the magma to raise the roof rather than form a stock (Hogan et al. 1998) or simply flow laterally until escaping the trap.

Emplacement depths for individual layers within the Elba laccoliths can be approximated from calculations on a stratigraphic basis, that is by measuring overburden thickness in cross-sections. These emplacement depths, however, are not just equal to the thickness of the present overburden, because, once the original section has been restored, original depths have to be calculated, taking into account the emplacement sequence and the likely amount of erosion. The total thickness of material lost by erosion since the beginning of magmatic activity has been calculated at 800 m, based on a rough estimate of the present mean erosion rate for Italy (0.1 mm yr^{-1}; Branca & Voltaggio 1993).

The Monte Capanne pluton intruded the base of the laccolith sequence. Its emplacement depth, however, is not simply the sum of laccoliths and their host-rock thicknesses, since the Monte Capanne pluton punched through several layers of Portoferraio porphyry in western Elba, as demonstrated by the strong mylonitic foliation developed in the Portoferraio porphyry close to the contact with the pluton on both its NE and SW sides (Fig. 3). After accounting for erosion, the calculated emplacement depth for the Monte Capanne pluton is about 4.5 km, a value in agreement with peak conditions of contact metamorphism produced by the Monte Capanne intrusion (0.1–0.2 GPa; Dini et al. 2002).

Geometry of the complex The western Elba igneous complex is composed of several sheet-like intrusions with extensively known roofs and floors, and a larger pluton with the roof exposed locally, but no known floor. All the sheet-like bodies of Orano porphyry and a few sheets of Portoferraio and San Martino porphyries cut across the structures of the host rock, and thus are dykes. The main bodies of Capo Bianco aplite, Portoferraio porphyry, and San Martino porphyry all occur subparallel to the planar structures of the host meta-ophiolite, metasediments, and flysch.

Based on detailed mapping (Fig. 3), maximum thickness values of the nine most significant intrusive layers, along with thicknesses of the intervening and overlying strata, were determined by measurement in cross-section (Fig. 5; Rocchi et al. 2002). Since layers either thicken or thin as they project below the surface, these values are invariably equal to or less than the true maxima. Thickness values of individual layers vary over an order of magnitude, from 50 to 700 m. The thickness of the Monte Capanne pluton is estimated at 2 km on the basis of preliminary magnetic modelling (O. Faggioni, pers. comm.).

Diameter values for each layer were approximated from the maximum horizontal exposed length. Layers exposed in central Elba have north–south strikes with lengths between 2.4 and 10 km, while in western Elba they have NE–SW strikes and lengths between 1.6 and 9.3 km (Rocchi et al. 2002). This latter length has been determined after reconstructing the layer before it was deformed and cross-cut by the Monte Capanne pluton: this reconstruction is supported by the occurrence of mylonitic foliation at the NE and SW contacts with the pluton (Fig. 3). Lengths perpendicular to strike are assumed to be similar, since Corry (1988) reports differences generally less than two times. The maximum diameter of individual layers varies nearly an order of magnitude, from 1.6 to 10 km (Rocchi et al. 2002). All nine of the main Elba laccolith layers have large aspect ratios (diameter/thickness), varying from 12 to 33.

Volumes were calculated assuming two end-member shapes: a laccolith approximated by a spherical bowl with height equal to maximum thickness, and a sill approximated by a cylinder with height equal to maximum thickness. The actual volume is likely between the two values; therefore, we assume the best volume estimate to be the average of the two calculations. Volumes of individual layers vary more than an order of magnitude, from 1.3 to 30 km^3 (Rocchi et al. 2002).

Relationships between the dimensional parameters of the layers constituting the Elba laccolith complex have been explored (Rocchi et al. 2002), and a significant exponential/power-law correlation has been found between thickness (T) and diameter (L). On a log T v. log L diagram this correlation results in a linear fit ($r^2 = 0.93$; Fig. 7) with equation $T = 0.026\ (\pm 0.006)\ L^{1.36(\pm 0.14)}$. Furthermore, all these layers plot in the horizontal elongation field, and only the two biggest individual layers fit the L–T relationships reported by McCaffrey & Petford (1997) for laccolith shapes.

Emplacement of the Elba laccoliths as layers that thin toward their edges added more than 2 km thickness to the upper crust above a common magmatic centre. Major dykes, such as the E–W-trending San Martino dykes in western Elba (Fig. 3), occur beneath individual units, while small vertical dykes occur between individual layers of each intrusive unit. Therefore, each intrusive unit is interpreted as a multilayer laccolith. The overall geometry resulting for the Elba subvolcanic complex is one of a nested multi-layer, multi-pulse Christmas-tree laccolith sequence.

Figure 9 presents a panel of cartoons illustrating construction of the pluton–laccolith complex, culminating with a 'snapshot' of the resulting dome structure. The oldest intrusion is the Capo Bianco laccolith (about 8.0 Ma), with a lower level at a strong discontinuity in Complex IV, and an upper level at the Complex IV–V interface. The geometry and shape of the almost coeval Nasuto microgranite is not represented, owing to its dismemberment by the closely following intrusion of the Portoferraio laccolith. The latter consists of two thin horizons and one major level occurring inside Complex IV, along with one major horizon emplaced at the Complex IV–V interface. The successive intrusion is represented by the San Martino laccolith (about 7.2 Ma); feeder dykes for the San Martino laccolith are preserved in western Elba, while the main intrusions were exclusively in central Elba within Complex V. This period of dome construction by laccolith growth was followed by the emplacement of the Monte Capanne pluton, whose roof is found very close to the Complex IV–Complex V contact. It is this ancient thrust surface on which much of the Central Elba fault movement is thought to have later occurred following the last pulse of igneous activity in western Elba, namely the emplacement of the east–west-trending Orano dyke swarm (about 6.8 Ma).

Filling of the laccoliths The growth of an intrusion can occur in various modes, depending on the magma-supply rate, the depth of emplacement, the availability of crustal magma traps, and the mechanisms by which country rock makes room. In the case of magmas emplaced in the upper 3–4 km of the crust, lifting the overburden is a probable mechanism for making space for the intrusion, as demonstrated in some limiting cases where overburden uplift has actually been documented (Corry 1988). How this space is filled, however, is a matter of debate recently addressed by Rocchi *et al.* (2002). Relevant data commonly collected to solve this problem are represented by the internal structure related to magma flow and the final shape that an intrusion acquired, assuming that the shape is the result of its growth history.

The scale-invariant distribution of the shapes of laccoliths (McCaffrey & Petford 1997) supports a common mechanism of formation. Very similar tabular shape, coupled with scale-invariant distribution of these shapes, has also been documented for plutons (McCaffrey & Petford 1997; Cruden & McCaffrey 2001), lending support to the idea that similar growth mechanisms exist for both laccoliths and plutons. A commonly acknowledged growth mechanism (Johnson & Pollard 1973; McCaffrey & Petford 1997; Cruden & McCaffrey 2001) is a

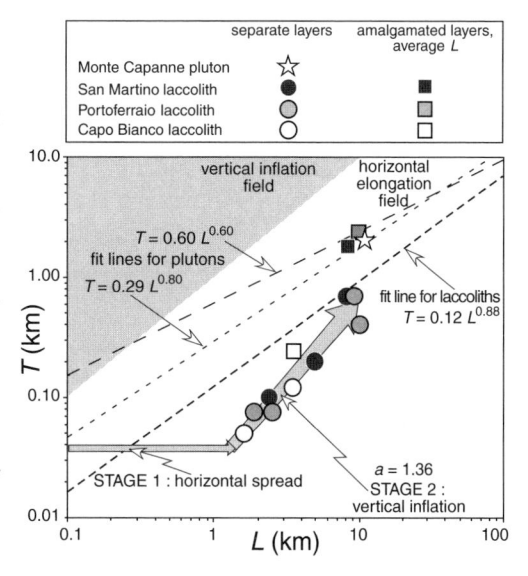

Fig. 7. Log T v. log L diagram, where T = maximum thickness of individual laccolith layers and L = horizontal length of individual laccolith layers (Rocchi *et al.* 2002). Also reported are: the fit lines for the laccoliths' and plutons' aspect ratios, and the horizontal elongation (white) and the vertical inflation (grey) fields of McCaffrey & Petford (1997). The dimensional parameters of Monte Capanne pluton (O. Faggioni, unpublished data) are shown for comparison.

two-stage filling story: first, the magma spreads laterally at the emplacement level until a layer almost as wide as the future intrusion is formed a aspect ratio; then the thin intrusion thickens by upward inflation.

These speculations are based on dimensional parameters reported for laccoliths and plutons that are always obviously measured for intrusions that have stopped filling. In this respect, the Elba individual laccolith layers are thought to be only part of a complete laccolith, thus offering a great opportunity to see and measure intermediate stages between horizontal spreading and vertical filling. Indeed, the steep slope ($a > 1$) in the log T v. log L plot (Fig. 7) is interpreted as evidence of laccolith growth frozen in the vertical inflation stage (Rocchi *et al.* 2002). This could be ascribed to insufficient magma supply, but it has to be noted that the cumulative thickness for the Capo Bianco laccolith layers almost fits the laccolith power-law line, and the cumulative thicknesses for both Portoferraio and San Martino laccolith layers fit the pluton's power-law line (Fig. 7). This suggests that the latter two appear as Christmas-tree laccoliths only because magma was not able to coalesce in a single reservoir, likely owing to the excess of crustal magma traps available at Elba that promoted magma injection in a new horizontal plane rather than continued filling of an already partially filled layer.

Tectonic splitting of the intrusive complexes

The portions of the Elba igneous complex cropping out in western and central Elba have not preserved their original emplacement geometry. All the intrusive units were emplaced within the tectonic Complexes IV and V, when they were stacked above the present western Elba (Fig. 8). Then, shortly after the intrusion sequence was completed, the upper part of the igneous–sedimentary complex was tectonically translated eastwards along the Central Elba Fault (CEF), leaving the lower part to be found in western Elba while the upper part came to rest in central Elba (Fig. 9a). Following this eastward translation, a 'west side up' movement occurred along the Eastern Border Fault with a throw of 2 to 3 km (Fig. 9b through d). This history is supported by:

1. the low-angle CEF at the base of the slice consisting of igneous layers and their host rocks in central Elba
2. shear fabrics, including east–west striations, in the fault mélange of the CEF;
3. fragments of rocks typical of western Elba in the footwall mélange;
4. the match of petrographic and geochemical features of the intrusive units in central and western Elba;
5. the strongly preferred east–west orientation of Orano dykes and their restriction to the northern areas in both western and central Elba.

The minimum amount of displacement along the central Elba Fault (CEF) is constrained to about 8 km by the distance from the pluton's aureole eastward to its leading edge where fragments of hornfels from that aureole occur in the fault mélange.

The timing of displacement on the CEF is constrained by

1. the occurrence of fragments of Monte Capanne hornfels in the footwall mélange of CEF some 8 km east of the nearest outcrop of the thermal aureole, indicating that movement on the CEF occurred after contact metamorphism linked to the Monte Capanne pluton (c. 6.8 Ma);
2. the matching distribution of Orano dykes in western and central Elba (Fig. 3), suggesting

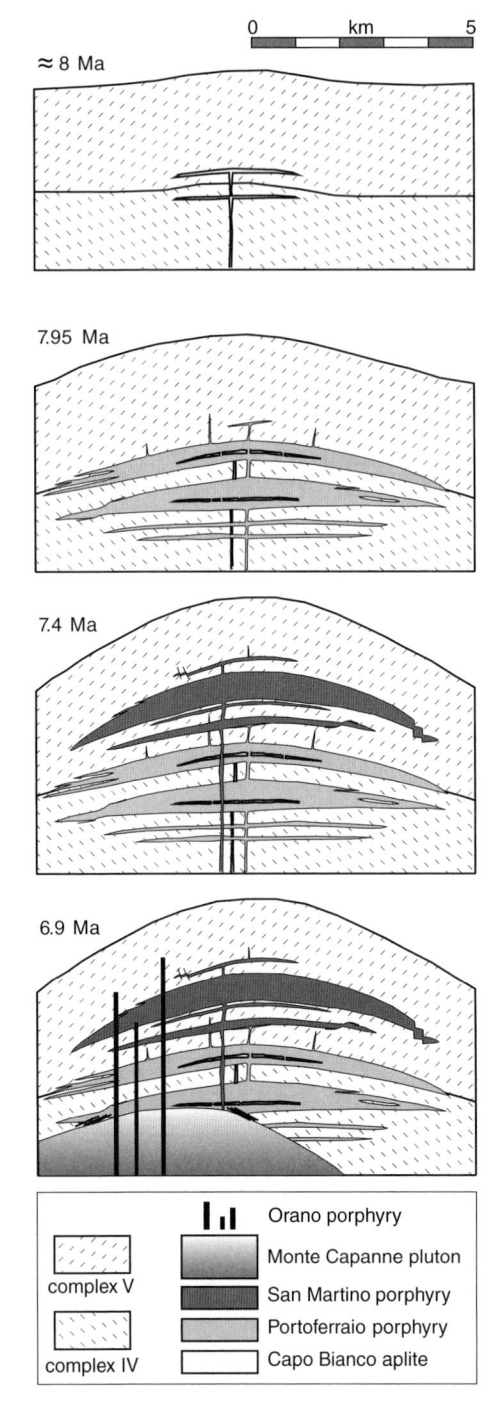

Fig. 8. Rise of the nested Christmas-tree laccolith complex: emplacement sequence of the intrusive units, building up the western Elba intrusive complex. The Nasuto microgranite intrusive event is not reported, owing to the small size of that intrusion, which is dismembered by the later Portoferraio porphyry.

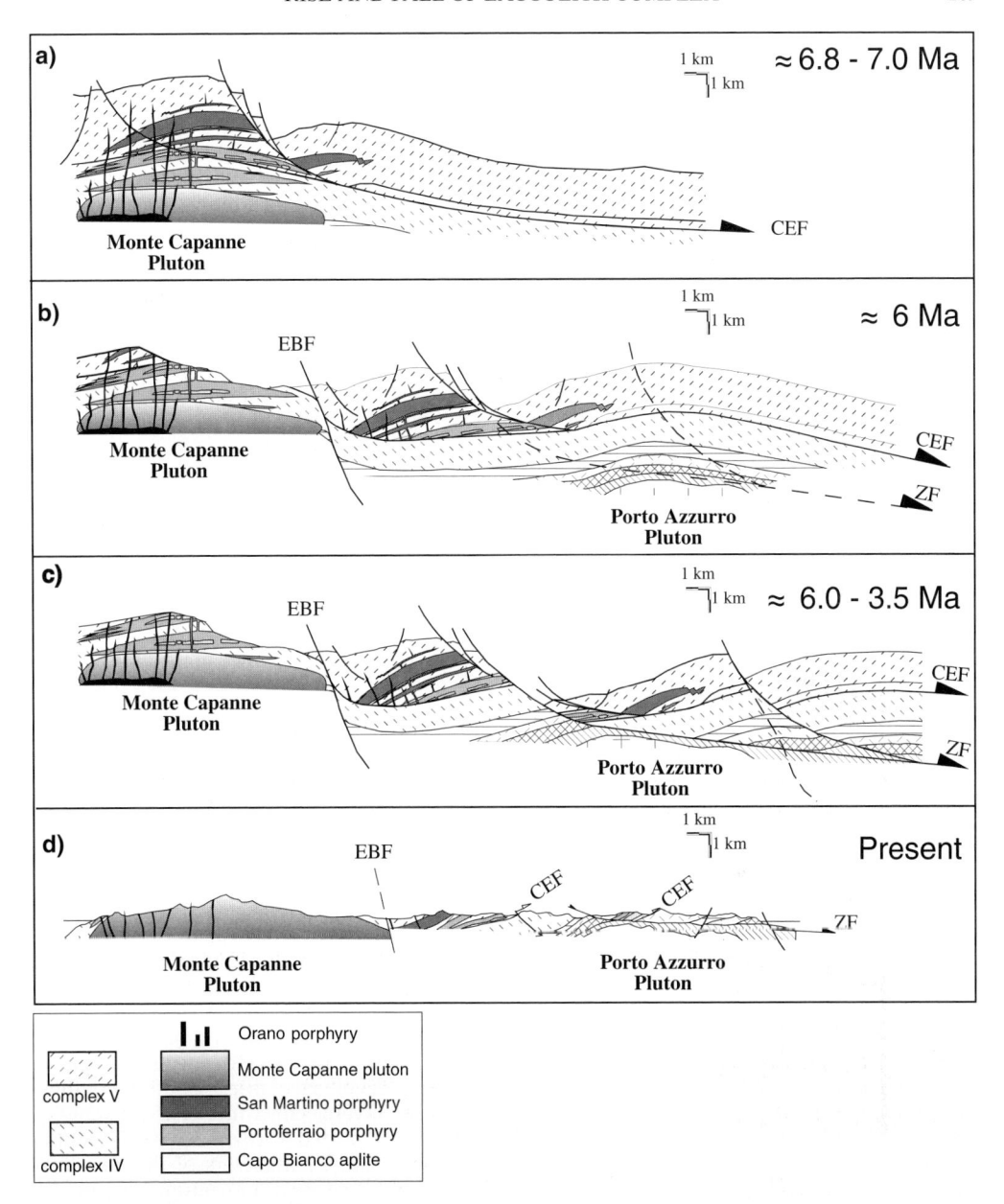

Fig. 9. Fall of the nested Christmas-tree laccolith complex: unroofing of the Monte Capanne pluton, showing progressive stages of décollement.

that the Orano dykes were truncated and translated by the movement along the CEF, that therefore took place after their emplacement at 6.85 Ma.

The timing of this scenario is further constrained by the occurrence on mainland Tuscany – some 50 km to the east – of abundant cobbles and boulders of tourmaline-bearing Capo Bianco aplite and Portoferraio porphyry in conglomerates deposited very close to the end of the Messinian (Bossio et al. 1993; Marinelli et al. 1993; Testa & Lugli 2000). Indeed, Capo Bianco aplite and Portoferraio porphyry layers were concentrated just above (and presumably below) the CEF, and the most logical mechanism to expose them without exposing the overlying San Martino porphyry units was by erosion of tilted layers (today's tilt of laccolith layers is about 30°; Fig. 6h).

The eastward displacement of the upper part of the complex is at least partly linked to gravitational instability. In about 1 million years, a 2700-m thick tectonostratigraphic section was inflated by the addition of at least 2400 m of laccolithic intrusions, leading to a total thickness for the new section of about 5000 m. A dome with a 10 km diameter and a height of 2.5 km, was produced with a surface slope of about 25° (assuming an originally flat surface). We envision that emplacement of the Monte Capanne pluton beneath this dome caused oversteepening and triggered the main eastward displacement of the upper section. Once significant movement began, transfer of the load from above Monte Capanne towards central Elba promoted movement on the east-dipping Eastern Border Fault as the unloaded pluton rose and the thickened central Elba section subsided. Final movement on the Eastern Border Fault took place entirely in a brittle regime, truncating the Central Elba fault that has since been eroded in western Elba and lies almost completely buried below central Elba.

Evidence from experimental analogue models (Merle & Vendeville 1995; Roman-Berdiel et al. 1995) and structural observations of natural examples (Reeves 1925; Gucwa & Kehle 1978) shows that laccolith-type magmatic intrusions can produce stresses large enough to induce thrusts and folds in the adjacent sedimentary rocks. In particular, the most efficient process seems to be gravity gliding, during which layer-parallel compression can result as rocks glide away from the topographic high created by laccolithic intrusion.

The rate of displacement is constrained by the time between onset of the movement along the Central Elba fault (c. 6.8 Ma) and the time when cobbles were deposited (close to the end of Messinian, i.e. before 5.3 Ma). Allowing for erosion and transport, a maximum estimate for the time available for the main movement on the Central Elba Fault is less than 1.5 million years. Thus, the eastward translation of at least 8 km occurred at an average rate in excess of 5–6 mm yr^{-1}. This movement rate is higher than movement rates reported for detachment faults (Stockli et al. 2001), while it is consistent with rates associated with gravity gliding (Fletcher & Gay 1971), possibly triggered by magma emplacement (Merle & Vendeville 1995).

A sequence of events similar to those described above also occurred for the tectonic evolution and pluton exhumation in eastern Elba (Keller & Pialli 1990; Pertusati et al. 1993), where the Zuccale Fault displaced a 2-km wide slice from the front edge of the Central Elba Fault along with part of the contact-metamorphic aureole of the Porto Azzurro pluton (Keller & Pialli 1990; Pertusati et al. 1993). Therefore, the movement on the Zuccale Fault occurred after emplacement of the upper Pliocene Porto Azzurro pluton (5.9 Ma; Maineri et al. 2003).

The common history of western–central and eastern Elba is presented schematically in Figure 10, and can be summarized as follows. Initially, dominantly subhorizontal movement on a low-angle detachment fault with top-to-the-east sense of shear (Central Elba and Zuccale faults) translated the overlying rocks eastwards, trimming out part of the contact aureole of the pluton with which it is associated. Then, high-angle structures were activated mainly at the eastern edge of the pluton, i.e. the Eastern Border Fault east of the Monte Capanne pluton and offshore faults east of the Porto Azzurro pluton (Bortolotti et al. 2001; Eastern Elba Faults (EEF)). In this scenario, pluton emplacement occurred before the main faulting, and promoted its activation. This inference challenges the hypothesis that eastward shear above the pluton was synchronous with pluton emplacement, accommodating the magmatism and controlling the level of emplacement of the Monte Capanne and Porto Azzurro plutons (Daniel & Jolivet 1995; Jolivet et al. 1998).

Conclusions

The favourable tectonic conditions on Elba Island allowed a detailed study of a 7–8-km thick crustal section. Detailed mapping of intrusions and their host rocks in western and central Elba led to the reconstruction of nine shallow-level intrusive layers, their shapes, intrusive sequence and emplacement depths. Combined, these intrusive layers constitute an outstanding example of nested Christmas-tree laccoliths. The dimensional parameters of the laccolithic layers yield general indications about the mechanisms of emplacement and growth of these sheet-like intrusions, which only in some cases completed the vertical inflation stage of their growth. Additionally, the analysis of this complex indicated that the site of magma emplacement was essentially governed by the crustal anisotropies behaving as magma traps, while no role has been recognized for the attainment of the magma neutral buoyancy level. Finally, the domal structure created in western Elba by laccolith emplacement was decapitated

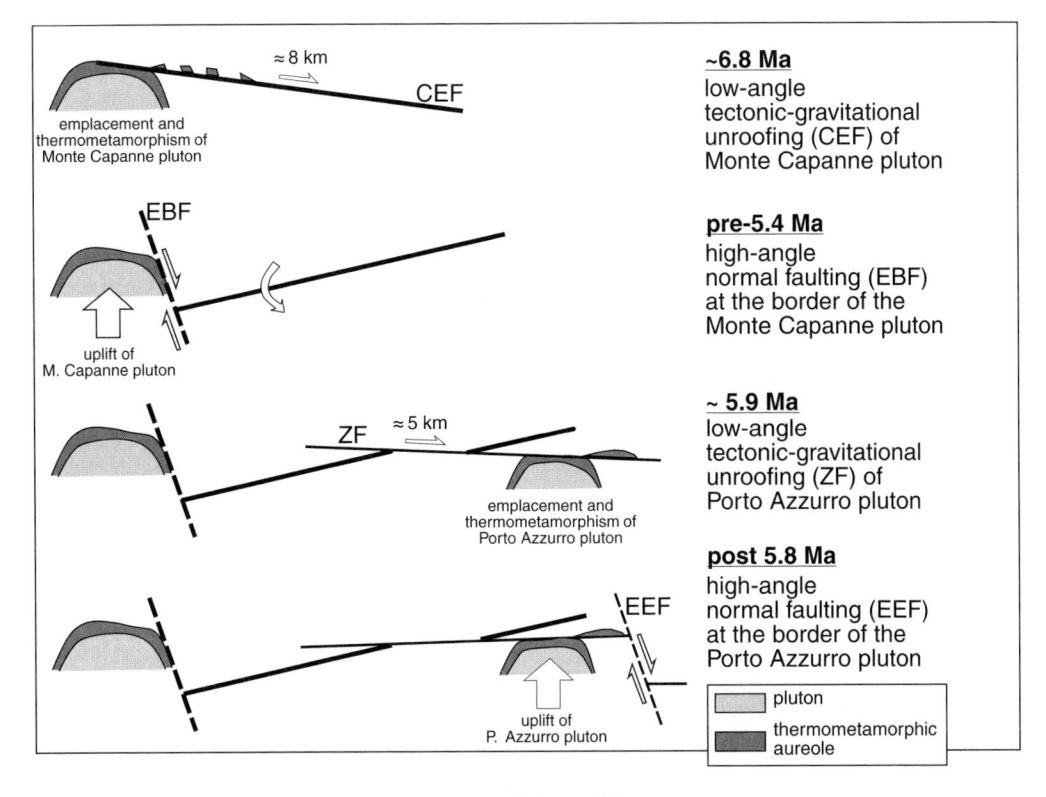

Fig. 10. Schematic summary of Late Miocene fault activity on Elba.

as a result of oversteepening, following emplacement of the Monte Capanne pluton into its basal portion.

This work is dedicated to the memory of G. Marinelli and L. Trevisan, who five decades ago drew attention to the geology of Elba Island and put forward stimulating hypotheses on magma genesis, ore genesis and tectogenesis regarding Elba geology. D. Wise provided criticism and comments on an earlier version of the manuscript. The thorough reviews of K. McCaffrey and N. Petford helped to improve the manuscript. The whole work was carried out with the financial support of MURST and CNR grants and with funding from Norwich University.

References

AMÉGLIO, L. & VIGNERESSE, J.-L. 1999. Geophysical imaging of the shape of granitic intrusion at depth; a review. *In*: CASTRO, A., FERNANDEZ, C. & VIGNERESSE, J.-L. (eds) *Understanding Granites: Integrating New and Classical Techniques.* Geological Society, London, Special Publications, **158**, 39–54.

BELLINCIONI, D. 1958. Rapporti tra 'argille scagliose ofiolitifere' flysch e calcare nummulitico nell'Elba centrale. *Bollettino della Società Geologica Italiana*, **77**, 112–132.

BENN, K., ODONNE, F. & DE SAINT-BLANQUAT, M. 1998. Pluton emplacement during transpression in brittle crust: new views from analogue experiments. *Geology*, **26**, 1079–1082.

BORTOLOTTI, V., FAZZUOLI, M., PANDELI, E., PRINCIPI, G., BABBINI, A. & CORTI, S. 2001. Geology of the central and eastern Elba Island, Italy. *Ofioliti*, **26**, 97–150.

BOSSIO, A., COSTANTINI, A., LAZZAROTTO, A., LIOTTA, D., MAZZANTI, R., MAZZEI, R., SALVATORINI, G. & SANDRELLI, F. 1993. Rassegna delle conoscenze sulla stratigrafia del Neoautoctono toscano. *Memorie della Società Geologica Italiana*, **49**, 17–98.

BOUCHEZ, J.-L., HUTTON, D.H.W. & STEPHENS, W.E. 1997. Granites: from Segregation of Melts to Emplacement Fabrics, Kluwer, Dordrecht, 3–10.

BOUILLIN, J.-P., BOUCHEZ, J.-L., LESPINASSE, P. & PECHER, A. 1993. Granite emplacement in an extensional setting; an AMS study of the magmatic structures of Monte Capanne (Elba, Italy). *Earth and Planetary Science Letters*, **118**, 263–279.

BRANCA, M. & VOLTAGGIO, M. 1993. Erosion rate in badlands of central Italy: estimation by radio-caesium isotope ratio from Chernobyl nuclear accident. *Applied Geochemistry*, **8**, 347–445.

BROWN, M. 2001. Crustal melting and granite magmatism: key issues. *Physics and Chemistry of the Earth A*, **26**, 201–212.

BRUNET, C., MONIÉ, P., JOLIVET, L. & CADET, J.P. 2000. Migration of compression and extension in the Tyrrhenian Sea, insights from $^{40}Ar/^{39}Ar$ ages on micas along a transect from Corsica to Tuscany. *Tectonophysics*, **321**, 127–155.

CASTRO, A., FERNANDEZ, C. & VIGNERESSE, J.-L. 1999. Understanding granites: integrating new and classical techiniques. *Geological Society, London, Special Publications*, **168**, 278 pp.

CONTICELLI, S., BORTOLOTTI, V., PRINCIPI, G., LAURENZI, M.A., D'ANTONIO, M. & VAGGELLI, G. 2001. Petrology, mineralogy and geochemistry of a mafic dike from the Monte Castello, Elba Island, Italy. *Ofioliti*, **26**, 249–262.

CORRY, C.E. 1988. Laccoliths – mechanics of emplacement and growth. *Geological Society of America, Special Papers*, **220**, 110 pp.

CRUDEN, A.R. & MCCAFFREY, K.J.W. 2001. Growth of plutons by floor subsidence: implications for rates of emplacement, intrusion spacing and melt-extraction mechanisms. *Physics and Chemistry of the Earth, A*, **26**, 303–315.

DANIEL, J.-M. & JOLIVET, L. 1995. Detachment faults and pluton emplacement: Elba Island (Tyrrhenian Sea). *Bulletin de la Société Géologique de France*, **166**, 341–354.

DEINO, A., KELLER, J.V.A., MINELLI, G. & PIALLI, G. 1992. Datazioni $^{39}Ar/^{40}Ar$ del metamorfismo dell'Unità di Ortano-Rio Marina (Isola d'Elba): risultati preliminari. *Studi Geologici Camerti, Volume Speciale 1992/2, CROP 1–1A*, 187–192.

DINI, A., INNOCENTI, F., ROCCHI, S., TONARINI, S. & WESTERMAN, D.S. 2002. The magmatic evolution of the laccolith–pluton–dyke complex of the Elba Island, Italy. *Geological Magazine*, **139**, 257–279.

FLETCHER, P. & GAY, N.C. 1971. Analysis of gravity sliding and orogenic translation. Discussion. *Geological Society of America Bulletin*, **82**, 2677–2682.

GUCWA, P.R. & KEHLE, R.O. 1978. Bearpaw Mountains rockslide, Montana, USA. *In*: VOIGHT, B. (ed.) *Rockslides and Avalanches, Natural phenomena*. Elsevier, Amsterdam. 393–421.

HOGAN, J.P. & GILBERT, M.C. 1995. The A-type Mount Scott Granite sheet: importance of crustal magma traps. *Journal of Geophysical Research*, **B8**, 15 779–15 792.

HOGAN, J.P., PRICE, J.D. & GILBERT, M.C. 1998. Magma traps and driving pressure: consequences for pluton shape and emplacement in an extensional regime. *Journal of Structural Geology*, **20**, 1155–1168.

INNOCENTI, F., WESTERMAN, D.S., ROCCHI, S. & TONARINI, S. 1997. The Montecristo monzogranite (Northern Tyrrhenian Sea, Italy): a collisional pluton in an extensional setting. *Geological Journal*, **32**, 131–151.

JOHNSON, A. & POLLARD, D.D. 1973. Mechanics of growth of some laccolithic intrusions in the Henry Mountains, Utah, I. Field observations, Gilbert's model, physical properties and flow of the magma. *Tectonophysics*, **18**, 261–309.

JOLIVET, L., DANIEL, J.M., TRUFFERT, C. & GOFFÉ, B. 1994. Exhumation of deep crustal metamorphic rocks and crustal extension in arc and back-arc regions. *Lithos*, **33**, 3–30.

JOLIVET, L. *ET AL*. 1998. Midcrustal shear zones in postorogenic extension: Example from the northern Tyrrhenian Sea. *Journal of Geophysical Research*, **103**, 12 123–12 160.

KELLER, J.V.A. & PIALLI, G. 1990. Tectonics of the Island of Elba: a reappraisal. *Bollettino della Società Geologica Italiana*, **109**, 413–425.

MCCAFFREY, K.J.W. & PETFORD, N. 1997. Are granitic intrusions scale invariant? *Journal of the Geological Society of London*, **154**, 1–4.

MAINERI, C., BENVENUTI, M., COSTAGLIOLA, P., DINI, A., LATTANZI, P., RUGGIERI, C. & VILLA, I.M. 2003. Sericitic alteration at the La Crocetta mine (Elba Island, Italy): interplay between magmatism, tectonics, and hydrothermal actvity. *Mineralium Deposita*, **38**, 67–86.

MALINVERNO, A. & RYAN, W.B.F. 1986. Extension in the Tyrrhenian Sea and shortening in the Apennines as results of arc migration driven by sinking of the lithosphere. *Tectonics*, **5**, 227–245.

MARINELLI, G. 1955. Le rocce porfiriche dell'Isola d'Elba. *Atti della Società Toscana di Scienze Naturali, Serie A*, **LXII**, 269–418.

MARINELLI, G. 1959. Le intrusioni terziare dell'isola d'Elba. *Atti della Società Toscana di Scienze Naturali, Serie A*, **LXVI**, 50–253.

MARINELLI, G., BARBERI, F. & CIONI, R. 1993. Sollevamenti neogenici e intrusioni acide della Toscana e del Lazio settentrioale. *Memorie della Società Geologica Italiana*, **49**, 279–288.

MERLE, O. & VENDEVILLE, B. 1995. Experimental modelling of thin-skinned shortening around magmatic intrusions, *Bulletin of Volcanology*, **57**, 33–43.

PERRIN, M. 1975. L'Ile d'Elbe et la limite Alpes–Apennins: données sur la structure géologique et l'évolution tectogénétique de l'Elbe alpine et de l'Elbe apennine. *Bollettino della Società Geologica Italiana*, **94**, 1929–1955.

PERTUSATI, P.C., RAGGI, G., RICCI, C.A., DURANTI, S. & PALMERI, R. 1993. Evoluzione post-collisionale dell'Elba centro-orientale. *Memorie della Società Geologica Italiana*, **49**, 297–312.

PETFORD, N., CRUDEN, A.R., MCCAFFREY, K.J.W. & VIGNERESSE, J.-L. 2000. Granite magma formation, transport and emplacement in the Earth's crust. *Nature*, **408**, 669–673.

POLI, G. 1992. Geochemistry of Tuscan Archipelago granitoids, central Italy: the role of hybridization processes in their genesis. *Journal of Geology*, **100**, 41–56.

REEVES, F. 1925. Shallow folding and faulting around the Bearpaw Mountains. *American Journal of Science*, **10**, 187–200.

ROCCHI, S., WESTERMAN, D.S., DINI, A., INNOCENTI, F. & TONARINI, S. 2002. Two-stage laccolith growth at Elba Island (Italy). *Geology*, **30**, 983–986.

ROMAN-BERDIEL, T. 1999. Geometry of granite

emplacement in the upper crust: contributions of analogue modelling. *In*: CASTRO, A., FERNANDEZ, C. & VIGNERESSE, J.-L. (eds) *Understanding Granites: Integrating New and Classical Techniques*. Geological Society, London, Special Publications, **158**, 77–94.

ROMAN-BERDIEL, T., GAPAIS, D. & BRUN, J.P. 1995. Analogue models of laccolith formation. *Journal of Structural Geology*, **17(9)**, 1337–1346.

SERRI, G., INNOCENTI, F. & MANETTI, P. 1993. Geochemical and petrological evidence of the subduction of delaminated Adriatic continental lithosphere in the genesis of the Neogene–Quaternary magmatism of central Italy. *Tectonophysics*, **223**, 117–147.

SPEAR, F.S. & CHENEY, J.T. 1989. A petrogenetic grid for pelitic schists in the system SiO_2–Al_2O_3–FeO–MgO–K_2O–H_2O. *Contributions to Mineralogy and Petrology*, **101**, 149–164.

STOCKLI, D.F., LINN, J.K., WALKER, J.D. & DUMITRU, D.A. 2001. Miocene unroofing of the Canyon Range during extension along the Sevier Desert Detachment, west central Utah. *Tectonics*, **20**, 289–307.

TESTA, G. & LUGLI. 2000. Gypsum–anhydrite transformation in Messinian evaporites of central Tuscany (Italy). *Sedimentary Geology*, **130**, 249–268.

THOMPSON, A.B. 1974. Calculation of muscovite–paragonite–alkali feldspar phase relations. *Contributions to Mineralogy and Petrology*, **44**, 173–194.

TRACY, R.J. & FROST, B.R. 1991. Phase equilibria and thermobarometry of calcareous, ultramafic and calcareous rocks, and iron formations. *In*: KERRICK, D.M. (ed.) *Contact Metamorphism: Reviews in Mineralogy*. Mineralogical Society of America, 207–289.

TREVISAN, L. 1950. L'Elba orientale e la sua tettonica di scivolamento per gravità. *Memorie dell'Istituto di Geologia dell'Università di Padova*, **16**, 1–30.

VOM RATH, G. 1870. Die Insel Elba. *Zeitschrift der Deutschen Geologischen Gesellschaft*, **XXII**, 591–732.

WESTERMAN, D.S., INNOCENTI, F., TONARINI, S. & FERRARA, G. 1993. The Pliocene intrusions of the Island of Giglio (Tuscany). *Memorie della Società Geologica Italiana*, **49**, 345–363.

WIEBE, R.A. & COLLINS, W.J. 1998. Depositional features and stratigraphic sections in granitic plutons: implications for the emplacement and crystallization of granitic magma. *Journal of Structural Geology*, **20**, 1273–1279.

Formation of saucer-shaped sills

A. MALTHE-SØRENSSEN[1], S. PLANKE[2], H. SVENSEN[1] & B. JAMTVEIT[1]

[1]*Physics of Geological Processes, Department of Physics, University of Oslo, Box 1048 Blindern, N-0316 Oslo, Norway (e-mail: malthe@fys.uio.no)*
[2]*Volcanic Basin Petroleum Research AS, Forskningsparken, Gaustadalleen 21, N-0349 Oslo, Norway*

Abstract: We have developed a coupled model for sill emplacement in sedimentary basins. The intruded sedimentary strata are approximated as an elastic material modelled using a discrete element method. A non-viscous fluid is used to approximate the intruding magmatic sill. The model has been used to study quasi-static sill emplacement in simple basin geometries. The simulations show that saucer-shaped sill complexes are formed in the simplest basin configurations defined as having homogeneous infill and initial isotropic stress conditions. Anisotropic stress fields are formed around the sill tips during the emplacement due to uplift of the overburden. The introduction of this stress asymmetry leads to the formation of transgressive sill segments when the length of the horizontal segment exceeds two to three times the overburden thickness. New field and seismic observations corroborate the results obtained from the modelling. Recent fieldwork in undeformed parts of the Karoo Basin, South Africa, shows that saucer-shaped sills are common in the middle and upper parts of the basin. Similar saucer shaped sill complexes are also mapped on new two- and three-dimensional seismic data offshore of Mid-Norway and on the NW Australian shelf, whereas planar and segmented sheet intrusions are more common in structured and deep basin provinces.

The emplacement of magmatic sills in sedimentary basins has major implications on the development and structure of the basins. The introduction of hot melt into the sedimentary sequence will cause heating, expulsion of pore fluids, and associated metamorphic reactions. The solidified and cold sills will subsequently influence the basin rheology, strength, and permeability structure. These magmatic processes will also influence the petroleum prospectivity through enhanced maturation, and formation of migration pathways (Schutter 2003).

Sheet intrusions, such as dykes, sills, and laccoliths, are important for rapid magma transport in the Earth's crust (e.g. Rubin 1995). Dykes are near-vertical magma-filled fractures driven primarily by the buoyancy of hot magma. Sills are dominantly layer-parallel sheets with transgressive segments. An impressive Jurassic sill complex is found in the Karoo Basin, South Africa, including extensive saucer-shaped sills (e.g. Du Toit 1920; Bradley 1965; Chevallier & Woodford 1999). A review of sill geometries and emplacement mechanisms is given by Francis (1982). Laccoliths are layer-parallel intrusions with a flat bottom and a dome-shaped top. The term was introduced by Gilbert (1877) based on studies in the Henry Mountains, Utah. Laccoliths in the Henry Mountains and elsewhere in the North America have subsequently been studied by, for example, Johnson & Pollard (1973) and Corry (1988). Sill–sediment boundaries represent high-impedance contrasts, and sill complexes are therefore well imaged on seismic reflection data. Recently, a number of studies on the 2D and 3D geometries of sill complexes have been made, in particular along the volcanic margins of the NE Atlantic (Skogseid *et al.* 1992; Berndt *et al.* 2000; Smallwood & Maresh 2002).

The wide range of sill complex geometries found in the field and from seismic data has generally not been addressed in theoretical studies, which have concentrated on the 2D shape of single dykes or sills, mostly in two dimensions (Pollard & Johnson 1973; Lister & Kerr 1991). The main focus of this work has been on the flow of viscous magma and the interplay between the viscous fluid flow and fracturing. In particular, various models on the effects of viscous pressure-drop along the sill and the fracturing of fluid-filled cracks have been discussed (Lister & Kerr 1991). The limitations of linear elastic fracture mechanics in studies of the propagation of an individual dykes or sills have also been addressed (Rubin 1993; Khazan & Fialko 1995).

Various authors have addressed mechanisms responsible for the dyke to sill transition, but with few quantitative modelling results to

From: BREITKREUZ, C. & PETFORD, N. (eds) 2004. *Physical Geology of High-Level Magmatic Systems.*
Geological Society, London, Special Publications, **234**, 215–227. 0305-8719/04/$15.00
© The Geological Society of London 2004.

support the ideas. The studies of laccolith and sill intrusions in the Henry Mountains (Johnson & Pollard 1973; Pollard & Johnson 1973) showed how linear elasticity theory could be applied to study the shape and development of sheet intrusions. Further theoretical studies of the interplay between structure, stress heterogeneities, and dyke and sill shapes (Pollard *et al.* 1975) have further shown that linear elasticity theory can bring important insights into dyke and sill emplacement without a full treatment of the effects of viscous magma or the detailed processes near the propagating dyke tip.

However, the advances in the study of flow in a single dyke have little significance for understanding the complex interaction between the intruding sill and the surrounding sediments, interactions between different sills, and the internal dynamics during emplacement. Effects such as magma-tectonism and inflation–deflation cycles have not previously been addressed by theoretical fluid-mechanical models.

In this paper, we focus on the interactions between sill emplacement and sediment deformation. We have developed a numerical model that can be used to study complexities of sill emplacement in heterogeneous materials. Our main emphasis here is to find the simplest possible model that contains only the essential physical processes needed to reproduce the most important phenomena observed without describing the whole process in detail.

Sill emplacement modelling

For dykes propagating upwards through the crust, the main driving forces are buoyancy, due to density difference between magma and host rock, and magma overpressure in the reservoir (Lister & Kerr 1991). It has been suggested that the injection of sills from dykes are driven by a pressure head due to the overshooting of dykes (Francis 1982), or that sills are fed laterally by dykes (Chevallier & Woodford 1999). In this article we will focus on the sill emplacement mechanism without considering how magma is supplied in detail. This is a reasonable assumption when studying several aspects of sill emplacement, such as the final sill geometry and tectonic effects of sill emplacment, in particular because these have proved very difficult if not impossible to infer the injection point of a sill complex based on geometries found in seismic data, or from field data. The focus is on providing the simplest possible model that contains the necessary mechanisms to provide an explanation for particular, observed characteristics of sill emplacement.

The sill emplacement is a coupled process where the flowing magma deforms and affects the surrounding matrix, and the deformation and fracturing of the matrix affects the flow of magma. A model of sill emplacement must take into account this fully coupled process: the magma should be represented by a fluid moving in a geometry determined by the deformation of the surrounding matrix, which will deform, and possibly change its material properties, due to the magma emplacement process.

We model this coupled process by representing the matrix as an elastic material and the magma as a fluid. The assumption of elasticity is reasonable at large length-scales and at shallow depths, that is at depths less than 5 km (Atkinson 1984). The elastic material can be represented by a discrete element model. Such models are well suited to study coupled processes, such as fracture-enhanced transport in eclogitization processes (Jamtveit *et al.* 2000) and hydro-fracturing processes (Flekkøy *et al.* 2002). Discrete element models also provide a simple, physical framework for studying dynamic fracture processes in which the surface geometry changes continuously during deformation. Discrete element models are also equally well suited for quantitative calculations, as other discretization approaches to continuum elasticity theory (Monette & Anderson 1994).

We initially make the assumption that the magma can be modelled as a non-viscous fluid with a constant pressure. This is a reasonable assumption for a slow deformation process, in which the viscous forces are small, and the rate of deformation due to magma injection is small compared to the time-scale of elastic equilibration in the host rock. It is also a reasonable assumption for a solidifying magma when the flow velocities are very small. The use of non-viscous fluid to model the magma implies that there is no pressure drop from the inlet along the length of the sill, and there are no shear forces on the boundaries. However, the actual thickness of a sill will depend on the viscosity, and therefore also on the rate of injection and the rate of solidification. Viscous effects are also important to understand the rate of emplacement, but may not be similarly important for modelling the shape of the sill or the deformation of the host rock during the emplacement. The effects of viscosity can, however, be studied in detail when the location of the injection point is known, and viscous effects can be approximated by a gradual change in the pressure in the sill away from the inlet.

Discrete element modelling

The sedimentary basin is modelled as a linear elastic material using a discrete element model

(DEM). The DEM model is a discrete realization of the linear elastic continuum formulation of the problem (Flekkøy et al. 2002). However, the DEM model deviates from the continuum formulation in the choice of boundary conditions. In the DEM approach, the elastic material is modelled as a network of elements – spheres – connected by elastic bonds in the form of springs or beams.

We have demonstrated that the equilibrium configuration of a triangular lattice of linear springs is equivalent to the equilibrium solution to the linear elasticity problem for an isotropic material (Flekkøy et al. 2002). Thus, as a first-order approach we have used a triangular lattice with springs to model the elastic sediments. The forces acting on a particle can be decomposed in an interaction force, f_i^n, due to inter-particle interactions, and interaction with boundaries, gravity, f_i^g, and external forces, f_i^e, such as the force transfer from the magma:

$$f_i^n = f_i^n + f_i^g + f_i^e. \quad (1)$$

For a network of springs, the force due to inter-particle interactions is:

$$f_i^n = \sum_j k_{i,j}\left(\left(|x_i - x_j|\right) - l_{i,j}\right) u_{i,j} \quad (2)$$

where the sum is over all connected neighbours j, x_j is the position of node j, $l_{i,j}$ is the equilibrium length for the connection between nodes i and j, and $u_{i,j}$ is a unit vector pointing from the centre of node i to the centre of node j.

The spring constant $k_{i,j}$ may vary locally for a heterogeneous material. This corresponds to local variations in the Young's modulus for the material. Typically, we assume that the material has homogeneous elastic properties, that is, $k_{i,j} = k$. However, the method presented here can easily be extended to model heterogeneous elastic properties.

The gravitational term f_i^g describes the gravitational force on the particle. For a particle of mass density ρ_i, the gravitational force in a two-dimensional system is:

$$f_i^g = \rho_i \pi r_i^2 w_z g u_y \quad (3)$$

where r_i is the radius of particle i, and w_z is the thickness in the third dimension. The mass density for the particle is related to the mass density of the elastic material ρ_m through the porosity ϕ of the packing: $\rho_m(1 - \phi) = \rho_i$.

The external force f_i^e is due to the coupling of the elastic material to the magma flow. For the case of a non-viscous magma, the only coupling force is the pressure gradient, given by the pressure (P) in the magma. However, the method presented here is easily extended to include both normal and shear forces in viscous magma. The pressure force acts in the direction of the local surface normal, which is calculated from the local configuration of elements as illustrated in Figure 1. This results in a force:

$$f_i^e = PA_i u_i^P \quad (4)$$

where P is the local pressure in the magma, A_i represents the local surface area, and u_i^P is the local surface normal to the sill–matrix interface.

The magma pressure P includes the effect of hydrostatic pressure in the magma with a given magma density ρ_s, $P = P_0 + \rho_s g(h - y)$, where h and P_0 corresponds to the level of injection.

For a system with small displacements, that is, a system which is at the linear elastic limit, we need only consider deviations from the lithostatic pressure. In this model, the effect of gravity is included in a lithostatic term for the stresses, and the hydrostatic pressure term in the magma pressure depends only on the difference in densities: $P = P_0 + \Delta\rho g(h - y)$. This method is used for the calculations, and all illustrations display the calculated stresses minus the lithostatic pressure. However, the gravitational term will be included in the equations in order to illustrate its role in non-isotropic and non-homogeneous systems.

The equations should be non-dimensionalized in order to show how the simulation parameters relate to real-world parameters. We introduce the non-dimensional parameters: $x = x'l$ and $k_{i,j} = k'_{i,j}k$, where l is the length corresponding to a diameter of the simulation particle. The spring constant normalization factor k can be related to the Young's modulus of the material for a triangular lattice of nodes (Flekkøy et al. 2002):

$$k = \frac{\sqrt{3}}{2} E w_z \quad (5)$$

where $E = E_0 2/\sqrt{3}$ is the Young's modulus of the material, and w_z corresponds to the thickness of the two-dimensional sample used in the simulation. It can be demonstrated that the value of w_z does not have any significance for simulation parameters or stresses, and is only needed for dimensional consistency. The surface area is rescaled using both l and w_z: $A_i = w_z l A'_i$.

The resulting equation for the force on particle is thus:

$$f_i = E_0 w l \left\{ \sum_j k'_{i,j}\left(\left|\vec{x}'_i - \vec{x}'_j\right| - l'_{i,j}\right)\vec{u}_{i,j} \quad (6)\right.$$

$$\left. - \frac{\pi \rho \lg}{E_0}(r'_i)^2 \frac{\rho_i}{\rho}\vec{u}_z + \frac{P}{E_0} A'_i \vec{u}_i^P \right\} \quad (7)$$

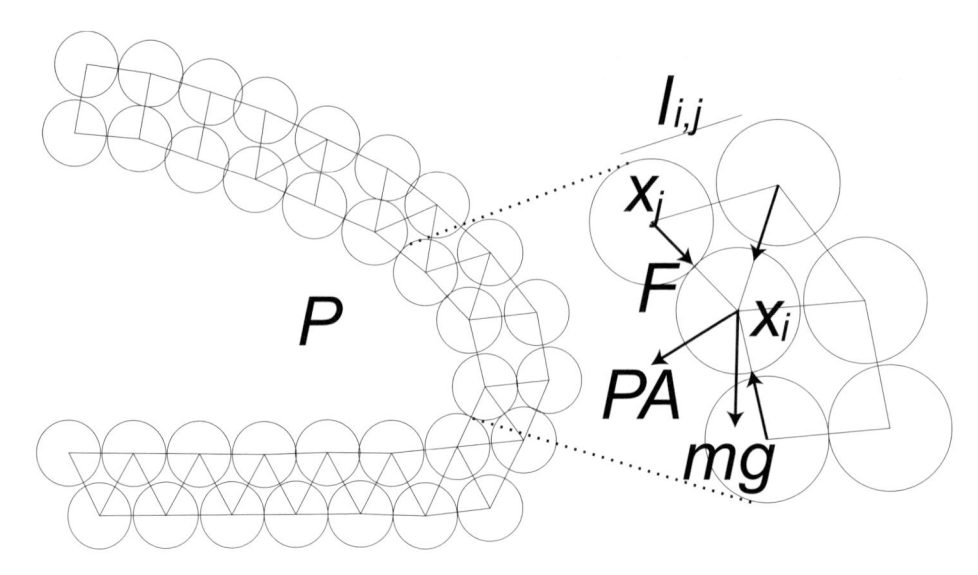

Fig. 1. Illustration of coupled DEM model for sill emplacement studies. The elastic material is modelled by a network of interconnected springs drawn as lines between the circular particles. The springs are initially placed on a triangular network. The sill is represented as a moving fracture, illustrated as the region without springs. A pressure P acts from the magma in the sill on the surrounding particles. The illustration on the right illustrates the forces acting on a particle i. The forces are: contact forces due to connections to other particles, calculated as a spring force proportional to the elongation of the springs; a gravitational force mg, and a pressure force PA_i, where A_i is the local surface area of the sill-magma boundary associated with particle i. For a given pressure P, the equilibrium configuration of the elastic network is found.

Simulations are performed in the non-dimensional co-ordinate system, and then scaled back to real-world quantities using the scaling relations. Stresses (σ'_{ij}) measured in simulation co-ordinates are rescaled to real co-ordinates according to $\sigma_{ij} = E_0 \sigma'_{ij}$.

The equilibrium configuration for the system can be found by standard relaxation techniques. We have used a specially tailored method based on successive over-relaxations (Allen 1954). This method uses extra relaxations close to fracture tips, and is particularly efficient for fracturing systems, where most of the deformation occurs near fractures and fracture tips.

The system is constrained by elastic boundaries in the form of elastic walls. The wall is modelled as a linear spring, so that the force on a particle due to the wall interaction is proportional to the distance from the particle to the wall. For a particle in contact with a wall parallel to the x-axis at $y = y_w$, the force from the wall on particle i is:

$$f_i = \begin{cases} k_w(y_i + r_i - y_w)\boldsymbol{u}_y & y_i + r_i - y_w > 0 \\ 0 & y_i + r_i - y_w \leq 0 \end{cases} \quad (8)$$

where r_i is the radius of particle i, and $\boldsymbol{u}_y$ is a unit vector along the y-axis, and k_w is the spring constant for the wall interaction.

As fracturing is not accounted for in linear elasticity theory, we have simulated its effects through the irreversible removal of a bond if the stress in the bond exceeds a maximum threshold value (σ^c), corresponding to the material strength. The material strength is assumed to be homogeneous on large scales, but to fluctuate around average values on the scale of individual particles. The variations correspond to variations in micro-crack densities and lengths, and other defects and variations which are always present in disordered materials. This local heterogeneity is described by a distribution of breaking thresholds, which corresponds to a distribution of the tensile strength of the material at a particular scale given by the scale of the elements. It can be demonstrated (Walmann et al. 1996; Malthe-Sørenssen et al. 1998a, b) that a normal distribution of breaking thresholds in this model reproduces fracture patterns observed in laboratory experiments and field studies. The average value of the breaking threshold may be related to the stress intensity factor (Flekkøy et al. 2002). The material behaviour can be tuned from a brittle material for narrow distribution of breaking threshold, to a more ductile behaviour for a wider distribution. We have extensive experience from simulating geological systems using this model, and this experience was used in selecting realistic distributions of breaking thresholds. Typical values for the breaking threshold correspond to breaking strains $\varepsilon^c = 2 \times 10^{-3}$ to 2×10^{-2} (Fyfe et al. 1978). A value corresponding to a strain of 2×10^{-3} has been used in the simulations presented here.

The simulation procedure corresponds to a quasi-static driving mechanism of sill emplacement. Initially, the breaking strength of a short line is reduced by a factor of 10 in order to initialize sill emplacement in the horizontal direction. Sill injection is simulated by gradually increasing the pressure P_0 until a fracture occurs in the surrounding matrix. A bond is removed, and the fluid is allowed to fill the new volume. This reduces the pressure in the fluid according to linear relationship between change in fluid volume and pressure, with a magma compressibility corresponding to the elastic stiffness of the surrounding matrix. This computational scheme ensures that the sill propagates in small steps, and not as a runaway brittle fracture. A new equilibrium configuration is found for the elastic material for the pressure P_0. The procedure is then repeated by slowly increasing the pressure. Snapshots of the sill geometry and the stresses around it illustrate the growth of the sill and the emplacement process.

Modelling results

The coupled model is used to study the mechanisms for the formation of saucer-shaped sills in an undeformed, homogeneous basin. The basin is represented as an initially homogeneous material with an isotropic lithostatic stress. The sill was emplaced at a depth h corresponding to the level of neutral buoyancy for the magma. The width of the simulated region of the basin was 40 km, and typical values for the depth h were 1 to 5 km. Other material parameters were selected within the range of typical properties of a sedimentary basin: $E = 2$ GPa, $\Delta\rho = 0.25$ Mg m^{-3}, $\varepsilon^c = 2 \times 10^{-3}$ (Atkinson 1984).

The developing geometry of the sill is illustrated in Figure 2 for a sill emplaced at at depth $h = 1.8$ km. Initially, the sill extends linearly, retaining its original orientation. In this regime, illustrated in Figure 2a, the sill is short compared to the thickness of the overburden. The shape of the sill and the stress field extending from the tips of the sill is approximately symmetrical, and the sill emplacement is not affected by the asymmetrical boundary conditions.

When the sill reaches a length of approximately the thickness of the overburden, the shape of the sill starts to become asymmetrical. This is illustrated by a plot of the position of the upper and lower sill surfaces shown in Figure 3. As a consequence, the stress in front of the sill tips becomes asymmetrical. Figure 2b illustrates the stress field immediately before the sill starts bending upwards.

The sill branches upwards due to the asymmetry of the stress field caused by the uplift of the overburden. This is illustrated in Figure 2c. Initially, the sill propagates steeply upwards; however, the ascent is limited by the reduction in pressure inside the sill, due to extension above its level of neutral buoyancy. Further development of the sill is shown in Figure 2d. We observe that the sill starts to develop an outer sill with lower inclination angle as the intrusion climbs further.

The formation of the saucer-shaped sill in the initially homogeneous basin can therefore be explained as an effect of an asymmetrical stress field generated by the sill intrusion itself and by the elastic interaction of the sill intrusion with the surrounding matrix. We have performed a series of simulations at various emplacement depths. A plot of the width of the horizontal part of the sill, w, as a function of the intrusion depth, h, is shown in Figure 4. The extent of the flat region of the sill, w, increases with depth due to increasing overburden. This observation is consistent with the cross-over from sill- to laccolith-like behaviour discussed by Johnson & Pollard (1973).

Discussion

Saucer-shaped sill complexes are frequently observed in unstructured sedimentary basins. The saucer-shaped sheet intrusions generally have a flat, circular central region, terminated at climbing sheet segments that cross-cut the sediments. Several saucers can be emplaced close to each other, interpenetrating and generating a complex interconnected pattern of sills.

Saucer-shaped sills are one of the most common structures observed in Karoo Basin, South Africa. This largely undeformed foreland basin is infilled with clastic sediments, which accumulated from the Late Carboniferous (310 Ma) (Vissner 1997) through the Mid-Jurassic (185 Ma) (Catuneaunu 1998). Voluminous extrusive and intrusive complexes were emplaced in the Karoo Basin at c. 183 Ma (Duncan et al. 1997).

The saucer-shaped sills control to a large extent the geomorphology of the landscape (Du Toit 1920; Chevallier & Woodford 1999). Figure 5a shows a sketch of the typical morphology of a ring structure. The Golden Valley (Fig. 5b) near

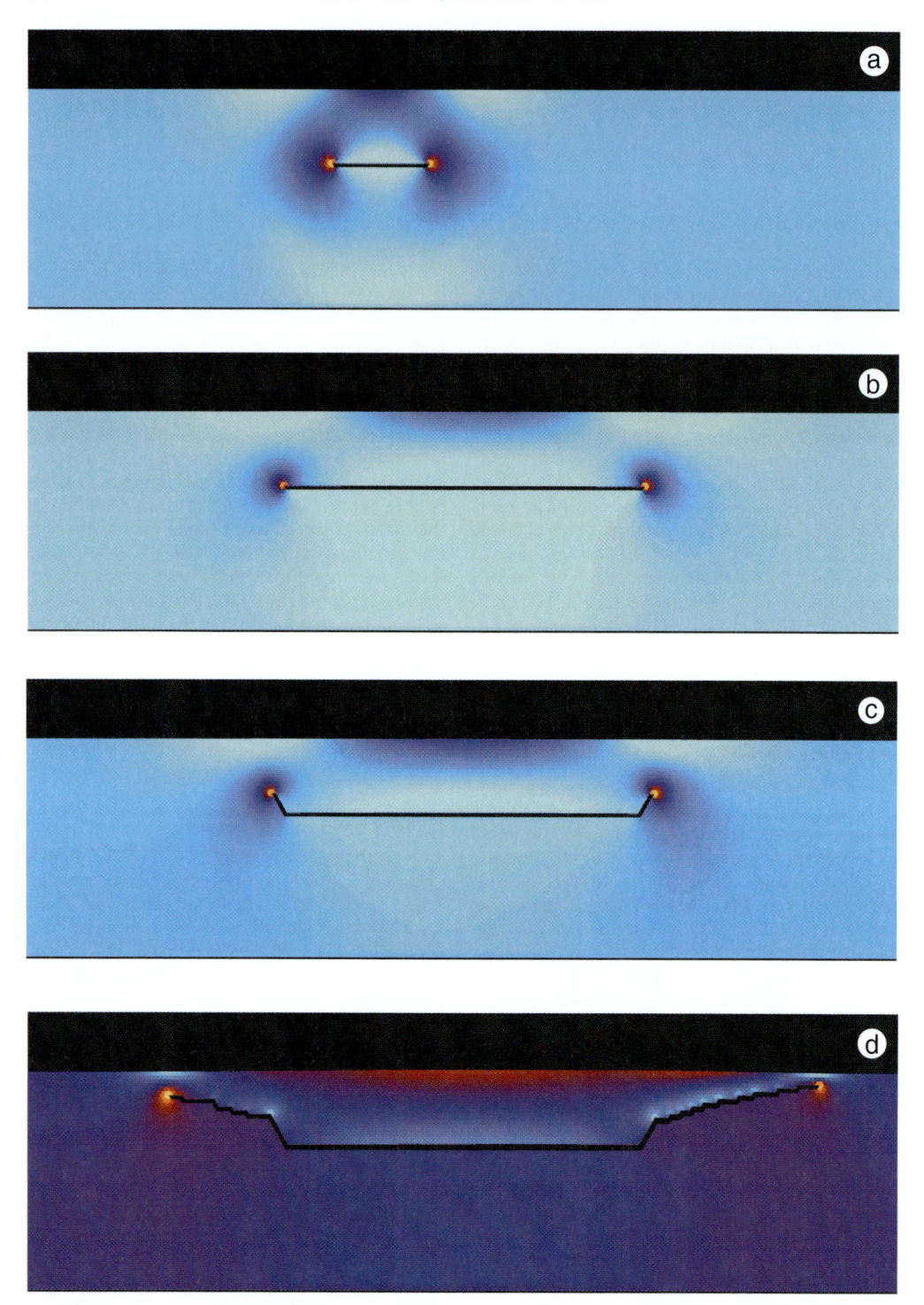

Fig. 2. Snapshots from a simulation of sill emplacement in a 40-km wide model. Only the central 20 km of the model is shown in the pictures. Figures (a) to (d) are snapshots showing various stages in the same simulation. The sill was emplaced at a depth of 1.8 km. The sill initially propagates horizontally, ((b) and (a)) but deflects upwards as the effect of the uplift of the overburden becomes important ((c) and (d)).

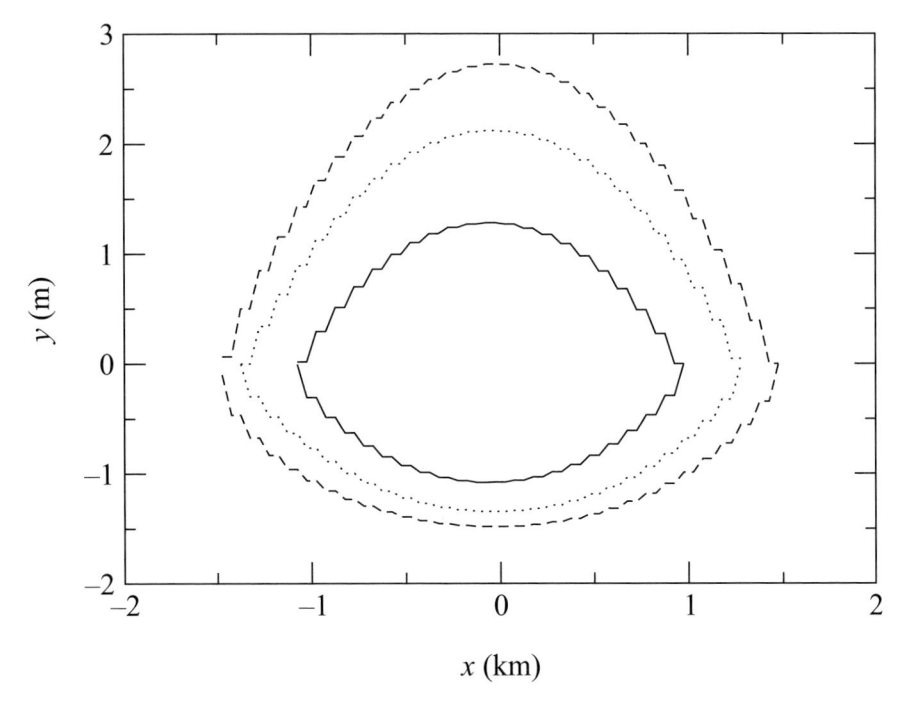

Fig. 3. The vertical position, y, of the top and bottom surfaces of the sill as a function of the spatial position along the sill, x, for several stages in the development of the sill. The shape of the sill becomes asymmetrical when it reaches a length approximately equal to the thickness of the overburden (1.8 km).

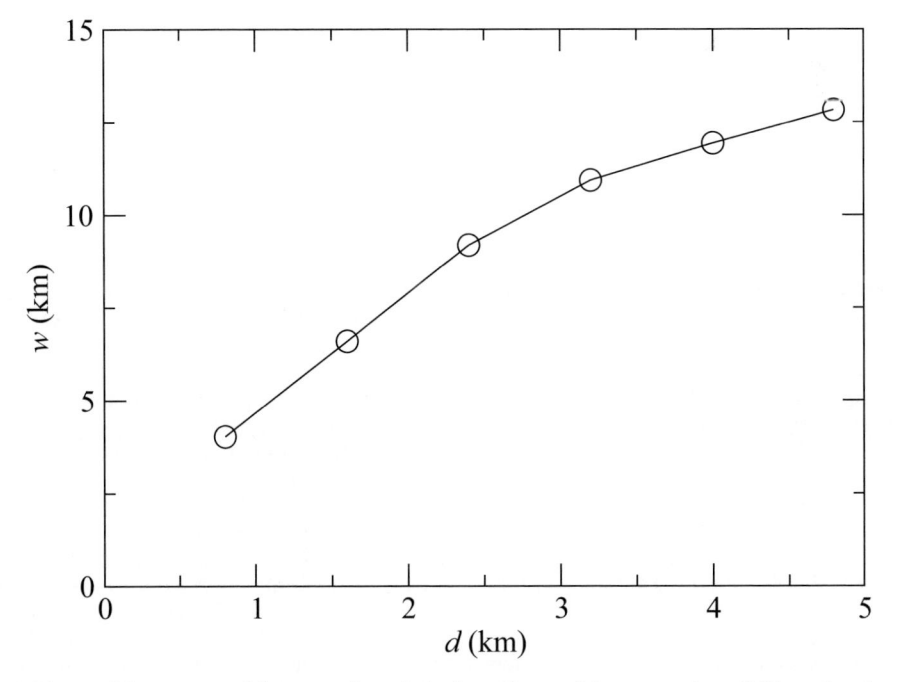

Fig. 4. The spatial extent, w, of the approximately horizontal part of the saucer-shaped sill as a function of emplacement depth, h, for simulations in a 40-km wide basin.

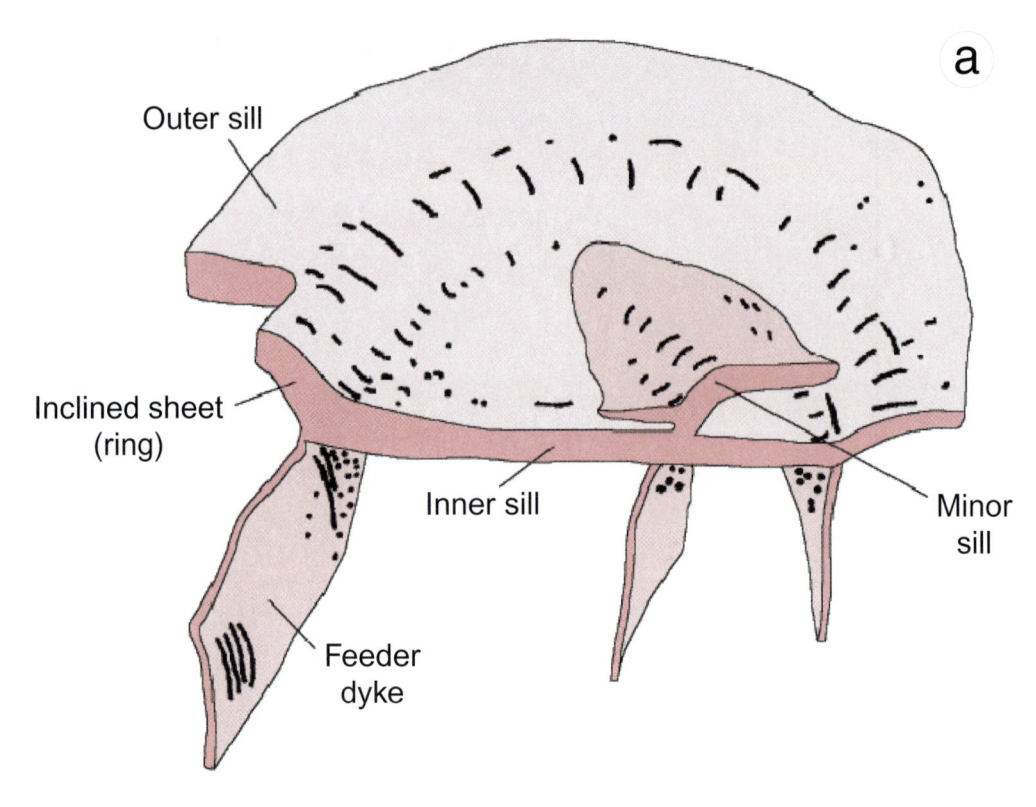

Fig. 5. (a) Sketch of the morphology of saucer-shaped sills and ring complexes in the Karoo Basin (after Chevallier & Woodford 1999). (b) Mosaic of aerial photographs of the Golden Valley sill. The Golden Valley is a classic saucer-shaped intrusion with a flat-lying central part and steep flanks. The saucer is 19 km by 10 km, with up to 350 m of elevation difference. (c) A photo of the Golden Valley sill seen from the outer southern part of the valley. Note that the inclined segments are formed by transgressive dolerites, and that the sedimentary sequence is approximately horizontal.

Queenstown is a large erosional basin defined by a saucer-shaped dolerite intrusion. The inclined rims are exposed as topographical highs that form an elliptical outcrop easily recognized on maps and aerial photographs. On the ground, the transgressive segments of the sill can clearly be seen to cross-cut the dominantly horizontally layered sedimentary strata (Fig. 5c). Similar saucer-shaped sill complexes are found regionally in the upper and middle parts of the Karoo Basin, whereas the sills in the lowermost sequences are dominantly subhorizontal.

Saucer-shaped sill intrusions are also commonly observed on volcanic rifted margins. Continental breakup between Greenland and Eurasia near the Palaeocene–Eocene transition was associated with voluminous extrusive and intrusive volcanism (Berndt et al. 2000). We have recently mapped the distribution of intrusions on more than 150 000 km of 2D seismic profiles and on 3D seismic data in the Vøring and Møre basins. The style of sill complexes clearly varies, with saucer-shaped intrusions dominating the undeformed basin segments.

The Gleipne Saddle in the outer Vøring Basin is located in a basin province with little pre- and post-magmatic structuring. Figure 6 shows a 3D image of a saucer-shaped high-amplitude seismic reflection interpreted as a sill intrusion. The size and geometry of this event is very similar to the geometry of the Golden Valley sill in the Karoo Basin (Fig. 5).

Saucer-shaped sills are also observed on the NW Australian shelf, e.g. on the Exmouth Plateau (Fig. 7). Here, the magmatism is associated with continental breakup between Australia and Greater India during the Valanginian (Symonds et al. 1998). A saucer-shaped sill intrusion can clearly be seen cross-cutting Triassic strata, locally causing deformation of the

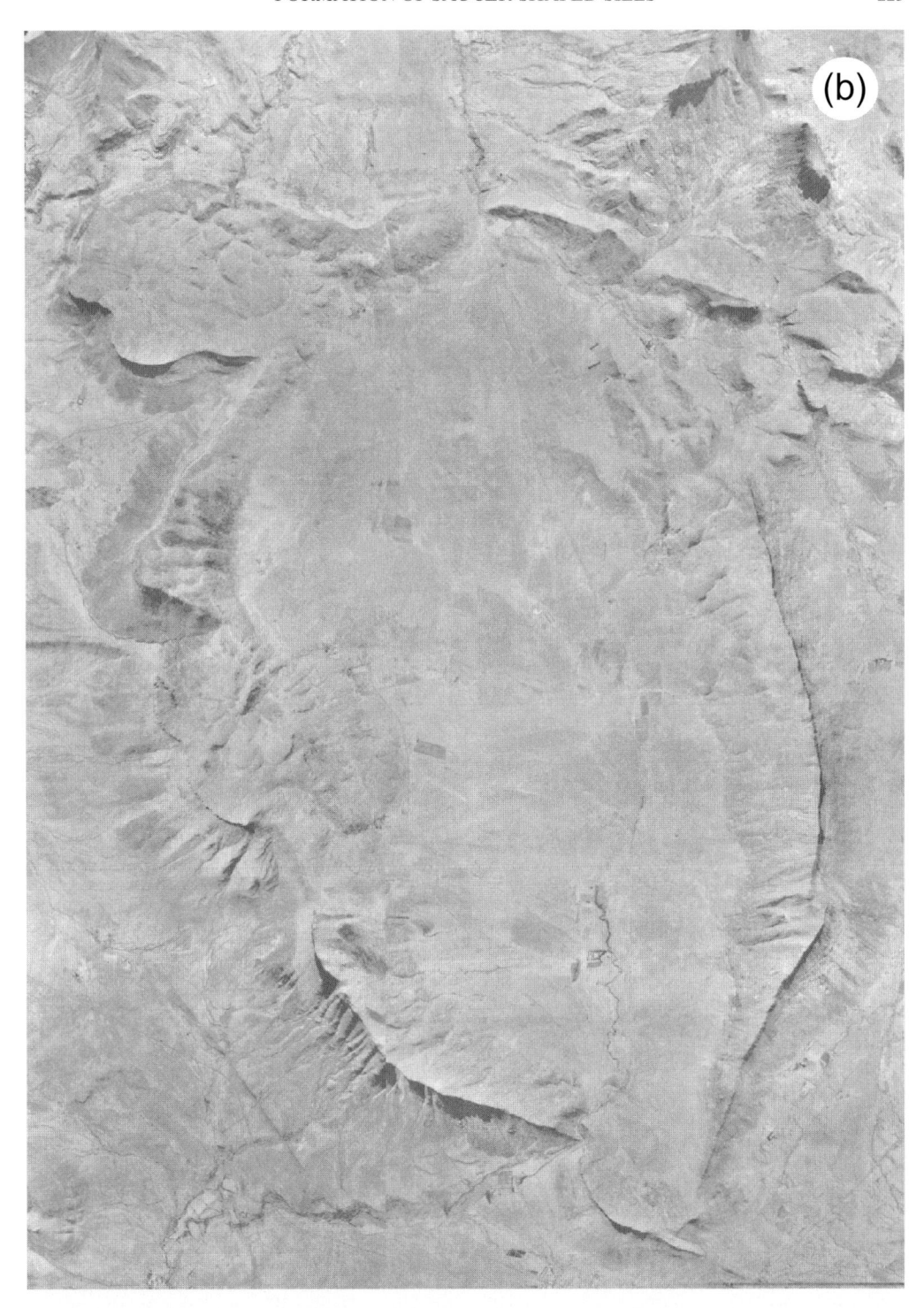

Fig. 5. *continued.*

Fig. 5. *continued.*

overburden sediments (magma-tectonism) (Fig. 7).

Field and seismic data from mainly undeformed volcanic basin segments suggest that the saucer-shaped geometry is a fundamental shape for sheet intrusions. The coupled model illustrates a physically plausible mechanism for the formation of saucer-shaped sills in homogeneous basins. Previously, the idea that the uplift of the overburden could lead to the formation of dykes leading upwards from the sill tips has been discussed and illustrated by laboratory experiments (Pollard & Johnson 1973). An alternative explanation is that the sill follows the surface of neutral buoyancy, and that this surface is shaped as a saucer (Bradley 1965). While saucer-shaped isostress surfaces may exist in some basin geometries, and may indeed be quite common on larger scales in saucer-shaped basins, the explanation of Bradley (1965) is not relevant for the formation of saucer-shaped sills in unstructured basins, because this explanation relies on the sill following a pre-existing structure in the form of a saucer-shaped density profile, which, in general, would not be present in unstructured basins.

Our model demonstrates in a self-consistent framework that sufficiently large sills emplaced into consolidated, elastic materials, should develop into saucer-shaped geometries. Shallow sills intruding into unconsolidated sediments will not produce the long-range stress fields necessary for the formation of saucers, and we therefore expect sills to be disordered and more similar to lava flows at this scale. The model also explains why deeper sills are mainly layer-parallel: these sills are not long enough to start curving upwards (Fig. 4).

While the model is strictly two dimensional, the fundamental mechanism presented here is equally valid in three dimensions. Boundary effects play a more important role in two-dimensional elasticity, because perturbations decay logarithmically in two dimensions. Also, the effective elastic stiffness of the overburden in the two-dimensional model is expected to be significantly lower than in a cylindrical symmetrical three-dimensional model. The functional

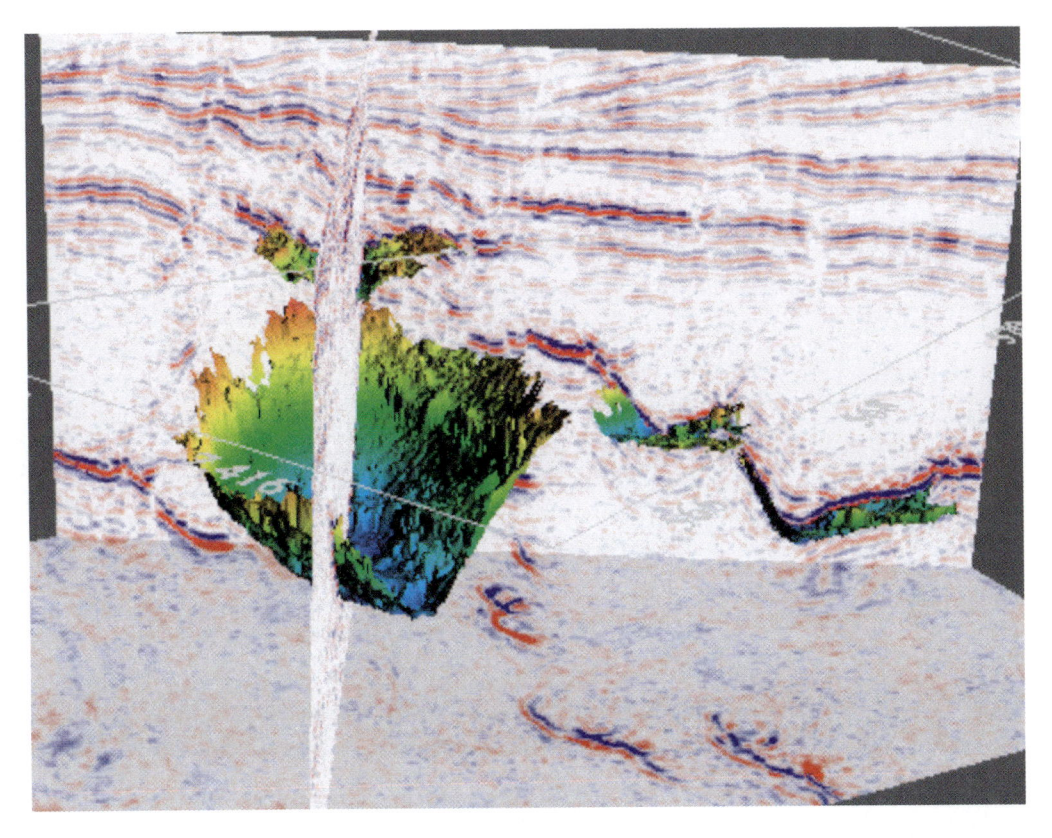

Fig. 6. 3D visualization of a saucer-shaped high-amplitude reflection on the Gleipne Saddle, Outer Vøring Basin. The event is interpreted as a sill reflection, with a geometry and size similar to the Golden Valley sill. The high-amplitude event is auto-tracked and plotted using the VoxelVision software. The main saucer is 3 km by 4 km, with approximately 500 m of elevation difference.

form of the relation between saucer extent and emplacement depth therefore cannot be compared directly with field and seismic data. However, the qualitative dependence is expected to be correctly represented.

Conclusions

We have developed a new, coupled numerical model for emplacement of magma in sedimentary basins. The host rock is modelled as an elastic material using a discrete element method where fracturing is accompanied by removal of elements. The magma is modelled as a non-viscous fluid.

Numerical modelling of sill emplacement in the simplest scenario – non-viscous magma emplaced in an undeformed, homogeneous basin – demonstrates that saucer-shaped sills are formed spontaneously and represent a fundamental shape. The saucer-shaped structure is generated due to interactions between the sill and the overburden. The emplacement process introduces stress heterogeneities in the basin, which subsequently influence the geometry of the emplaced sill.

The transgressive nature of sill emplacement is a fundamental effect that should be taken into consideration when studying sill emplacement. The saucer shape is a fundamental shape in shallow sills: our model indicates that sills will be flat unless their extent exceeds the depth by a factor of two to three, depending on details of the material properties of the overburden.

There is good agreement between the modelling results and observational data from seismic interpretation of the Vøring and Møre basins, offshore Mid-Norway, and on the NW Australian Shelf, and field studies in the Karoo Basin, South Africa. The model provides a unifying explanation for the observation of saucer-shaped intrusions in shallow to intermediate

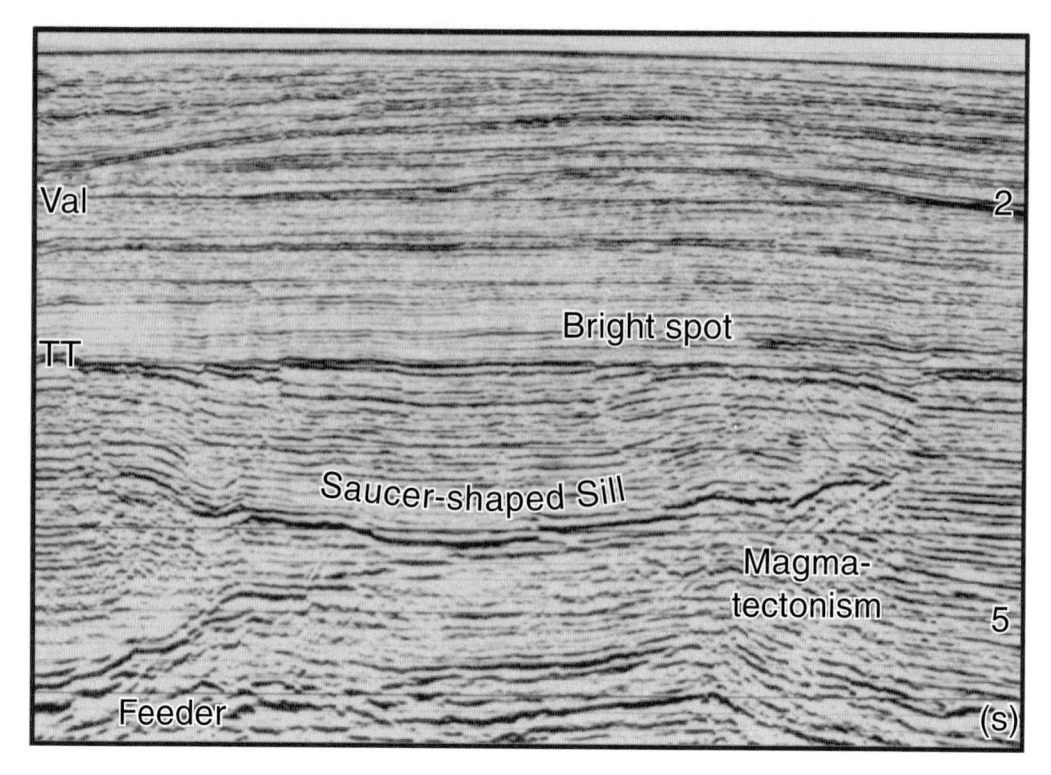

Fig. 7. Saucer-shaped high-amplitude event interpreted as a magmatic sill cross-cutting Triassic sedimentary strata on the Exmouth Plateau, Australian NW Shelf. Modified from Symmond et al. (1998). The width of the saucer is 27 km.

emplacement depths, and provides a framework in which more complicated sill emplacement geometries and more complicated emplacement dynamics can be studied.

We gratefully acknowledge access to seismic data from TGS–NOPEC, and financial support from the sponsors of the 'Petroleum Implications of Sill Intrusion' project, and the Norwegian Research Council. L. Chevallier, Council of Geoscience, South Africa, and G. Marsh, Rhodes University, South Africa have provided valuable information and fieldwork guidance. The VoxelVision interpretation was done by S. Johansen of the Bridge Group.

References

ALLEN, D.M.D.G. 1954. *Relaxation Methods*. McGraw-Hill, New York.

ATKINSON, B.K. 1984. Subcritical crack growth in geological materials. *Journal of Geophysical Research*, **89**, 4077–4114.

BERNDT, C., SKOGLY, O.P., PLANKE, S., ELDHOLM, O. & MJELDE, R. 2000. High-velocity breakup-related sills in the Vøring Basin off Norway. *Journal of Geophysical Research*, **105**, 28 443–28 454.

BRADLEY, J. 1965. Intrusion of major dolerite sills. *Transactions of the Royal Society of New Zealand, Geology*, **3**, 27–55.

CATUNEAUNU, O., HANCOX, P.J. & RUBIDGE, B.S. 1998. Reciprocal flexural behavior and contrasting stratigraphies: a new basin development model of the Karoo retroarc foreland system, South Africa. *Basin Research*, **10**, 417–439.

CHEVALLIER, L. & WOODFORD, A. 1999. Morphotectonics and mechanism of emplacement of the dolerite rings and sills of the western Karoo, South Africa. *South African Journal of Geology*, **102**, 43–54.

CORRY, C.E. 1988. Laccoliths: mechanisms of emplacement and growth. *Geological Society of America, Special Papers*, **220**, 1–110.

DUNCAN, R.A., HOOPER, P.R., REHACEK, J., MARSH, J.S. & DUNCAN, A.R. 1997. The timing and duration of the Karoo igneous event, southern Gondwana. *Journal of Geophysical Research*, **102**, 18 127–18 138.

DU TOIT, A.I. 1920. The Karoo dolerites. *Transactions of the Geological Society South Africa*, **33**, 1–42.

FLEKKØY, E.G., MALTHE-SØRENSSEN, A. & JAMTVEIT, B. 2002. Modeling hydrofracture. *Journal of Geophysical Research*, **107**, 10 259–10 272.

FRANCIS, E.H. 1982. Magma and sediment – I: emplacement mechanism of late Carboniferous

tholeiite sills in the northern Britain. *Journal of the Geological Society of London*, **139**, 1–20.
FYFE, W.S., PRICE, N. & THOMPSON, A.V. 1978. *Fluids in the Earth's crust*. Elsevier Science, New York.
GILBERT, G.K. 1877. *Geology of the Henry Mountains, in the U.S. Geographical and Geological Survey of the Rocky Mountain Region*, U.S. Government Printing Office, Washington, D.C., USA.
JAMTVEIT, B., AUSTRHEIM, H. & MALTHE-SØRENSSEN, A. 2000. Accelerated hydration of the Earth's deep crust induced by stress perturbations. *Nature*, **107**, 10 259–10 272.
JOHNSON, A.M. & POLLARD, D.D. 1973. Mechanics of growth of some laccolithic intrusions in the Henry Mountains, Utah, 1. *Tectonophysics*, **18**, 261–309.
KHAZAN, Y.M. & FIALKO, Y.A. 1995. Fracture criteria at the tip of fluid-driven cracks in the earth. *Geophysical Research Letters*, **22**, 2541–2544.
LISTER, J.R. & KERR, R.C. 1991. Fluid-mechanics models of crack propagation and their application to magma transport in dykes. *Journal of Geophysical Research*, **96**, 10 049–10 077.
MALTHE-SØRENSSEN, A., WALMANN, T., FEDER, J., JØSSANG, T., MEAKIN, P. & HARDY, H.H. 1998a. Simulation of extensional clay fractures. *Physical Review E*, **58**, 5548–5564.
MALTHE-SØRENSSEN, A., WALMANN, T., JAMTVEIT, B., FEDER, J. & JØSSANG, T. 1998b. Modeling and characterization of fracture patterns in the Vatnajökull glacier. *Geology*, **26**, 931–934.
MONETTE, L. & ANDERSON, M.P. 1994. Elastic and fracture properties of the 2-dimensional triangular and square lattices. *Modelling and Simulation in: Materials Science and Engineering*, **2**, 53–73.
POLLARD, D.D. & JOHNSON, A.M. 1973. Mechanics of growth of some laccolithic intrusions in the Henry Mountains, Utah, 2. *Tectonophysics*, **18**, 311–354.
POLLARD, D.D., MULLER, O.H. & DOCKSTADER, D.R. 1975. The form and growth of fingered sheet intrusions. *Geological Society of America Bulletin*, **86**, 351–363.
RUBIN, A.M. 1993. Tensile fracture of rock at high confining pressure: implications for dike propagation. *Journal of Geophysical Research*, **98**, 15 919–15 935.
RUBIN, A.M. 1995. Propagation of magma-filled cracks. *Annual Review, Earth Planetary Sciences*, **23**, 287–336.
SCHUTTER, S.R. 2003. *Hydrocarbon Occurrence and Exploration in and Around Igneous Rocks*. Geological Society, London, Special Publications, **214**.
SKOGSEID, J., PEDERSEN, T., ELDHOLM, O. & LARSEN, B. 1992. Tectonism and magmatism during NE Atlantic continental break-up: the Vøring margin. *Geological Society, London, Special Publications*, **68**, 305–320.
SMALLWOOD, J. & MARESH, J. 2002. *The Properties, Morphology and Distribution of Igneous Sills: Modelling, Borehole Data and 3d Seismic from the Faroe–Shetland Area*. Geological Society, London, Special Publications, **197**.
SYMONDS, P.A., PLANKE, S., FREY, Ø. & SKOGSEID, J. 1998. Volcanic development of the Western Australian continental margin and its implications for basin development. *In*: PURCELL, P.G. & PURCELL, R.R. (eds) *The Sedimentary Basins of Western Australia 2, Proceedings of the Petroleum Exploration Society of Australia, Symposium, Oxford*, Petroleum Exploration Society of Australia, Perth, Australia, 33–54.
VISSNER, J.N.J. 1997. Geography and climatology of the Late Carboniferous to Jurassic Karoo Basin in the south-western Gondwana. *South African Journal of Geology*, **100**, 233–236.
WALMANN, T., MALTHE-SØRENSSEN, A., FEDER, J., JØSSANG, T., MEAKIN, P. & HARDY, H.H. 1996. Scaling relations for the lengths and widths of fractures. *Physical Review Letters*, **77**, 5393–5396.

Table 1. *List of symbols*

ϕ	Porosity of particle packing	i	Particle index
ε^c	Critical strain – strain at which a spring breaks	$k_{i,j}, k$	Spring constant
ρ_m	Matrix mass density	k_w	Spring constant for wall interaction
ρ_i	Particle mass density	$k'_{i,j}$	Non-dimensional spring constant
ρ_s	Magma mass density	$l_{i,j}$	Equilibrium length for spring from i to j
$\Delta\rho$	Density difference between magma and matrix	l	Physical length, typical equilibrium length
σ'_{ij}	Non-dimensional stress tensor	m	Mass of particle
σ_{ij}	Stress tensor	P	Magma pressure at height y
σ^c	Critical stress – stress at which a spring breaks	P_0	Magma pressure at point of injection
A_i	Contact area between magma and particle	r_i	Radius of particle i
E	Young's modulus of matrix	$u_{i,j}$	Unit vector from node i to node j
E_0	Young's modulus of a spring		Surface normal for the sill–sediment interface
	Interparticle force on particle i	u_y	Unit vector along the y-axis
	Force on particle i due to gravity	w	The width of the horizontal part of the sill
	Magma-pressure force on particle i	w_z	Thickness of the two-dimensional sample
	Net force on particle i	x_j	Position of node j
g	Acceleration of gravity	x'	Non-dimensional spatial position
h	Sill emplacement depth	y_w	Position of vertical wall, i.e. elastic boundary

Sill complex geometry and internal architecture: a 3D seismic perspective

KEN THOMSON

School of Geography, Earth and Environmental Sciences, University of Birmingham, Edgbaston, Birmingham B15 2TT, UK
(e-mail: K.Thomson@bham.ac.uk)

Abstract: Seismic volume visualization techniques demonstrate that saucer-shaped sill complexes consist of a series of radiating principal flow units rising from an inner saucer and fed by principal magma tubes. Such flow units contain smaller scale secondary flow units, each being fed by a secondary magma tube branching from the principal magma tube. This pattern is repeated down to scales of approximately 100 m with successively smaller flow units being fed from magma tubes repeatedly branching from higher order tubes. The data demonstrates that each sill complex is independently fed from a centrally located point source, that sills grow by climbing from the centre outwards and that peripheral dyking from the upper surface is a common feature. These features suggest a laccolith emplacement style involving peripheral fracturing and dyking during inner saucer growth and thickening.

Crucial to validating models for the emplacement of sills is a detailed understanding of their complete geometry and internal architecture. Even in classic outcrop examples (e.g. the Karoo sills of South Africa; Du Toit 1920) the amount of 3D structural information is limited by the available exposure and, consequently, the true 3D complexity in such areas can, in many cases, be underestimated. Given these limitations the increasing availability of 3D seismic data provides a powerful tool with which to investigate sill geometry (e.g. Thomson & Hutton 2004). With typical spatial resolutions of 25 metres, such data-sets allow the accurate and complete description of sill geometries at a level of detail comparable to many outcrop-based studies. In addition, advances in seismic volume visualization (particularly opacity rendering) allow their internal architecture and features relatable to magma-flow directions to be observed (Thomson & Hutton 2004). Using 3D seismic data from the North Rockall Trough, this paper will examine the geometries of sill complexes, the magma-flow patterns within such bodies and their relationships to volcanic features.

Using a combination of conventional seismic picking and opacity rendering, the seismic data from the North Rockall Trough demonstrate that there are two dominant sill types within the region, both exhibiting complex internal architectures. The common structural form is radially symmetrical, and consists of a saucer-like flat inner sill at the base with an arcuate inclined sheet connecting it to a gently inclined outer rim

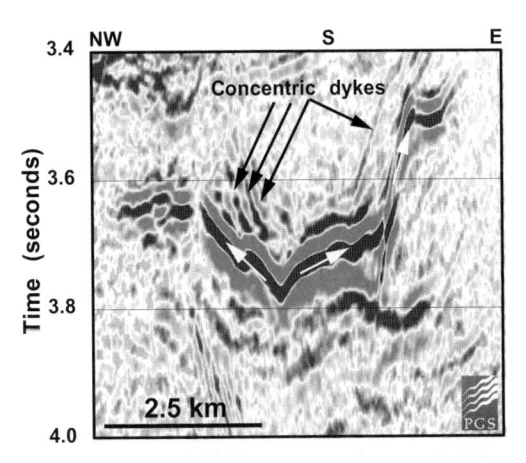

Fig. 1. Seismic section across a typical radially symmetrical sill complex, with evidence for multiple concentric dykes rising from the upper surface of the inner saucer, producing the distinctive 'ring-in-ring' pattern seen in Figure 2. Note that the thickest section occurs in the middle. Magma-flow directions derived from opacity rendering are indicated in white. For location see Figure 2.

(Fig. 1). Such geometries have previously been documented using 3D seismic data from the Faeroe–Shetland Basin (Davies *et al.* 2002; Smallwood & Maresh 2002) and in classic onshore localities such as the Karoo of South Africa (Du Toit 1920; Chevalier & Woodford 1999). The largest, seismically observed, fully

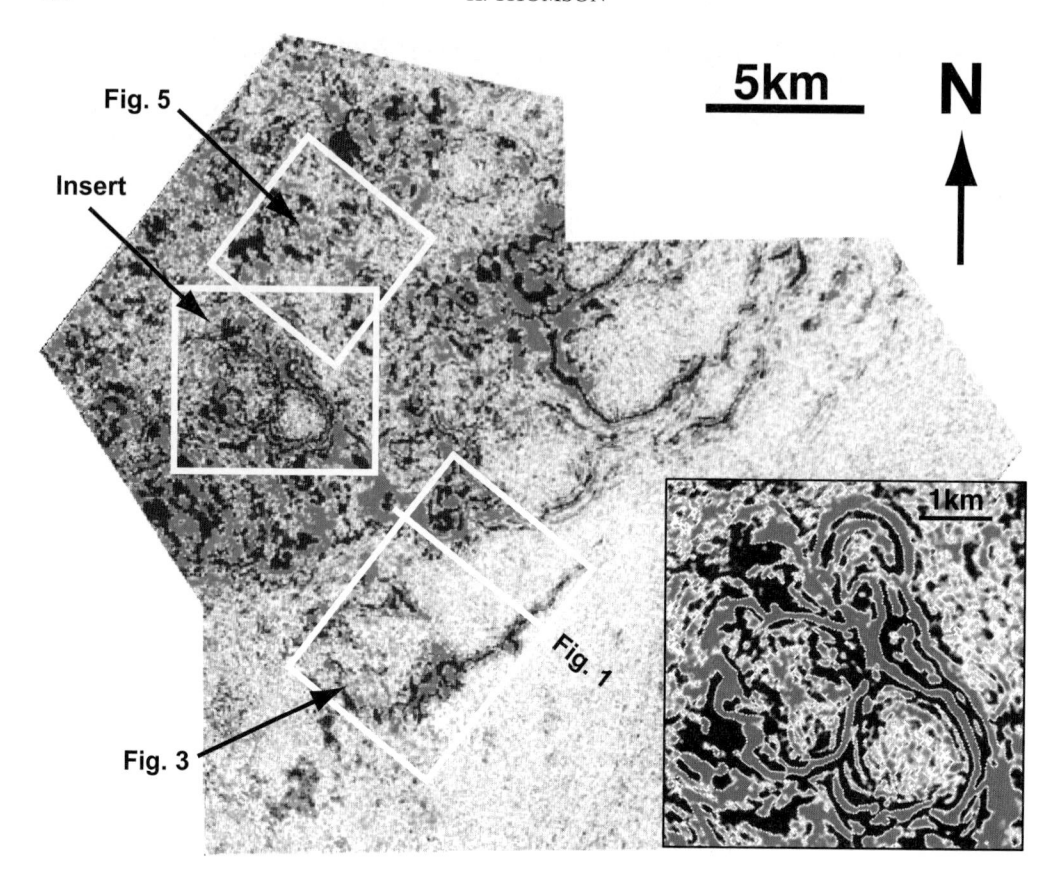

Fig. 2. Time slice from the North Rockall Trough 3D seismic volume. Insert shows the distinctive 'ring-in-ring' pattern due to concentric dyking and the cuspate geometry of the inclined sheets.

developed sills in the North Rockall Trough displaying this geometry are roughly radially symmetrical, approximately 5 km wide (cf. Chevalier & Woodford 1999) and appear to form the fundamental building blocks for sill intrusion across the region. Rising from the upper surface of each inner saucer, a number of thin concentric dykes can also be observed (Fig. 1); producing the distinctive 'ring-in-ring' pattern previously documented for the Karoo sills (Fig. 2; Chevalier & Woodford 1999). Within a sill complex, the seismic data also demonstrate that the inclined sheet consists of a series of concave-inwards/convex-outwards segments, providing the cuspate geometry also seen in the Karoo sills (Fig. 2; Chevalier & Woodford 1999). Such sill complexes can appear as isolated bodies, but commonly occur in close proximity to each other and consequently merge to produce hybrid geometries involving multiple inner saucers, inclined sheets and outer rims.

Figure 3 is a 3D volume visualization of a hybrid sill complex, in which the country rock has been made transparent through the application of an opacity filter. Again, each sill complex can be shown to consist of a thick inner saucer connected to the variable thickness outer rim by an inclined sheet. However, further details of the internal structure can also be seen. The inclined sheet – generally the thinnest part of the sill complex – contains thicker linear regions radially distributed around the inner saucer. From these linear thickenings, smaller-scale linear features can be seen branching off to form a dendritic network that extends from the periphery of the inner saucer through the inclined sheet and into the outer rim. These branching relationships can be interpreted as magma-flow channels within the sheet and, as the branches always terminate towards the sill periphery, and can be traced back through parent channels to the inner saucer, this is interpreted to mean that sill complexes grow radially outwards from the inner saucer. This implies

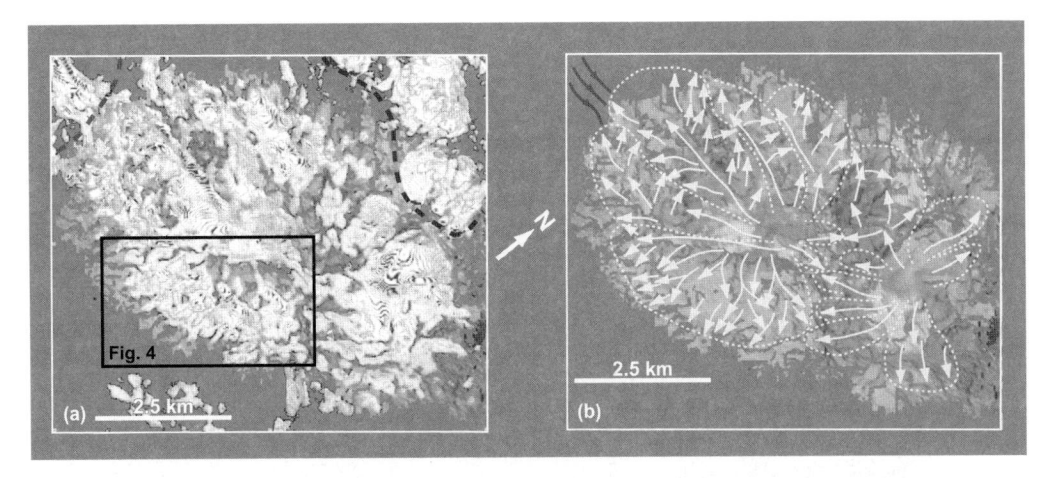

Fig. 3. (a) Plan view of a 3D seismic volume containing a hybrid sill complex in which the low-amplitude 'country rock' has been made transparent. For location, see Figure 2. The artificially illuminated plan view of the hybrid sill complex used in (b) is also shown for ease of comparison. The rendered pattern illustrates the branching nature of sill complexes sourced from two thick inner saucers. The area to the north of the northern dashed line contains rendered material lying above the hybrid sill body, and consequently can be ignored. The area to the west of the western dashed line contains a rendered sill body lying below the hybrid sill body of interest. (b) Artificially illuminated plan view of the hybrid cill complex in (a). The solid lines mark the inferred position of the magma tubes, with the arrowheads indicating the flow directions. The dashed lines mark the limits of primary flow units.

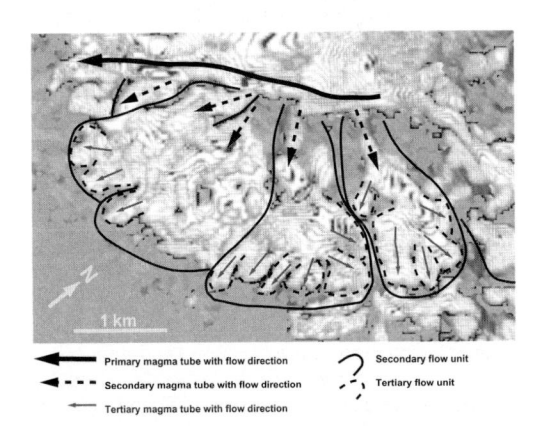

Fig. 4. 3D seismic volume image of part of a primary flow unit within the hybrid sill complex shown in Figure 3. Secondary magma tubes, feeding secondary flow units, can be seen branching from the primary magma tube. Secondary flow units can also be seen to subdivide into lower-order (tertiary) flow units.

that the sill complexes consist of a series of primary flow units rising from the inner saucer and radiating outwards and upwards. The primary flow unit is fed by a primary magma tube, generally elongated in the direction of propagation (i.e. towards the sill complex periphery) and has a slightly concave upwards cross-section normal to the direction of propagation (Figs 3 & 4). This latter geometry accounts for the cuspate geometry of the inclined sheets (Fig. 2). Primary magma tubes bud secondary magma tubes that feed secondary flow units that propagate at an angle of approximately 45° to their parent primary magma-tube flow/propagation direction (Figs 3 & 4). As with primary flow units, secondary flow units also display a slightly concave-upwards cross-section. Secondary flow units derived from the same primary flow unit are generally subparallel. This pattern of subdividing flow units is replicated down to scales of approximately 100 m (the effective limits of seismic resolution), with successively smaller flow units being fed from magma tubes repeatedly branching from higher-order tubes (Fig. 4).

Elongate, bilaterally symmetrical sill complexes can also be found within the North Rockall Trough. Unlike the radially symmetrical sill complexes, these bodies are in close proximity and likely directly related to volcanic centres (Fig. 5). They occur as a series of radiating sills originating from the volcanic centre or from the periphery of the ring fault/dyke. Their geometric form is an elongate, concave-upwards trough – similar in general form to the primary and secondary flow units seen in radially symmetrical

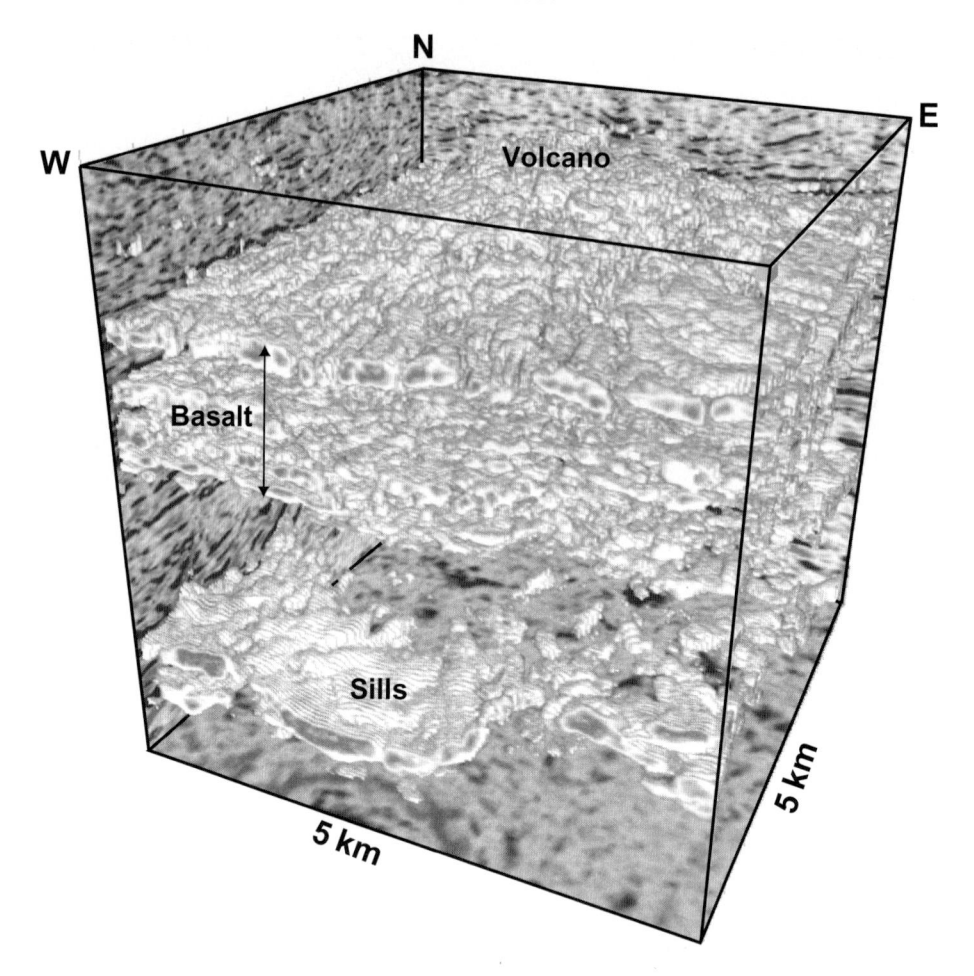

Fig. 5. 3D opacity-rendered image of a volcano within the North Rockall Trough and the underlying concave-upwards, trough-like, sill complexes beneath it. In order to make the limits of the volume more readily recognizable, the NW and NE walls contain conventional seismic sections. Similarly, the volume floor contains a time slice. For location see Figure 2.

sill complexes. With a general tendency to climb away from the volcanic centre from which they originate, these elongate sill complexes can also be seen to branch as they propagate away from the volcanic centre, with individual buds (or flow units) also having a general concave-upwards profile.

References

CHEVALIER, L. & WOODFORD, A. 1999. Morph-tectonics and mechanism of emplacement of the dolerite rings and sills of the western Karoo, South Africa. *South African Journal of Geology* **102**, 43–54.

DAVIES, R., BELL, B.R., CARTWRIGHT, J.A. & SHOULDERS, S. 2002. Three-dimensional seismic imaging of Paleogene dike-fed submarine volcanoes from the northeast Atlantic margin. *Geology*, **30**, 223–226.

DU TOIT, A.L. 1920. The Karoo dolerite of South Africa: a study of hypabyssal injection. *Transactions of the Geological Society of South Africa*, **23**, 1–42.

SMALLWOOD, J.R. & MARESH, J. 2002. The properties, morphology and distribution of igneous sills: modelling, borehole data and 3D seismic data from the Faeroe–Shetland area. *In:* JOLLEY, D.W. & BELL, B.R. (eds) *The North Atlantic Igneous Province: Stratigraphy, Tectonic, Volcanic and Magmatic Processes.* Geological Society, London, Special Publications, **197**, 271–306.

THOMSON, K. & HUTTON 2004. Geometry and growth of sill complexes: insights using 3D seismic from the North Rockall Trough. *Bulletin of Volcanology*, **66**, 364–375.

Hydrothermal vent complexes associated with sill intrusions in sedimentary basins

BJØRN JAMTVEIT[1], HENRIK SVENSEN[1], YURI Y. PODLADCHIKOV[1] & SVERRE PLANKE[1,2]

[1]*Physics of Geological Processes (PGP), University of Oslo, PO Box 1048 Blindern, 0316 Oslo, Norway (e-mail: bjorn.jamtveit@geo.uio.no)*
[2]*Volcanic Basin Petroleum Research, Oslo Research Park, Gaustadalleen 21, 0349 Oslo, Norway*

Abstract: Subvolcanic intrusions in sedimentary basins cause strong thermal perturbations and frequently cause extensive hydrothermal activity. Hydrothermal vent complexes emanating from the tips of transgressive sills are observed in seismic profiles from the Northeast Atlantic margin, and geometrically similar complexes occur in the Stormberg Group within the Late Carboniferous–Middle Jurassic Karoo Basin in South Africa. Distinct features include inward-dipping sedimentary strata surrounding a central vent complex, comprising multiple sandstone dykes, pipes, and hydrothermal breccias. Theoretical arguments reveal that the extent of fluid-pressure build-up depends largely on a single dimensionless number (Ve) that reflects the relative rates of heat and fluid transport. For $Ve \gg 1$, 'explosive' release of fluids from the area near the upper sill surface triggers hydrothermal venting shortly after sill emplacement. In the Karoo Basin, the formation of shallow (< 1 km) sandstone-hosted vents was initially associated with extensive brecciation, followed by emplacement of sandstone dykes and pipes in the central parts of the vent complexes. High fluid fluxes towards the surface were sustained by boiling of aqueous fluids near the sill. Both the sill bodies and the hydrothermal vent complexes represent major perturbations of the permeability structure of the sedimentary basin, and are likely to have long time-scale effects on its hydrogeological evolution.

Large igneous provinces, such as the Northeast Atlantic igneous province and the Karoo igneous province in South Africa, are characterized by the presence of an extensive network of sills and dykes embedded in sedimentary strata. The thermal and hydrological effects of these intrusions on surrounding sedimentary rock strata are of considerable interest not only to the research community, but also to companies carrying out petroleum exploration in volcanic basins.

Previous field studies of sills emplaced in porous sedimentary rocks and loosely consolidated sediments describe fluidization of initially consolidated sediments near sill contacts and expulsion of large volumes of pore-waters towards the surface, thus creating a hydrothermal system (Grapes *et al.* 1973; Einsele 1982; Krynauw *et al.* 1988). Major hydrothermal effects triggered by sills have also been inferred from seismic profiles from the sedimentary basins along the Northeast Atlantic margins (Svensen *et al.* 2003). Figure 1 shows our interpretation of a seismic profile in the north-central Vøring Basin on the mid-Norwegian margin. Vertical structures starting at sill tips reach the palaeosurface, where they terminate in eye-like structures interpreted to represent ancient hydrothermal eruption centres similar to mud volcanoes (Planke *et al.* 2003; Svensen *et al.* 2003, 2004).

In this paper we describe and characterize hydrothermal vent complexes from the Karoo Basin in central South Africa. These structures are geometrically very similar to the subsurface vent complexes identified on seismic reflection data in sedimentary basins along the NE Atlantic margins. It is demonstrated how high fluid pressures, fluidization processes, and channelized flow play a major role in restructuring the sediments around sills, affecting both the short- and long time-scale hydrological evolution of the system. We finally present a model quantifying the conditions required for the build-up of fluid pressures sufficient to trigger hydrothermal vent complex formation above sills, in the case where the pore fluid is pure water.

From: BREITKREUZ, C. & PETFORD, N. (eds) 2004. *Physical Geology of High-Level Magmatic Systems.*
Geological Society, London, Special Publications, **234**, 233–241. 0305-8719/04/$15.00
© The Geological Society of London 2004.

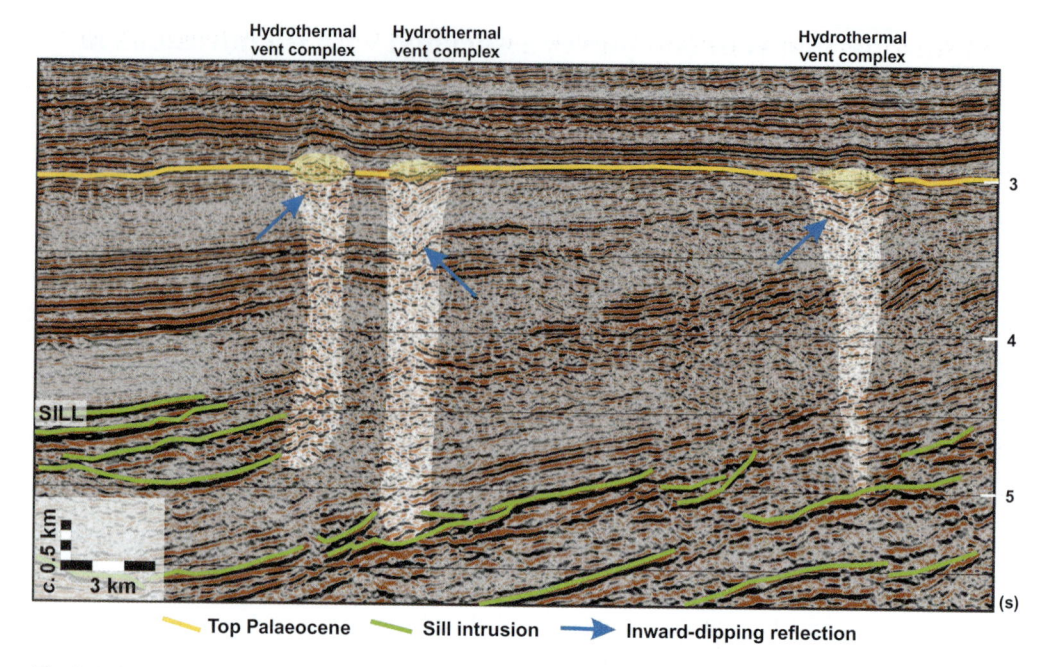

Fig. 1. Seismic expression of three hydrothermal vent complexes cutting Palaeocene and Cretaceous clastic sedimentary strata in the north-central Vøring Basin. High-amplitude reflections are interpreted as mafic sills based on their morphology, seismic shadowing effects, aeromagnetic and well data. Note that the sedimentary strata dip into the vent (blue arrows). The vent complexes are associated with a positive perturbation of the topographic relief at the palaeosurface (yellow ellipses). These vent eye-like structures are interpreted as 'mud-volcanoes' and may reach several kilometres in diameter.

Hydrothermal vent complexes in the Karoo Basin

The Karoo Basin in South Africa is a sedimentary basin covering more than half of South Africa. The basin comprises up to 6 km of sedimentary strata, capped by at least 1.4 km of basaltic lava, and is bounded by the Cape Fold Belt along its southern margin (Smith 1990; Johnson et al. 1997). The sediments accumulated from the Late Carboniferous to the Mid-Jurassic, in an environment ranging from marine (the Dwyka and Ecca groups) to fluvial (the Beaufort and parts of the Stormberg Group) and aeolian (upper part of the Stormberg Group) (Catuneanu et al. 1998). At the time of lava eruption, an up to 400-m thick sequence of fine-grained sand had been deposited by aeolian, fluvial and lacustrine processes (Smith 1990). The sedimentary rocks comprising the SE parts of the basin were gently folded during phases of the Cape Orogeny (278–215 Ma; Catuneanu et al. 1998), whereas the rest of the basin is essentially undeformed.

Both southern Africa and Antarctica experienced extensive volcanic activity in early Jurassic times (183 ± 1 Ma; Duncan et al. 1999). Sills and dykes are present throughout the sedimentary succession, and locally make up 70% of the basin volume (Rowsell and de Swardt 1976). Horizontal sill intrusions preferentially occur in the deep basin sequences (the Ecca Group), whereas transgressive sheets and rings are common in intermediate sequences (the Beaufort Group) (Chevallier & Woodford 1999; Woodford et al. 2001). The magmatic material in the upper formations of the Stormberg Group is dominantly present as dykes.

Hundreds of vent complexes have been identified in the Stormberg Group (Du Toit 1904; Keyser 1997; Dingle et al. 1983), but have previously received limited attention. The complexes range from being almost purely volcanic to being almost entirely filled by sedimentary material (Du Toit 1904, 1912; Gevers 1928; Seme 1997; Dingle et al. 1983; Woodford et al. 2001). The hydrothermal vent complexes have also been termed diatremes, volcanic necks, and breccias, and, since the pioneer work of Du Toit, have been interpreted as results of phreatic or phreatomagmatic activity (e.g. Gevers 1928; Coetzee 1966; Taylor 1970; Seme 1997; Woodford et al. 2001).

However, the presence of juvenile magmatic material in the sediment-dominated hydrothermal vent complexes is minor (Seme 1997). The focus in this paper is on the sediment-dominated hydrothermal vent complexes in the Molteno–Rossouw area in the central parts of the Karoo Basin (Fig. 2A).

The hydrothermal vent complexes in the Molteno–Rossouw area are observed as erosional remnants in the subhorizontal Clarens and Elliot formations (upper Stormberg Group). They comprise an inner and an outer zone with structurally modified strata (Fig. 2B). The outer zone is characterized by flexured and inward-dipping (up to 45°) sedimentary sequences. The inner zone represents circular topographic highs with diameters up to 300–400 metres (Fig. 2B). Characteristic features of the inner zones include sediment breccias with clay and sandstone fragments in a sandstone matrix, and cross-cutting sandstone dykes and pipes (Fig. 2C, 2D & 2E). Brecciated sediments are volumetrically important in most of the sediment-dominated hydrothermal vent complexes. Sharp cross-cutting relations, with deformed border zones, are common between sandstone pipes and the brecciated sediments. The field observations indicate that brecciation took place before the intrusion of sediment pipes and dykes. Together, these structures represent unambiguous evidence that these vent complexes represent conduits for overpressured fluids, fragmented and fluidized sediments that at least partly originated at deeper stratigraphic levels (at least from the level of the Molteno Formation).

Modelling the fluid pressure evolution around shallow sill intrusions

Studies of geothermal systems suddenly heated from below distinguish an early stage of isothermal pressurization from a later period of heating and hydrothermal circulation (Delaney 1982). Previous numerical models for the early (prior to the onset of convection) heat and fluid flow around cooling magmatic intrusives assume stable one-dimensional removal of fluid away from the sill contact (cf. Delaney 1982; Litvinovskiy et al. 1990; Podladchikov & Wickham 1994). However, if the fluid pressure exceeds the lithostatic load pressure, the fluid flow away from the heat source will be associated with matrix deformation and fluid channelling.

The conditions necessary to generate fluid overpressures sufficient to cause fluid channeling (i.e. hydrothermal venting) *shortly after* sill emplacement can be constrained by simple scaling relationships for diffusive transport. At the early stages, the width of the overpressured (fluid pressure exceeding hydrostatic) zone (H_{OP}) is given by the characteristic length scale for hydraulic diffusion:

$$H_{OP} \approx 2\sqrt{k_{fluid} t} \quad (1)$$

where $k_{fluid} = \dfrac{k}{\beta \phi \mu_{fluid}}$ is hydraulic diffusivity, $\beta \approx 10^{-8}$ Pa^{-1} is the effective fluid and pore compressibility, ϕ is porosity, k is permeability, μ_{fluid} is fluid viscosity, and t is the time since (instantaneous) sill emplacement. A similar scaling relation expresses the boiling zone thickness:

$$H_{boil} \approx 2\sqrt{k_T t} \quad (2)$$

where $k_T \approx 10^{-6}$ m^{2}s^{-1} is the heat diffusivity. Therefore, the boiling front velocity can be expressed as:

$$V_{boil} \approx \sqrt{\dfrac{k_T}{t}} \quad (3)$$

Note that at the early stages of sill cooling the fluid is heated by the rock, and not vice versa. At these early stages, advective heat transport is insignificant (due to the small volume of pore-fluids stored initially in the rocks), and permeability does not have a first-order effect on the boiling zone thickness, not even when its effect on fluid pressure is taken into account (cf. Delaney 1982; Podladchikov & Wickham 1994). At the later hydrothermal circulation stages, advection becomes important due to the large volume of fluids circulating through the same rock, and fluids start to dominate the heat transport and to 'heat the rocks'. In such advection-dominated convecting systems, some fluid pathways are near adiabatic, which gives rise to extensive two-phase regions (the latent heat of vaporization of water is very large and prevents adiabatic vaporization by decompression (cf. Ingebritsen & Sanford 1998, p. 200)). On the contrary, at the early stages, the boiling two-phase zone is restricted to a narrow front mainly controlled by conductive heat transfer in the solid.

The hot vapour may move faster through the boiling front but, after coming in contact with cold rocks, it quickly condenses back into water without significant influence on the boiling front velocity. Delaney (1982) concluded that '. . . the properties of the steam region do not play a significant role in determining the maximum pressure'. Since we are interested in the

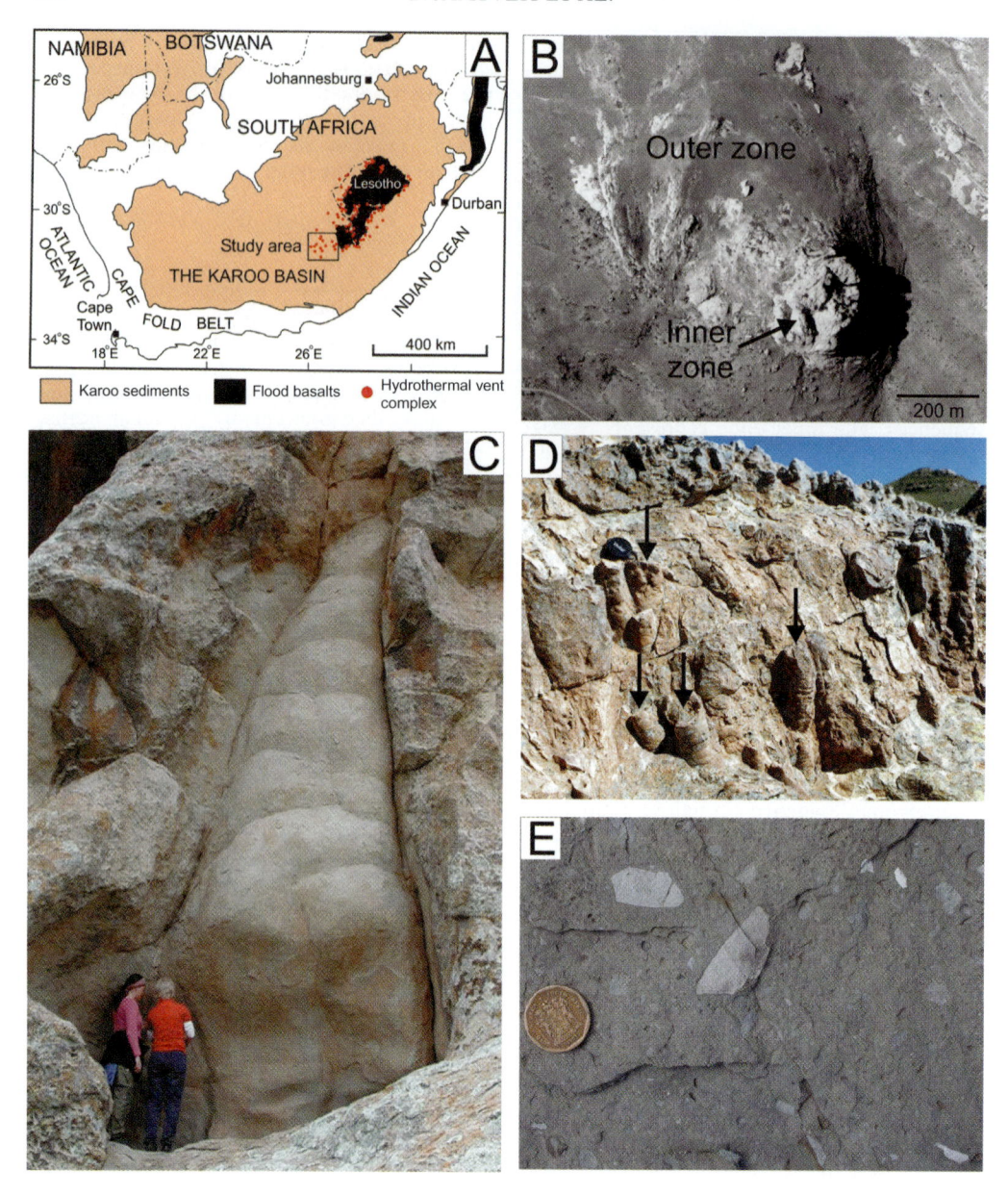

Fig. 2. (**A**) The Karoo Basin, with the Lesotho flood basalts and the locations of abundant shallow hydrothermal vent complexes. (**B**) Aerial photo of hydrothermal vent complex from the Molteno–Rossow area. The vent complex comprises an inner zone seen as a vegetation-free erosional remnant, and an outer zone with inward dipping sedimentary strata. The surrounding sedimentary layering is subhorizontal. (**C**) Sandstone dyke in hydrothermal vent complexes. The dyke contains abundant carbonate cement and erodes more easily than the surrounding vent sandstone. (**D**) Numerous small (up to 30 cm in diameter) sandstone pipes (position shown by arrows) are present within the inner zone of vent complex. (**E**) Brecciated sediment from a hydrothermal vent complex. Clasts are mostly of sedimentary origin (sandstone, claystone and siltstone), although a minor component of juvenile magmatic material is present. The sediment clasts are believed to have formed during fluidization of a mixture of fragmented sediments from the Elliot and Clarens formations, and fine-grained sand from the Clarens Formation.

maximum fluid pressure achieved during the early pressurization stage and for the cases of our interest H_{boil} is much smaller than H_{OP}, we can employ simple cold-water properties for the porous flow outside the narrow boiling zone. The Darcian fluid flux (Q_D) out of the boiling zone is approximately:

$$Q_D \approx \frac{k}{\mu_{fluid}} \frac{P_{fluid} - P_{hyd}}{H_{OP}} \quad (4)$$

where ($P_{fluid} - P_{hyd}$) is the difference between fluid pressure and hydrostatic pressure (the overpressure). Initially, the overpressure is small, thus $Q_D \ll V_{boil}$, leading to massive production of steam. Thus the fluid pressure will rise to a slowly decaying value characterized by a quasi-static flux balance:

$$\rho Q_D = \phi \Delta \rho_{boil} V_{boil} \quad (5)$$

where ρ and $\Delta \rho_{boil} (\approx \rho)$ are the density of the cold water and the density difference between cold water and supercritical steam. Solving for non-hydrostatic fluid pressure build-up we obtain:

$$P_{fluid}^{max} \approx P_{hyd}\left(1 + 2\frac{\Delta \rho_{boil}}{\rho \beta P_{hyd}}\sqrt{\frac{k_T}{k_{fluid}}}\right) \quad (6)$$

In order to quantify the proximity of this pressure build-up to venting we define the dimensionless parameter **Ve** as follows:

$$Ve = \frac{P_{fluid}^{max} - P_{hyd}}{P_{hyd}} \approx 2\frac{\Delta \rho_{boil}}{\rho \beta P_{hyd}}\sqrt{\frac{k_T}{k_{fluid}}} \quad (7)$$

Assuming $\Delta \rho_{boil} \approx \rho/2$ and substituting parameters yields:

$$Ve = \frac{1}{\beta P_{hyd}}\sqrt{\frac{k_T}{k_{fluid}}} \approx \frac{1}{10^7 Z}\sqrt{\frac{\mu_{fluid} k_T}{k \beta}} \\ \approx \frac{10^{-7}}{Z\sqrt{k}} \quad (8)$$

where Z is the intrusion depth in km. $Ve \ll 1$ corresponds to the situation where a sill is emplaced in an environment that is sufficiently permeable to prevent significant fluid pressure build-up because pressure diffusion is fast compared to the rate of pressure production. In contrast, for shallow emplacement depths, $Ve \gg 1$ leads to a fluid-pressure increase and a 'blow-out' situation when the fluid pressure exceeds the lithostatic pressure. At this stage, overpressured fluids will create their own permeability by deforming the overlying rocks and generating fluid release structures, i.e. hydrothermal vent complexes. Figure 3 shows under what depth and permeability conditions boiling-driven venting may occur, based on eqn (8). Note that, for pure water, boiling-driven venting will not occur at depths exceeding approximately 1.1 km, because the lithostatic pressure at such depths exceeds the critical pressure of water. This is consistent with the confinement of Karoo Basin hydrothermal vent complexes to the shallow-level strata below the Lesotho basalts (Fig. 2A). The results of the simple 1D model presented above are in full agreement with a recently developed fully coupled 2D numerical model, including the complete thermodynamics of H_2O (Podladchikov *et al.* in prep.).

Given the range of permeabilities, as well as the spatial permeability variations in natural systems, it is clear that the permeability is *the* controlling parameter. For a given field scenario, the total uncertainty in the estimated **Ve**-value due to all other parameters will be orders of magnitude less than the uncertainty due to poorly constrained permeability. It is important to realize that the relevant permeability in this case is the bulk permeability for the sedimentary rocks on a scale comparable to H_{OP}. This is likely to be controlled by the permeability of the least-permeable sedimentary strata in the sedimentary sequence near the sills.

In the three-dimensional world, the pressure build-up is largest near the highest point of the sill surface (i.e. near the tip of the sill for the gently climbing sills in the shallow parts of the Karoo Basin). For high **Ve**-numbers, boiling will boost the fluid pressure shortly after sill emplacement. In reality, boiling and significant rises in fluid pressure may already take place during emplacement.

The analysis above is not constrained to the situation where the porous fluid is pure H_2O. In the pure H_2O case, the **Ve**-number will be reduced at depths corresponding to the critical point of water, due to the growing density of steam $\Delta \rho_{boil} < \rho$. However, in a sedimentary basin setting one can easily imagine deeper hydrothermal venting processes driven by boiling of saline solutions, pressure build-up mechanisms related to gas generation from organic-rich sediments, devolatilization reactions (e.g. Litvinovskiy *et al.* 1990), retrograde boiling of crystallizing magma (Podladchikov & Wickham 1994), or increasing **Ve**-numbers with depth resulting from a permeability reduction. This may explain the presence of the deep roots of the hydrothermal vent complexes from the Mid-Norwegian volcanic margin (Fig. 1).

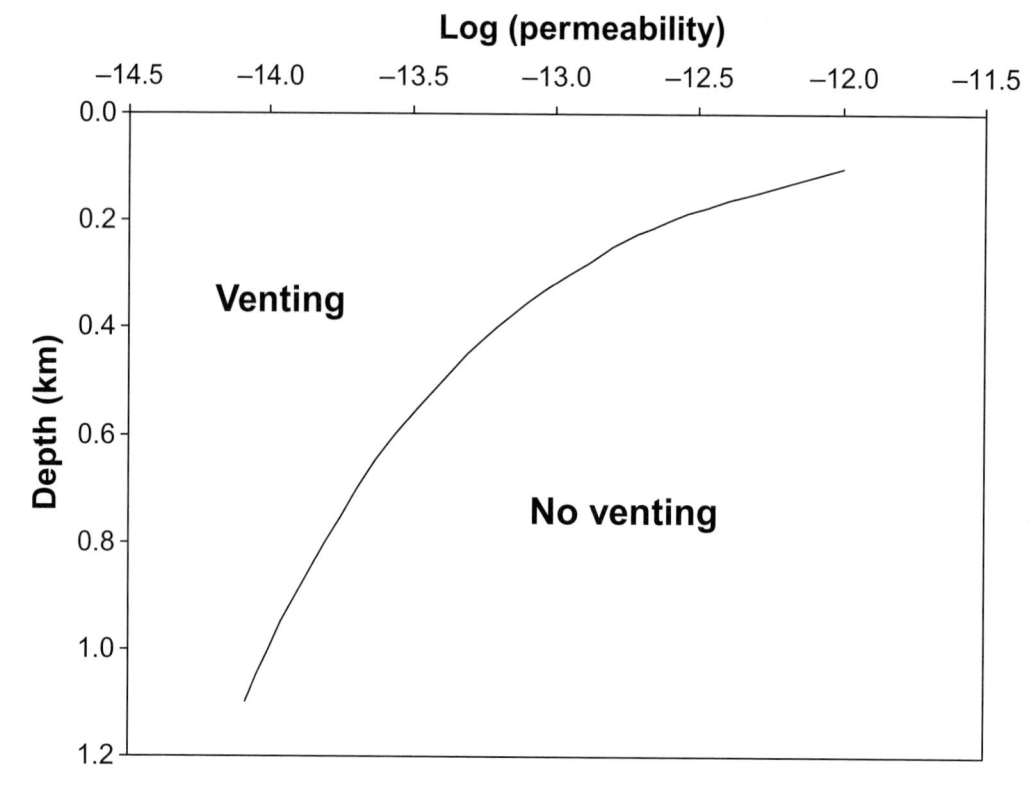

Fig. 3. Depth of intrusion versus log permeability of country rock for a *Ve*-value of 1, based on eqn (8). Venting may occur in the region where depths are shallow and/or permeabilities are low. In the region to the right, fluid pressure will leak off too rapidly for venting to occur. At depths exceeding c. 1.1 km, boiling does not occur in pure water because the lithostatic pressure exceeds the critical pressure. Hence, no venting is expected.

Discussion

Field observations from the Karoo Basin demonstrate that the semi-circular hydrothermal vent complexes readily identified from aerial photos (Fig. 2B) represent restructured conical bodies typically having a surface diameter of hundreds of metres. By analogy with seismic profiles from the Mid-Norwegian volcanic margin (Fig. 1), we argue that these hydrothermal vent complexes are spatially connected to the termination of sill intrusions deeper in the basin.

We propose a model where overpressure generated from boiling of pore fluids near the sill–sediment interface is the driving mechanism for venting. The most obvious source of fluid overpressure in the shallow-level Karoo sediments is boiling of aqueous pore fluids. This will result in a short-lived and gas-driven system able to brecciate consolidated sediments in the source region and fluidize sand closer to the surface. In contrast to previous models applicable to hydrothermally driven piercement structures in sedimentary basins (e.g. Du Toit 1904, 1912; Grapes et al. 1973; Navikov & Slobodskoy 1979; Lorenz 1985), our model does not depend on the presence of a dyke network or emplacement-related fracturing in the source region, or on extensive mixing between magma and sediments. Juvenile volcanic material is present only sporadically in the sandstone-dominated complexes, suggesting phreatic activity without disintegration of sill dolerite or major restructuring of the sill–sediment interface.

Formation of hydrothermal vent complexes related to overpressured fluids derived from close to the sill–sediment contact is supported by ongoing aureole studies around sills in the Beaufort Group. Abundant sediment dykes cutting dolerite sills are evidence for high fluid pressures in contact-metamorphic aureoles, where fluidization and brecciation of sandstone beds below sills cause back-veining following cooling and thermal contraction (Walker & Poldervaart 1949; van Biljon & Smitter 1956; Planke et al. 2000).

Our modelling results indicate that fluid discharge vents triggered by sill intrusions may have formed very shortly after sill emplacement, probably while most of the sill was still molten. An important consequence of the rapid evolution of the system is that the formation of hydrothermal vent complexes mainly depends on the dimensionless number *Ve*. The sensitivity both to the detailed geometry of the system and possible uncertainties in the control parameters is negligible for the short time-scale evolution of the system.

Analogue experiments of fluid flow through sand beds have produced vent structures very similar to those observed in the Karoo Basin, including inward-dipping strata towards subvertical pipes of fluidized sand (Woolsley 1975). More recently, experiments on gas flow through beds of ultrafine cohesive powders showed how gas flow through porous beds leads to a sequence of different migration mechanisms, depending on gas flux (Li *et al.* 1999). For low gas fluxes, the gas flows through the stagnant porous medium, as in normal Darcian flow. However, with increasing gas flux, channel-ways form through the particle beds and, for very high fluxes, the overpressured gas lifts the roof and ultimately causes disaggregation and partial fluidization of the overlying porous wedge.

In the Karoo Basin, the fluid pressure must have been very high during initial vent formation and subsequently dropped as the system 'ran out of steam'. Thus an initial stage where the overpressured fluids literally lifted the roof and caused the observed disruption, disaggregation and brecciation of the initially well-stratified sediments may have been followed by a period where reduced fluid pressure pushed the system into a state where release of fluid and fluidized sand occurred through smaller discrete pipes or channels. The venting process produced local topographic depressions that could have acted as sites for aeolian and lacustrine sediment deposition (cf. Holzförster *et al.* 2002). A model of the vent complex formation consistent with field observations, modelling and the above-mentioned analogue experiments, is outlined in Figure 4.

Drilling of vent structures both in the western and eastern Karoo has shown that these structures represent local aquifers and thus a 'permanent' local perturbation of the hydrological properties of the Karoo Basin (Woodford *et al.* 2001). The long time-scale utilization of hydrothermal vent complexes for basin fluids is furthermore emphasized by the close association between seep-carbonate deposits and underlying vent structures (Svensen *et al.* 2003).

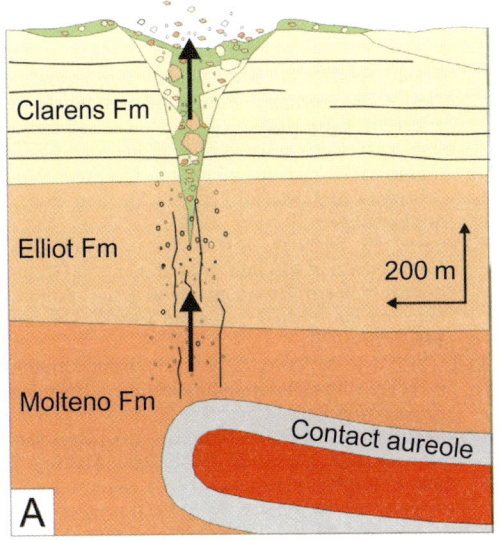

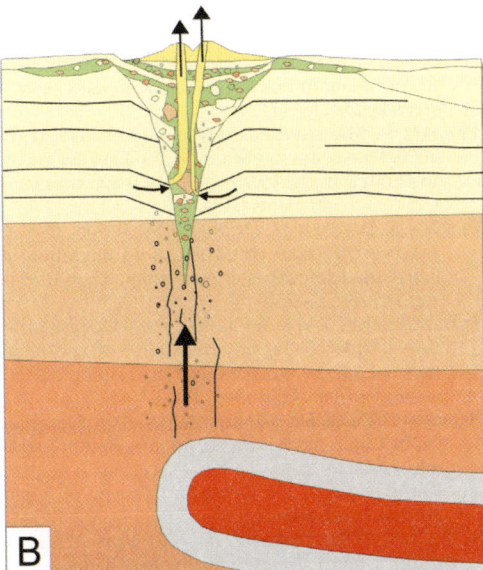

Fig. 4. Schematic evolution of hydrothermal vent complexes, based on data from the Karoo Basin, seismic data from the Norwegian Sea, and numerical modelling. The labelling of key units and the approximate scale applies to the Karoo Basin. (**A**) Boiling of pore fluids causes fluid pressure build-up and formation of a cone-shaped hydrothermal vent complex. The initial fluid expulsion is associated with fragmentation and generation of hydrothermal breccias. The erupted material may have formed crater rim deposits, but these have not been identified in the field, due to the level of erosion. (**B**) Subsequently, large (several metres across) pipes of fluidized sand cross-cut the brecciated rocks. Smaller hydrothermal pipes (centimetre-size) form during later reworking of the cone structure, following a reduction in fluid pressure gradients.

This study was supported by the Norwegian Research Council through Grant No. 113354-420 to the *Fluid–Rock Interactions* Strategic University Program, and by Volcanic Basin Petroleum Research AS. Continuous support by TGS-Nopec and the industrial participants on the 'Petroleum Implications of Sill Intrusions' is gratefully acknowledged. Finally, we thank G. Marsh (Rhodes University) and L. Chevallier (Council for Geoscience, Republic of South Africa) for discussions and valuable support during our trips to South Africa; J. Connolly (ETH–Zürich) for many useful comments; and A. Malthe-Sørenssen, S. A. Lorentzen, C. Haave (University of Oslo), E. Eckhoff (VBPR, Oslo), S. Polteau (Rhodes University), and S. Piazolo (University of Mainz) for the company and assistance during fieldwork.

References

CATUNEAUNU, O., HANCOX, P.J. & RUBIDGE, B.S. 1998. Reciprocal flexural behaviour and contrasting stratigraphies: a new basin development model for the Karoo retroarc foreland system, South Africa. *Basin Research*, **10**, 417–439.

CHEVALLIER, L. & WOODFORD, A.C. 1999. Morphotectonics and mechanism of emplacement of the dolerite rings and sills of the western Karoo, South Africa. *South African Journal of Geology*, **102**, 43–54.

DELANEY, P.T. 1982. Rapid intrusion of magma into wet rocks: groundwater flow due to pore pressure increases. *Journal of Geophysical Research*, **87**, 7739–7756.

DINGLE, R.V., SIESSER, W.G. & NEWTON, A.R. 1983. *Mesozoic and Tertiary Geology of Southern Africa*. A.A. Balkema, Rotterdam.

DUNCAN, R.A., HOOPER, P.R., REHACEK, J., MARSH, J.S. & DUNCAN, R.A. 1997. The timing and duration of the Karoo igneous event, Southern Gondwana. *Journal of Geophysical Research*, **102**, 18 127–18 138.

DU TOIT, A.L. 1904. Geological Survey of Elliot and Xalanga, Tembuland. *Annual Report Geological Comm. Cape of Good Hope for 1903*, **8**, 169–205.

DU TOIT, A.L. 1912. Geological Survey of part of the Stormbergen. *Annual Report Geological Comm. Cape of Good Hope for 1911*, **16**, 112–136.

EINSELE, G. 1982. Mechanism of sill intrusion into soft sediment and expulsion of pore water. *Deep Sea Drilling Project, Initial Reports*, **64**, 1169–1176.

GEVERS, T.W. 1928. The volcanic vents of the Western Stormberg. *Transactions of the Geological Society of South Africa*, **31**, 43–62.

GRAPES, R.H., REID, D.L. & MCPHERSON, J.G. 1973. Shallow dolerite intrusions and phreatic eruption in the Allan Hills region, Antarctica. *New Zealand Journal of Geology and Geophysics*, **17**, 563–577.

HOLZFÖRSTER, F., HOLZFÖRSTER, H.W. & MARSH, J.S. 2002. New views on the Jurassic Clarens Formation: volcano-sedimentary interaction. *16th International Sedimentological Congress Abstract Volume*, p. 159.

INGEBRITSEN, S.E. & SANFORD, W.E. 1998. *Groundwater in Geological Processes*. Cambridge University Press, 340 pp.

JOHNSON, M.R. ET AL. 1997. The foreland Karoo Basin, South Africa. *In:* R.C. SELLEY (ed.) *African Basins. Sedimentary Basins of the World, 3.* Elsevier, Amsterdam.

KEYSER, N. 1997. *Geological Map of the Republic of South Africa and the Kingdoms of Lesotho and Swaziland. 1:1,000,000*. Council for Geoscience, South Africa.

KRYNAUW, J.R., HUNTER, D.R. & WILSON, A.H. 1988. Emplacment of sills into wet sediments at Grunehogna, western Dronning Maud Land, Antarctica. *Journal of the Geological Society, London*, **145**, 1019–1032.

LI, H., HONG, R. & WANG, Z. 1999. Fluidizing ultrafine powders with circulating fluidized bed. *Chemical Engineering Science*, **54**, 5609–5615.

LITVINOVSKIY, B.A., PODLADCHIKOV, Y.Y., ZANVILEVICH, A.N. & DUNICHEV, V.M. 1990. Melting of acid volcanites in contact with a shallow basite magma. *Geokhimiya*, **6**, 807–814 (translated).

LORENZ, V. 1985. Maars and diatremes of phreatomagmatic origin: a review. *Transactions of the Geological Society of South Africa*, **88**, 459–470.

NAVIKOV, L.A. & SLOBODSKOY, R.M. 1979. Mechanism of formation of diatremes. *International Geology Review*, **21**, 1131–1139.

PLANKE, S., MALTHE-SØRENSSEN, A. & JAMTVEIT, B. 2000. Petroleum implications of sill intrusions: an integrated study. *VBPR/TGS-NOPEC Commercial Report*, 43 plates (A3).

PLANKE, S., SVENSEN, H., HOVLAND, M., BANKS, D.A. & JAMTVEIT, B. 2003. Mud and fluid migration in active mud volcanoes in Azerbaijan. *Geo-Marine Letters*, **23**, 258–268.

PODLADCHIKOV, Y.Y. & WICKHAM, S. 1994. Crystallization of hydrous magmas: calculation of associated thermal effects, volatile fluxes, and isotopic alteration. *Journal of Geology*, **102**, 25–45.

ROWSELL, D.M. & DE SWARDT, A.M.J. 1976. Diagenesis in Cape and Karoo sediments, South Africa and its bearing on their hydrocarbon potential. *Transactions of the Geological Society of South Africa*, **79**, 81–145.

SEME, U.T. 1997. *Diatreme deposits near Rossouw, north Eastern Cape: sedimentology–volcanology and mode of origin*. Geology Honours Project, Department of Geology, Rhodes University, Grahamstown, South Africa.

SMITH, R.M. 1990. A review of stratigraphy and sedimentary environments of the Karoo Basin of South Africa. *Journal of African Earth Sciences*, **16**, 143–169.

SVENSEN, H., PLANKE, S., JAMTVEIT, B. & PEDERSEN, T. 2003. Seep carbonate formation controlled by hydrothermal vent complexes: a case study from the Vøring Basin, the Norwegian Sea. *Geo-Marine Letters*, **23**, 351–358.

SVENSEN, H., PLANKE, S., MALTHE-SØRENSSEN, A., JAMTVEIT, B., MYKLEBUST, R., RASMUSSEN, T. & REY, S.S. 2004. Release of methane from a volcanic basin as a mechanism for initial Eocene global warming. *Nature*, **429**, 542–545.

VAN BILJON, W.J. & SMITTER, Y.H. 1956. A note on the occurrence of two sandstone dykes in a Karoo dolerite sill near Devon, south-eastern Transvaal. *Transactions of the Geological Society of South Africa*, **59**, 135–139.

WALKER, F. & POLDERVAART, A. 1949. Karoo dolerites of the Union of South Africa. *Bulletin of the Geological Society of America*, **60**, 591–706.

WOOLSLEY, T.S. 1975. Modeling of diatreme emplacement by fluidization. *Physics and Chemistry of the Earth*, **9**, 29–42.

WOODFORD, A.C. ET AL. 2001. *Hydrogeology of the Main Karoo Basin: Current Knowledge and Research Needs*. Water Research Commission Report, **860**, Pretoria, 310 pp.

Experimental constraints on the mechanics of dyke emplacement in partially molten olivines

S. VINCIGUERRA[1], X. XIAO[2] & B. EVANS[2]

[1] *Osservatorio Vesuviano – Istituto Nazionale di Geofisica e Vulcanologia, Via Diocleziano 328, 80124, Naples, Italy (e-mail: vinciguerra@ov.ingv.it)*

[2] *Department of Earth, Atmospheric and Planetary Sciences, Massachusetts Institute of Technology, 77 Massachusetts Avenue, Cambridge, MA 02139, USA*

Abstract: We investigated the mechanics of basalt dyke emplacement during laboratory deformation experiments at up to 300 MPa confining pressure and temperatures up to 1200 °C. Experiments have been conducted on two-phase samples of a basalt (MORB) dyke and a matrix of 90% San Carlos Olivine plus 10% MORB. No migration of MORB is observed when samples have been hot isostatically pressed. Conversely, significant diffusion of basalt into the matrix is found when a deviatoric stress is applied. Creep (80–160 MPa) and strain-rate experiments (from 3×10^{-4} to 5×10^{-5} s^{-1}) induced melt propagation up to 50% of the initial dyke length/width ratio. The kinematics of deformation are essentially plastic and show a strong dependence on the load applied and the dyke geometry. Local pressure drops, due to dilatancy, are suspected to have enhanced melt migration.

In considering the mechanics of dyke emplacement the details of interactions between the viscous stresses, country-rock deformation and the role of fractures are poorly understood. Modelling macroscopic effects of distributed damage, due to either ductile and/or brittle creep rupture, requires mechanical parameters for stress–strain distributions and material strength degradation. The growth of microcracks or granular flow may produce degradation of the material ahead of the crack-tip, by reducing the elastic coefficients, enhancing the formation of an evolving damage zone, called the 'process zone' (Delaney et al. 1986; Rubin 1993; Agnon & Lyakhovsky 1995; Meriaux et al. 1999). For basaltic dykes in partially molten systems, the melt/rock viscosity ratio may be sufficient for dyke growth to be treated simply as a purely elastic process. In this case, flow can be treated as a diffusional process in a poroelastic medium, as long as the dykes are large enough to grow 'critically', acting as leading pathways, according to linear elastic fracture mechanics (Rubin 1998). This view is supported by field observations on thin dykes (widths of 0.5–5 m) that extend over considerable distances (1–10 km). However, rapid magma injection is required in order to prevent solidification (Delaney & Pollard 1982; Spence & Turcotte 1985).

A complete analysis of the problem requires simultaneous solution of the coupled equations governing host-rock deformation and magma flow (Lister 1990; Rubin 1993). Here we analysed, in the laboratory, the mechanics of dyke emplacement. This was done by quantifying the migration of 100% basalt dykes into a matrix of 90% San Carlos olivine and 10% MORB, following rock deformation experiments. The experiment used a triaxial high-pressure, high-temperature Paterson apparatus under different load and temperature conditions.

Experimental details

The mechanics of dyke emplacement have been investigated, using a Paterson-type gas apparatus, whose main feature is a servo-controlled loading frame with pressure vessel, operating at high-pressure and high-temperature conditions (Fig. 1). Lithostatic conditions are imposed by a gas pressure medium, which ensures a strictly hydrostatic environment up to 500 MPa. Axial deformation is then provided by a loading actuator (maximum load 100 kN) and measured by an internal load cell, which permits sensitive measurements of axial load and displacement. A three-zone internal furnace permits heating and cooling of the sample (up to 1300 °C), providing a uniform high temperature of the sample area. Temperature was measured using a Pt/Rh–Pt thermocouple. The main pressure vessel is a steel cylinder, with its axis held vertical and a

From: BREITKREUZ, C. & PETFORD, N. (eds) 2004. *Physical Geology of High-Level Magmatic Systems.* Geological Society, London, Special Publications, **234**, 243–249. 0305-8719/04/$15.00
© The Geological Society of London 2004.

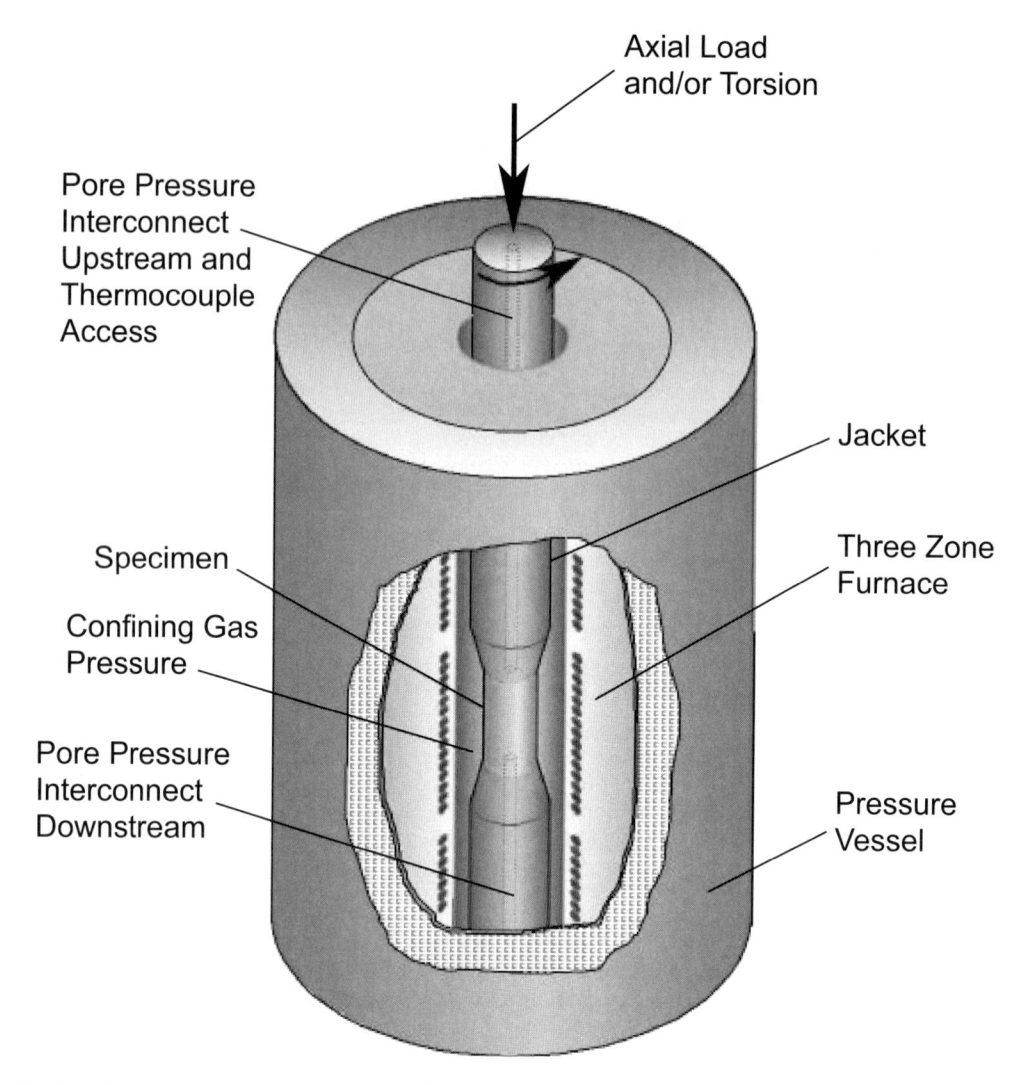

Fig. 1. Paterson-type gas apparatus for triaxial deformation experiments under high-pressure–high-temperature conditions. The main elements are the confining pressure vessel, the three-zone internal furnace, the sample set-up and the axial load.

working volume of c. 1195 cm³ (65 mm diameter and 360 mm length). The cylindrical sample has an initial size of 10 mm diameter × 20 mm length.

Experiments were conducted on isostatically hot-pressed aggregates of olivine + basalt. Fine-grained San Carlos olivine powder (< 35 μm) was mechanically mixed with 10 wt % of mid-ocean ridge basalt (MORB), an olivine tholeiite, with a grain size of <15 μm. The compositions of San Carlos Olivine + MORB, and their rheology, in terms of partially molten system dynamics, are well known (Hirth & Kohlstedt 1995a,b and references therein).

In order to obtain the basalt dykes, 100 wt % MORB samples were cold-pressed into Ni cans (26 mm long, 11.6 mm OD, 10 mm ID) with a uniaxial pressure of 100 MPa and air dried for 12 hours at 425 K, in a f_{O_2} controlled-atmosphere furnace. Cold-pressed samples were then isostatically hot-pressed (HIP) at 1473 ± 5 K and 300 MPa for six hours. After the samples were hot pressed, the jackets were dissolved with a mixture of nitric and hydrochloric acids (aqua regia), in order to prevent any chemical reaction with the sample. Dykes 7 mm long, 3 mm wide and 2 mm thick were then machined from the samples, by using cutters and surface grinders. Dykes were located at one end of a new Ni can, which was then filled with olivine + MORB powder. Next, the cold and hot pressing procedures were performed. Jackets were again dissolved with aqua regia. Samples were then cut in half, longitudinally, along the maximum dyke width, to prepare polished sections for optical and chemical analyses.

In preparation for deformation experiments, samples were ground and jacketed in sleeves of Ni. The Ni sleeves were then inserted into Fe jackets, in which the ceramic and hardened steel deformation pistons were also enclosed, to produce the deformation assembly. The assembly was then dried at 425 K in an f_{O_2} controlled-atmosphere furnace.

Deformation experiments were conducted at a temperature of 1473 ± 5 K; a confining pressure of 300 MPa; a differential stress of 80 to 160 MPa; and strain rates of 3×10^{-4} to 5×10^{-5} s^{-1}. Initially, experiments were conducted with a constant piston displacement rate. For one set of experiments, at fixed displacement rate, a total strain of c. 12–13% of the initial dimension was reached. Another suite of experiments was performed, switching to a constant load of 80–160 MPa (i.e. creep tests), with the load applied for 45 minutes.

After removal from the apparatus, the sample jackets were dissolved again using aqua regia. The sample ends were ground until the dyke was visible. Samples were then cut and polished.

Results

As a first step we measured the dimensions of dykes both after HIP and after deformation experiments. No appreciable (greater than 0.5 mm) dimensional change could be measured for HIP, while changes of 1–4 ± 0.5 mm in length and 1–2 mm in width were found for deformed samples.

In order to evaluate whether/how melt migration occurred from the 100% MORB dyke into the 90% San Carlos Olivine + 10% MORB, we needed a reliable measurement of melt concentration. Microstructural observation of hot-pressed and deformed aggregates was carried out, by means of optical and chemical analyses. The chemical analyses of the two materials used are reported in Table 1.

For the optical analysis we used a SEM to provide back-scattered electron images analysis, that shows atomic number contrasts, while in

Table 1. *Chemical analyses of starting material*

	MORB	Olivine + 10% MORB
SiO_2	49.93	41.47
TiO_2	1.47	0.14
Al_2O_3	15.55	1.43
Cr_2O_3	0.02	–
FeO	9.17	7.69
MnO	0.01	0.12
NiO	–	0.28
MgO	9.02	47.48
CaO	11.42	1.07
Na_2O	3.05	0.28
K_2O	0.20	0.02
P_2O_5	0.14	0.02
Cr_2O_3	0.02	–

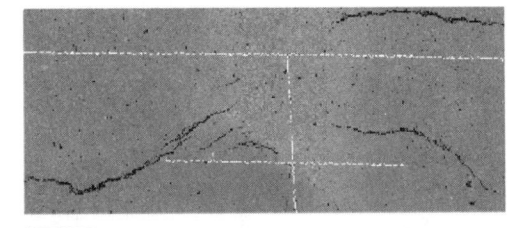

Fig. 2. Example of the area investigated. Profiles are performed both perpendicularly and along the dyke. Extra profiles are performed in the tip region.

terms of chemical analysis we mapped heavy-element concentration, through X-ray fluorescence measurements. Since we were interested in understanding whether melt diffusion occurred from the dyke into the matrix, we performed measurements both parallel and perpendicular to the dyke direction (Fig. 2).

The microstructural image (Fig. 3) analysis was performed by calculating the number of pixels in each of the different grey scales. This procedure allows calculation of the percentage of the area occupied by the olivine and the melt (i.e. MORB). Macroscopically, two types of melt were recognized: one darker (with the same composition of the MORB dyke) and the other lighter (forming thin sheets around the olivine crystals). The chemical analyses showed that the light-coloured melt has a higher content of Mg and a lower content of Fe and Na, with respect to the initial MORB composition. This indicates that some chemical reaction between the melt and olivine crystals has occurred. It is likely that MORB originally located in the matrix reacted

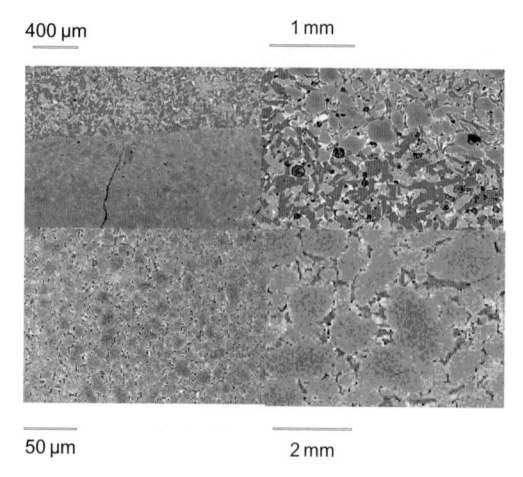

Fig. 3. (**a–b**) Different magnifications of the dyke–matrix contact and (**c–d**) matrix structure.

Fig. 4. Plot of melt percentage vs. distance relative to the right side of profiles performed perpendicularly to the dyke direction, after constant stress experiments performed for 45 minutes. Unfilled squares indicate 130 MPa experiments; filled squares indicate 70 MPa.

with the olivine crystals, whilst the darker melt migrated from the dyke. However, a change in the chemistry of the melt does not affect our microstructural analysis, since we are primarily interested in any overall relative changes of melt content. Such changes are the real markers of suspected migration processes, rather than in the absolute composition of melt.

Profiles along and perpendicular (Fig. 4) to the dyke show that melt content increases as a function of the stress applied. By considering the dyke-matrix contact as 0 on the x-axis, and referring to the melt content in the matrix, MORB ranges from 12–18% for 70 MPa constant stress applied, to 14–22% for 130 MPa in the first 2 mm, gradually reaching the background value (10%). The melt shows a higher content with respect to the background up to c. 4 mm away from the dyke, for profiles along the dyke, and c. 2 mm away, for profiles perpendicular to the dyke. Only beyond these distances does the melt go back to its background matrix content. This indicates the absence of further diffusion.

As a further investigation, we used an X-ray spectrometer that allows us to measure heavy-element concentrations. Iron and Calcium were the candidates for this analysis. However, the Fe content varies from c. 2.2 wt % in the olivine + MORB matrix to c. 2.8 wt % in the MORB dyke. Thus, the content difference is almost within the instrumental error (c. 0.1). Calcium is much more reliable, since its content is between c. 0–0.02% in the olivine matrix and c. 4.5% in the MORB dyke. Its concentration can be estimated from the number of Ca counts per unit of time. Calcium shows a marked variation in content from the dyke and the dyke–matrix contact (2500–3000 counts), to the matrix (<500 counts), for isostatically hot-pressed samples (Fig. 5a). Conversely, Ca content decreases smoothly from the dyke to the matrix for samples that were also deformed (Fig. 5b). In fact, Ca drops from 2500 counts in the dyke to 1000 counts at the dyke–matrix contact, and then to c. 750 from 0 to 2 mm away from the contact. Only beyond 2 mm does Ca register less than 500 counts. This pattern clearly further indicates that melt migration occurred only when axial deformation was applied.

Finally, we investigated the dyke-tip region in order to evaluate whether higher degrees of melt migration had occurred. We compared full Ca profiles of deformed samples perpendicular to the dyke (upper profile in Fig. 2) to profiles perpendicular to the dyke tip (lower profile in Fig. 2). By plotting the average of Ca content measured, it appears that the dyke-tip profiles define a slightly narrower region around the dyke propagating deeper into the matrix, with respect to the central part of the dyke (black lines). Thus, the hypothesis that deformation experiments induce higher stress concentration in the dyke-tip region seems to be supported by the chemical analyses.

Discussion and conclusions

We investigated the mechanics of dyke emplacement, in the laboratory by studying the migration of 100% MORB 'dykes' into a matrix of 90% San Carlos olivine and 10% MORB.

Microstructural observations (back-scattered electron images and X-ray fluorescence analyses) of the melt distribution around the dykes indicates that no significant melt migration

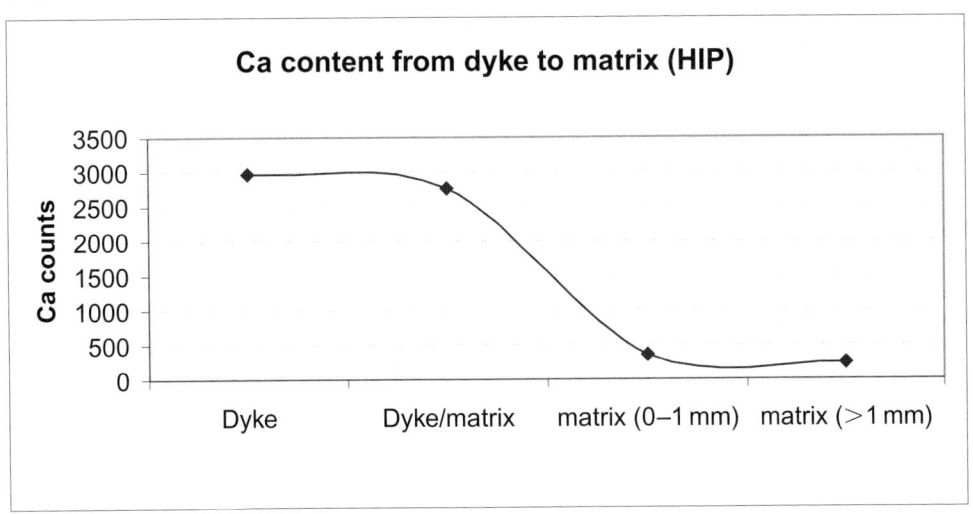

(a)

Ca content from dyke to matrix (HIP)

(b)

Ca content from dyke to matrix (after deformation)

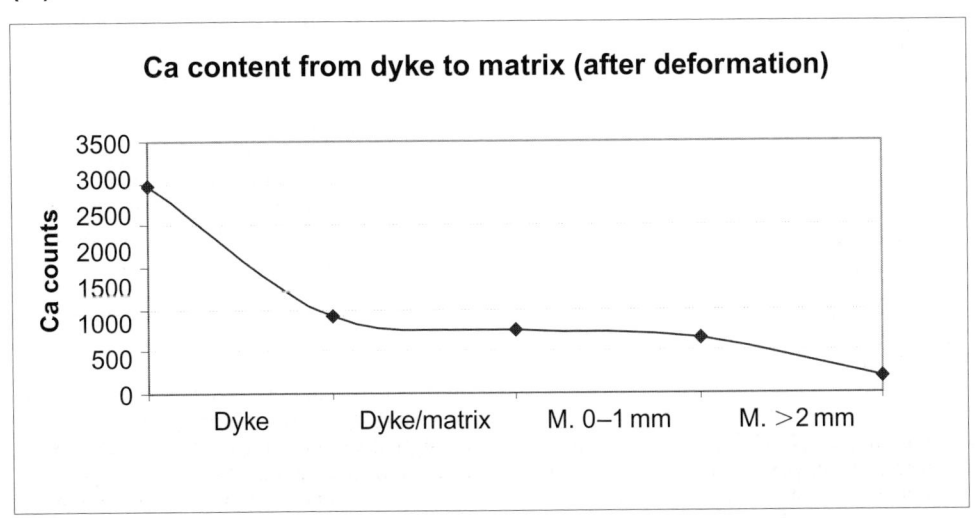

Fig. 5. (a) Ca profile (right side) from dyke to matrix for HIP samples and (b) deformed samples. Again we plot one hand, as the melt content is symmetrical with respect to the dyke position.

occurred after six hours of HIP at 1473 K with a confining pressure of 300 MPa (Fig. 7).

Conversely, significant dependence between melt migration and stress applied has been found. Deformation experiments, in terms of creep and constant displacement rate, respectively, at constant stresses (80–160 MPa) for 45 minutes and at constant strain rates ranging from 3×10^{-4} to 5×10^{-5} s^{-1}, with a total strain of c. 12–13% induced (Fig. 7) melt propagation up to 50% of the initial dyke dimensions.

The highest stress concentrations are produced at the near-dyke-tip region, where the superposition of the stress perturbation, due to the dyke intrusion and the elastic dilation of the host rock near the tip cavity, may produce a drop in the host-rock pore pressure that can play an important role in enhancing melt migration.

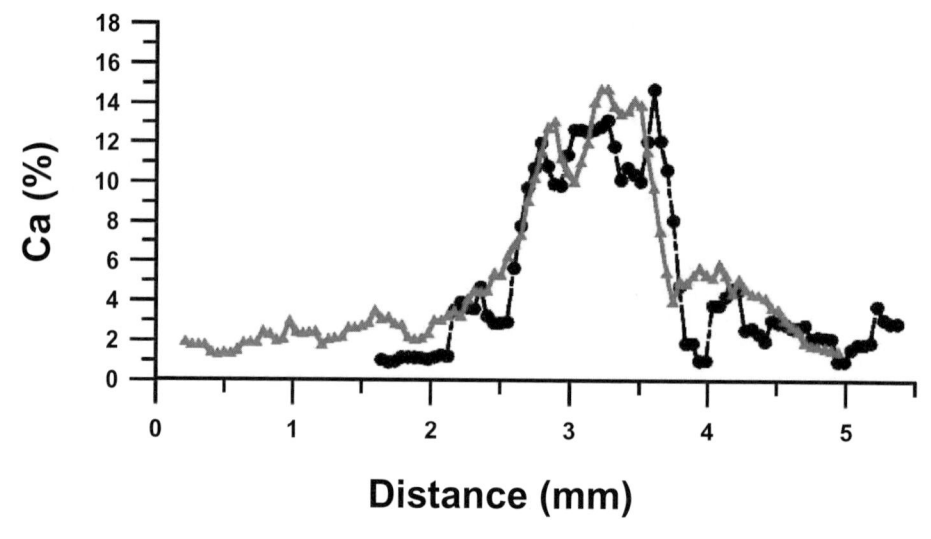

Fig. 6. Full Ca profiles performed perpendicularly to the dyke (black line), and to the dyke tip region (red lines).

(a) Hot-isostatically-pressed

(b) After triaxial deformation

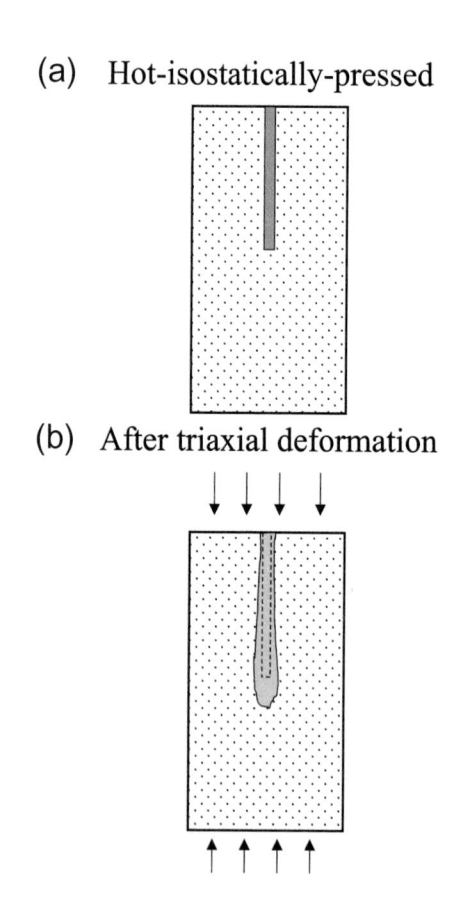

Fig. 7. Conceptual sketch of dyke–matrix mechanical behaviour for (**a**) hot-isostatically-pressed samples and (**b**) triaxially deformed samples.

In this view, two competing processes might favour melt migration: porous flow and dyke propagation. The first process appears evident, in agreement with recent theoretical studies indicating that porous flow between the dyke and the process zone (Rubin 1993) and the viscous deformation of the host rock (Lyakhovsky et al. 1997) are important migration mechanisms. On the other hand, sporadic fracture formation does not show repeatability, and does not seem to be related to either a dyke's geometry or its propagation. The causes of fracturing should be sought in quenching due to the fast cooling rate (30 K per minute), rather in the existence of a real 'damage' or process zone as modelled, which would require a massive presence of distributed micro-cracks necessary for strain softening or material strength degradation. However, the dependence found in our experiments, between stress applied and melt migration, supports numerical simulation results of magma propagation (Agnon & Lyakhovsky 1995; Lyakhovsky et al. 1997). In damaged host rock, magma invades the damaged zone once a critical damage value is exceeded. Under the simulated conditions of our experiments we cannot find evidence of degradation of the material by damage ahead of the crack-tip. However, the response to loading suggests that it is more plausible that the kinematics of migration are strongly influenced by plastic mechanisms and occur at some critical stress, favoured by a pressure-drop during dilatancy.

G. Di Toro is warmly thanked for some optical analyses. P. Romano and L. Pappalardo are thanked for X-ray fluorescence analyses. G. Hirth and J. Renner are thanked for fruitful discussions.

References

AGNON, A. & LYAKHOVSKY, V. 1995. Damage distribution and localization during dyke intrusion. *In*: BAER, G. & HEIMANN, A. (eds) *Physics and Chemistry of Dykes*, Balkema, Rotterdam, 65–78.

DELANEY, P.T. & POLLARD, D.D. 1982. Solidification of basaltic magma during flow in a dike. *American Journal of Science*, **282**, 856–885.

DELANEY, P.T., POLLARD, D.D., ZIONY, J.I. & MCKEE, E.H. 1986. Field relations between dikes and joints: emplacement processes and paleostresses analysis. *Journal of Geophysical Research*, **91**, 4920–4938.

HIRTH, G. & KOHLSTEDT, L.D. 1995a. Experimental constraints on the dynamics of the partially molten upper mantle: deformation in the diffusion creep regime. *Journal of Geophysical Research*, **100**, 1981–2001.

HIRTH, G. & KOHLSTEDT L.D. 1995b. Experimental constraints on the dynamics of the partially molten upper mantle 2. Deformation in the dislocation creep regime. *Journal of Geophysical Research*, **100**, 15 441–15 449.

LYAKHOWSKY, V., BEN-ZION, Y. & AGNON, A. 1997. Distributed damage, faulting and friction, *Journal of Geophysical Research*, **102**, 27 635–27 649.

MERIAUX, C., LISTER, J.R., LYAKOWSKY, V. & AGNON, A. 1999. Dyke propagation with distributed damage of the host rock. *Earth and Planetary Science Letters*, **165**, 177–185.

RUBIN, A. 1993. Tensile fracture of rock at high confining pressure: implications for dike propagation. *Journal of Geophysical Research*, **98**, 15 919–15 935.

RUBIN, A. 1998. Dike ascent in partially molten rock. *Journal of Geophysical Research*, **103**, 20 901–20 929.

SPENCE, D.A. & TURCOTTE, D.L. 1985. Magma-driven propagation of cracks. *Journal of Geophysical Research*, **90**, 575–580.

Index

andesites, Flechtingen–Roßlau Block, Magdeburg 51–66
Apennines, Gavorrano intrusion laccolith 151–61
Armorica, Zelezniak Hill, [S]206[s]Pb–[S]238[s]U zircon dating 71–3
aureoles, Gavorrano intrusion 155–6
Australian NW Shelf, Exmouth Plateau 226

Black Mesa bysmalith, Henry Mountains, Utah 163–73
    construction rate 163–73
    geological setting and previous work 164–5
    thermal numerical simulations 166–70
Bohemian Massif, Intra-Sudetic Basin, Poland 5–11
Brescian Prealps *see* Montecampione Group, S Alps
bysmalith construction rate, Black Mesa, Utah 163–73

caldera-related ring complexes 126–42
    vs laccolith complexes 126
    schematic cross-sections 145
Carboniferous–Permian transition 5–11, 14–15
cauldron subsidence, Rallier-du-Baty nested ring complex 144–6
Central Europe, Permo-Carboniferous examples 13–31
    geotectonic setting at Carboniferous–Permian transition 14–15
    laccolith complexes
        magmatism in strike slip-settings 13–14
        strike-slip control 23–9
        types 19
    Variscan intramontane strike-slip basins 15–19
    see also Ilfeld; Saale; Saar–Nahe Basins
Christmas-tree laccolith complex 195–213
Colorado Plateau, laccolith complexes 26, 27, 163–73

deformation experiments, dyke emplacement in partially molten olivines 243–9
discrete element modelling (DEM), sill emplacement 217–27
Donnersberg intrusive–extrusive dome, Saar–Nahe Basin 15, 17–25, 81–3, 96–102, 108–11
Donnersberg type laccolith complexes 15, 17–25, 96–9
dyke emplacement
    mechanics in partially molten olivines 243–9
    experimental details 243–5
    results 245–6
    mid-ocean ridge basalt (MORB) 243–9

Elba Island, Christmas-tree laccolith complex 195–213
    construction of intrusive complex 205–10
    geological outline 196–200
    intrusive sequences 200–5
    Monte Capanne pluton 197–8, 204–10
    petrographic and chronological features summarized 201
Exmouth Plateau, Australian NW Shelf, saucer-shaped sill 226

Flechtingen–Roßlau Block, Magdeburg 51–66
    emplacement scenario of andesitic magmas 63–5
    Flechtingen Sill Complex (FSC) 54–9
    Late Palaeozoic volcano-sedimentary succession 52–4
    lower andesites 59–61
    structure, lithology and emplacement processes 54–9
    upper andesites 61–3
fluid–rock interactions, offshore Norway, hydrothermal vents 233–41
fumaroles, Kerguelen Archipelago 142

gas apparatus, Paterson-type 243–4
Gavorrano intrusion, Tuscany 151–61, 196
Germany *see* Flechtingen–Rosslau Block; Ilfeld; Saale; Saar–Nahe Basins
granodioritic intrusions, Kaczawa Mountains, Poland 67–74
groundwater
    and diagenesis 112–13
    and emplacement of volcanic bodies 113–14

Halle Volcanic Complex 19–23
Hartz Mountains, Ilfeld Basin 15–17
hawaiites 139
heat flow, groundwater and diagenesis 112–13
Henry Mountains
    laccoliths, theoretical emplacement depth 179–80
    see also Black Mesa bysmalith
Hungary, Ság-hegy peperitic lava lake-fed sills
hydrothermal vent complexes associated with sill intrusions 233–41
    Karoo Basin 234–5
    modelling fluid pressure evolution 235–8

Ilfeld Basin, Hartz Mountains 15–17
Indian Ocean S *see* Kerguelen Archipelago
Intra-Sudetic Basin, Poland 5–11
    online geology and volcanic evolution 5–6
    subvolcanic intrusions 6–8
intrusion shape, power-law distribution 158
intrusive–extrusive domes, Saar–Nahe Basin
    Donnersberg 15, 17–25, 81–3, 96–102, 108–111

Himmelberg dome 95–6
Kreuznach dome 100–1
Kuhkopf laccolith and dome 99–100
Lemberg dome 101–2
Nohfelden dome 102
Wilzenberg dome 95
Italy *see* Elba; Montecampione group, Southern Alps; Tuscany

Joliot–Curie caldera, Rallier-du-Baty nested ring complex 134

Kaczawa Mountains, Sudetes, Poland, Zelezniak Hill 67–74
  geochemical features and age 69–71
  geological setting/map 67–9
  geometry of intrusive complex 71
  petrography 67–9
  zircon [S]206[s]Pb–[S]238[s]U dating 71–3
Karoo Basin
  hydrothermal vent complexes associated with sill intrusions 234–5
  saucer-shaped sill formation 219–25
Kerguelen Archipelago, Rallier-du-Baty nested ring complex 126–42
  chronological sequence of igneous rocks 135
  field notes
    country rocks 128–31
    Dome Carva volcano 140–1
    external trachytic domes and associated pumice deposits 131
    Joliot–Curie caldera 134
    northern and southern ring complexes 131–4
    Pic Saint-Allouarn volcano 141–2
    plutonic arcuate dykes 133
    plutonic ring dykes and cone sheets 132–8
    satellite cupolas 133
    Table de l'Institut caldera volcano 138–40
    thermal activity 142
  time durations and dimensions 142–6
    cauldron subsidence 144–6
    stoping process 144
    tilting and doming of basaltic wallrocks 143–4

laccolith complexes 13–14, 15–19, 23–9
  Christmas-tree geometry 19, 26, 208
  defined 215
  Donnersberg type 15, 17–25, 81–3, 96–102, 108–11
  emplacement depth 178–80
  Halle type 25
  intrusion geometry 181–8
  locations
    Colorado Plateau 26, 27, 163–73
    Elba Island 195–213
    Gavorrano intrusion 151–61
    Halle Volcanic Complex 19–23
    Kaczawa Mountains, Poland 67–74
    Saar–Nahe Basin 15–19, 92–6, 108–11
    Utah, Colorado 26, 27
  and magmatism 13–14
  punch(ed) geometry 19, 26, 183
  vs ring complexes 126

strike-slip basins 13–14
strike-slip control on evolution 23–9
volume estimates 180–1
Little Hungarian Plain *see* Ság-hegy

maar–diatreme volcanoes, Saar–Nahe Basin 80–8, 106–7
magma–water interactions 33–50
mid-ocean ridge basalt (MORB), dyke emplacement 243–9
modelling
  fluid pressure evolution, hydrothermal vent complexes 235–8
  saucer-shaped sill formation 215–27
  strike-slip basins 28
Monte Capanne pluton 197–8, 204–11
Montecampione Group, S Alps 175–93
  emplacement depth 178–80
  geological setting 176–8
  geometry and relations with host group 181–9
  volume estimates 180–1
Monticello fault, Gavorrano area 153, 154–5
More Basin, offshore Norway 222–5
mugearites 139

Nohfelden intrusive–extrusive dome, laccoliths, cupolas, intrusive–extrusive domes, Saar–Nahe Basin 102
North Rockall Trough sill seismic data 229–32
Norway (offshore)
  More Basin 222–5
  Voring Basin 222–5, 233–41

Pannonian Lake, Hungary 36
Paterson-type gas apparatus 243–4
[S]206[s]Pb–[S]238[s]U zircon dating, Zelezniak Hill, Armorica 71–3
Peninsula Tuff Cone, California 38–40
peperites
  Flechtingen–Rosslau Block 59
  Ság-hegy peperitic lava lake-fed sills 33–50
Permo-Triassic, central S Alps 175
phreatomagmatic pyroclastic units, peperitic lava lake-fed sills 36–8
pluton emplacement
  rate and mechanism, Black Mesa bysmalith 163–73
  tilting, Elba Island 195–213
  see also Kerguelen Archipelago; Tuscany
Poland *see* Intra-Sudetic Basin; Kaczawa Mountains, Sudetes
Porto Azzurro pluton 200, 209, 211

Rallier-du-Baty nested ring complex 126–42
  see also Kerguelen Archipelago
ring complexes, caldera-related 126–42
Rockall, North Rockall Trough sill seismic data 229–32

Saale Basin 24
Saar–Nahe Basin 15–19, 75–124

INDEX

dykes 88, 107
formation of late Variscan basins 76–7
groundwater and emplacement of volcanic bodies 113–14
growth of subvolcanic and volcanic bodies 114
heat flow, groundwater and diagenesis 112–13
ignimbrite 112
laccoliths, cupolas, intrusive–extrusive domes 92–102
  Bauwald laccolith 92–4
  Donnersberg intrusive–extrusive dome 15, 17–25, 81–3, 96–102, 108–11
  Herrmannsberg cupola 94–5
  Himmelberg dome 95–6
  Konigsberg laccolith 94
  Kreuznach intrusive–extrusive dome 100–1, 105
  Kuhkopf laccolith and intrusive–extrusive dome 99–100
  Lemberg intrusive–extrusive dome 101–2
  Nohfelden intrusive–extrusive dome 102
  Obermoschel cupola 94
  Palatinate Anticline cupolas 94
  Potzberg cupola 95
  Selberg cupola 94
  Waldbockelheim cupola 92
  Wilzenberg dome 95
late orogenic compression and extension in Variscan collision belt 76–7
lava flows 102–3, 111
maar–diatreme volcanoes 80–8, 106–7
Nahe caldera 104–6
Prims and Nahe syncline ignimbrite 104, 112
sills 88–92, 107
tephra deposits 103–4, 111–12
Ság-hegy peperitic lava lake-fed sills 33–50
  corrugation zones 38–40
  fluidization halo 41–4
  geological setting 35–6
  lava lake-fed intrusions 41–4
  phreatomagmatic pyroclastic units 36–8
saucer-shaped sill formation 215–27
  discrete element modelling 216–19
  sill emplacement modelling 216
sedimentary basins
  hydrothermal vent complexes associated with sill intrusions 235–8
  modelling 216–19
Servino Formation 176–89
sills
  3D seismic perspective 229–32
  depth range of intrusion 107
  emplacement depth 178–80

emplacement modelling 216
hydrothermal vent complexes associated 233–41
peperitic lava lake-fed sills 36–8
propagation mechanisms 107
saucer-shaped sill formation 215–27
volume estimates 180–1
South Africa, Karoo Basin
  hydrothermal vent complexes associated with sill intrusions 234–5
  saucer-shaped sill formation 219–25
Southern Alps *see* Montecampione Group
stoping process 144
strike-slip basins
  control on evolution of laccolith complexes 23–9
  laccolith complexes 13–14
  simple model 28
  Variscan 15–19
Sudetes *see* Intra-Sudetic Basin; Kaczawa Mountains

thermal numerical simulations, Black Mesa bysmalith, Utah 166–70
trachytes 139
transtensional basin systems 13–31
Tuscany, Gavorrano intrusion 151–61
  fault systems 154–5
  geological framework 152–5
  shape 156–8
  thermal metamorphism and hydrothermal alteration 155–6
Tyrrhenian Basin 152
Tyrrhenian Sea 196

Utah *see* Henry Mountains, Black Mesa bysmalith

Variscan belt 68
  Intra-Sudetic Basin, Poland 5–11
  Kaczawa Mountains 67–74
  Saar–Nahe Basin 76–7, 108–111
Verrucano Lombardo 176–89
Voring Basin, offshore Norway
  Gleipne Saddle, sills 222–5
  hydrothermal vents 233–41

Walbrzych Trough, Intra-Sudetic Basin, Poland 7–8

Zelezniak Hill *see* Kaczawa Mountains, Poland
zircon [S]206[s]Pb–[S]238[s]U dating, Armorica 71–3